T0186478

Discrete Mathematics
with Ducks

Second Edition

Discrete Mathematics with Ducks

Second Edition

sarah-marie belcastro

serious mathematics treated with levity

CRC Press
Taylor & Francis Group
Boca Raton London New York

CRC Press is an imprint of the
Taylor & Francis Group, an **informa** business

A CHAPMAN & HALL BOOK

CRC Press
Taylor & Francis Group
6000 Broken Sound Parkway NW, Suite 300
Boca Raton, FL 33487-2742

First issued in paperback 2020

© 2019 by Taylor & Francis Group, LLC
CRC Press is an imprint of Taylor & Francis Group, an Informa business

ISBN 13: 978-0-367-57070-5 (pbk)
ISBN 13: 978-1-138-05259-8 (hbk)

This book contains information obtained from authentic and highly regarded sources. Reasonable efforts have been made to publish reliable data and information, but the author and publisher cannot assume responsibility for the validity of all materials or the consequences of their use. The authors and publishers have attempted to trace the copyright holders of all material reproduced in this publication and apologize to copyright holders if permission to publish in this form has not been obtained. If any copyright material has not been acknowledged please write and let us know so we may rectify in any future reprint.

Except as permitted under U.S. Copyright Law, no part of this book may be reprinted, reproduced, transmitted, or utilized in any form by any electronic, mechanical, or other means, now known or hereafter invented, including photocopying, microfilming, and recording, or in any information storage or retrieval system, without written permission from the publishers.

For permission to photocopy or use material electronically from this work, please access www.copyright.com (http://www.copyright.com/) or contact the Copyright Clearance Center, Inc. (CCC), 222 Rosewood Drive, Danvers, MA 01923, 978-750-8400. CCC is a not-for-profit organization that provides licenses and registration for a variety of users. For organizations that have been granted a photocopy license by the CCC, a separate system of payment has been arranged.

Trademark Notice: Product or corporate names may be trademarks or registered trademarks, and are used only for identification and explanation without intent to infringe.

**Visit the Taylor & Francis Web site at
http://www.taylorandfrancis.com**

**and the CRC Press Web site at
http://www.crcpress.com**

Deduckation: To my frolleagues and colleagues for supporting this book through the first edition, and trusting my vision for this material… and especially to Tom Hull, who first pointed out that writing this text was inevitable, and then helped me with parts of it—both editions. He has been a sounding board and support structure for this project as well as the rest of my ever-changing life.

Contents

Preface for Instructors and Other Teachers

1 About This Book

Discrete Mathematics with Ducks is intended for a sophomore-level audience (including some first-year students) to support a gentle course so that students who find mathematics and proofs and abstraction challenging can still succeed. However, I am mindful that classes including weaker students almost always contain stronger students as well, and so I include more challenging problems with every topic and activity. Additionally, I am mindful that different institutions and different faculty members at those institutions have different ideas about what pace and amount of material is appropriate for a lower-level class. For this reason, I have included Bonus sections for those instructors who wish to have faster pacing. The Bonus sections can be used as fodder for take-home exams or projects as well as for students who just want to know more about a topic outside of class. The material in the text is not new; my contributions are a curation of curriculum, a tone of text, and a philosophy of pedagogy.

The guiding pedagogical principle behind the organization of *Discrete Mathematics with Ducks* is that students can discover many ideas, concepts, theorems, and proofs for themselves with a bit of guidance. Where I see an engaging way for them to do this, I have written Try This! sections that are sets of problems that allow students to construct fundamental parts of the material. However, I also believe that students are likely to miss a detail here or there in their work and, more importantly, that as beginning mathematicians they need reinforcement for their newfound learning. For this reason, I follow sets of discovery problems with sections that explain the relevant material and give both examples and details. I outline some ways instructors might capitalize on this organization of the material in Section 2; instructors who want to gain experience in discovery-based teaching will particularly want to read Section 2.2.

The guiding pedagogical principle behind the style and tone of this text is pretty silly. I mean that literally: I believe that students are more likely to absorb mathematics that is presented in a goofy way. Bizarre situations help students separate the abstraction of underlying mathematics from the presentation of a problem and thus give students practice in recognizing the mathematical essence of problems they find in other contexts. Students who are enjoying the weirdness of problem presentations are also focusing on the mathematics. It's easier to remember a zany concept setup than to recall a straightforward statement. And there's no reason to be serious when there's an opportunity to have fun!

There is also a hidden agenda in my structuring of this material. (I guess it won't be hidden anymore after you read this paragraph.) I think it is hard to learn mathematical techniques without a surrounding context. Attempting to do so is sort of like opening a toolbox and simply holding up each tool, describing its function, and then passing the tools around the audience. Without a carpentry project, it is difficult to build a reliable mental library of situations in which each tool is useful. So, in this text I use discrete mathematics as the context via which proof is introduced. Similarly, tools such as set theory, logic, and functions are companions to the basic combinatorics and graph theory that are introduced at the start of the text.

Discrete mathematics is a growing area of mathematics that is used throughout industry, so I think a discrete mathematics text should function as an introduction to and survey of the field and its myriad possibilities. Faculty who specialize in discrete mathematics are housed in mathematics, applied mathematics, or computer science (depending on the institution, and they may show up in multiple departments as well). They do combinatorics, graph theory, geometry, and optimization. In this text, I have attempted to balance combinatorics and graph theory topics that lead naturally to use in computer science with those that lead naturally to mathematics investigations. This is so that students will have a taste of the many flavors discrete mathematics has and thus of the paths discrete mathematics can take. As part of this flavor tasting, I have tried to introduce optimization topics where possible. Hopefully, this diverse introduction to the field will excite students into desiring further study of discrete mathematics.

The ACM (Association for Computing Machinery) Special Interest Group on Computer Science Education (SIGCSE) guidelines suggest curricula for computer-science-focused and mathematics-focused discrete mathematics courses. *Discrete*

Mathematics with Ducks gives overviews of the topics and techniques listed by SIGCSE and then reinforces them throughout the course by applying them to discrete mathematics topics and problems. The content and approach of the text comply with the SIGCSE guidelines in this way.

I made some decisions about the inclusion and exclusion of content and terminology that are potentially controversial and so should be disclosed (or at least mentioned) here. There are certain terms (including *predicate* and *combinations*) that I avoided using because they only arise in mathematical subspecialties or teaching contexts and are not used throughout mathematics or even mathematics courses. I deliberately separated the treatment of recursion from the treatment of induction so that students would have time to internalize the idea of induction before linking it to recurrences. It's easy for beginning students to get bogged down in the study of formal logic, so I minimized its treatment here. Finally, in this second edition I have chosen to use gender-neutral pronouns, with *ze* as the third-person singular, *hir* as the corresponding possessive, and *Mx.* as an honorific; I avoid the singular "they" as it is often confusing (and thus imprecise!) in writing. (Historical figures and personal acquaintances of the author are referred to by their publicly disclosed genders.)

The chapters on probability, cardinality, and number theory are included because many instructors want or need to teach that material as part of a discrete mathematics course (and because they are beautiful mathematics), but they are not central to discrete mathematics as a subfield and so I have placed them outside of the themes around which the book is organized. Similarly, an introduction to computational complexity has been included by popular demand for big-O notation. For probability, I chose to emphasize expected value and downplay the techniques of counting/proportions because, on one hand, expected value is central to discrete probability and, on another hand, students have already learned and used counting techniques elsewhere in the book. Cardinality is treated via a play, to avoid the potential dryness of a formal treatment. For number theory, I selected a sampler that would show different flavors of number theory—and hopefully whet students' appetites, as I think every mathematics student should study number theory for an entire semester. Still, the chapter includes almost enough background to justify the workings of the RSA algorithm (only one bit of it is black-boxed). Computational complexity is introduced via a sequence of abstractions from measuring algorithmic efficiency and performance, so that big-O (and Θ and Ω) notation arises as formal ways of describing algorithm complexity estimates.

2   How to Use This Book

First things first: please don't interpret anything written in this section as prescriptive. I'm not attempting to tell you how to use the book (despite the title of this section), but instead, inviting you to think about what will work best for you and offering suggestions that you may take seriously or toss aside at your whim. Different classroom techniques are effective for different instructors, and in this you must find your own way.

 Second things second: each chapter is designed to take one week of class time and contains a mixture of discovery activities, expository text, in-class exercises, and homework problems. At the end of each expository section, there are elementary exercises labeled as Check Yourself problems and signaled by the marginal pencil-toting duck shown here (as are all sections of problems students should attempt); these are placed at the ends of sections rather than inside the section to prompt students to review soon after reading a section. Additionally, almost every chapter contains bonus material for enrichment or fast-paced classes, and all chapters contain guides to further study. The chapters are organized into three themes (background, combinatorics, and graph theory), with four additional chapters (on probability, cardinality, number theory, and computational complexity). The first chapter introduces both combinatorics and proof, and the third chapter introduces graph theory, so there is thematic foreshadowing within the first theme. All chapters after the first five assume knowledge of the first five chapters. (A detailed disclosure of dependencies appears in Section 3.)

My advice for how to deal with the ebb and flow of course pacing is to strictly adhere to a one-chapter-per-week schedule (choose topics for your syllabus accordingly!), and in this way achieve breadth rather than depth in student acquisition of the material. It may be tempting (especially for the first few chapters!) to have students work through the entirety of the activities in a chapter and to review the reading and… but in this way, one can get dragged down into belaboring material. Instead, leave some material for student study. In fact, while there are three classes' worth of material presented in each chapter, they are underpinned by only two substantial classes' worth of activities. Students' pacing will not be uniform; if your class gets behind your intended schedule, no harm will be done if you skimp on some chapters by reducing the work to two days.

On the topic of course mechanics, I assign daily readings (specified in each chapter's Instructor Notes section) and elementary practice problems (under the

Check Yourself label). I also assign homework each week, taken from the Problems section of the previous week's material. The purpose of this timing is to allow concepts some time to sink in and to prompt students to review. For a course of this level, I give both in-class and take-home exams. In-class exams are composed mostly of computational or simple problems (like Check Yourself exercises), but with a few end-of-chapter-level problems thrown in as well. My take-home exams often involve end-of-chapter-level problems and sets of problems from Bonus sections. I have included a selection of additional problems at the end of each of the central themes from which you may draw exam and review problems.

Discrete Mathematics with Ducks is written to be ideal for instructors who like active learning. If you have no experience with active learning techniques but would like to try some, then read on, for I'll give a short introduction below. However, if you don't give a flying figwhistle for that frippery-frappery, this book can work for you too. There are lots and lots of problems for a lecturer to use as examples (particularly in the Try This! sections) and homework assignments. All instructors should be aware that many sections throughout the text begin with Hey! You! warnings (indicated by the marginal stop-sign-holding duck shown here) that caution students not to read portions of the text before working on the relevant in-class activities. Make sure to tell your students whether or not to honor the don't-read-ahead edicts.

At the end of each chapter, I have included a section entitled Instructor Notes. This gives a breakdown of how I would (and how you might) conduct two to three classes on the chapter material. If you have other ideas on how to use the chapter, try them! And if they work well, please do share them with me. For those who like to partition course material among groups of students who present topics, the Try This! sections can function as projects. If you prefer that individual students make presentations, then you may wish to steer weaker students away from material in chapters where Try This! sections come before informational sections.

All links mentioned in the text, as well as GeoGebra files mentioned in the text, are available electronically at http://www.toroidalsnark.net/dmwdlinksfiles.html.

2.1   A Start on Discovery-Based Learning

My personal implementation of discovery-based learning in the classroom rests on having students collaborate in small groups to do mathematics. In order to have enough class time for group work, I require that students attempt to read relevant material in advance. For this book, I have taken special care to make the reading as elementary as possible so that students are able to read it and learn from it. (Let's hope I have succeeded in this endeavor.)

I am regularly asked how I get students to read a textbook. The pat answer is that I assign students to read the book, and expect students to read the book, and then they do so. However, others tell me that they also assign and expect reading but that students do not do it. Observers of my classes tell me that the difference is that I truly hold students responsible for reading the book. I do not repeat material in class that they could have learned by reading. At most, I give a review or an interactive example or exercise at the start of class. I recommend that if you lecture over book material, do so briefly. How much time is spent on lecture-like activities depends on you; I spend less than 15 minutes per class on lecture-like activities.

When instructors step away from lecturing and turn over some control of the class to students, they often feel as though they are not covering as much material as they would be covering if they lectured. Coverage is mainly an illusion whether or not we lecture; however we structure our classes, and whatever material we believe we transmit to students, we have no actual control over what material enters or is synthesized within students' minds. The main shift is in our perspective.

Most of the Try This! sets of in-class exercises will likely take longer than a single class period to complete. This is intentional; my expectation is that the students will not collectively finish all of the problems. I have tried to include enough problems, and some difficult problems, so that strong/advanced/speedy students will have things to think about while other students soldier on. Additionally, having more problems means that students have some choice in what they work on. For the most part, Try This! problems do not need to be done sequentially.

It takes a while to gauge in advance how long it will take your students to work a problem. With many subjects and texts, an instructor will allocate additional class time to incomplete in-class activities, or will lecture on remaining problems. For this text, I recommend a different approach. If it takes your students an hour to do three problems, then I advise you to accept that the students will only experience discovery for those three problems. Just continue with the course; after a class period's worth of experimentation, students can read about the material in the reinforcing sections. (Everything important is contained in the reading.) Because this is a survey course, it is more important that students gain exposure to a variety of concepts than that they fully master many of them.

Finally, nowadays many students have experience with collaborative learning from high school. In-class group work may seem less foreign to them than it does to you. Here are some practical tips on how to conduct group work. As with all advice I give, use that which works for you and discard that which rings false.

2.2   Details of Conducting Group Work

Begin by breaking students into groups. I suggest having three to five students per group. There are many ways to allocate students to groups; one is to count the students off by ⌈number of students/number of students per group⌉. Clumping students by first letter of first name, or first letter of last name, or month of birthday will work as well. Have students move so that all group members can see each others' faces and share papers and books. For the first few instances of group work, ask students to start by introducing themselves to each other.

Try to achieve a different partition in each class meeting for the first few weeks so that students get to know each other and experience a variety of each others' learning styles, strengths, and weaknesses. After the first couple of class meetings, you will be able to simply give a command ("Get into groups!" or "Clump up!") and the students will automatically rearrange themselves physically in preparation for the activity.

Once students have been partitioned, have them turn to the appropriate page of the text and tell them to work on the problems together. Walk slowly through the classroom, circulating among the groups. For the first few minutes, just listen. If they are collectively silent for more than two minutes—time this or count it out, because it feels like a long time to the instructor, and it sometimes takes students a while to digest the problem statements—then remind the class that they should be talking to each other. The sound level in the classroom usually rises quite quickly after such a reminder.

For the bulk of the collaborative learning time, move from group to group. When visiting a group, listen to see what they are saying to each other; look to see what they are writing down. If they are making errors, step in to gently correct them. (If this is your first experience doing group work, you may find that the students understand far less than they seemed to when you lectured. The difference is, I suspect, not in the students' level of understanding, but in your awareness of the students' level of understanding. There is little chance when lecturing to interrogate the students' understanding, but a great chance to do so when involved in group work; now is your opportunity to directly intervene in your students' learning process.) If one student seems separated from the rest of the group, remind the rest of the group to include that student. (How you do this will depend on your personality. I often make melodramatic statements like "Poor X! Ze is all alone in the wilderness over here....") If students seem to be working independently, encourage them to collaborate by trying to get a conversation going. For me, it works well to squat, so I'm physically on their level, and ask what they're thinking about. If the students seem to be stuck, ask them to tell you where they're stuck and what's getting in their way. Then give a small hint and promise you'll be back with more if that doesn't unstick them quickly. Then, move on to the next group.

At some point in this process, students or groups will start raising their hands to ask you questions. It's a good idea to outline a circuit of the class in your mind so that you can systematically visit every group in turn. Going to see a group who have raised hands can throw this off, so make sure to do something like reversing direction or returning to your previous place in the circuit. Otherwise, you may discover that you have lost track of the progress of a group or two.

Another dynamic you may encounter is that some students will be social instead of academic (do discourage this, unless they're taking a one-minute break from otherwise hard work), and some students may work on other material (for your class or for other classes) during this time. Everyone has a different philosophy on how to deal with this. Personally, I am not offended if students work on other material; they have busy lives. If a student is doing well in class, I allow hir to work on whatever ze wants to, as long as ze is not distracting hir group or slowing them down.

Some students are sharper than others, and with discrete mathematics in particular, there can be groups where some students struggle while others quietly do twice the number of problems one expects. I allow this as long as the stronger students give hints to the weaker students, and I try to make sure the stronger students don't run out of problems to do. I have readied the text for this eventuality by building extra problems into the Try This! activities, but you may wish to also prepare a list of end-of-chapter problems that are suitable add-ons or extensions just in case you have some extra-fast students.

Especially when groups work at different rates, or when some groups are composed of students of divergent abilities, it is difficult to know when to declare the activity done. I find it useful to set an approximate deadline by which time I think it will be appropriate to move on. Still, it regularly happens that students are working productively at that time, and then you must decide whether to let the students continue or cut off the activity. I recommend erring in the direction of cutting students' work short rather than allowing it to drag on; over time, you will develop a sense from "reading" the groups as to when it is best to have them keep working and when not.

In order to help students achieve closure on a set of problems, and so that you can set their work in a larger context, devote some time at the end of class to large-group discussion. Announce that they should stop working in groups (my cue is usually some variant on "Let's talk!") and ask them to summarize work on a particular problem or collection of problems. (Sometimes I survey them as to which problems they have completed before requesting this summary.) In practice, this can mean that each group gives a presentation at the board, or it can mean that a few students speak from their seats, or some combination of the two. Once students are used to this practice, I can use a cue such as, "Tell me about what you did," and students will know what to do. At the start of a term, I ask, "How did you approach problem X and what result did you get?" or "How does the proof for number Y go?" and students will volunteer responses. If I saw a group taking a notable approach when circulating among the students, I may say, "Z and W are going to brief us on their work on problem N." Because everyone has thought about the problems, the summaries are usually quick and elicit many nods.

Encourage different groups to describe their work on different problems in order to spread around the practice of mathematical speech. I think it is effective to reflect students' speech back to them in slightly more formal (and completely mathematically correct) language so that they understand the correctness of their conclusions and can improve their communication of mathematics over time. Such reflection can be followed by stating or reminding students what the point of the problem was and (briefly!) how it relates to the larger study in which you are communally engaged. Now, this sounds quite involved, but should only take ten minutes or so—perhaps not all problems are discussed, but only those for which different groups took different approaches, or perhaps very few details are mentioned.

When you first try this type of group work, it's common to feel either distant or over-involved. By distant, I mean that you might feel shy about intervening with groups, or feel as though you're just walking around and listening and not doing anything. By over-involved, I mean that you end up spending lots of time with each group, essentially walking the group through the exercise, and not visiting every group as a result. And you might find that you're not sure what to say when you hear students being stuck in ways you didn't expect, or making errors you weren't aware students could make, etc. That's okay. This type of classroom activity takes lots of practice. My personal foibles are (1) I tend to get absorbed in going from group to group and listening and helping and answering questions and then suddenly there isn't enough time to discuss the problems as a large group (drat!), and (2) if I'm tired, I tend to float instead of listening carefully, or instead allow students to get off topic.

3 Chapter (and Bonus-Section) Sequencing and Dependencies

Roughly speaking, Chapters 1–5 are needed for most of the rest of the book, and within the Combinatorics and Graph Theory themes, the chapters are mildly sequentially ordered. More conceptually, Chapter 1 includes an introduction to proof and all other chapters use proof; Chapter 2 includes set notation, logical thinking, and proof by contradiction, all of which are used in the remainder of the text.

Some instructors who use material from Chapters 14–17 place it after the Theme I–Theme III material, and others place the additional material mid-semester, for example between Themes I and II or between Themes II and III.

The following chapter dependency list is given by direct use of content.

Chapter 2 needs:	Chapter 1
Chapter 3 needs:	Chapter 1, Chapter 2
Chapter 4 needs:	Chapter 1, Chapter 2, Chapter 3
Chapter 5 needs:	*nothing* (Bonus needs Chapter 3)
Chapter 6 needs:	Chapter 1, Chapter 3, Chapter 4 (Bonus needs Chapter 5)
Chapter 7 needs:	Chapter 1, Chapter 3, Chapter 6 (Bonus needs Chapter 5)
Chapter 8 needs:	Chapter 1, Chapter 4, Chapter 5, Chapter 6
Chapter 9 needs:	Chapter 4, Chapter 6, Chapter 8 (Bonus needs Chapter 1, Chapter 3, Chapter 5)
Chapter 10 needs:	Chapter 1, Chapter 3, Chapter 5, Chapter 6 (Bonus needs Chapter 7 Bonus)
Chapter 11 needs:	Chapter 3, Chapter 4, Chapter 6, Chapter 10
Chapter 12 needs:	Chapter 3, Chapter 5, Chapter 10 (Bonus 3 needs Chapter 7)
Chapter 13 needs:	Chapter 3, Chapter 4, Chapter 5, Chapter 10, Chapter 11
Chapter 14 needs:	Chapter 1, Chapter 3, Chapter 7 (Bonus needs Chapter 6)

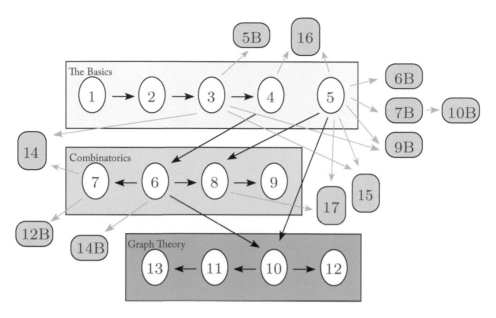

Figure 1. In this chapter dependency chart, Bonus sections are designated by "B."

Chapter 15 needs: Chapter 1, Chapter 2, Chapter 3, Chapter 5

Chapter 16 needs: Chapter 1, Chapter 2, Chapter 3, Chapter 4, Chapter 5

Chapter 17 needs: Chapter 5, Chapter 8

Indirect chapter dependency information is given graphically in Figure 1.

Preface for Students and Other Learners

1 About This Book (and about Learning Mathematics)

In my experience, it is difficult to learn mathematical techniques without a surrounding context. Attempting to do so is like attending a seminar wherein the presenter opens a toolbox, holds up each tool and describes its function, and then passes the tools around the audience. Without a carpentry project, it is difficult to build a reliable mental library of situations in which each tool is useful. So, in this text I use discrete mathematics as the context via which you learn about proving mathematical statements.

I believe that you can discover many interesting mathematical ideas, and even theorems and proofs, with a bit of guidance. In that vein, I have written Try This! sections that are sets of problems that allow you (usually collaboratively) to construct fundamental parts of the material. However, I also know that everyone misses a detail here or there and, more importantly, that as beginning mathematicians you will want to verify that your discoveries are correct! For this reason, I follow sets of discovery problems with sections that explain related material and give both examples and details.

The style and tone of this text is, let's face it, pretty silly. I mean that literally: I believe that you are more likely to absorb mathematics that is presented in a goofy way. Completely strange and unrealistic problem setups help you to separate the abstraction of the underlying mathematics from the presentation of a problem, and thereby give you practice in recognizing the mathematical essence of problems you may find in other contexts. I think that if you are laughing about a problem presentation, it aids you in focusing on the mathematics. It's easier to remember a weird introduction to a concept than to recall a straightforward statement. And there's no reason to be serious when there's an opportunity to have fun!

The only way to truly learn mathematics is by *doing* it and *practicing* it, not by *observing* it. It's just like any other skill: If you want to learn how to dance, you watch someone do and explain some steps, and then you practice those steps. To become really good, you usually have to engage in auxiliary activities, such as stretching or weightlifting (note that these are also practice). If you want to learn how to write, you write every day and try different forms of writing and ask for feedback on your work. You may also study grammar or read examples of excellent writing. That's general commentary, but now you need to know how to proceed with discrete mathematics.

2   How to Use This Book

This book is designed to help you learn discrete mathematics through a mix of discovery-based activities and the more traditional read-text-and-then-do-problems technique. You may be wondering how I chose which topics would be initiated via discovery-based activities and which would be introduced via text; here is the answer. Whenever I knew of a way that students could come upon ideas reasonably quickly themselves, I wrote problems and activities that would direct you along that way. Topics for which my experience has been that students will not readily reinvent the relevant ideas have gotten the I'll-just-tell-you-about-it treatment.

You may also be wondering, *Why* did I decide to write the book in this way? When people discover ideas for themselves, they tend to retain those ideas longer and to understand them more deeply than if someone else revealed the ideas. So where possible, I have provided discovery activities. At the same time, it's easy to miss part of a relevant idea when you're thinking about it on your own. Therefore, I've also written about the ideas that you should discover for yourself. *Please,* dear reader, do not read this text until you've worked through the discovery activities! The sections that might spoil your fun begin with a Hey! You! alert and are accompanied by the stop-sign-holding duck shown here. (By the way, the discovery activities are titled Try This! and are accompanied by a pencil-toting duck, also shown here, and the same duck signifies other activities and sets of problems throughout the text.)

Maybe you want to know more about what discovery activities are before you do them. Some of them are intended to have you create good definitions. (Definitions are written the way they are because they're useful for talking about ideas or proving theorems. It's easy to make bad definitions—too vague, too restrictive, insufficiently relevant—if you're not careful.) Some are designed to help you construct a theorem and proof at the same time. Some are structured so you will work with many examples, become familiar with the topic, and generate intuition.

You might be used to solving problems that are direct applications of previous text, as, for example, in an algebra class where a new type of factoring is introduced and then 30 problems are given that use this type of factoring. That will only sometimes be the case here, and there will only be a few such problems at a time. In this text, problems are of the following kinds:

- direct applications of the text (should only take a minute or so);

- indirect applications of the text, where it will not be instantly clear how to apply the text;

- creative thinking, where the ideas in the text will be useful but not in a specific way;

- extension problems, where you'll be trying to extend the ideas in the text to other situations;

- discovery problems, where you won't have text to rely on and just need to solve the problems by thinking.

Please don't be intimidated by these problems. You can do it! And lots of guidance is given; maybe you won't even notice how creative and critical your thinking is.

Warning: some problems take a long time to solve. By "a long time," I mean "more than an hour." There are some other problems that take maybe a minute or so to solve. You can probably recognize these because they are marked as problems to check your reading comprehension. But feel good about yourself if you can solve any end-of-chapter problem in under ten minutes. (This means that you should start your homework earlier than the night before it's due. A bonus to starting early is that you can be more efficient—spreading out the time you work on homework will allow your brain to percolate and soak and produce clearer thinking.)

It's good practice to write up solutions to problems. At first, you should write out solutions to *everything*. This will be a huge pain, but well worth it because in this way you will gain the skills needed to write difficult solutions when you encounter them later. As your expertise increases, you can start writing solutions only to not-immediate-to-solve problems. If you get *really* good, you can just write up difficult problems!

One way to do this course would be to have a notebook in which you record your work on the Check Yourself problems and on the Try This! problems, for later reference. Essentially, this would create a solutions manual for the book as you go but also would contain things that didn't work, etc. (Maybe you could lightly X out those pages).

By the way, all links and GeoGebra files mentioned in the text are available electronically at http://www.toroidalsnark.net/dmwdlinksfiles.html. A final note: this textbook uses gender-neutral pronouns, with *ze* as the third-person singular and *hir* as the corresponding possessive, and *Mx.* as an honorific. (We do not use "they" in the singular as it can be confusing in writing.) Exceptions are made for historical figures and people of the author's personal acquaintance whose gender identification is public.

2.1   How to Use This Book in a Class

If you are using this book in conjunction with a multi-student course, then certainly you should use it as your instructor advises. In the absence of advice, refer to Section 2.2 and substitute "classmate" for "buddy who is also self-studying."

2.2   How to Use This Book for Self-Study

If you are using this book for self-study, just read it cover to cover and follow the instructions given in the text. There are sections to read before working on Try This! problems, and these end with Check Yourself problems. You probably could have answered these questions immediately after reading about the corresponding concepts, but a page or so later you may need to review a definition or idea. That's why they've been placed later—to help you reinforce what you've just read. There are also sections for which there is no pre-trying-problems reading but where the problems are designed to help you discover ideas. These are also labeled Try This! If you can't solve the Try This! problems after an hour or two, it's time to read the corresponding text and then Try This! again. It's best to have a buddy—you might find one through The Art of Problem Solving (see http://www.artofproblemsolving.com)—who is also self-studying and with whom you can discuss your ideas. Working with a buddy is effective for Try This! problems, and there are some Check-Yourself Challenges in the book that ask you to generate your own examples; a buddy can help you determine whether your examples fit the criteria you desire. Then, of course, there are end-of-chapter problems. These are not marked with difficulty levels because what's easy for one person is difficult for another (and vice versa). However, there are some problems marked as Challenges (beware).

3   Tips for Reading Mathematics

Generally mathematics is much more difficult to read than fiction (this is probably not too surprising) or many kinds of nonfiction such as history books or instruction manuals (maybe not surprising in practice but perhaps surprising in theory). The reason is that mathematics is conventionally written in as concise a way as possible, both in the sense that symbols substitute for some words or phrases and in the sense that verbosity is avoided.

Hopefully these difficulties have been reduced by the friendly manner in which this text is written. However, the author is keenly aware that first, it is rarely possible to sufficiently reduce reading difficulties, and second, she has probably failed to achieve her ideals. So, here are some tips to help you through.

- ✤ Reading a sentence or paragraph or chapter multiple times is quite helpful. Here, "multiple" should be interpreted as some number in the three to seven range, especially if you don't think you understood every detail the first time through.

- ✤ If you feel like you just don't "get" some idea after reading supporting text a few times, go read some other source. Then come back to this one. Exposure to multiple perspectives helps you synthesize ideas.

- ✤ Do *not* skip words (… unless you are intentionally skimming). Every word in a math book is important and there for a reason. If you are having trouble understanding a sentence (or paragraph or problem) try reading aloud, even if (especially if) you feel silly. It does help. One of the most common causes of stuckness when solving a problem is having not read the problem statement carefully enough.

- ✤ It's useful to have some scratch paper and a writing implement nearby when reading. That way you can do calculations, attempt problems fully (not in your head), record questions and ideas that occur to you, try examples, note a definition or two that you keep having to look up, etc.

🦆 Oh, speaking of attempting problems, don't try to do problems in your head. I'm serious about this: I know you think that you should be able to, but almost no one can. And there's no reason you should—our brains are made for thinking, not memory, so use the paper as your recording device and free up your brain for thinking.

🦆 Don't believe any of the mathematical claims made in a book or paper without verifying them. For yourself. Yes, really. You can do it! On the other hand, don't let this bog you down; sometimes you just need to read the next sentence or two in order to clear things up. (This happens to the author all the time.) Don't get more than a paragraph or so ahead of that point where you last understood what was going on, unless you're *completely* stuck on that paragraph. In that case, mark it as something you need to go back to, and proceed in the hopes that life will improve.

🦆 Be active, not passive, while reading. That means that you should try to answer any questions raised in the text and try to solve any problems posed, and definitely don't trust the author's claims. Yes, that last repeats a point made in the previous bullet point. That's because it's super-important.

🦆 Ask yourself questions. (Do you sense a theme? Good.) For example, what are the main points of the section/chapter you just read? Can you generate your own examples of the newest definitions? What are some situations to which a recently stated theorem will apply? If there seem to be extraneous words, read again—why are they there? (Remember that mathematicians rarely use excess words.) Is a new concept similar to a concept you already understood? If so, how?

🦆 After a first read-through, read nonlinearly. That is, flip forwards or backwards to follow your own train of thought. Perhaps you will need to review an earlier concept or look up the meaning of a symbol, and in the process you generate a new question that sends you to yet another page. That's fine. Use your scratch paper to note your original goal so you don't forget it while following interesting mental tangents.

🦆 Read only for 20 minutes at a time (unless you lose track of time and keep going because you're having a great time). Then take a 5-minute break to let your brain absorb and process behind the scenes, and start reading carefully again.

4   Problem-Solving Prompts

Questions to ask yourself when you're stuck:

- ❧ Do I truly understand what the problem is asking? Maybe I need to read it aloud or look up some of the terms.

- ❧ Am I using the constraints introduced in the statement of the problem? They're probably there for a reason. Related question: am I using the criteria given in definitions of terms used in the problem?

- ❧ Is there a super-easy or even trivial example I can work through? This often helps to make sure you understand the setup of the problem. Try using 0 or 1 or $n = 0$ or $n = 1$.

- ❧ Is there a diagram I could draw? That might help.

- ❧ Can I break this down into a set of smaller or simpler problems?

- ❧ Is this problem related to any theorems I know? Or does it look similar to any examples I've seen?

- ❧ Am I sure that all the statements I've written down are correct (both in terms of reasoning and of symbolic manipulation)?

- ❧ Is the statement of the problem correct or true? Maybe I should be looking for a counterexample.

(Faculty readers will recognize this section as inspired by George Polya's *How to Solve It*.)

5   Tips for Writing Mathematics

So. About writing things up. There is not a single correct way to write, and as you write mathematics you will develop a mathematical writing voice of your own. Your first goal is always to communicate mathematics to a reader. Presumably that reader does not have identical comprehension of the particular mathematics to yours, or ze wouldn't need to read your write-up! This is particularly true with peers who are stuck on problems and seek your help. But you probably often have a second goal, namely, to communicate your *understanding* of mathematics. Many students who are being graded on problems or proofs mistakenly believe (or temporarily fool themselves into thinking) that the idea is to indicate an answer or a basic reason that a theorem is true and that the instructor/teacher/grader/professor will see this as a verification of task completion. Not so. The instructor/teacher/grader/professor wants to verify that you (a) have understood the mathematical material, (b) have been able to solve this related problem, and (c) can clearly communicate your understanding. Trust me—no one who chose to use this textbook would have any lower standard.

While we're talking about this, let me point out that writing mathematics well requires a lot of practice. Try not to get miffed if you're asked to rewrite something. It's not necessarily a problem with your understanding (although it might be) but instead with clarity (i.e., a lack thereof) in your communication. Just take a deep breath and remember that if a reader didn't understand something you wrote in your proof, then your communication has not been sufficient to convey that point. (Okay, now you're thinking, *So, when I don't understand something in this book, I can blame the author because she didn't communicate clearly!* As a mathematical beginner, it's more likely that you're not used to reading mathematics than that the mathematical writing is unclear. See the Tips for Reading Mathematics given in Section 3. But also, the author is human and sometimes does write unclearly no matter how many times she revises. She's sorry in advance if her writing happens to be incompatible with your brain at some point in this book.)

Some of the following tips may only make sense to you after you have begun writing proofs, and others will only sink in after you have practiced writing mathematics for a while, so you may wish to revisit this section regularly.

- 🦆 Make sure to define new terms and symbols as you introduce them and qualify them appropriately. For example, just because you use p doesn't mean everyone will know it's a prime.

- 🦆 Structure a solution as you would a paper (except your solution will hopefully not be as long). The introduction usually consists of a restatement of the problem, the body consists of a discussion and solution of the problem, and the concluding statement either places the solution in context or verifies that your argument has proved what you set out to prove.

🐦 End every statement with a period. A mathematical expression is part of a sentence; a statement has a verb and therefore is a sentence and therefore should end in a period.

🐦 Try not to have too many symbols appear in a row; insert words between them. For example, "$2+3 = 5\ 2+3+10 = 5+10 = 15$" would make more sense as "We know that $2+3 = 5$ and can add 10 to each side to obtain $2+3+10 = 5+10 = 15$."

🐦 Try not to begin a sentence with mathematical notation. If you have ended a previous sentence with notation, then it will be confusing. Plus, the capitalization can be worrisome; for example, when beginning a sentence with a it seems one should write A, except that A is a different beast.

🐦 Never use a pronoun without an antecedent. No one will know what you're talking/writing about. For example, don't write (or even say) "It's 5." *What*'s 5? If we had been faced with $10-x = 5$, it would not be clear whether you meant "$10-x$ is 5" or "$x = 5$".

🐦 Check to see that your written solution addresses the original question or proves the original statement. (This is related to making sure your solution contains a concluding statement.)

🐦 Be careful not to write the way you work. This has two meanings: one is that one's exposition can always be improved. The other is that often one works backwards when solving a problem or finding the path to a proof. But don't write that way! Often such text begins with (statement) $=^?$ (other statement) and proceeds to change each side of the proposed equation until a definite equality results. It ends up reading, "If statement S is true, then (fill in some steps here), so $1 = 1$ and we are done." But of course, $1 = 1$ whether or not S is true, so it does not need a proof to accompany it.

🐦 Read over your solution or proof after you're done. You might notice a flaw in your reasoning, or find that you need to add justification for a statement, or that you could say something in a way shorter or much clearer way.

Acknowledgments

First, I must acknowledge the four ducks who lived next door to me years ago, who are mentioned throughout the book. They didn't seem to lie, but they did sometimes lay eggs that became my breakfast. My favorite was the courageous grey duck, who regularly escaped her enclosure to root for bugs in the nearby field portion of my backyard. She (photo below) is the model for the grey duck shown in the book; she disappeared on April 4–5, 2011.

The large white duck (pictured on the title page and many other places in this book) was always happy to eat spinach stems and random greens from my garden; she was taken on June 30, 2011, most likely by a fox.

Thanks to Daniel Pinkwater for writing the hilarious auxiliary text *Ducks!* that, together with the neighbor ducks, inspired the theme for this book... and thanks to my parents for not lying (see auxiliary text). Thanks also to D. P. for writing *Lizard Music* (my favorite of his books) and the other 82 (at last count) D. P. books I own. Indeed, I couldn't have written this text, or lived my life the way I have, without them.

Thanks to Dylan Shepherdson, who first suggested I write a textbook based on the curriculum I designed for a class (and who had better darned well use this textbook now that it's here for him to use)!

My thanks go also to Alice and Klaus Peters, who felt that this textbook would be a useful contribution to the range of discrete mathematics textbooks. Klaus, sadly, died after the first edition of *Discrete Mathematics with Ducks* was published, but before his death he encouraged me to continue enhancing the text.

I would additionally like to acknowledge Alice and Klaus for their work in creating A K Peters, Ltd., a business and line of books that stood apart from the rest of the industry. Their books are known to be of high quality, beautifully made, and of interesting content. Alice and Klaus made possible the publication of many volumes that other publishers would have avoided or ruined; they structured all aspects of publishing to be excellent experiences for authors; and, it's sad that their unique environment no longer exists in the publishing world.

Along those lines: to editor Charlotte Byrnes, thank you for your many efforts on behalf of both editions of this book. I am *so* grateful that I could work with you, and work with you again; you make the behind-the-scenes aspects of production as much fun as—actually *more* fun than—*writing* a book!

Doug Shaw, thank you for your suggestions on improving the first edition before it was published, both in terms of the content and in terms of the writing. Also, thank you for modeling good teaching and mentoring me in the ways of active learning for lo these many years. The potential hot-air balloon ride continues to be duly noted.

Heather Ames Lewis, thank you for your support of this book and in particular for your happy email of October 2013, which lives in my inbox to this day to be reread on occasion.

Finally, I'm thankful to my personal feline Pantalaimon, who sat next to me or on me during much of the work on the second edition.

And now for some content-specific acknowledgments:

Dave Perkins, thank you for lots of great suggestions and for testing this book using our shared vision for how it could be used. Dave contributed Problems 19, 7, and 4 in the Problems on the Theme of the Basics and Problem 25 in Chapter 1 and, I'm sure, also some other problems that I forgot to note after incorporating them into the text.

Thanks go out to David Cox and Dana Rowland for finding many of my errors in the preliminary edition. Dana is a master at finding subtle inconsistencies; thank goodness she read my text! Dana wisely suggested that I create Problem 5 in the Check Yourself exercises for Section 4.2 and that I vertex-label graphs used in Chapter 12. She is also the genesis of the example in Section 14.5.1 of computing the expected cost to the insurer of a basic health insurance policy. David gave the text an extremely careful reading, and I thank him for the many interesting and valuable discussions about the teaching of discrete mathematics that resulted. David contributed Problems 33 and 34 in the Potential Practice Proof Problems as well as several set theory problems, and also Problems 39 and 41 in the Problems on the Theme of the Basics (and probably a few more elsewhere that I forgot to note).

There are explanations of cultural references and acknowledgment of outside sources scattered throughout the book in *Credit where credit is due* paragraphs of *Where to Go from Here* sections. My use of this phrase comes from the name of a New York City community credit union cofounded by one of my college classmates, Mark Levine.

I'm grateful to Neil J. A. Sloane for creating the Online Encyclopedia of Integer Sequences and making it such a useful website.

Thanks to Harriet Pollatsek for telling me about the Luminous Nose restaurant, for providing resources on discrete mathematics teaching, and for lots of great conversations about education (and other topics).

Daddy, your influence is directly felt on pages 249, 350, and 301.

Rachel Shorey, recall that the result of your running joke about Mr. B. is on page 503. Ben Eisen, thanks for sending me the *Elder Sign* dice info.

Section 3.3.3 was inspired by Josh Greene and Ari Turner (thank you both). A new and excellent approach to Section 8.7 was suggested by an anonymous reviewer. Section 8.6 was derived from notes written by Tom Hull (thank you), who also provided the idea for Problem 1 in Section 10.6. The exposition of Chapter 14 is based on notes taken in Adam Marcus's introductory probability lectures in 2007; thank you, Adam, for teaching me this material. Tamara Veenstra helped me think through the Try This! sections of Chapter 16.

Thanks to Tom Hull for many fruitful conversations about writing Bonus Check-Yourself problems throughout the text, though particularly for conversations about Chapters 4, 11, and 13.

Thanks to Karl Schaffer for suggesting the permutahedron activity of Chapter 12 and the idea of introducing the Euler φ function in Chapter 16, to Goran Konjevod for alerting me to the existence of [5] (without which I could not have written Chapter 17!), to Max Woerner Chase for giving permission to print his "Induction Song" lyrics, to Jillian Bakke for providing chemical information used in Chapter 13, and to Sean Kinlin for drawing street intersections from memory (on scratch paper) for my use in Chapter 13. And *of course* (much more!) thanks to Sean Kinlin for sharing my life, for cooking my dinner, and for being unfailingly supportive and funny (even if that includes a few too many bad puns).

I'm grateful to the unnamed students at Merrimack College, Amherst College, Western New England College (now Western New England University), and Luzerne County Community College who used the preliminary edition of this text, and to the named students Dana Fry, Marcy Rattner, Yang Song, and Dara Zirlin of Mount Holyoke College, as well as faculty Jessica Sidman, Heather Lewis, and Eric Clark, who used the first printing of this text; each collection of students+instructor found typos that none of the others did.

Also thanks to the people who notified me of errata in the first edition: Nicholas Zoller, Inna Pivkina, Michael Paul Cosentino, Tom Hull, Dana Rowland, Kevin Lillis, Nick Mariano, Sue Metzger, Doug Shaw, Ben Baumer, Elliott Warkus, Corrine Yap, Sam Oshins, Claudia Drury, and Xingchen Xu (the last five of whom were in my excellent Fall 2013 Discrete Mathematics class at Sarah Lawrence College).

Penultimately, thanks to the many students I've had over the years who responded so positively to serious mathematics treated with levity; you showed me that this is definitely the way to go. Thank you, thank you, *thank you*.

And finally, thank you, reader! I'm happy that you've decided to take a look at this book.

Affiliations: My work on both editions of this text has benefited from my appointments as Fellow of the Centre for Textiles and Conflict Studies and as Research Associate at Smith College.

Technical credits: Figures for the first edition were originally made with ACD Canvas X. New figures were made, and most original figures were edited, in Eazy-Draw, a technical illustration program for the Macintosh. Dave Mattson, the creator, gives excellent and prompt user support. A few images were output from *Mathematica*, and the photos of next-door-neighbor ducks were edited in Photoshop by Tom Hull.

Part I
Theme: The Basics

Chapter 1

Counting and Proofs

1.1   Introduction and Summary

Our introduction to discrete mathematics will begin with some problems. You should make a significant effort to solve these problems before proceeding further—it will help if you can meet with others to work collaboratively, but you can also solve them on your own (it'll just take longer). Working these problems will allow you to discover some basic principles (the sum and product and pigeonhole principles) of counting.

   The problems are followed by some reinforcing text that will make sure you acquire all the needed details. Please do not read it until after you have worked through the problems! It might spoil your fun.

   After that, things go downhill. (Just kidding.) One of the themes in this textbook is learning how to prove things, and we'll start by discussing proof and counterexample right after the reinforcing text about the sum and product principles. Finally, we will tell you everything you desire to know about the pigeonhole principle.

1.2   Try This! Let's Count

Even though you have no experience with discrete mathematics yet, just jump in here—these problems do not require any prior knowledge and are great to discuss with classmates. Do not be alarmed if you do not finish the entire set within a single class period.

1. At WEBS, America's Yarn Store, there are two aisle displays of sale yarns, six aisle displays of closeout yarns in the back warehouse, one aisle display of Grandpa's Garage Sale yarns in the back warehouse, and one display shelving unit of über-clearance $2/ball yarns in the retail area. To how many display areas can you go in WEBS to buy yarn that is not full price?

2. A group of friends goes out for single-scoop ice-cream cones. There are sugar cones, cake cones, and waffle cones. But there are only five flavors of ice cream left (peppermint, hoarhound, chocolate malt, gingerbread, and squirrel). How many cone/ice cream combinations can be ordered?

3. At this ice-cream store, ice-cream scoops are stored right in the ice-cream containers between uses. At least how many ice-cream scoops must be in use if two of them have to be stored in the same flavor ice-cream container?

4. A server at the Luminous Nose restaurant goes to this same ice-cream store but decides to get a triple-decker cone. The stacking of scoops on the cone is important: a cone with peppermint atop two scoops of squirrel tastes different than a cone with two scoops of squirrel atop a scoop of peppermint, so an order of peppermint-squirrel-squirrel is different from an order of squirrel-squirrel-peppermint. How many possible triple-decker ice-cream orders are there?

5. Some people heading to a party stop by the ice-cream store to buy quarts of ice cream. How many orders of three quarts could they make? What if the three flavors have to be different? What if no one will agree to order squirrel ice cream?

6. Four teams are attending a local Ultimate Frisbee meet. If each team plays each other team exactly once, how many games are played?

7. Some of the Ultimate Frisbee players decide to form temporary teams in an arbitrary way. They put royal blue and lime green armbands into a bag, and each player closes hir eyes and grabs an armband to see which temporary team ze'll be on. How many armbands need to be grabbed in order to ensure that one of the teams has at least two players? How many armbands need to be grabbed in order to assure that one of the teams has at least seven players?

8. Some Ultimate Frisbee meet attendees saunter over to the Healthy Snack Box Machine, where they each choose one of five kinds of fruit, one of three herbal teas, and one of six flavors of wrap sandwich to get packed in a box. How many possible snack boxes are there?

9. Let's generalize Problem 6 to a regional Ultimate Frisbee tournament where there are n teams attending. Teams are assigned numbers (1 through n) when they register. As before, each team will play each other exactly once.

(a) How many games does Team 1 play?

(b) How many games does Team 2 play? Wait, that counts the Team 1 versus Team 2 game twice. How many not-yet-counted games does Team 2 play?

(c) Keep going. How many "new" (uncounted) games does Team i play?

(d) How many games are played in total?

10. Let's also generalize Problems 2 and 3 to a more reasonable ice-cream store. There are still three kinds of cones (the usual), but now there are k flavors of ice cream.

(a) How many different single-scoop ice-cream cones can be ordered?

(b) How many ice-cream scoops must be in use if two of them have to be stored in the same flavor ice-cream container?

11. **Terminology alert:** We write finite sets as lists of their members (also called elements). For example, $\{2,3,5,7\}$ is an excellent set. So is $\{1,4\}$. These sets are *disjoint* because they have no members in common. On the other hand, $\{1,2\}$ is not disjoint from either $\{2,3,5,7\}$ or $\{1,4\}$. The *union* of two sets A,B (or many sets $A,B,\ldots,N$) is a set containing all members of A and of B (and of $C,\ldots,N$). The union of the three sets listed so far is $\{1,2,3,4,5,7\}$.

(a) How many elements are in the union of two disjoint finite sets?

(b) How many members does a union of finitely many disjoint finite sets have?

(c) Are the previous two questions related to any of the previous problems?

(d) How many members does the union of n disjoint sets, each with m elements, have?

12. **Another terminology alert:** We call the notation (a,b) an *ordered pair* and (a,b,c) an *ordered triple* (and yes, we call $(a,b,\ldots,n)$ an *ordered n-tuple*). Generally, the first member of the pair (or triple, etc.) is from some set A, and the second member of the pair (or triple, etc.) is from some set B, etc. If A has m elements and B has k elements, how many ordered pairs can be formed from A and B? Is this related to any of the previous problems?

1.3   The Sum and Product Principles

Hey! You! Don't read this unless you have worked through the problems in Section 1.2. I mean it!

There are two principles that underlie most of the problems you worked in Section 1.2. Here they are, stated formally.

Oh, wait, we need to define one piece of notation first. The number of elements in a set A is denoted by $|A|$.

> **The sum principle.**   The number of elements in a finite number of disjoint finite sets $A, B, \ldots, N$ is the sum of their sizes $|A| + |B| + \cdots + |N|$.

You might think of this visually as in Figure 1.1.

Problem 1, about WEBS, used the sum principle directly. Problems 6 and 9, about counting the number of games played in an Ultimate Frisbee tournament, used the sum principle by subdividing a set into smaller disjoint sets.

Here is another example.

Example 1.3.1.   An employer offers ten days of paid vacation, three paid sick days, and four paid personal days per year. How many days can one not work and still get paid? The sets V of vacation days, S of sick days, and P of personal days are disjoint, so by the sum principle the total number is $|V| + |S| + |P| = 10 + 3 + 4 = 17$.

We will eventually get around to figuring out what to do if our sets aren't disjoint, but you'll have to wait for Chapter 7 for that. (There will be lots of other interesting things to think about in the meantime!)

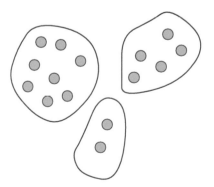

Figure 1.1.   The number of elements in the union of these three disjoint sets is $8 + 5 + 2 = 15$.

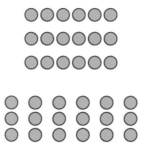

Figure 1.2. There are 18 elements total in three six-element sets, or in six three-element sets.

Wait, in order to continue, we need a second piece of notation! Ordered n-tuples $(a, b, \ldots, n)$ are members of the set denoted $A \times B \times \cdots \times N$, called the *Cartesian product* of the sets $A, B, \ldots, N$.

> **The product principle.** The number of elements in the Cartesian product of a finite number of finite sets $A \times B \times \cdots \times N$ is the product of their sizes $|A| \cdot |B| \cdot \cdots \cdot |N|$.

You might think of this visually as in Figure 1.2, where we might consider $|A| = 6$ and $|B| = 3$. The same Cartesian product can be grouped as $|B|$ copies of A, or as $|A|$ copies of B.

The product principle can also be formulated as making a collection of decisions or as putting choices in slots. For example, counting the number of ways to decide first which of r rooms to enter and then which of the c chairs to sit in and then which of p pencils to pull out of a case uses the product principle to see that there are $r \cdot c \cdot p$ ways. Similarly, making license plates that start with three numbers and end in BAT uses the product principle to see that any ten digits can be placed into each of the three number slots for a total of 1,000 license plates.

Problems 2, 4, and 8, about counting single-scoop and triple-decker ice-cream orders and Healthy Snack Boxes orders, used the product principle. The subproblem of Problem 11 that asked for the number of elements in the union of n disjoint sets, each with size m, could be solved using the sum principle by adding m to itself n times. Or, notice that adding m to itself n times is exactly what it means to multiply $n \cdot m$ (geometrically, make an $n \times m$ grid of elements), and solve the problem using the product principle.

Notice that we apply both the sum and product principles by letting sets stand in for something else—for example, the flavors of ice cream can be represented by the elements of a five-element set I (for ice cream), as can the types of cone by the elements of a three-element set C (for cone). This is a specific case of a more general counting technique of using stand-ins. For example, instead of counting pets in a shelter, one could count paws and divide by four. (It might be a good idea to check that each pet retains all four paws, as otherwise, modifications will need to be made to the total.) Sometimes we will let one set stand in for another set in our attempts to count. This will work as long as we know how the sizes of the sets in question are related to each other.

Here is another example that uses the product principle.

Example 1.3.2. The Restaurant Quatre-Étoile offers prix fixe meals only. (That means you pay a fixed amount and get a k-course meal, where k usually varies between three and five. It's pronounced "pree fix.") Their menu allows a choice of appetizers, a choice of main dishes, and a choice of desserts. We could view the menu as three sets: A has members Escargot Sampler, Quichelets, Puff Pastry Plantain Purses, and French Fries; M has members Veal Medallions with Infant Carrots, Foie Gras Falafel with Fig Fondue, Caviar-Crusted Croutons with Consommé, Zucchini Stuffed with Okra and Mushrooms, and Filet Mignon with Hard-Boiled Onions; and, D has members Ice Cream with Chocolate-Covered Grasshoppers under Mint Sauce, and Eight-Layer Orange-Glazed Pound Cake with Ganache Filling. The product principle says that the number of different meals that could be ordered is $|A| \cdot |M| \cdot |D| = 4 \cdot 5 \cdot 2 = 40$.

Hey, can we use the sum and product principles together in one problem? Yes, it does happen.

Example 1.3.3. A debit-card company, DCC Corp., decides that in order to increase security, it will allow three-digit and five-digit personal identification numbers (PINs) in addition to the usual four-digit PINs. Informally, we notice that there are ten choices for each digit, so there are 10^3 possible three-digit PINs, 10^4 possible four-digit PINs, and 10^5 possible five-digit PINs, for a total of 111,000 possible PINs for DCC customers. In terms of sets, we let T be the set of possible three-digit PINs, F be the set of possible four-digit PINs, and V be the set of possible five-digit PINs, so that by the sum principle the total number of possible PINs is $|T| + |F| + |V|$. But we don't know the sizes of T, F, and V. Here is where the product principle comes in: T is secretly the Cartesian product of three sets T_1, T_2, T_3, each corresponding to one of the digits of a three-digit PIN. So $|T| = |T_1 \times T_2 \times T_3| = |T_1| \cdot |T_2| \cdot |T_3|$. We know that $|T_1| = |T_2| = |T_3| = 10$. Likewise, F is the Cartesian product of four ten-element sets and V is the Cartesian product of five ten-element sets. Combining these observations, we obtain $|T| + |F| + |V| = (|T_1| \cdot |T_2| \cdot |T_3|) + (|F_1| \cdot |F_2| \cdot |F_3| \cdot |F_4|) + (|V_1| \cdot |V_2| \cdot |V_3| \cdot |V_4| \cdot |V_5|) = 10^3 + 10^4 + 10^5 = 111{,}000$.

Both the sum and product principles are only stated for finite sets and for finite numbers of sets. If you are interested in learning a little bit about infinite sets, look at Chapter 15.

You might notice at this point that we haven't said anything about ordering quarts of ice cream (as in Problem 5). Ha! This problem involves more advanced counting ideas, and we will address their governing principles in Chapters 6 and 7.

Check Yourself ───

1. Gelly Roll pens come in 6 solid colors of fine point and 11 of medium point, 10 moonlight colors, 10 shadow colors, 12 stardust colors, and 14 metallic colors. (Not kidding.) How many different Gelly Roll pens are there?

2. When redeeming a prize coupon, you may choose one of six charms and *either* one of three carabiners *or* one of two bracelets. How many different prize choices could you make?

3. **Challenge:** Invent your own problem that uses both the sum principle and the product principle.

───

1.4   Preliminaries on Proofs and Disproofs

In order to begin our study of careful reasoning and how to communicate our thoughts, we have to know the meanings of the words most commonly used in the reading and writing of mathematics.

Definition 1.4.1 (a clump of 'em). A *definition* is a precise statement of the meaning of a term. (Think dictionary, but better.) A *conjecture* is a statement proposed to be true and made on the basis of intuition and/or evidence from examples. (You already made some conjectures when you worked the problems in Section 1.2, and you'll make many more before this book is through.) A *theorem* is a statement that can be demonstrated to be true. A *proposition* is… well, some people use it as a smallish theorem, and others use it as a theorem offered (proposed) to the reader. A *lemma* is a small theorem, usually stated and proven in the process of proving a regular-size theorem. A *corollary* is a statement whose truth follows directly (or almost directly) from a related theorem. A *proof* is a justification of the truth of a statement using reasoning so rigorous that the argument compels assent.

Notice that definitions are precise; they are precise not only so that one can distinguish between similar concepts but also because they are used as references for rigorous reasoning. Definitions are not arbitrary (even though they often seem that way). Instead, a definition comes about because someone needs it either to shorten communication or to help justify an idea. (Do you think that our definition of "definition" fulfills the criteria to be a definition?) Often a definition lists criteria that must be checked in order for the definition to be fulfilled; be on the watch for such criteria, as they are the key to using a definition as part of a proof.

Here are a few examples of definitions—you may already be familiar with these ideas.

Definition 1.4.2. An integer $n > 1$ is *prime* if the only positive divisors of n are n and 1.

By this definition, 3 is prime because $3 \cdot 1 = 1 \cdot 3 = 3$, and there is no different possible factorization into positive integers.

Definition 1.4.3. A number is *even* if it is evenly divisible by 2. Equivalently, a number m is even if $m = 2k$ for some integer k. A number m is *odd* if $m = 2k + 1$ for some integer k.

The number 64 is even because $64 = 2 \cdot 32$. However, the number 3 is not even because $3 = 2 \cdot \frac{3}{2}$ and $\frac{3}{2}$ is not an integer, but $3 = 2 \cdot 1 + 1$ so the number 3 is odd.

Definition 1.4.4. A *binary number* is a number expressed using only the digits 0 and 1, with counting proceeding as $1, 10, 11, 100, 101, 110, \ldots$ and with places representing powers of two, increasing to the left and decreasing to the right.

Thus, the number 64 is not binary because it uses digits other than 0 and 1, but 101 and 101011.101 can be binary numbers or decimal numbers. The binary number 101 represents $2^2 + 0 + 2^0 = 5$ in decimal notation, and the binary number 101011.101 represents $2^5 + 0 + 2^3 + 0 + 2^1 + 2^0 + 2^{-1} + 0 + 2^{-3} = 43.625$ in decimal notation. (Remember that $2^{-1} = \frac{1}{2}$ and $2^{-3} = \frac{1}{8}$.)

Example 1.4.5. Let's look at the numbers 5, 28, and 10. Because $5 \cdot 1 = 1 \cdot 5 = 5$ and there are no other integer factorizations of 5, it is prime. It is also odd because $5 = 2 \cdot 2 + 1$. On the other hand, $28 = 14 \cdot 2$ is even, and not prime; it is also not binary because it uses digits other than 0 and 1. We can rewrite $28 = 16 + 8 + 4$ $= 2^4 + 2^3 + 2^2$, so its binary representation is 11100. The number 10 could be the decimal number 10 or the binary number 10 representing the decimal number 2 (which, by the way, is both even and prime).

You probably made some conjectures when you were working through the problems in Section 1.2. The creation of conjectures is a most important process in mathematics, so we will be concerned with it throughout this text. It is part of what makes mathematics an art rather than a collection of facts or rules. When you encounter a new problem or concept, you should generate and explore some examples. This in turn will help you generate ideas, and then you can notice patterns and say what you think is true (and that's a conjecture!). Practice this process often. Start now by examining these data:

$$24 = 5 + 19 = 7 + 17 = 11 + 13.$$
$$8 = 3 + 5.$$
$$38 = 19 + 19 = 7 + 31.$$

- ✦ What property do the numbers on the left-hand sides of the equations have in common?

- ✦ What property do the numbers on the right-hand sides of the equations have in common?

- ✦ Come up with three more examples that fit this pattern.

- ✦ Do you think the pattern always holds?

- ✦ What is your conjecture? (We will revisit this later.)

One of the skills you must learn as a mathematician is making conjectures, and another is determining whether your conjectures—and those conjectures others share with you—are true. In that vein, you will often be asked to prove statements that are true, but sometimes you will be asked to prove statements that are false. (It is not possible to successfully prove a false statement.)

Here are some examples of theorems.

Example 1.4.6 (of theorems). Three. Yes, three theorems are here.

- ✦ Every natural number greater than 1 has a unique factorization into prime numbers.

- ✦ Suppose n teams play in a tournament. Then for each team to play each other team exactly once, there need to be $(n-1) + (n-2) + \cdots + 1$ games, which is equal to $n(n-1)/2$ games.

🐦 If a natural number is expressed in both binary and decimal forms, the binary number will have at least as many digits as its decimal equivalent.

You proved the first part of the second theorem above, and you may prove the accompanying formula in Chapter 4. A proof must be convincing in the logical sense, but it need not explain why a theorem is true or provide insight as to why the theorem is true. Those are both devoutly to be wished, of course. A proof must compel assent and, in order to do so, must communicate ideas to the reader or listener. Does this mean that a proof must be intelligible to anyone who reads or listens? In some sense, yes—if someone doesn't believe your proof, then it is inadequate. But the reader/listener must make a reasonable effort to understand, by translating symbols and checking definitions of unfamiliar terms.

Our next example requires a new bit of notation: $a \in A$ means that a is an element of the set A.

Example 1.4.7 (of a proof). Let us prove a special case of the product principle. We would like to show that for finite sets A and B, the number of elements in $A \times B$ is $|A| \cdot |B|$. First note that by definition, the elements of $A \times B$ are ordered pairs (a_i, b_j), where $a_i \in A$ and $b_j \in B$. For each element $a_i \in A$, there are $|B|$ pairs $(a_i, \star)$ because there are $|B|$ different ways to put an element of B in the $\star$ slot. Now, there are $|A|$ elements of A, so the total number of pairs is (number of elements of A) · (number of pairs formable with one element of A) $= |A| \cdot |B|$. We have reached the desired conclusion, so we are done!

The simplest proof technique is direct proof. Here is how to do it.

Template for a direct proof:

1. Restate the theorem in the form *if (conditions) are true, then (conclusion) is true*. Most, but not all, theorems can be restated this way. (For example, some are secretly *(conditions) are true if and only if (conclusions) are true*, a structure you will learn about in Section 2.3.1.)

2. On a scratch sheet, write *assume (conditions) are true* or *suppose (conditions) are true*.

3. Take some notes on what it means for (conditions) to be true. See where they lead.

4. Attempt to argue in the direction of *(conclusion) is true*.

5. Repeat attempts until you are successful.

6. Write up the results on a clean sheet, as follows.

  🐦 Theorem: (State theorem here.)

  🐦 Proof: Suppose (conditions) are true.

  🐦 (Explain your reasoning in a logically airtight manner, so that no reader could question your statements.)

  🐦 Therefore, (conclusion) is true. (Draw a box or checkmark or write Q.E.D.—the abbreviation of *quod erat demonstrandum*, Latin for "which was to be demonstrated"—to indicate that you're done.)

Admittedly, there is a lot of grey area in just how one should argue in the direction of *(conclusion) is true.* This is where the creativity and art of proof come in. However, having a structure to work within is very helpful. Working backwards from (conclusion) is sometimes helpful as part of the attempts, as long as the results are presented "forwards." One must be careful to avoid the temptation to start with the conclusion and work backwards and then hand that in as a finished proof; the steps have to be reversible and presented in the appropriate order. (One must also avoid the temptation to give a few examples and call it a day. That is *not* a proof.) Let's think through a simple direct proof.

Example 1.4.8 (of a direct proof). Let us show that if n is an even number, then for any integer k, the number kn is even. We have been given the statement in if-then form, so we may suppose that n is an even number. Our desire is to find a way to show that kn is even. What do we know about even numbers? Well, the definition of an even number says that it is a multiple of 2, so that means that $n = 2m$ for some integer m (in fact, it's the integer $\frac{n}{2}$). We can substitute this into the expression we want to know about, kn, to see that $kn = k2m$. Aha! This is also $2(km)$ and that means the expression is a multiple of 2, and so it is even. We are done (except for writing it up nicely, which we leave to you in order to save some typing).

Usually direct proofs are not so simple. Sometimes they are much longer; sometimes they require a number of cases. Also, in the wide world of mathematics, people don't usually name the proof techniques they're using (except for induction, which we'll learn about in Chapter 4). Thus, it is rare to find the sentence, "We proceed by direct proof" written outside of an introduction to proof writing.

Then there is the problem of dealing with false statements. What are you supposed to do if you are given a proposition to prove and it turns out to be false? Well, here's the deal. First, you have to figure out that the proposition is false. (It's not a bad idea to suspect that any statement you're asked to prove might be false.) If a statement is false, you know why: you've found a particular case in which it is untrue, also known as a *counterexample*. So that's all you have to do … state that counterexample.

Example 1.4.9 (of a counterexample). *Proposition:* If n is even, then $2n - 5$ is also even.

This proposition is false, because 4 is an even number but $2 \cdot 4 - 5 = 8 - 5 = 3$ is odd and thus not even.

An excellent reference for learning about proof techniques and proof writing is *Book of Proof* by Richard Hammack. It has tons of examples and elementary exercises and is freely available online [12].

Example 1.4.10 (of an open problem). *Proposition:* If $n > 2$ is even, then n can be written as the sum of two primes.

Maybe you conjectured this on page 11. Surprise! This is a famous statement known as the *Goldbach conjecture*. No one knows whether this proposition is true or not! However, most people think it is very likely that the Goldbach conjecture is true—it has been verified for numbers up to $3 \cdot 10^{18}$. (Christian Goldbach (1690–1764) made the conjecture in 1742 as part of a correspondence with Euler, who figures prominently in Chapter 12.)

Check Yourself ───────────────────────────────

1. Prove that if n is even, then n^2 is even.

2. Prove that if n is odd, then $n^2 + 5n - 3$ is also odd.

3. **Challenge:** Invent your own false proposition and accompany it with a counterexample.

1.5   Pigeons and Correspondences

We will have much discussion of sets and subsets in Chapter 2, but for now we will define subsets so that we can count them. A *subset* A of a set B is a set all of whose members are also members of B. For example, $\{1, 4, duck\}$ is a subset of

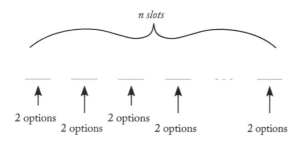

Figure 1.3. By the product principle, there are 2^n ways to fill in these slots.

$\{1,2,3,4,5,duck\}$, but $\{duck,egg\}$ is not. We would like to count the number of subsets of a finite set with n elements, so we will do it more than once, in different ways.

The first way excellently uses the product principle. It also uses the idea of one-to-one correspondence. This is the idea behind converting any counting problem (call this one Problem 1) into another counting problem (perhaps called Problem 2): if the items counted in Problem 1 are in one-to-one correspondence with the items counted in Problem 2, then there are the same number of items counted in each problem. But before discussing this excellent way of counting subsets, let's do an example.

Example 1.5.1. What are all the subsets of $\{egg,duck\}$? Certainly $\{egg\}$ and $\{duck\}$ are subsets. Also, $\{egg,duck\}$ is a subset of itself (the elements are the same), and the empty set (denoted $\emptyset$) is also a subset. In fact, the empty set is a subset of every set, though in a rather boring way. So in total, $\{egg,duck\}$ has four subsets.

Consider these subsets as follows. Each subset corresponds to a way of filling in two blanks ___ ___. The first blank either has *egg* or doesn't, and there are two options there. The second blank either has *duck* or doesn't, and there are two options there. The product principle says there are $2 \cdot 2 = 4$ subsets in total.

From this, one can abstract that if a set E has n elements, one of which is *egg*, then half of the subsets of E contain the element *egg*. Each subset of E corresponds to a way of filling in n blanks, as indicated in Figure 1.3. The first blank either has *egg* or it doesn't, and there are two options there. Likewise, each other blank either has its assigned element or it doesn't, and each has two options. The product principle says there are $2 \cdot \cdots \cdot 2 = 2^n$ subsets in total. We will revisit this argument in another context (graph theory) in Chapter 10.

Here is a related way to count the subsets of an n-element set. We assign a 1 or 0 to each set element, depending on whether it is or is not in the given subset (much like filling in or leaving a blank). This produces a one-to-one correspondence between subsets and strings of binary digits (called *binary strings*). We again use the set $\{egg, duck\}$ as an example. As shown in the table below, we convert each subset to a binary string.

Subset	Binary String	Decimal Number	Counting Number
$\emptyset$	00	0	1
$\{duck\}$	01	1	2
$\{egg\}$	10	2	3
$\{egg, duck\}$	11	3	4

We can read each binary string as representing a binary number and then convert each such binary number to decimal (base 10). However, the smallest of those decimal numbers is 0, which is not useful for counting, so we add 1 to each of the decimal numbers. This effectively produces another one-to-one correspondence, between binary strings and counting numbers. (By the way, one-to-one correspondences are more formally known as bijections, and we will discuss them in more detail in Chapter 3.)

This reduces our original question to "How many binary strings are there with n digits?" We might note that the largest binary number represented by an n-digit binary string is $111\ldots1$ (n ones). Now, there is one more binary string than there are numbers counting up to $111\ldots1$ (n ones) because we need to include the string $000\ldots0$ (n zeroes). So, we can simply add 1 to the decimal equivalent of $111\ldots1$ (n ones). Also, we could add 1 before or after converting to decimal, so let's do it before and get $1000\ldots0$, or 1 followed by n zeroes. That's 2^n—ta da!

Each way of counting the number of subsets of a set is a different proof of

Theorem 1.5.2. *A set with n elements has 2^n subsets.*

It is useful to have different proofs of the same theorem because they give different understandings of, or different perspectives on, the mathematics involved. Hidden in the above proofs is the following.

Fact. If two sets A and B are in one-to-one correspondence, then they have the same size.

Yes, you probably knew this, but it is worth stating explicitly so that you will remember it when it is useful. How exactly will it be useful? Well, we will focus in Chapters 6 and 7 on a few types of counting problems—and most other counting

Figure 1.4. Pigeons sitting in pigeonholes.

problems can be solved by creating one-to-one correspondences with those famil-iar problems. So trying to find one-to-one correspondences is a skill you'll want to acquire over time. There are related facts about the sizes of sets that are *not* in one-to-one correspondence, and we'll find those in Chapter 3.

Here is another explicit statement of a fact that you probably already know.

The pigeonhole principle. If you have more pigeons than pigeonholes, then if every pigeon flies into a hole, there must be a hole containing more than one pigeon. (See Figure 1.4.)

Really. Not kidding, it is actually called the pigeonhole principle by pretty much everyone; this is not a silly name invented for this book, unlike some other names you will find here. How on earth are these pigeons relevant?

Example 1.5.3. Suppose you have a bag of pigeons, some grey and some black. More classically, suppose you have a drawer full of grey socks and black socks. How many pigeons/socks must you grab in order to be sure you have two of the same color? One is clearly not enough, and will only ask to drive the bus; two might be enough if you're lucky, but you could also get one grey and one black; but three gives the guarantee that even if the first two were grey and black, respectively, the third must be either grey or black and thus be the same color as one of the first two pigeons/socks.

While we are making explicit things you know (but might not have stated out-right), here is

How to apply the pigeonhole principle:

1. Figure out what represents the pigeons. In Example 1.5.3, these were both pigeons and socks.

2. Figure out what represents the pigeonholes. In Example 1.5.3, the pigeon-holes are pigeon/sock colors.

3. Figure out how pigeons correspond to holes. In Example 1.5.3, a pigeon flies into a hole that matches its color.

Sometimes it is not obvious how to apply the pigeonhole principle, and in such cases the explicit instructions will be useful. Let's do a more complicated example.

Example 1.5.4. Did you know that in San Francisco, at least five people have the same number of hairs on their heads? Wow, that's gnarly. Intuition first:

The population of San Francisco is at least 870,000, according to https://www.census.gov/quickfacts/fact/table/sanfranciscocitycalifornia,US/PST045216.

According to various unreferenced sources on "teh intarwebs," the average person has 100,000 hairs on hir head; those who are naturally blonde average 140,000 hairs. It seems that 180,000 would be a reasonable upper bound for the number of hairs on a human head, but let's be safe and use 200,000 as an upper bound.

Then, $4 \cdot 200,000 < 800,000 < 870,000$, and so it looks like we have more than four people per hair-number. With pigeons:

Pretend that each resident of San Francisco is a pigeon.

And, pretend that there is a set of pigeonholes numbered from 1 to 200,000.

Even if the first 200,000 pigeons fly into different holes, and then the next 200,000 pigeons each fly to a hole containing only one pigeon, and then the next 200,000 pigeons each fly to a hole containing only two pigeons, and then the *next* 200,000 pigeons each fly to a hole containing only two pigeons, there will be four pigeons in each hole and there are at least 70,000 pigeons who still need holes. Thus, there must be some hole that houses at least five pigeons, and therefore there are at least five San Franciscans with the same number of hairs on their heads.

By the way, this kind of argument is known as an *existence proof*. That's because we know the five people exist, but we don't know who they are. (This is also called a *nonconstructive proof*, in contrast to a *constructive proof*, in which we would explain how to find the five people.) Some people find existence proofs unsatisfying. Oh, well.

Example 1.5.4 used a variant on the pigeonhole principle, namely

The generalized pigeonhole principle. If you have more than k times as many pigeons than pigeonholes, then if every pigeon flies into a hole, there must be a hole containing more than k pigeons.

Notice that when proving Theorem 1.5.2, we used a one-to-one correspondence; the pigeonhole principle is essentially using the *lack* of a one-to-one correspondence. (After all, if there were a one-to-one correspondence, there would be the same number of pigeons as pigeonholes.) We will investigate some formal details along these lines in Chapter 3. Similarly, the generalized pigeonhole principle is essentially using the lack of a *many*-to-one correspondence.

For a final example, we will use Theorem 1.5.2 together with the pigeonhole principle.

Example 1.5.5. Given any list of 25 numbers, each of which has at most five digits, two subsets of the list have the same sum. Again, intuition first:

Any one of the numbers is less than 100,000, so the sum of all 25 of them is less than 2,500,000. Therefore any *subset* of the 25 numbers also has sum less than 2,500,000. (We will ignore the empty set, even though it is a subset of the numbers.)

To find the lowest sum possible, consider the case of a subset that's just the number 00001. It has sum 1. Therefore, there are at most 2,500,000 different possible sums the subsets could have.

Now, how many subsets are there? We know this from Theorem 1.5.2—there are $2^{25} = 33,554,432$ possible subsets. (Actually, because we have ignored the empty set, we are only considering $2^{25} - 1 = 33,554,431$ subsets.)

There are *way* more subsets than sums, so two of the subsets must have the same sum. In terms of pigeons, we represent the subsets by pigeons and the subset-sums by pigeonholes; a pigeon flies to the pigeonhole labeled with its subset's sum.

Check Yourself ————————————————————————————

1. List all the subsets of $\{egg, duck, goose\}$. How many are there? How many of them contain *egg*? … *duck*? … *goose*?

2. Consider a standard deck of cards with suits hearts ($\heartsuit$), spades ($\spadesuit$), clubs ($\clubsuit$), and diamonds ($\diamondsuit$), and values 2–10, jack, queen, king, and ace. How many cards must you deal out before being assured that two will have the same suit? How many must you deal out before being assured that two will have the same value?

3. **Challenge:** Invent your own counting question that can be answered using the pigeonhole principle.

1.6   Where to Go from Here

This chapter contained a very basic introduction to enumerative combinatorics, the science of counting. To learn more, consult Chapters 6–9 (and then see where those chapters direct you!).

More specifically, binary numbers and strings are used throughout computer science as ways of representing data in computers. Sets and subsets are treated extensively in Chapter 2. The study of number properties such as *even*, *odd*, and *prime* is part of the larger field of number theory, of which we will encounter more in Chapter 5 and which is addressed in Chapter 16. One-to-one correspondences are studied at length in Chapter 3. We will address more proof techniques in Chapters 2 and 4, and Richard Hammack has written the lovely *Book of Proof* [12] for further study.

Yes, this is a brief section, but that's because this is the most introductory chapter! In other chapters we will give more information and advice.

Credit where credit is due: Most of the problems in Section 1.2 were inspired by [3], and several problems in Section 1.9 were inspired by [1]. In Section 1.2, WEBS is a real store (see www.yarn.com) and the Luminous Nose is a real restaurant in Japan (or at least that's what I'm told the Luminous Nose building is). Example 1.5.3 refers to *Don't Let the Pigeon Drive the Bus* by Mo Willems. Bonus Check-Yourself Problem 9 was suggested by Doug Shaw; Bonus Check-Yourself Problem 1 was inspired by colleagues at the Centre for Textiles and Conflict Studies. In Section 1.9, the grape-nut burgers in problem 1 are Jim Henle's recipe, Problem 5 references an old internet joke from the age of modems, and Problems 30–32 were inspired by Karl Schaffer's notes thereon.

1.7   Chapter 1 Definitions

disjoint sets: Sets with no elements in common.

union of sets: The union of two sets A, B (or many sets $A, B, \ldots, N$) is a set containing all members of A and of B (and of $C, \ldots, N$).

subset: A subset A of a set B is a set all of whose members are also members of B.

definition: A precise statement of the meaning of a term. (Think dictionary, but better.)

conjecture: A statement proposed to be true and made on the basis of intuition and/or evidence from examples.

theorem: A statement that can be demonstrated to be true.

proof: A justification of the truth of a statement using reasoning so rigorous that the argument compels assent.

proposition: A smallish theorem, or a theorem offered (proposed) to the reader.

lemma: A small theorem, usually stated and proven in the process of proving a regular-size theorem.

corollary: A statement whose truth follows directly (or almost directly) from a related theorem.

counterexample: A particular case in which a statement is untrue. For example, 3 is a counterexample to the statement *all numbers are even*.

prime number: An integer $n > 1$ whose only positive divisors are n and 1.

even number: A number evenly divisible by 2. Equivalently, a number m is even if $m = 2k$ for some integer k.

odd number: A number m is odd if $m = 2k + 1$ for some integer k.

binary number: A number expressed using only the digits 0 and 1, with counting proceeding as $1, 10, 11, 100, 101, 110, \ldots$ and with places representing powers of two, increasing to the left and decreasing to the right.

1.8 Bonus Check-Yourself Problems

Solutions to these problems appear starting on page 593. Those solutions that model a formal write-up (such as one might hand in for homework) are to Problems 2, 4, and 6.

1. A Timbuk2 custom messenger bag comes in four sizes, has 46 options for the left-panel and center-panel and right-panel fabrics, 18 different binding options, 27 logo colors, 11 liner colors, three options for pocket style, two handednesses, and 47 different options for the strap pad. (Really, not kidding—these numbers came from the Timbuk2 website in October 2014.) How many different custom messenger bags could one order?

2. Prove that the product of any three odd numbers is also odd.

3. Takeo, a paper store in Tokyo, has walls lined with coded drawers. Each code designates a type of paper. One such drawer is 2Q08. If the first entry has to be 1, 2, or 3 (there are only three walls with drawers), the second is a letter, and the last two are numbers, then how many drawers could Takeo have?

4. You want to buy an electric car. The Chevy Volt comes in eight colors (red, brown, grey, pale blue, two blacks, two whites), offers three kinds of wheels, and has five kinds of interiors (two cloth, three leather). The Tesla comes in nine colors (black, two whites, two greys, brown, red, green, blue), and gives a choice of three roof styles (one is glass), four wheel styles, four seat colors, four dashboard prints, and three door-trim colors. There are three versions of the Nissan Leaf (S, SV, SL), each of which comes in seven colors (two whites, two greys, red, blue, black). How many different choices of car do you have?

5. Prove, or find a counterexample: the sum of two consecutive perfect cubes is odd.

6. How many four-digit phone extensions have no 0s and begin with 3?

7. In 2016, there were 3,945,875 live births in the US. (Source: http://www.cdc.gov/nchs/fastats/births.htm.) Did there have to be two of these births within the same second?

8. How many length-8 binary strings have no 0s in the fourth place?

9. You receive a choose-your-own-adventure certificate for a jewelry store! The deal is that you get to pick one of eight precious gems, and either a ring or a bracelet to put it in. There are three possible ring styles and six possible bracelet styles.

   (a) How many possible prizes are there?

   (b) How did you answer the previous question? If you used the product principle first, re-answer the question using the sum principle first.

(And if you used the sum principle first, re-answer the problem using the product principle first.)

   (c) On closer look, you realize that neither the ruby nor the emerald would look good on the bracelet. How many prizes are still possible?

10. I have a lot of stuff in my stuff-holder: six ball-point pens, a silver star wand, three teal signature pens, a bronze-yellow colored pencil, five liquid ink pens, three mechanical pencils, a highlighter, six permanent markers, seven gel pens, a Hello Kitty lollipop, two markers, three wooden pencils, a 3-inch-long pen, a calligraphy marker, a pen shaped like a cat, and a pair of left-handed office scissors.

How many writing utensils do I have in the stuff-holder?

1.9   Problems That Use Counting or Proofs

Even when a problem statement doesn't explicitly say that you must explain your reasoning, you still should give some justification for your answer—even if it's just a few words.

1. Bruno Burger's specialty is, you guessed it, burgers. They offer four different burger patties (chicken, fish, soy, and grape-nut) with your choice of seven vegetables (onions, lettuce, tomato, kale, red onions, zucchini, and eggplant). How many patty-with-a-vegetable burgers can be ordered?

2. The Supreme Bruno is any patty-with-a-vegetable burger plus a condiment (choose from Worcestershire sauce, wasabi sauce, or mustard); you can also have cheese, or not. How many Supreme Brunos could be ordered?

3. Prove that the sum of two even numbers n_1 and n_2 is also even.

4. Prove that the sum of two odd numbers n_1 and n_2 is even.

5. You are assigned to communicate with a truly ancient computer. You must do this by telephone by shouting binary digits over the line, in clumps of eight digits. How many different eight-digit binary strings are there to shout?

6. A local creperie offers sweet crepes and savory crepes. A sweet crepe could have any fruit (banana, strawberry, mango, apple, lemon) and any syrup

(nutella, chocolate, caramel, honey). A savory crepe could have any vegetable (broccoli, mushroom, spinach) and any protein (turkey, cheese, prosciutto). How many different crepes are on the menu?

7. Prove that every binary number n that ends in 0 is even.

8. Prove that every odd number n ends in 1 in its binary representation.

9. Scary Clown offers a Sad Meal containing a sandwich, a salad, a dessert, and a drink. (They are not mixed together in the box.) There are 11 types of sandwiches, 3 types of salads, and 5 different kinds of desserts. A person with low standards for food could eat a different Sad Meal every day for three years. So how many drinks are possible choices for a Sad Meal?

10. Prove, or find a counterexample: the difference of two consecutive perfect squares is odd.

11. Every US coin is stamped with the year in which it was minted. How many coins do you need to have in your pocket to be assured that at least two of them have the same last digit? How many do you need to be assured that at least two have the same first digit?

12. Prove, or find a counterexample: the sum of two perfect squares is even.

13. In order to keep track of circulation numbers, the library asks you to note on a form, when you leave the library, which combinations of 15 subject areas and of 8 types of material (books, current journals, databases, bound journals, videotapes, microfilm, microfiche, DVDs) you used. How many possible ways are there to fill in a line on the form?

14. (Still about the library) Of course, not every combination is realistically possible, as the library does not hold materials in every type for every discipline. If the library has six types of material for each discipline, how many possible ways are there to fill in a line on the form?

15. (And more about the library) More realistically, some disciplines use materials in more differing forms than others. Let's look at just a few disciplines. The Dance holdings are in videotape, DVD, current journals, bound journals, and books. The Math holdings are in books, current journals, databases, bound journals, videotapes, and microfilm. The Computer Science holdings are in books, databases, and DVDs. Ancient Studies holdings are just bound journals, videotapes, microfilm, and microfiche. How many possible ways are there to fill in a line on the form for these four disciplines?

16. Prove, or find a counterexample: the sum of two primes is even.

17. At Chicago O'Hare International Airport, there are an average of 1,185 direct flights per day (source: http://www.flychicago.com/ohare/myflight/direct/pages/default.aspx). Prove that at least two of these flights must take off within 90 seconds of each other.

18. Prove that if n is even, then $(-1)^n = 1$.

19. How many different seven-digit phone numbers are there?

20. How many different seven-digit phone numbers begin with 231- and contain no 9s?

21. Is the product of two odd numbers even or odd? Prove it.

22. Let us try to strengthen the result in Example 1.5.5.

    (a) Does a list of distinct five-digit numbers of length 20 have the property that there must be two subsets of the list with the same sum?

    (b) What is the smallest list of distinct five-digit numbers such that there must be two subsets of the list with the same sum?

23. Prove that if n is any integer, then $3n^3 + n + 5$ is odd. (Suggestion: do one case for n odd and one case for n even.)

24. A cold-footed centipede has a drawer filled with many, many socks. And yes, that centipede does have 100 feet. If the centipede only owns green and brown socks, how many must it pull from the drawer in the dark of the morning to be assured that it has a matching set for all of its feet (100 socks of the same color)? What if the centipede also owns polka-dotted socks? What if the centipede's drawer has many, many socks of k different colors?

25. **Challenge:** What if the centipede wants 50 (possibly different) matching pairs, one pair for each pair of feet? Consider first a centipede who only owns green and brown socks, then a centipede who also owns stripey socks, and then a centipede who owns k different colors of socks.

26. Let us propose an alternate definition for prime numbers: *An integer $n \geq 1$ is prime if the only positive divisors of n are n and 1.* Which numbers are prime under this definition? Why has the mathematical community chosen *not* to use this definition?

27. Let us propose an alternate definition for prime numbers: *An integer $n > 1$ is prime if the only positive divisor of n is n.* Which numbers are prime under this definition? Why has the mathematical community chosen *not* to use this definition?

28. The Red Dot company sells laser pointers in three colors (red, green, purple) and two lengths (keychain, pencil). The green and purple laser pointers only come with regular tips, but there is also a Fancy Tip option for the red laser pointers. How many options are there for Red Dot laser pointers?

29. You buy a Scheepjes Catona Colour Pack, which contains 109 mini-skeins in rainbow colors that "create a collection of mercerized cotton that will dazzle and delight you." However, you only have eight rainbow-color bins (red, orange, yellow, green, blue, indigo, violet, and neutral) to store them in. Prove that some bin must contain at least 14 mini-skeins.

30. Magic Trick! You challenge a friend to choose seven different natural numbers in the range 1–12. You claim that (and so you should be able to prove that)...

    (a) ... two of your friend's seven numbers sum to 13.

    (b) ... two of your friend's seven numbers have a difference of 6.

    (c) ... two of your friend's seven numbers have a difference of 3.

31. Magic Trick #2! This one requires *two* friends. One of your friends picks eight different numbers in the range 1–20. The second friend chooses one of these eight numbers to remove. You claim that there are two subsets of the remaining numbers that have the same sum. Prove your claim!

32. In Massachusetts there are a *lot* of Dunkin Donutses. It is not particularly unusual to find five Dunkin Donutses within one square mile. Must there be two of them within $\frac{3}{4}$ of a mile of each other? What about within $\frac{1}{2}$ mile of each other?

33. In the author's current house, the foyer has a two-shelf bookshelf, the living room has two six-shelf bookshelves, the dining room has a two-shelf bookshelf, and the pantry has a shelf full of books. In addition, there are three other shelves in the dining room that have books on them, and there are three shelf-like surfaces in the living room that hold piles of books. How many different shelves could be holding the book *Keeping Ducks and Geese* by Chris and Mike Ashton?

34. The Jinhao Shark fountain pen comes in 12 different colors and two different nib styles. How many different shark-headed pens could one own?

35. Prove that if $a(ab + 1)$ is odd, then one of a, b must be odd and the other must be even.

1.10   Instructor Notes

Most of the notes for instructors given in this text are simply descriptions of what I do in class and how I think about it. You should do what works for you in your classroom; feel free to ignore any advice I give that does not apply to your situation. However, I hope that some of this commentary is of use.

The first week of the semester often has less class time than most other weeks. It is feasible to spend only one or two class meetings on this chapter. The first class meeting must of course begin with some orientation (such as introductions and/or syllabus review) but the bulk of the class can be spent with students working in groups on problems from Section 1.2. It is worth reassuring students that even though they have no experience yet with discrete mathematics, these are problems they can approach just by thinking about them; discrete math is a natural way of thinking.

Working in groups on the first class day has the advantages of setting a collaborative and interactive tone early on and having classmates meet each other (this is especially useful for first-year class members). I advise counting the students off as $1, 2, \ldots, \lfloor \frac{n}{4} \rfloor$ (where n is the number of students present), asking the 1s to collect in one area, the 2s in another, etc., and reminding students to introduce themselves to each other before starting work. Be sure to reserve ten minutes at the end of class to discern which problems have been completed by all groups and elicit verbal explanations of their solutions from group representatives.

To reinforce what students learned in class, have them read Sections 1.3–1.5 for the next class, specifically assigning the Check Yourself problems.

A good warmup for the second day of class is asking the students to share their Check Yourself Challenge responses. This may lead to a discussion of other Check Yourself problems if students have questions about them. Then have the class walk the instructor through their choice of proof that a set of size n has 2^n subsets, and use this as an opening to discuss how to turn an argument into a written proof. Generally, students will have different ways of explaining the same proof, and generally, they will not have very precise language this early in the semester. It may take one-half to one hour for students to hash out this simple proof, even with prompts from the instructor. This is a valuable exercise for them to learn how much work is involved after solving a problem in order to submit homework. If any time remains in class, ask whether students have questions over the reading or Check Yourself problems, and then revisit problems not completed or discussed the first day. (If you have lots of extra time, have the students start in on the Counting Exercises—but be sure to save a few to have them write up as homework!)

In assessing your students after the first week of classes, remember that they will not have been able to master basic proof techniques yet—they're just starting! Mastery will come over a period of weeks as they practice proof writing in multiple mathematical contexts. In case you desire (now or later) additional basic proof problems for your students, a selection of them is provided in Section TI.2.

Chapter 2

Sets and Logic

2.1 Introduction and Summary

Sets and logic are the fundamentals that underlie all of mathematics, not just discrete mathematics. However, a discrete mathematics course is a customary place to address them directly. Sets are collections of objects. Logic is a formal way of describing reasoning. We will both describe and construct sets, and we will develop truth tables as a way to use logic on compound statements. Logical tools are available for when we have trouble figuring out how to reason precisely using English.

Both sets and logic come with a lot of notation. In order to do anything interesting with either sets or logic, you need to be familiar with that notation. (In the case of logic, we will not use the notation very often after this chapter.) Hence, this chapter has a lot of reading that you must complete before you can get on with the discovery and *doing* of related mathematics. It may feel a bit tedious; sorry. Break it up into smaller chunks to aid focus and retention.

This chapter also contains our first introduction to the interesting proof technique of contradiction (and to the less interesting, but super-useful, proof technique of double-inclusion). Proof by contradiction basically works by hypothesizing that a theorem is false (say "suppose not!") and then obtaining a statement that is clearly false (such as $0 = 1$).

Try not to be intimidated by the amount of unfamiliar material in this chapter. We will be working with logical thinking and proof techniques all semester, and you are not expected to fully grasp them yet. The intent of this chapter is to give you the ideas and terminology so you can work to master the ideas as you use them in context. You will probably want to reread parts of this material later in the course to assist in that endeavor.

2.2 Sets

Sets are ubiquitous in mathematics (and in life!). The definition of the word *set* has a long and sordid history, full of confusions such as whether a set is allowed to contain itself. We will be a bit imprecise here and give more of a description than a definition.

Definition 2.2.1 (of set). A *set* contains *elements*. The elements must be distinct, but their order does not matter. There may be finitely many or infinitely many elements in a set. Elements can be words, objects, numbers, or other sets (i.e., basically anything).

When an element a is a member of a set A, we denote this by $a \in A$ (and read it aloud as "a is in A" or "a is an element of A"). The notation $a_1, a_2 \in A$ means that both a_1 and a_2 are elements of A. Often, sets are denoted by capital letters, and their elements are denoted by related lowercase letters.

Example 2.2.2 (of your favorite sets). The sets most commonly used in discrete math are

- the natural numbers, $\mathbb{N} = \{1, 2, 3, \ldots\}$,

- the binary digits, $\mathbb{Z}_2 = \{0, 1\}$,

- the integers, $\mathbb{Z} = \{\ldots, -2, -1, 0, 1, 2, \ldots\}$.

Beware that some people (many computer scientists and some mathematicians) think that $0 \in \mathbb{N}$, perhaps because computer scientists often start counting with zero instead of with one. In order to have consistency with mathematical induction (see Chapter 4), we disagree with this view. Instead, we refer to the set $\{0, 1, 2, 3, \ldots\} = \mathbb{W}$ as the whole numbers (but we refer to it rarely).

Example 2.2.3 (of other sets). The set $\{1, 2, 3\}$ is the same set as $\{2, 3, 1\}$. Similarly, $\{\ldots, -6, -4, -2, 0, 2, 4, \ldots\}$ is the same infinite set as $\{0, 2, -2, 4, -4, \ldots\}$. (The dots indicate that the established pattern keeps on going.) By some definitions, $\{1, 1, 2, 3\}$ is not a set because elements are repeated, but in this text we will simply consider $\{1, 1, 2, 3\}$ as an inefficient expression of the set $\{1, 2, 3\}$. On the other hand, $\{1, \{1, 2, 3\}, 3\}$ is a perfectly fine (and well-expressed) set. The set with no elements $\{\}$ is often denoted $\emptyset$ and called the *empty set* or the *null set*. It is different from $\{\{\}\} = \{\emptyset\}$, which contains one element (the empty set). A set of four duck heads is shown in Figure 2.1.

Figure 2.1. The elements of the set $\{dh_1, dh_2, dh_3, dh_4\}$ are duck heads.

This is an appropriate moment to recall that $|A|$ denotes the number of elements in a set, also called its *size* or its *cardinality*. We will only consider the cardinality of finite sets here, and if you are interested in infinite sets, you should look at Chapter 15. Here are a few examples: $|\{1,2,3\}| = 3$; $|\{\{1,2,3\}\}| = 1$; $|\{\{1,2,3\}, \mathbb{N}\}| = 2$. Do not confuse set cardinality with absolute value, even though they use the same notation; one applies to sets and the other to numbers, so there is no conflict.

2.2.1 Making New Sets from Scratch

So far, we have described a set by listing all its elements. Most of the time we instead describe the pattern that the elements follow. For example, $2\mathbb{Z} = \{k \in \mathbb{Z} \mid k$ is even$\} = \{\ldots, -4, -2, 0, 2, 4, \ldots\}$. The first expression is read as "two zee is the set of k in zee such that k is even," or as "two zee is the set of integers k such that k is even," or as "two zee is the set of all integers that are even." Another way of writing this same set dispenses with the word "even": $2\mathbb{Z} = \{k \in \mathbb{Z} \mid k = 2\ell$ for some $\ell \in \mathbb{Z}\}$. Here we have substituted the definition of *even* for the word "even."

Example 2.2.4. The set $\{a_1 a_2 a_3 \mid a_i \in \mathbb{Z}_2\}$ is the set of all three-digit binary strings $\{000, 001, 010, 011, 100, 101, 110, 111\}$. Similarly, $\{a_1 a_2 a_3 a_4 \mid a_i \in \mathbb{Z}_2, a_1 = 1, a_3 = 0\}$ is the set of all four-digit binary strings with first digit 1 and third digit 0, or $\{1000, 1001, 1100, 1101\}$. The set $\{(a, b) \mid a \in 2\mathbb{Z}, b \in \{0, 1, 2\}\}$ is the set of all ordered pairs where the first component is an even integer and the second component is 0, 1, or 2.

Basically, we write sets in the form {type of elements | condition(s)}. Often the type of elements will include a restriction to some set.

2.2.2   Finding Sets inside Other Sets

Recall from Chapter 1 that if we have two sets A and B, then A is a subset of B if every element of A is also an element of B. Let's say it again:

Definition 2.2.5. If A and B are sets, then A is a *subset* of B if every element of A is also an element of B. We denote this relationship as $A \subset B$.

Technically, the symbol $\subset$ means that A is a *proper* subset, so that there is at least one element in B that is not in A, but we will be loosey-goosey with our usage and allow $A \subset B$ to mean that A is perhaps equal to B. The symbol $\subseteq$ is used to indicate that perhaps A and B are equal, and the symbol $\subsetneq$ indicates that A and B are definitely not equal. (Do not confuse $\subsetneq$ with $\not\subset$, which means that A is *not* a subset of B!) Notice that $\emptyset \subset A$ for any set A—because all zero of the elements in $\emptyset$ are also elements of A! Every set contains some nothingness.

Example 2.2.6 (of flavors of subsets and non-subsets). We start with $A = \{2k \mid k > 0, k \in \mathbb{Z}\}$, the even natural numbers; $A \subset \mathbb{N}$ and, in fact, $A \subsetneq \mathbb{N}$. In binary land, $\{1\} \subset \mathbb{Z}_2$ and $\{0,1\} \subseteq \mathbb{Z}_2$ but $\{2\} \not\subset \mathbb{Z}_2$. Less commonly seen are the equivalent statements $\mathbb{Z}_2 \supset \{1\}$, $\mathbb{Z}_2 \supseteq \{0,1\}$, and $\mathbb{Z}_2 \not\supset \{2\}$. We could have instead written $1 \in \mathbb{Z}_2, 2 \notin \mathbb{Z}_2$ for the first and last of those statements (do you see why?).

A related concept is that of the *power set* $\mathscr{P}(A)$ of a set A. It is the set of all subsets of A. (You know from Theorem 1.5.2 that if A is finite, then $|\mathscr{P}(A)| = 2^{|A|}$.) We will not use this concept very often, but it is worth mentioning because other sources you encounter in your mathematical life will expect you to recognize it.

The notion of subset allows us to define the idea of set complement. We denote the complement of A by $\overline{A}$, though other people use notations like A^C or A' (that last one is silly because the symbol $'$ is used for so many other things, but still, you should be warned).

Definition 2.2.7. If $A \subset B$, then $\overline{A} = B \setminus A$, all the elements of B that are not in A, is called the *complement* of A relative to B. (This is sometimes written as $B - A$.)

So if you see the symbol $\overline{A}$, know that there is secretly a B out there that you must know about in order to understand what $\overline{A}$ is. Sometimes the *universe* is temporarily redefined as a particular set (instead of the universe we live in) and it takes the place of B for all sets $A_1, A_2, \ldots, A_n$ in a discussion. (By the way, if there are several sets under discussion, we may refer to them as the first set or A_1 (pronounced "A-one"), the second set, the nth set, etc.). We can think of a set complement as a way of removing one set from another.

Example 2.2.8 (of complements). As a small example, note that $\{1,3,5,7\} \setminus \{1,5\}$ $= \{3,7\}$. Now let B be the set of four-digit binary strings. Then $B \setminus \{a_1 a_2 a_3 a_4 \mid a_i \in \mathbb{Z}_2, a_1 = 1, a_3 = 0\} = \{0000, 0001, 0010, 0011, 0100, 0101, 0110, 0111, 1010,$ $1011, 1110, 1111\}$.

The notation $B \setminus A$ can be extended to situations where A is not a subset of B; in these cases, we interpret $B \setminus A$ to mean $B \setminus (\text{elements of } A \text{ in } B) = B \setminus (A \cap B)$. For example, $\{1,3,5,7\} \setminus \{1,5,6\} = \{3,7\}$. We simply remove any elements of B that are elements of A.

2.2.3   Proof Technique: Double-Inclusion

There is a simple way to show that two sets are equal (if in fact they are), and it has a special name because it is used so frequently. You may deduce that name from the title of this section. To show that $A = B$, show first that $A \subset B$ and then show that $B \subset A$. This means that A is included in B and B is included in A and thus arises the term *double-inclusion*.

Of course, it might be useful to understand how to show that $A \subset B$ (or $B \subset A$) in order to execute a double-inclusion proof. A technical way to think about $A \subset B$ is with the statement *if $a \in A$, then $a \in B$*. So a formal inclusion proof proceeds as follows:

- Let a be any element of A.

- (Reasoning, statements.)

- Therefore, $a \in B$, and so $A \subset B$.

Example 2.2.9. Two different expressions can describe the same set. Let us show that two descriptions of the set of even numbers are equivalent. To that end, let $E_1 = \{k \in \mathbb{Z} \mid k = 2\ell \text{ for some } \ell \in \mathbb{Z}\}$ and let $E_2 = \{2r + 6 \mid r \in \mathbb{Z}\}$. First, we will show that $E_1 \subset E_2$. Let e be any element of E_1. Then $e = 2\ell$ for some $\ell \in \mathbb{Z}$. If we let $r = \ell - 3$, then $e = 2\ell = 2(r + 3) = 2r + 6$, where $r \in \mathbb{Z}$, and therefore $e \in E_2$. Now, we will show that $E_2 \subset E_1$. Let t be any element of E_2. Then $t = 2r + 6$, where $r \in \mathbb{Z}$. Setting $\ell = r + 3$, we have that $t = 2r + 6 = 2(r + 3) = 2\ell$, where $\ell \in \mathbb{Z}$, and therefore $t \in E_1$. Because $E_1 \subset E_2$ and $E_2 \subset E_1$, we conclude that $E_1 = E_2$.

2.2.4 Making New Sets from Old

The most common operations on sets are the three defined here.

Definition 2.2.10. The *union* of sets A and B is a set $A \cup B$ containing all the elements in A and all the elements in B (with any duplicates removed). Similarly, the union of sets $A_1, A_2, \ldots, A_n$ is $A_1 \cup A_2 \cup \cdots \cup A_n = \bigcup_{i=1}^{n} A_i$ and contains all elements in the A_i (with any duplicates removed). Dealing with infinitely many sets is a little bit trickier and depends on how many there are (see Chapter 15 for more on this), but for now we'll say that $\bigcup_{i=1}^{\infty} A_i$ and $\bigcup_{i \in \mathbb{N}} A_i$ are the same.

Example 2.2.11. Let $A = \{egg, duck, 3, 4\}$ and let $B = \{duck, goose, 7, 8\}$. Then $A \cup B = \{egg, duck, goose, 3, 4, 7, 8\}$.
   Let $A_i = \{i\}$. Then $\bigcup_{i=1}^{\infty} A_i = \mathbb{N}$.

Definition 2.2.12. The *intersection* of sets A and B is a set $A \cap B$ containing every element that is in both A and B. Similarly, the intersection of sets $A_1, A_2, \ldots, A_n$ is $A_1 \cap A_2 \cap \cdots \cap A_n = \bigcap_{i=1}^{n} A_i$ and contains only elements that are in all of the A_i. We may sometimes take infinite intersections as in $\bigcap_{i=1}^{\infty} A_i$ and $\bigcap_{i \in \mathbb{N}} A_i$.

Example 2.2.13. With A and B and A_i defined as in Example 2.2.11, $A \cap B = \{duck\}$ and $\bigcap_{i=1}^{\infty} A_i = \emptyset$.

   Two sets A and B are called *disjoint* if $A \cap B = \emptyset$. We now have enough notation to give a super-formal way of restating the sum principle.

> **Theorem 2.2.14.** *If $A_1, \ldots, A_n$ are disjoint finite sets, then* $|A_1 \cup \cdots \cup A_n| = |A_1| + \cdots + |A_n|$.

   That is perhaps the most boring way to state the sum principle (can you think of a more boring way?), so we will not generally use it. It is, however, worth noting that almost every mathematical statement can be rewritten to use formal set language; and, it is also worth noting that this often borifies a given statement. (**Definition:** $bor \cdot i \cdot fy$, to intensify the level of boringness something has.) At the same time, we informally use set theory in our daily lives; for example, red-headed women are the intersection of the set of redheads and the set of women. Most of the time, we don't even notice that we're using set theory, but if you listen to conversations and look in the media, it's all over the place (albeit implicitly).

Definition 2.2.15. The *Cartesian product* of sets A and B is a set $A \times B$ containing all possible ordered pairs where the first component is an element of A and the second component is an element of B. In other words, $A \times B = \{(a,b) \mid a \in A$ and $b \in B\}$. Likewise, the Cartesian product $A_1 \times A_2 \times \cdots \times A_n$ is the set of all n-tuples $(a_1, a_2, \ldots, a_n)$ where $a_i \in A_i$.

Example 2.2.16 (of Cartesian products). The set $\{duck, goose\} \times \{egg\} = \{(duck, egg), (goose, egg)\}$. When the empty set is involved, there's a trick; $\{5, 7, 9, 11\} \times \emptyset = \emptyset$ because there are no possible ordered pairs with the second component from the empty set. Binary strings of length two are formally $\mathbb{Z}_2 \times \mathbb{Z}_2 = \{0, 1\} \times \{0, 1\} = \{(0, 0), (1, 0), (0, 1), (1, 1)\}$. This is sometimes abbreviated as $(\mathbb{Z}_2)^2$. Likewise, binary strings of length n are formally $\mathbb{Z}_2 \times \mathbb{Z}_2 \times \cdots \times \mathbb{Z}_2 = (\mathbb{Z}_2)^n$.

2.2.5 Looking at Sets

The most common example of a Cartesian product is that the real plane $\mathbb{R}^2$ is secretly $\mathbb{R} \times \mathbb{R}$, as shown in Figure 2.2. ($\mathbb{R}$ is shorthand for the real numbers.)

Figure 2.3 shows two other examples of Cartesian products.

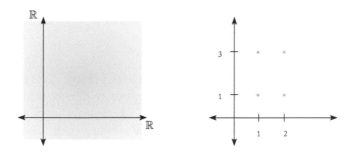

Figure 2.2. At left, $\mathbb{R}^2$; at right, $\{1,2\} \times \{1,3\}$.

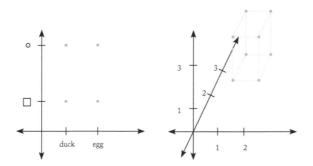

Figure 2.3. At left, $\{duck, egg\} \times \{\square, \circ\}$. At right, $\{1,2\} \times \{1,3\} \times \{2,3\}$. Although the set looks as though it is misplaced, it is not. (Grey lines are added to help locate the points in space but are not part of the set.)

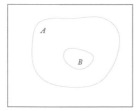 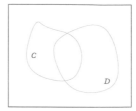

Figure 2.4. Each Venn diagram shows the relationship between two sets. Note that $B \subset A$ but no subset relationship exists between C and D.

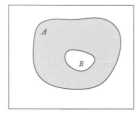 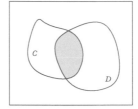

Figure 2.5. At left, $A \setminus B$, the part of A that does not include B, is shaded; at right, $C \cap D$, the overlap between C and D, is shaded.

We need ways of visualizing larger and more abstract sets. The usual method is called a *Venn diagram*, in which we draw a big box to denote the universe and then blobs to represent sets. Here are a couple of examples, shown in Figure 2.4. Those are pretty boring because they simply show two sets each. The information provided by the Venn diagrams is what kind of subset relationship (if any) exists between the two sets. Let's indicate some new sets that are derived from the old sets—in Figure 2.5 we shade the results of performing set operations on our old sets. This process extends to some fancy shaded diagrams when we have three sets and multiple set operations, as in Figure 2.6. In Figures 2.7–2.9, we show how to find these same sets using hatching.

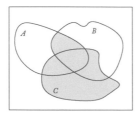 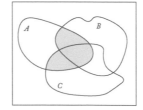 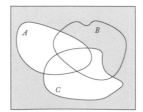

Figure 2.6. From left to right, $(A \cap B) \cup C$, $A \cap (B \cup C)$, and $\overline{A \cup C}$.

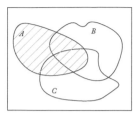 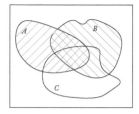 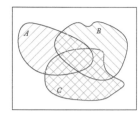

Figure 2.7. From left to right, A, $A \cap B$, and $(A \cap B) \cup C$.

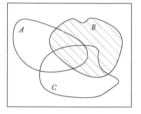 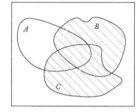 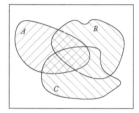

Figure 2.8. From left to right, B, $B \cup C$, and $A \cap (B \cup C)$.

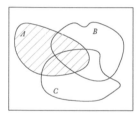 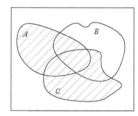 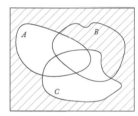

Figure 2.9. From left to right, A, $A \cup C$, and $\overline{A \cup C}$.

To exhibit $(A \cap B) \cup C$, we look within the parentheses. We start at left in Figure 2.7 by hatching A. Because we want $A \cap B$, we use a different hatching for B so that $A \cap B$ is crosshatched. Then, to demonstrate the union with C, we crosshatch C to match.

To exhibit $A \cap (B \cup C)$ in Figure 2.8, we again look within the parentheses. We start by hatching B. Because we want $B \cup C$, we use the same hatching on C as on B. In contrast, we want to intersect this set with A, so we use a different hatching on A so that the intersection is crosshatched.

The first step in showing $\overline{A \cup C}$ is to hatch A at left in Figure 2.9. To show $A \cup C$, we use the same hatching on C as on A. Finally, what we want to exhibit is $\overline{A \cup C}$, so we apply hatching on the remainder of the diagram and erase the previously applied hatching.

If you would like to practice using Venn diagrams, here are three online resources that you will likely find helpful.

- 🦆 http://demonstrations.wolfram.com/InteractiveVennDiagrams/: This software lets you click on parts of a two- or three-set Venn diagram to shade them, and then it shows the set notation for the corresponding set and its complement.

- 🦆 http://randomservices.org/random/apps/VennGame.html: This applet has you click on a set-notation description and then shades the corresponding regions of a Venn diagram.

- 🦆 http://math.uww.edu/~mcfarlat/143venn.htm: This "quiz" applet has 15 different symbolic descriptions of sets. You have to figure out which regions on the corresponding Venn diagrams should be shaded, and mousing over a nearby diagram will show the correct shading.

Check Yourself

There may seem to be a lot of these problems, but each one is quick to do.

1. List the elements of $\{z \in \mathbb{Z} \mid -10 \le z < 10\}$.

2. Write the set $\{2,4,6,8,10\}$ as a set of elements subject to a condition.

3. What is the cardinality of the set $\{duck,\emptyset,\{duck,egg\},\{duck,\{duck,egg,\emptyset\}\}\}$?

4. Is $\{3,6,13,67\} \subset \{67,4,53,5,13,6\}$?

5. List the elements of $\mathscr{P}(\{-1,5,20\})$.

6. Let $A = \{5,6,7,8,9,23\}$, $B = \{6,7,9,456,3.142\}$, and $C = \{7,4,8,2.3,\pi,6\}$. List the elements of ...

   (a) ... $A \cup B$.

   (b) ... $B \cap C$.

   (c) ... $A \setminus C$.

7. Let $D = \{6.53,42,1,hat\}$ and $F = \{0,-2\}$. List the elements of ...

   (a) ... $D \times F$.

   (b) ... $F \times D$.

   (c) ... $D \times D$.

   (d) ... $\emptyset \times F$.

8. Draw a visual representation of the set $\{1,2,3\} \times \{4,5\}$.

9. Make a Venn diagram that represents $\{1,2,3,4,5,6\} \cap \{4,5,6,7,8,9\}$.

10. **Challenge:**

    (a) Invent three sets of your own.

    (b) Find a different way to write each of the sets (for example, list the elements, or describe what the elements have in common using set notation).

    (c) Make a Venn diagram showing the relationships between your three sets.

2.3 Logic

Regular old English communication is not very precise, and many sentences have more than one interpretation. The reason logical notation and language have developed is so that there can be no question as to what a statement is intended to convey. The word "logic" is used to refer to an area of mathematics as well as a type of thinking. In all of mathematics, we use logical thinking, and we use the notation and language of the area of mathematics known as logic when less formal communication does not serve us well.

The basic component of logical language is the *statement*, which is a sentence that is either true or false. (To say that in a snooty way, a statement has a truth value from the set $\{true, false\}$.) Here is a non-statement: "Be a blue-footed booby." That sentence is an imperative; likewise, questions are not statements. Similarly, "$\{-3,0,2\} \setminus \{0,1\}$" is not a statement because it lacks a verb; it is only an expression.

Example 2.3.1 (of statements). Here are a few statements.

- The December 2009 issue of *Mathematics Magazine* has 78 pages.

- $32 - 6 = 16$.

- $\{1,5,7\} \cap \{1,2,8\} = \{2\}$.

- There is a one-to-one correspondence between four-digit binary strings and the corners of a four-dimensional cube.

In logic, we don't care about whether a statement is true or whether it is false. (Reread Example 2.3.1 with this in mind!) Our intent will be to examine the relationships between statements when they are combined in certain ways. We care about the roles that statements play rather than their validity or truth value. Thus, logical language omits the details of statements by referring to them with variables (usually P or Q or R), so that one can stick *any* statements into the templates that result. This simultaneously makes logical language useful and more difficult to read.

2.3.1 Combining Statements

There are just a few constructions used in logic to combine statements, called *connectives*. They are as follows:

- ✤ *and* is the verbal analogue to set intersection, so P-and-Q is only true if both P and Q are true;

- ✤ *or* is the verbal analogue to set union, so P-or-Q is true whenever either P or Q is true;

- ✤ *not* makes a true statement false and makes a false statement true; it gives a statement its opposite meaning;

- ✤ *implies* means that one statement is a consequence of the other; it is also written as *if-then* and is called a *conditional* statement.

Example 2.3.2 (of a very compound statement). Consider the statement *if $x \in \mathbb{Z}$ and $x < 2.7$ then x is negative or $x \in \{0, 1, 2\}$*. The implication combines the substatements *$x \in \mathbb{Z}$ and $x < 2.7$* and *x is negative or $x \in \{0, 1, 2\}$*. Each of those has two substatements of its own; the *and* has substatements *$x \in \mathbb{Z}$* and *$x < 2.7$*, and the *or* has sub-statements *x is negative* and *$x \in \{0, 1, 2\}$*. Then, note that the statement under consideration is true. (If we changed $x \in \{0, 1, 2\}$ to $x \in \{0, 1\}$, then it would be false.)

Example 2.3.3 (of ambiguity without parentheses). Consider the statement *$x \in \mathbb{Z}$ and $x < 3.6$ or $x > 628.3$*. Does it mean *($x \in \mathbb{Z}$ and $x < 3.6$) or $x > 628.3$*, or does it mean *$x \in \mathbb{Z}$ and ($x < 3.6$ or $x > 628.3$)*? The number $x = 1,002.7$ is described by the first statement but not the second statement. The number $x = -23$ is described by both statements. When we combine statements, we must be careful that the resulting statements are unambiguous, and so we must use enough parentheses.

Now we will be completely precise: we will define each of the connective terms using a *truth table*. As the name indicates, a truth table is a table that lists the truth values of a statement. Here is a silly and useless truth table:

$$\begin{array}{c|c} P & P \\ \hline T & T \\ F & F \end{array}$$

This can be read aloud as *when P is true, P is true; when P is false, P is false.* See? It is indeed useless.

We will now define *and* (denoted $\wedge$), *or* (denoted $\vee$), and *not* (denoted $\neg$) using serious and useful truth tables.

P	Q	$P \wedge Q$		P	Q	$P \vee Q$		P	$\neg P$		P	Q	P xor Q
T	T	T		T	T	T		T	F		T	T	F
T	F	F		T	F	T		F	T		T	F	T
F	T	F		F	T	T					F	T	T
F	F	F		F	F	F					F	F	F

Looking at these truth tables, we can see that there is a difference between the usual English use of *or* and the formal logical use of *or*. After dinner, a host might ask, "Would you like coffee or tea?" (The answer "neither" corresponds to the line in the truth table where P and Q are both false.) The intent is to offer *either* coffee *or* tea, not both—regular English *or* is actually *exclusive or*, abbreviated *xor*. We have given a bonus truth table for xor above. Notice that the number of rows in a truth table depends on the number of statements involved. We need 2 rows for P, 4 for P, Q, 8 for P, Q, R, 16 for P, Q, R, S, and so forth, so that we can have all possible combinations of true and false.

Example 2.3.4. We will make a truth table for $(P \wedge Q) \vee R$.

P	Q	R	$P \wedge Q$	$(P \wedge Q) \vee R$
T	T	T	T	T
T	T	F	T	T
T	F	T	F	T
T	F	F	F	F
F	T	T	F	T
F	T	F	F	F
F	F	T	F	T
F	F	F	F	F

Sometimes we can ignore a few rows of a truth table: if we have particular statements corresponding to $P, Q, R, \ldots$, and we know that one of the statements is true (or, likewise, false), then we only need the rows of the truth table corresponding to that truth (or falsehood). Let us suppose that Q stands for the statement *the sun is plaid*. This is clearly false, so we could just write

P	Q	R	$P \wedge Q$	$(P \wedge Q) \vee R$
T	F	T	F	T
T	F	F	F	F
F	F	T	F	T
F	F	F	F	F

Translating between logic and set notations.   There is a correspondence between set and logic notations, particularly when the logical statements are about sets. The elements for which the statement $P \wedge Q$ holds are those in the set $A = \{x \mid P$ is true for $x\}$ *and* the set $B = \{x \mid Q$ is true for $x\}$, and together those elements form the set $A \cap B$. Similarly, the elements for which the statement $P \vee Q$ holds are those in the set $A = \{x \mid P$ is true for $x\}$ *or* the set $B = \{x \mid Q$ is true for $x\}$, and together those elements form the set $A \cup B$. In this sense, $\wedge$ (or *and*) for statements corresponds to $\cap$ for sets, and $\vee$ (or *or*) for statements corresponds to $\cup$ for sets. The analogy for the connective *not* is a bit subtler; elements for which $\neg P$ holds are those not in the set $A = \{x \mid P$ is true for $x\}$, but then where *are* they? For this to make sense, we must make reference to a universe set U so that the elements not in A are those in $\overline{A}$, the complement of A relative to U.

Example 2.3.5 (of combining set and logic notations).   We can describe the set $A_1 \cap (A_2 \cup A_3)$ as $\{x \mid x \in A_1 \cap (A_2 \cup A_3)\}$. Via a set of equivalences, we can turn it into another set:

$$\{x \mid x \in A_1 \cap (A_2 \cup A_3)\} = \{x \mid x \in A_1 \text{ and } x \in (A_2 \cup A_3)\}$$
$$= \{x \mid x \in A_1 \text{ and } x \in (A_2 \text{ or } A_3)\}$$
$$= \{x \mid (x \in A_1 \text{ and } x \in A_2) \text{ or } (x \in A_1 \text{ and } x \in A_3)\}$$
$$= \{x \mid (x \in A_1 \cap A_2) \text{ or } (x \in A_1 \cap A_3)\}$$
$$= \{x \mid (x \in A_1 \cap A_2) \cup (x \in A_1 \cap A_3)\}$$
$$= (A_1 \cap A_2) \cup (A_1 \cap A_3).$$

Cool!

Next is *implies* (denoted by $\Rightarrow$). We read $P \Rightarrow Q$ as "P implies Q" or as "If P, then Q." Implication can be seen from different perspectives; when we are writing a proof, $P \Rightarrow Q$ needs justification, and we consider P and Q as separate statements, with $\Rightarrow$ standing in for the chain of argumentation that forms the bulk of a proof. In a logical context, $P \Rightarrow Q$ is a single statement that has truth values defined by the following truth table.

P	Q	$P \Rightarrow Q$
T	T	T
T	F	F
F	T	T
F	F	T

This might seem a little weird. Or, more precisely, the last two lines of the table might seem a little bit weird. How can $P \Rightarrow Q$ be true if P is false? Consider a practical and pleasant example, namely, the statement *if you go to the party, then you will get some candy.* If you don't go to the party, you don't expect to get any candy, but you might get some anyway from some other source. But it's still true that if you *did* go, you'd get candy, so even though you don't go to the party, the implication still holds; the promise made to you is true.

There are many equivalent ways of writing implication, which is lovely but sometimes confusing. The statement $P \Rightarrow Q$ is usually read as *P implies Q* or as *if P then Q* but can also be read as *P only if Q* and *P is sufficient for Q* to hold. On the other hand, $Q \Rightarrow P$ can also be read as *P if Q* (see, if Q then …) and *P is necessary for Q*. Let's look again at the statement *if you go to the party, then you will get some candy.* Here, P is *you go to the party* and Q is *you will get some candy*. We could restate the statement as *going to the party is sufficient for getting some candy*, or as *you go to the party only if you get some candy*, or also as *getting some candy is necessary when you go to the party.*

Now, check this out: we can combine truth tables. (Note that arrows do work the way they should, so $P \Leftarrow Q$ means "If Q, then P.")

P	Q	$P \Rightarrow Q$		P	Q	$P \Leftarrow Q$
T	T	T		T	T	T
T	F	F		T	F	T
F	T	T		F	T	F
F	F	T		F	F	T

P	Q	$P \Rightarrow Q$	$P \Leftarrow Q$	$(P \Leftarrow Q) \wedge (P \Rightarrow Q)$
T	T	T	T	T
T	F	F	T	F
F	T	T	F	F
F	F	T	T	T

It may not surprise you to learn that we abbreviate $(P \Leftarrow Q) \wedge (P \Rightarrow Q)$ as $P \Leftrightarrow Q$ and read that as "P if and only if Q." This is a fairly common kind of mathematical statement, used to show that two statements P and Q are logically equivalent. (More generally, any time two compound statements have the same truth tables, they are considered logically equivalent.) Some people are irritated by having to write out the words "if and only if" and abbreviate the phrase to *iff*. This statement type is called a *biconditional*. Additionally, even though we're not talking about proofs at the moment, it's worth pointing out that if you want to prove a biconditional statement you almost always have to split it into the two implications and prove them separately. (It's possible to string together a bunch of biconditionals, but that's hard. Don't bother.) We often write $(\Rightarrow)$ to indicate we'll prove that P implies Q and then write $(\Leftarrow)$ to indicate we'll prove that Q implies P ... and we start a new paragraph for each.

Advice. If you're new to the mathematical uses of *and*, *or*, *not*, and *implies*, then you might want to carry their truth tables around with you for a while until you internalize them.

Logic is related to our goal of learning proof crafting because there we need to produce rigorous and airtight reasoning. Well, using logical language certainly does that! When we aren't sure whether we're being rigorous enough, logic is here for us to fall back on. However, we don't want to resort to formal logic too often because it kills ease of communication. Plus, logical language is devoid of context—it doesn't care whether a given statement is true or false, but we do. And we want to convince others of that truth or falsehood.

On the other hand, logical notation is used in writing computer code, especially in creating conditionals (that's code-speak for if-then statements). For example, If[(a==b || a==0) && c < 5, c, 0] says *if $a = b$ or $a = 0$, and if c is less than 5, then return the value of c; otherwise, return 0*. It may seem like the major use of logic for computer scientists is knowing the notation so that code can be written, but it is important to understand logical equivalence so that code can be refined for speed increases. Hardware designers use circuitry that corresponds to logical connectives, so minimizing their number can have positive consequences for power consumption and manufacturing cost.

2.3.2   Restriction of Variables via Quantifiers

One can make—and in fact we have already made—statements that include variables, such as *k is even* or $x^2 - 3 = 1$. In these cases, whether or not the statement is true depends on what value the variable (here, k or x) has. The statement *k is even* is true only when k is even (duh) and $x^2 - 3 = 1$ is true only when $x = 2$ or $x = -2$ (slightly less duh). Notice that statements always have verbs in them (is this a "duh"?) so they differ from functions like k or $x^2 - 3$ that merely produce numbers. Sometimes people will refer to a variable-including statement as $P(k)$ instead of just P; we won't do that here because it confuses us and, therefore, potentially you as well.

The *quantifiers* "for all" (denoted $\forall$, which is sometimes colloquially referred to as "the upside-down A" by students who forget what it stands for) and "there exists" (denoted $\exists$, which is similarly sometimes colloquially referred to as "the backwards E") restrict the variables referred to in a statement. We can rewrite our two example variable statements using these quantifiers.

Example 2.3.6. The statement *for all even k, k is even* is certainly true, though *for all k, k is even* is false and *there exists k such that k is even* is true. Similarly, *for all x, $x^2 - 3 = 1$* is false, whereas *there exists x such that $x^2 - 3 = 1$* is true.

We can prove that last statement. Consider $x = 2$ and note that $2^2 - 3 = 1$, so there does exist an x such that $x^2 - 3 = 1$. This technique generalizes. Existence proofs can be done simply by giving an example: you've shown that the desired object exists! But this is the *only* time an example works as a proof.

Sometimes it would be more convenient if people used quantifiers in ordinary English. For example, in the common statement *every duck wants a cookie*, the speaker could mean that given any duck, it desires some cookie ($\forall d \in Ducks, \exists c \in Cookies$ such that d wants c), or the speaker could mean there exists a cookie that every duck wants ($\exists c \in Cookies$ such that $\forall d \in Ducks, d$ wants c). Notice that this exemplifies not only the vagueness of English but that placing quantifiers in different orders changes the meaning of a statement. So be careful!

Example 2.3.7. Consider the statement $\forall n \in 2\mathbb{Z}, \exists a, b \in \mathbb{Z}$ *such that $a = 2k_1 + 1$, $b = 2k_2 + 1$, and $n = b - a$.* This basically says that for every even integer, there exist two odd integers such that the even integer is the difference of the odd integers. This is a true statement; given any even integer n, the integer $a = n - 1$ will be odd, as will $b = n - 1 + n = 2n - 1$, and $b - a = 2n - 1 - (n - 1) = n$.

If we change the order of the quantifiers, we may obtain $\exists a, b \in \mathbb{Z}$ such that $a = 2k_1 + 1, b = 2k_2 + 1$, and such that $\forall n \in 2\mathbb{Z}, n = b - a$. This says that there exist two odd integers such that for every even integer, that even integer is the difference of the two odd integers. This is a false statement; no matter which two odd integers a, b are considered, they have a single difference $b - a$ that is even. Any other even integer, such as $b - a + 2$, cannot be the difference of a and b, so the statement does not hold for most even integers (let alone for *every* even integer).

2.3.3   Negation Interactions

Even professional mathematicians sometimes find negating statements to be some-what challenging. To be safe, take the English-mathematics version of a statement, substitute quantifiers (but don't go to full logic-speak), and then use the rules we will see here.

Example 2.3.8. ¬(*All ducks like cookies*) is logically equivalent to *there exists a duck who does not like cookies*. Unsurprisingly then, ¬(*some duck likes cookies*) is logically equivalent to *all ducks dislike cookies*. More mathematically, ¬(*for all integers k, k = 2.5*) is equivalent to *there exists an integer k such that k ≠ 2.5*.

Basically, if you have the statement ¬(∀ stuff), that converts to ∃¬(stuff), and if you have the statement ¬(∃ stuff), that converts to ∀¬(stuff). At least this reduces the problem of negating to a shorter statement, though (stuff) might have some more quantifiers hidden within it.

Example 2.3.9 (of wacky negations). Let's negate a couple of statements. Consider *for all ducks, there exists a cookie such that a tree weeps*. In logic notation, this becomes ∀ *ducks, ∃ a cookie such that a tree weeps*. Thus, the negation proceeds as ¬(∀ *ducks, ∃ a cookie such that a tree weeps*), which becomes ∃ *a duck, ¬(∃ a cookie such that a tree weeps)*, and then ∃ *a duck, such that ∀ cookies ¬(a tree weeps)*, ending with *there exists a duck such that for all cookies, no tree weeps*. Consider now *there exists an egg such that it cracks for all cooks*. Its negation is slightly simpler. We translate first to logical notation to achieve ∃ *an egg, such that it cracks ∀ cooks*. Its negation is ¬(∃ *an egg, such that it cracks ∀ cooks*), which becomes ∀ *eggs, ¬(∀ cooks it cracks)*, then ∀ *eggs, ∃ cooks ¬(it cracks)* and finally ∀ *eggs, ∃ cooks it does not crack*. This doesn't make much grammatical sense, so we reword it to read *for any egg, there exists a cook who cannot crack it*.

Negation plays nicely with other connectives, as follows.

DeMorgan's laws (logic version). $(\neg P) \vee (\neg Q)$ is logically equivalent to $\neg(P \wedge Q)$, and $(\neg P) \wedge (\neg Q)$ is logically equivalent to $\neg(P \vee Q)$.

Example 2.3.10. *No ducks and no chickens* is the same as *no ducks or chickens*.

People often think DeMorgan's laws are pretty obvious, but we have stated them here for completeness (as well as because sometimes they are needed when statements P and Q are elaborate). We will investigate another form of DeMorgan's laws in Section 2.4.

> **Negation and implication.** The statement $P \Rightarrow Q$ is logically equivalent to the statement $\neg Q \Rightarrow \neg P$.

Definition 2.3.11 (of implication relatives). We sometimes call $P \Rightarrow Q$ the *original statement* and always call $\neg Q \Rightarrow \neg P$ the *contrapositive* statement. Along these lines, $Q \Rightarrow P$ is the *converse* statement, and $\neg P \Rightarrow \neg Q$ is the *inverse* statement, and also the contrapositive of the converse statement. All four of these statements are known as *implications*.

Notice that an implication and its converse are usually not both true at the same time. For example, *if I am at the combination Pizza Hut and Taco Bell, then I am at the Pizza Hut* is always true, but *if I am at the Pizza Hut, then I am at the combination Pizza Hut and Taco Bell* is often false.

Should you wish to practice the use of logic notation, logical thinking, and truth tables, here are some resources.

🐦 http://demonstrations.wolfram.com/PropositionalLogicPuzzleGenerator/: You are shown some polygons along with a list of statements in logic notation. (The logic notation is not quite the same as used in this book, but there is a help option that explains it.) Each statement is marked as true or false. The challenge is that the polygons are not labeled but referred to in the statements as A, B, C, etc., and you get to match the labels with the polygons.

🐦 http://demonstrations.wolfram.com/LogicWithLetters/ and http://demonstrations.wolfram.com/2DLogicGameWithLetters/ and http://demonstrations.wolfram.com/LogicWithLogicians/: These puzzles do not use formal logic notation but give practice in logical thinking.

🐦 http://www.cs.utexas.edu/~learnlogic/truthtables/: After typing in a logical statement, you are given a corresponding blank truth table to fill in—it just has headers and a few beginning columns. You can choose whether to have your work checked entry by entry, or when you're done filling in the table. Warning: this applet uses a single arrow for *implies* instead of the double arrow we use in this text.

These problems take less time to do than they at first appear to take.

1. Let *P* represent the statement *Ximena is pretty*, *Q* represent *Ximena is quizzical*, and *R* represent *Ximena is a rugby player*. Write $(P \vee Q) \wedge R$ as an English sentence.

2. Write *Miyuki does not like kumquats, but ze likes pickles or daikon* in logic notation.

3. Rewrite *every cat drinks beer* as an implication.

4. **Challenge:** Come up with two examples of mathematical statements and two examples of mathematical non-statements.

5. Using truth tables, verify that the converse of a statement is not logically equivalent to the original statement. (Suggestion: make the columns P, Q, $P \Rightarrow Q$, and $Q \Rightarrow P$, and compare the last two columns.)

6. Write the contrapositive of the statement *if the maple tree is orange, then the scissors are closed.*

7. Using truth tables, verify that the statement *if I am at the combination Pizza Hut and Taco Bell, then I am at the Pizza Hut* is always true.

8. Negate the statement *there exists an even number n such that n < 10.*

2.4   Try This! Problems on Sets and Logic

These problems are intended to be discussed with peers. Some students find these problems quite challenging and others find them easy. Your eventual success in discrete mathematics is unlikely to be related to your feelings about this particular collection of problems.

1. What is the cardinality of $\{0, cat, \{dog\}, \{2.1, 6\}\}$? List all its subsets. (How many should there be?)

2. Formally negate the statement "You can fool all of the people all of the time."

3. List several elements of the set $E = \{x \in \mathbb{Z} \mid \frac{1}{2}x \in \mathbb{Z}\}$ and then give a simpler description of E.

4. Here are DeMorgan's laws, given in logic notation: $\neg(P \vee Q)$ is logically equivalent to $(\neg P) \wedge (\neg Q)$ and $\neg(P \wedge Q)$ is logically equivalent to $(\neg P) \vee (\neg Q)$.

(a) Express DeMorgan's laws using set notation.

(b) Prove DeMorgan's laws using truth tables.

(c) Prove DeMorgan's laws using Venn diagrams.

(d) Prove DeMorgan's laws using set-element notation. (Suggestion: use double-inclusion.)

(e) Can you state DeMorgan's laws for three or more sets?

(f) Does that give you any ideas for stating, using logic notation, DeMorgan's laws for three or more statements?

5. Let A be the set of even numbers from -6 to 6 (inclusive), and let B be the set of odd numbers from -6 to 6 (inclusive), living in the universe of integers from -10 to 10 (inclusive).

(a) List the elements of $\overline{B}$.

(b) What is $\overline{A \cup B}$?

(c) Describe $A \setminus B$ using fewer symbols.

6. Is $\neg(P \Rightarrow Q)$ logically equivalent to $P \wedge \neg Q$?

7. Let $A_k = \{0, 1, \ldots, k\}$. What is $\bigcup_{i=1}^{n} A_i$? How about $\bigcap_{i=0}^{n} A_i$?

8. Draw a Venn diagram representing $(A \cap B) \cap (A \cup C)$.

9. Is it true that $\exists m \in \mathbb{Z} \mid \forall n \in \mathbb{Z}, m = n + 5$?

2.5   Proof Techniques: Not!

After all that boring reading, you probably are sighing at the thought of dealing with more material in this chapter. But fear not! This is shorter and more interesting (really!).

We already know how to do a straightforward proof, by directly proving an implication $P \Rightarrow Q$: we assume P is true and then deduce that Q is therefore true. We already know one way to disprove $P \Rightarrow Q$: find a counterexample. Now we will use a single fact from logic to burst wide open the clouds surrounding proof and shine glowing rays of truth on the situation.

Remember from Section 2.3.3 that the contrapositive of a statement is logically equivalent to the statement itself. That means we could prove $(\neg Q) \Rightarrow (\neg P)$

instead! This is but a tiny step removed from doing a direct proof: here we assume $\neg Q$ and deduce that $\neg P$ is therefore true. In fact, you can use the template from Section 1.4 (page 12) by simply inserting $\neg Q$ for P and inserting $\neg P$ for Q.

Example 2.5.1. Let $n, m \in \mathbb{N}$. We will prove that if $n \cdot m$ is odd, then an $n \times m$ grid cannot be tiled with dominoes. (A grid is *tiled* if every square is covered exactly once.) The contrapositive of this statement is *if an $n \times m$ grid can be tiled with dominoes, then $n \cdot m$ is not odd.* So, suppose an $n \times m$ grid can be tiled with dominoes. There are a total of $n \cdot m$ squares, and every domino covers two squares. Therefore, the tiling uses $\frac{n \cdot m}{2}$ dominoes, and so $n \cdot m$ must be even. Therefore, $n \cdot m$ is not odd.

There is a related technique we can use—it is called *proof by contradiction* and it proceeds by assuming the statement we want to prove is false and obtaining a logical problem of some kind. For an oversimplified example, if we want to prove that $P \Rightarrow Q$, we would assume P is true and Q is false, and if we can show that Q false implies P false, then this contradicts our assumption that P was true. (Read *that* aloud three times...) You may astutely notice that this is actually proving the contrapositive. In this case, we might start by drafting a proof by contradiction, continue by discovering that we've proven the contrapositive, and write the clean version of the proof as a contrapositive proof.

More commonly when using proof by contradiction, the P in $P \Rightarrow Q$ is a compound statement containing several conditions (e.g., *if k is an integer, ℓ is even, and the moon is green*), and we will only contradict one part of P rather than proving the negation of P as a whole (e.g., showing that the moon is not green and thus deriving a contradiction).

Less common but still useful is assuming Q is false and deriving a contradiction unrelated to the statements under consideration—for example, showing that Q is false implies that 2 is an odd number.

Template for a proof by contradiction:

1. Restate the theorem in the form *if (conditions) are true, then (conclusion) is true.*

2. On a scratch sheet, write *suppose not*. Then write out (conditions) and the negation of (conclusion).

3. Try to simplify the statement of ¬(conclusion) and see what this might mean.

4. Attempt to derive a contradiction of some kind—to one or more of (conditions) or to a commonly known mathematical truth.

5. Repeat attempts until you are successful.

6. Write up the results on a clean sheet, as follows.

> 🦆 **Theorem:** (State theorem here.)
>
> 🦆 **Proof:** Suppose not. That is, suppose (conditions) are true but (conclusion) is false.
>
> 🦆 (Translate this to a simpler statement if applicable. Derive a contradiction.)
>
> 🦆 Contradiction!
>
> 🦆 Therefore, (conclusion) is true. (Draw a box or checkmark or write Q.E.D. to indicate that you're done.)

Example 2.5.2. We will prove that there are infinitely many powers of 2, i.e., $2^0, 2^1, 2^2, \ldots$. Suppose not. Then there are finitely many powers of 2; let the number of them be n. Therefore, we can sort them in increasing order of size. Consider the largest of these, k. Then 2^k is not one of the n powers of 2; it is larger than any of them because $2^k > k$. Therefore, there are at least $n+1$ powers of 2, which contradicts the supposition that there were only n of them.

Contradiction can also be used to disprove false statements. In this case, assume the statement is true and derive a contradiction.

Check Yourself ————————————————————————

1. Prove that if n^2 is odd, then n is odd. (Suggestion: try proving the contrapositive.)

2. Prove that if there are ten ducks paddling in four ponds, then some pond must contain at least three paddling ducks. (Suggestion: try contradiction.)

3. **Challenge:** Develop your own statement that can be proved by contradiction.

2.6   Try This! A Tricky Conundrum

Consider the following argument: *You must learn about sets or learn about logic if you go on to the next chapter. You did not learn about sets and did not go on to the next chapter. Therefore, you must not have learned about logic.*

1. Decide for yourself whether or not the conclusion is correct (that you must not have learned about logic). Make a note of this decision.

2. In a small group, exchange your decisions and share your reasoning (justify your decisions). Please collaborate from here on out.

3. Let's check our logic formally.

   (a) Dissect the first sentence and find three statements within it that you can label with letters.

   (b) Turn the first sentence into an expression using formal logic symbols.

   (c) Express the second and third sentences in formal logic symbols, too.

   (d) Make a (big) truth table that includes parts for each of the sentences and for the argument as a whole.

4. Compare the result of this truth table to your original idea. If they agree, explain how they are compatible. If they do not agree, find the source of the error.

5. If you have some time left over, work on these proofs.

   (a) For $n \in \mathbb{N}$, prove that if $n^3 + 6n^2 - 2n$ is even, then n is even.

   (b) Let $x \in \mathbb{R}$. Show that if $x^5 + 7x^3 + 5x \geq x^4 + x^2 + 8$, then $x \geq 0$.

   (c) Prove that an 8×8 chessboard with a square missing cannot be tiled with dominoes.

   (d) Prove that for n odd, an $n \times n$ chessboard missing its lower-right-hand corner can be tiled with dominoes.

2.7   Additional Examples

Example 2.7.1 (of manipulating set notation). Let $S_1 = \{q + 1 \in \mathbb{Z} \mid q = 2k$ for some $k \in \mathbb{Z}\}$, and let $S_2 = \{2r + 5 \mid r \in \mathbb{Z}\}$; we want to show that $S_1 = S_2$. First, we will show that $S_1 \subset S_2$. Let s be any element of S_1. Then $s = 2k + 1$ for some

$k \in \mathbb{Z}$. If we let $r = k - 2$, then $s = 2k + 1 = 2(r + 2) + 1 = 2r + 5$, where $r \in \mathbb{Z}$, and therefore $s \in S_2$. Now, we will show that $S_2 \subset S_1$. Let t be any element of S_2. Then $t = 2r + 5$, where $r \in \mathbb{Z}$. Setting $k = r + 2$, we have that $t = 2r + 5 = 2(k - 2) + 5 = 2k + 1$, where $k \in \mathbb{Z}$, and therefore $t \in S_1$. Because $S_1 \subset S_2$ and $S_2 \subset S_1$, we conclude that $S_1 = S_2$.

Example 2.7.2 (of Venn diagrams). We will exhibit $(A \cap \overline{B}) \cup (\overline{A} \cap B)$ using Venn diagrams.

We begin by looking within the parentheses. The first set of parentheses contains $A \cap \overline{B}$. We start at left in Figure 2.10 by hatching A. Because we want $A \cap \overline{B}$, we use a different hatching for $\overline{B}$ and then combine these so that $A \cap \overline{B}$ is crosshatched.

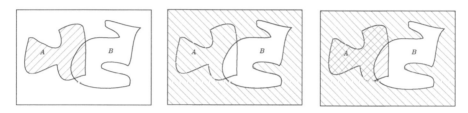

Figure 2.10. At left, A; in the middle, $\overline{B}$; at right, $A \cap \overline{B}$.

The second set of parentheses contains $\overline{A} \cap B$. We start at left in Figure 2.11 by hatching $\overline{A}$. Because we want $\overline{A} \cap B$, we use a different hatching for B and then combine these so that $\overline{A} \cap B$ is crosshatched.

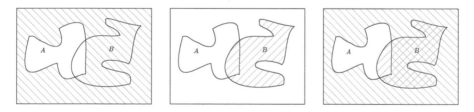

Figure 2.11. At left, $\overline{A}$; in the middle, B; at right, $\overline{A} \cap B$.

Finally, we combine these sets. We start at left in Figure 2.12 by showing $A \cap \overline{B}$, and in the middle we show $\overline{A} \cap B$. Because we want $(A \cap \overline{B}) \cup (\overline{A} \cap B)$, we display both at once using the same type of hatching.

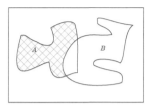 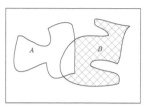 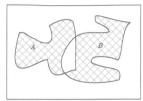

Figure 2.12. At left, $A \cap \overline{B}$; in the middle, $\overline{A} \cap B$; at right, $(A \cap \overline{B}) \cup (\overline{A} \cap B)$.

Example 2.7.3 (of breaking down a very compound statement). Consider the statement *if $x \in \mathbb{Z}$ and $x > -7.2$ then x is positive or $x \in \{0, -1, -2, -3, -4, -5, -6, -7\}$.* The largest logical substructure is the if-then implication, which combines the substatements $\langle x \in \mathbb{Z}$ *and* $x > -7.2 \rangle$ and $\langle x$ *is positive or* $x \in \{0, -1, -2, -3, -4, -5, -6, -7\} \rangle$. Each of those has two substatements of its own; the *and* has substatements $\langle x \in \mathbb{Z} \rangle$ and $\langle x > -7.2 \rangle$, and the *or* has substatements $\langle x$ *is positive* $\rangle$ and $\langle x \in \{0, -1, -2, -3, -4, -5, -6, -7\} \rangle$.

Example 2.7.4 (of evaluating statements with truth tables). Here is an argument someone might make: *The jelly bean is blue. Blue things are tasty. Therefore, the jelly bean is tasty.* Is this argument correct? We will represent *jelly bean* as J, *blue* as B, and *tasty* as T. Then *the jelly bean is blue* is really *if it is a jelly bean, then it is blue* or $J \Rightarrow B$. We can similarly write the other statements as $B \Rightarrow T$ and $J \Rightarrow T$. Surely, if $J \Rightarrow B$ and $B \Rightarrow T$, then $J \Rightarrow T$, right? Let's see…

J B T	$J \Rightarrow B$	$B \Rightarrow T$	$(J \Rightarrow B) \wedge (B \Rightarrow T)$	$J \Rightarrow T$	$((J \Rightarrow B) \wedge (B \Rightarrow T)) \Rightarrow (J \Rightarrow T)$
T T T	T	T	T	T	T
T T F	T	F	F	F	T
T F T	F	T	F	T	T
T F F	F	T	F	F	T
F T T	T	T	T	T	T
F T F	T	F	F	F	T
F F T	T	T	T	T	T
F F F	T	T	T	T	T

Yup, it's all true! Literally, all entries in the last column of the truth table are T—this means the implication, and therefore the argument, is correct.

Example 2.7.5 (of quantifier order mattering). Let $d, e \in \mathbb{Z}$. Consider the statement $\forall e, \exists d$ *such that* $d < e$. This true statement basically says that given an integer, we can find a smaller one. For example, given $e = -32$, we can find $d = -4{,}389$.

If we change the order of the quantifiers, our new statement is $\exists d, \forall e$ such that $d < e$. This statement says there is some integer such that every other integer is larger. That's not true!

(If you are (or have been) a student of calculus, compare this example to the formal (ε-δ) definition of limit.)

Example 2.7.6 (of wacky negations). Consider the statement *for all futons, there exists a duck such that stripes are in fashion.* In logic notation, this becomes $\forall$ *futons,* $\exists$ *a duck such that stripes are in fashion.* Thus, the negation proceeds as $\langle \neg(\forall$ *futons,* $\exists$ *a duck such that stripes are in fashion*$)\rangle$; $\langle \exists$ *a futon,* $\neg(\exists$ *a duck such that stripes are in fashion*$)\rangle$; $\langle \exists$ *a futon, such that* $\forall$ *ducks* $\neg$(*stripes are in fashion*$)\rangle$; ... and finally, $\langle$*there exists a futon such that for all ducks, stripes are not in fashion*$\rangle$.

2.8   Where to Go from Here

> Commandment.   Go back and reread the material on proof in Section 1.4.
> And (*grin*) reread Section 3 on how to read mathematics.

We will apply the concepts introduced in this chapter throughout the text, but logic will be particularly important in Chapter 5 when we study the construction of algorithms. The type of basic set theory introduced in this chapter is pervasive in and essential for all of mathematics and has a somewhat different flavor when used in courses based in continuous as opposed to discrete mathematics, such as real analysis and topology. If after working through the material in this chapter, you want to see more examples and have more elementary exercises to work, consult *Book of Proof* by Richard Hammack [12].

Venn diagrams are a source of much interesting investigation. If you try to draw a Venn diagram that represents four or more sets, you will quickly run into trouble showing all possible intersections. For a good survey of approaches to this problem, see http://www.combinatorics.org/Surveys/ds5/VennEJC.html, which also tells you more than you ever wanted to know about Venn diagrams—and includes a zillion references.

Set theory and logic are subfields of mathematics on their own, so there is a great deal to learn about each of these. (Sometimes they are lumped together as *foundations of mathematics*.) We will address a small bit of set theory in Chapter 15. You can take upper-level undergraduate courses on set theory and on logic; if you wish to self-study, *Sweet Reason: A Field Guide to Modern Logic* by Tom

Tymoczko and Jim Henle and *An Outline of Set Theory* by Jim Henle should be the resources you use first.

Within mathematics, set theory and logic are small subfields but are quite active. For example, the Association for Symbolic Logic sponsors sessions of research talks at national mathematics conferences. One famous result in the area is Gödel's incompleteness theorem, which basically says that in any logical system there are statements that cannot be proven to be true or shown to be false. Classical problems in foundations of math were often related to what set of axioms (assumptions or rules) is needed, or is best, for various statements to be true. Modern logic research involves making formal abstract models of other parts of mathematics in order to prove more powerful theorems.

Credit where credit is due: The first activity in Section 2.6 was adapted from an example in [8]; the first puzzle and the project in Section 2.10 were adapted from exercises in [1]. The example on page 45 references a song by Das Racist (find it on YouTube). Problem 12 in Section 2.12 includes a phrase from "Song for a Future Generation" by the B-52s. Four problems in the latter part of Section 2.12 were donated or inspired by Heather Ames Lewis.

2.9   Chapter 2 Definitions

set: A mathematical object that contains distinct unordered elements. There may be finitely many or infinitely many elements in a set.

element: Elements can be words, objects, numbers, or sets (i.e., basically anything).

empty set: The set with no elements. Also called the null set.

null set: The empty set.

cardinality: The number of elements in a set.

size: The cardinality of a set.

subset: A is a subset of B if every element of A is also an element of B.

proper subset: A is a proper subset of B there is at least one element in B that is not an element of A.

power set: The set of all subsets of A, denoted $\mathscr{P}(A)$.

set complement: If $A \subset B$, then $\overline{A} = B \setminus A$, all the elements of B that are not in A, is called the complement of A relative to B.

union: The union of sets A and B is a set $A \cup B$ containing all the elements in A and all the elements in B (with any duplicates removed). The union of many sets A_i contains all elements in the A_i (with any duplicates removed).

intersection: The intersection of sets A and B is a set $A \cap B$ containing every element that is in both A and B. The intersection of many sets A_i contains only elements that are in all of the A_i.

disjoint: Two sets A and B are called disjoint if $A \cap B = \emptyset$.

Cartesian product: The Cartesian product of sets A and B is a set $A \times B$ containing all possible ordered pairs where the first component is an element of A and the second component is an element of B. In other words, $A \times B = \{(a,b) \mid a \in A \text{ and } b \in B\}$. Likewise, the Cartesian product $A_1 \times A_2 \times \cdots \times A_n$ is the set of all n-tuples $(a_1, a_2, \ldots, a_n)$ where $a_i \in A_i$. The name *Cartesian* is derived from René Descartes (1596–1650).

Venn diagram: A picture in which a big box denotes the universe of things under consideration and blobs represent sets. Venn diagrams are used to show relationships between sets. Named after John Venn (1834–1923), who wrote influential works on logic and probability/statistics.

statement: A sentence that is either true or false; it is the basic component of logical language. (To say that in a snooty way, a statement has a truth value from the set $\{true, false\}$.)

connective: A logical construction used to combine statements.

truth table: A table that lists the truth values of a statement.

and: The verbal analogue to set intersection, so P-and-Q is only true if both P and Q are true; denoted $\wedge$. The corresponding truth table is shown in Figure 2.13.

or: The verbal analogue to set union, so P-or-Q is true whenever either P or Q is true; denoted $\vee$. The corresponding truth table is shown in Figure 2.13.

xor: "Exclusive or" means that one statement or the other is true, but not both. The corresponding truth table is shown in Figure 2.13.

not: This gives a statement its opposite meaning; denoted by $\neg$, it makes a true statement false and makes a false statement true. The corresponding truth table is shown in Figure 2.13.

implies: This means that one statement is a consequence of the other; denoted $\Rightarrow$. The corresponding truth table is shown in Figure 2.13.

if-then: A statement involving implication.

conditional: An if-then statement.

if and only if: "P if and only if Q" is denoted $P \Leftrightarrow Q$ and means that the statements P and Q are logically equivalent. The corresponding truth table is shown in Figure 2.13.

DeMorgan's laws: The logical rules for how *not* interacts with *or* and *and*. Named after Augustus DeMorgan (1806–1871).

iff: If and only if.

biconditional: An if-and-only-if statement.

quantifier: Quantifiers such as "for all" and "there exists" restrict the variables referred to in a statement.

implication: A statement of the form $P \Rightarrow Q$.

contrapositive: When $P \Rightarrow Q$ is the original statement, $\neg Q \Rightarrow \neg P$ is the contrapositive statement.

converse: When $P \Rightarrow Q$ is the original statement, $Q \Rightarrow P$ is the converse statement.

inverse statement: When $P \Rightarrow Q$ is the original statement, $\neg P \Rightarrow \neg Q$ is the inverse statement.

P	Q	$P \wedge Q$
T	T	T
T	F	F
F	T	F
F	F	F

P	Q	$P \vee Q$
T	T	T
T	F	T
F	T	T
F	F	F

P	Q	P xor Q
T	T	F
T	F	T
F	T	T
F	F	F

P	$\neg P$
T	F
F	T

P	Q	$P \Rightarrow Q$
T	T	T
T	F	F
F	T	T
F	F	T

P	Q	$P \Leftrightarrow Q$
T	T	T
T	F	F
F	T	F
F	F	T

Figure 2.13. The truth tables for *and*, *or*, xor, *not*, *implies*, and *if and only if*.

2.10 Bonus: Truth Tellers

One application of logical thinking is the class of truth-teller puzzles. The basic format for these is that some statements are made, and each speaker either always tells the truth or always lies. Your assignment is to figure out what's going on (either who is telling the truth or what the truth of the matter is). Such puzzles can be unraveled using truth tables or simply by using logical reasoning. Here we will give a few examples of how to use truth tables to resolve these puzzles.

Suppose you meet some ducks. It is known that a given duck either always tells the truth or always lies. (This is theorized to be the origin of the common expression "Ducks usually lie." See [21].)

Example 2.10.1. One duck says, "I am a truth-telling duck." Another duck quacks, "I am a lying duck." Can we determine anything about either duck's nature? Let us make a truth table to investigate. Let D represent the duck; it gets the value T if it is a truth-telling duck and the value F if it is a lying duck. The statement "D tells the truth" is true exactly when D is a truth-telling duck; the statement "D lies" is true exactly when D is a lying duck.

D	D tells the truth	D lies
T	T	F
F	F	T

That's the unvarnished truth of the situation. But, of course, a lying duck lies (duh)… and our truth table doesn't take that into account. So we modify the truth table to reveal what each type of duck would say in each situation—we swap T and F for the lying duck:

D	D tells the truth	D lies
T	T	F
F	~~F~~ T	~~T~~ F

We can now see that either sort of duck would say that it tells the truth, so we can determine nothing about the first duck. We also see that neither sort of duck would say that it lies, so the second "duck" must not be a duck at all.

Example 2.10.2. A pair of ducks approaches. One quacks, "Exactly one of us is a liar." The other says, "Both of us tell the truth." Huh! What is going on? Let's look at a truth table.

D_1	D_2	D_1 xor D_2 lies	$D_1 \wedge D_2$ tell the truth
T	T	F	T
T	F	T	F
F	T	T	F
F	F	F	F

Again, we modify the table to account for what lying ducks say, and remember that D_1 made the statement in the third column, whereas D_2 made the statement in the fourth column:

D_1	D_2	D_1 xor D_2 lies	$D_1 \wedge D_2$ tell the truth
T	T	F	T
T	F	T	~~F~~ T
F	T	~~T~~ F	F
F	F	~~F~~ T	~~F~~ T

Interestingly, we can only conclude that D_2 is a liar—the statements are consistent whether D_1 is a truth teller or a liar!

Puzzle 1. Amy finds a present on hir doorstep. Ze suspects it was left by either Rachel, Tess, or Nicol. Ze confronts each one.

Rachel: Not me! Tess knows you, and Nicol is your BFF.

Tess: I don't know you, and besides, I've been on vacation in Europe for the last several weeks. I didn't leave you a present.

Nicol: It wasn't me, but I did happen to see Tess and Rachel walking along the river together last week. It must have been one of them.

Let us assume that the present-giver is lying and the other two individuals are telling the truth. Who left Amy the present?

Puzzle 2.   In *Math Curse* [22], the main character has a strange experience at dinner. "While passing the mashed potatoes, Mom says, 'What your father says is false.' Dad helps himself to some potatoes and says, 'What your mother says is true.' ... Can that be true?" Figure out what is going on here... and if you have not already done so, read *Math Curse*. Your local public library surely has it in the picture-book section.

> Project: You are walking about and see some tasty-looking berries. You also meet a duck, which, like any duck, always lies or always tells the truth. You may ask the duck exactly one question. Explain why you will not definitely learn whether the tasty-looking berries are safe to eat by asking any of the following questions:
>
> 🦆 Are these tasty-looking berries safe for a human to eat?
>
> 🦆 Do you tell the truth?
>
> 🦆 Do you tell the truth and are these tasty-looking berries safe for a human to eat?
>
> 🦆 Do you tell the truth or are these tasty-looking berries safe for a human to eat?
>
> 🦆 If you tell the truth, then are these tasty-looking berries safe for a human to eat?
>
> 🦆 If these tasty-looking berries are safe for a human to eat, then do you tell the truth?
>
> 🦆 Do you tell the truth if and only if you lie?
>
> Design a single question to ask the unknown duck such that the answer will tell you whether the tasty-looking berries are safe to eat.

If you want to play with many, many, many more puzzles of this sort, consult a book by Raymond Smullyan. He has written lots of logic puzzle books—perhaps the first was *What Is the Name of This Book?*—and they are easy to find. If you prefer an electronic playground, here are a few sources of logic puzzles:

http://demonstrations.wolfram.com/KnightsKnavesAndNormalsPuzzleGenerator/,
http://demonstrations.wolfram.com/KnightsAndKnavesPuzzleGenerator/,
http://demonstrations.wolfram.com/AnotherKnightsAndKnavesPuzzleGenerator/.

All generate collections of statements. You decide which speakers are knights (who tell the truth) and which are knaves (who lie). The software has options to translate each statement into logic notation and to reveal the solution to each puzzle.

2.11 Bonus Check-Yourself Problems

Solutions to these problems appear starting on page 595. Those solutions that model a formal write-up (such as one might hand in for homework) are to Problems 7 and 9.

1. On an October 2014 visit to the CVS Minute Clinic, the check-in kiosk asked the question, "If you have a copay for today's visit, will you be paying for it with a credit or debit card?"

   (a) Identify the formal logic quantifiers and structure in this question.

   (b) The visit in question was for a flu vaccine, which does not require a copay. The kiosk gave options of *Yes* and *No*. How should the visitor have answered?

   (c) Can you find a simpler way to word the question clearly? (In other words, what *should* the kiosk question ask?)

2. There was a recent campaign slogan heard on the radio: *Not just Blue Cross Blue Shield of Massachusetts, but Blue Cross Blue Shield ... of you.* Why is this mathematically nonsensical for residents of Massachusetts?

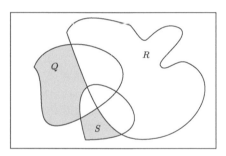

Figure 2.14. A Venn diagram of mystery.

3. Consider the Venn diagram in Figure 2.14.

   (a) Express the shaded area as a set using unions, intersections, and/or complements of the sets Q, R, and S.

   (b) Let $Q = \{k \in \mathbb{Z} \mid |k| \leq 10\}$, $R =$ even numbers, and $S = \{n \in \mathbb{N} \mid n$ is a perfect square$\}$. List the elements of the shaded area.

4. Let $A =$ multiples of 4, and $B =$ multiples of 6. Write $A \cap B$ as a set in the form { sets | conditions }.

5. Negate the statement $\forall\, n \in \mathbb{Z}, \exists\, y \in 2\mathbb{N}$ *such that* $n = y \cdot k$ *for some* $k \in \mathbb{Z}$. Is either the statement or its negation true?

6. Prove that $k \in \mathbb{Z}$ is positive if and only if k^3 is positive.

7. Make a truth table for $\neg(P \wedge Q) \wedge ((P \vee Q) \wedge R)$. Can you express this

statement (henceforth referred to as *aaaaaa!*) more simply?

8. Let $A = \{0, 1, 2\}$ and $B = \{1, 3, 5, 7\}$.
   (a) List the elements of $(A \times B) \cap (B \times A)$.
   (b) List the elements of $(A \setminus B) \times (B \setminus A)$.

9. Show that $(A \times B) \cup (C \times B) = (A \cup C) \times B$.

10. Show that $\{2k \mid k \in \mathbb{N}\} \cup \{4k + 1 \mid k \in \mathbb{W}\} \cup \{4k + 3 \mid k \in \mathbb{W}\} = \mathbb{N}$.

2.12   Problems about Sets and Logic

1. List the elements of $\{n \in \mathbb{N} \mid n^2 = 4\}$.

2. An excerpt from a 2010 Blue Cross Blue Shield survey: "Do **not** include care you got when you stayed overnight in a hospital. Do **not** include the times you went for dental care visits … In the last 12 months, **not** counting the times you needed care right away, how often did you get an appointment for your health care at a doctor's office or clinic as soon as you thought you needed?" What type of needed care is the question asking about? What is excluded? Can you find a simpler way to word the question clearly?

3. Another excerpt from a 2010 Blue Cross Blue Shield survey: "In the last 12 months, how often did your doctor or health provider discuss or provide methods and strategies other than medication to assist you with quitting smoking or using tobacco?" Analyze the connectives in the question. Are any or all of

them used in the same way we use them in mathematics?

4. Compute $|\{z \in \mathbb{Z} \mid z > -10, z^3 < 0\}|$.

5. Make a truth table for $P \wedge (\neg P \vee Q)$.

6. Write the set $\{1, 2, 4, 8, \dots\}$ without using dots.

7. Use Venn diagrams to indicate the even numbers less than ten.

8. Let $A = \{1, 2, 3\}$ and $B = \{2, 3, 4\}$. List the elements of …
   (a) … $(A \times A) \cap (B \times B)$.
   (b) … $(A \times B) \cup (B \times A)$.
   (c) … $A \times (A \setminus B)$.

9. Using truth tables, verify that the contrapositive and original statement are logically equivalent.

10. Again using truth tables, verify that the converse and inverse statements are logically equivalent.

11. Give a counterexample to the statement $|A \cup B| = |A| + |B|$.

12. Is the statement *if the moon is made of green cheese, then Aristotle is the President of Moscow* true or false?

13. Draw a Venn diagram that indicates $(A \cup B) \setminus C$.

14. Decide whether or not it is true that $(A \times B) \cup (C \times D) = (A \cup C) \times (B \cup D)$. If true, give a proof. If false, give a counterexample.

15. Show that if A and B are sets, then if $A \setminus B = \emptyset$, then $B \neq \emptyset$ (unless $A = \emptyset$).

16. Suppose R is false but that $(P \Rightarrow Q) \Leftrightarrow (R \wedge S)$ is true. Is P true or false? What about Q?

17. Could we rewrite the conditional
    ((c > 5 && b == a) || c >= 5)
    in a simpler way? If so, what is it? (Suggestion: use a truth table.)

18. Write this in English: $\forall k \in 3\mathbb{Z}, \exists S \subseteq \mathbb{N}, |S| = k$. (Is it true?) What is the negation of this statement? (Is the negation true?)

19. Prove that $n \in \mathbb{N}$ is odd if and only if n^2 is odd.

20. Prove that $\mathbb{Z} = \{3k \mid k \in \mathbb{Z}\} \cup \{3k + 1 \mid k \in \mathbb{Z}\} \cup \{3k + 2 \mid k \in \mathbb{Z}\}$.

21. Prove that there are infinitely many prime numbers. (Suggestion: try using contradiction.)

22. Show that $n \in \mathbb{N}$ is not divisible by 4 if and only if the binary representation of n ends in 1 or in 10. (Suggestion: use the contrapositive.)

23. Express $P \Rightarrow Q$ using $\neg$ and $\vee$ but not $\Rightarrow$. (Suggestion: play around with truth tables.)

24. Some of the pigeonhole principle proofs in Chapter 1 are secretly proofs by contradiction or proofs that use the contrapositive. Which ones?

25. On route I-91 near Springfield, MA, there was once a sign that said "WASH YOUR BOAT" (pause) "AFTER USE" (pause). Explain why you are complying with the sign if you do not own a boat. How does this relate to truth tables?

26. Compute the cardinality of the set ...

    (a) ... {*wiggle*, *worm*, *wiggle worm*}.

    (b) ... {*wiggle*, {*wiggle*}, {*worm*}, *worm*}.

    (c) ... {{{*wiggle*, *worm*}}}.

27. Let $A = \{(2,5), (-3,1), (4,2), (1,1), (0,1)\}$. List the elements in each of the following sets (or write $\emptyset$ if appropriate).

    (a) $\{(a_1, a_2) \in A \mid a_1 < a_2\}$.

    (b) $\{a_1 \mid (a_1, a_2) \in A \text{ and } a_1 > a_2\}$.

    (c) $\{a_2 \mid (a_1, a_2) \in A \text{ and } a_2 = 0\}$.

28. Let the universe be $U = \{x \in \mathbb{N} \mid x \leq 10\}$, and let $A = \{1,2,3,4,5\}$, $B = \{5,6,7\}$, and $C = \{1,6,9\}$. List the elements of ...

    (a) ... $\overline{A} \cup C$.

    (b) ... $(B \setminus C) \setminus A$.

    (c) ... $(A \cap B) \times C$.

29. Write the negation of *x is prime or x < 52*. (Don't say, "It's not true that ....")

30. Use a truth table to show that $((\neg p) \wedge q) \wedge (p \vee (\neg q))$ is a contradiction.

31. Write the negation of *for all integers x and y, the number $\frac{x-y}{5}$ is an integer.* (Don't say, "It's not true that ....")

32. Write each of the following statements using formal logic notation.

    (a) Even numbers are never prime.

    (b) Triangles never have four sides.

    (c) There are no integers a, b such that $a^2/b^2 = 2$.

(d) No square number immediately follows a prime number.

33. Write the contrapositive of *if $x^2 > 100$, then y has a sister.*

34. Carefully write out some of your results from Problem 4 of Section 2.4: State DeMorgan's laws for two sets using set notation, and prove them using set-element notation. Now state DeMorgan's laws for *n* sets.

35. Carefully write out more of your results from Problem 4 of Section 2.4: Prove DeMorgan's laws for two statements using Venn diagrams, being sure to include intermediate steps and complete sentences. Now state DeMorgan's laws for *n* statements.

36. Prove that if a natural number *n* is even, then $n - 1$ is odd …

    (a) … using a direct proof.

    (b) … by proving the contrapositive.

    (c) … using proof by contradiction.

37. Prove that *x* is even if and only if $4x^2 - 3x + 1$ is odd.

38. **Challenge:** Try to rewrite *this* sentence as a logical statement!! (That is, write it as a collection of short statements joined by logical connectives and quantifiers.) Can you write a simplified version of the next statement? *The following two categories of charitable organizations are not required to have a "Certificate of Solicitation": An organization that is primarily religious in purpose and falls under the regulations 940 CMR 2.00; or An organization that does not raise or receive contributions from the public in excess of $5,000 during a calendar year or does not receive contributions from more than ten persons during a calendar year, if all of their functions, including fundraising activities, are performed by persons who are not paid for their services and if no part of their assets or income inures to the benefit of, or is paid to, any officer or members (M.G.L. c. 68, s. 20).* (Source: http://www.mass.gov/ago/doing-business-in-massachusetts/public-charities-or-not-for-profits/soliciting-funds/overview-of-solicitation.html)

39. Write the set $\{\ldots, -8, -4, 0, 4, 8, \ldots\}$ without using dots.

40. Evaluate the statement $A \cap B = A \setminus B$. Is it true? If so, prove it. If not, find a counterexample and determine whether it is *always* false or whether there exist A, B for which the statement is true.

2.13   Instructor Notes

This chapter is written with the intent that students will read Sections 2.1, 2.2, and 2.3 and attempt the Check Yourself problems before the first class of the week. You may look at the amount of text/material in the chapter and think, "There's no way we can get through this much material in a week." If you expect mastery from the students, then yes, there's no way. But if you expect that the students will get the gist of the material, with little immediate recall and some details filled in over time, then a week is enough time (says the author from experience). The point of dumping all this material on the students at once, and quickly, is to de-emphasize background material while giving them surface familiarity with the concepts; they can then develop deeper familiarity over time as they use sets and logical thinking in other contexts. The practical effect is that students will need to look up notation and terminology and facts/theorems/truth tables all week and for some weeks to come.

Because set theory and logic involve so much new notation, and because different sources use different notation, it is worth exposing students to variances. Examples include denoting *such that* as s.t. or | or :, denoting the set $\{1,\ldots,n\}$ as $[n]$, using $-$ or $\setminus$ for set subtraction, noting that | can mean *divides* as well as *such that*, and denoting complementation by an overline versus a superscripted C versus a prime. Whatever notation you like to use, point it out to the students. Of course, you may not prefer the notation used in this book, and students are likely to encounter other notations in their mathematical lives; you may as well warn them now.

Such a discussion of notation is a good warmup for the first class of the week. There is a lot of reading in Sections 2.2 and 2.3, so it makes sense to follow a short warmup with a request for any questions over the reading or Check Yourself problems. After such a discussion, break students into groups to work on Section 2.4. The DeMorgan's law exercise is likely to take them quite a while, so it is unlikely that they will complete these problems in the remaining class time.

Ask the students to read Section 2.5 for the next class. You may want to devote some class time to further work on Section 2.4 before embarking on the activity in Section 2.6, and it's always good to ask whether there are questions over the reading or the Check Yourselfs. (Should those be pluralized as Check Yourselves?) It is likely that this activity will take most of a class period, if not all of it. My experience is that much of a third class meeting is needed to fully address all the problems.

A cheery warmup for a third day of class is to project the Greek alphabet (Google Images will produce a table to your liking) and go through the pronunciations and uses of the letters. Some are listed on page 641. Students like to share their prior knowledge as part of this discussion.

If you choose to include the Bonus Section 2.10 material in class, you might show your class a *Doctor Who* clip (from "The Pyramids of Mars") containing a truth-tellers problem; it is available at https://www.youtube.com/watch?v=W90s58LtYhk. (This tip courtesy of Tom Hull!) Beware that this may provide savvy students with significant clues for solving the final question of the Section 2.10 Project.

Finally, please remember that this chapter is an *overview* of set theory and logic and proof techniques. Students will practice using these ideas throughout the course and need not have mastered them just yet. Should you want to supplement this material with some additional basic proof problems, a few are provided in Section TI.2.

Chapter 3 🦆🦆🦆

Graphs and Functions

3.1   Introduction and Summary

We will combine an introduction to graphs (sets of dots connected in various ways) with a study of functions, taking as our primary example graph isomorphism. The same graph can be drawn in lots of different ways, and sometimes it is hard to tell that two drawings represent the same graph. A graph isomorphism is a function that turns one representation of a graph into another. The idea of isomorphism is ubiquitous in mathematics, so we will discuss how it is used with other mathematical objects as well.

We begin with basic material on functions, after which we link functions to sets and counting and the pigeonhole principle (see Chapters 2 and 1). Then we will do some exploratory exercises and follow this by playing some graph games. (Yes, really.) You will definitely want a game-playing partner for those activities!

Graphs are cool. They are the focus of Chapters 10–13, as well as the focus of the author's (pure mathematics) research. Graphs come with lots of terminology, not all of which is standard; this can be a bit tiresome, but graphs are fundamental to computer science and applied mathematics, so the terminology is well worth learning. This chapter gives common examples of graphs and investigates some simple graph properties.

The introductory graph material is followed by an introduction to isomorphism and then an activity on graph isomorphism. Finally, we explain the graph theory behind the game played near the beginning of the chapter.

3.2   Function Introdunction

Most people have a general idea of what a function is; it's like a machine where one puts something in the hopper at the top and gets something out of the slot on the side. But that doesn't help anyone figure out which mathematical items are functions and which are not! Therefore, we need a definition.

Definition 3.2.1. We call $f : A \rightarrow B$ a *function* when, given any element a of the set A as input, the function f outputs a unique element $f(a) = b \in B$. We sometimes say that f *maps* a to b and call f a *map*. In particular, a function is *well defined*: it satisfies the criterion that if $a_1 = a_2$, then $f(a_1) = f(a_2)$. (Sometimes people state the contrapositive of this criterion, namely, that if $f(a_1) \neq f(a_2)$, then $a_1 \neq a_2$; this may be easier to conceptualize but quite difficult to use when proving something is a function.)

The set A from which inputs are taken is called the *domain* and the set B from which outputs are selected is called the *target* or *target space*. The element $f(a)$ is called the *image* of the element a. The *range* of a function is all the elements of the target space that are mapped to by the function; that is, for $f : A \rightarrow B$, the range of f is $Range(f) = \{f(a) \mid a \in A\}$. We may intuit from Figure 3.1 the origins of these terms. A function is not just the rule for transforming elements of A into elements of B but also includes what sets A and B are. For example, $f : A \rightarrow B$ is a different function than $f : C \rightarrow B$. If $D \subset A$ and we want to talk about applying $f : A \rightarrow B$ just to elements of D, we write $f|_D$ to indicate that we are restricting the domain of f to D.

We sometimes need a word for a thing that given *i*nput, *p*roduces *o*utput, but is not necessarily well defined; we will use the word *gipo* for this purpose. (Notice that every function is a gipo, but not every gipo is a function.) All of the definitions given in the previous paragraph apply to gipos as well as to functions.

Let us examine a few examples and nonexamples of functions. (A nonexample is an example that does not fit the desired definition.) In Figure 3.1, we show domains on the left and targets on the right, with each gipo rule indicated by arrows. Notice that the input elements hang out in their domain and are sent by arrows to their target. (It's not literally the input elements in the target space—it's their images as seen through the lens of the gipo rule.)

Here are some aspects of the definition of function to which we should pay special attention:

- 🐦 A function f on a domain A has to be defined on every single element of A. If some of them are skipped, either it's not a function after all or else it *is* a function, but secretly defined on some subset of A.

- 🐦 It is not cool to have two outputs for one input. For example, consider a gipo f defined on the set S of length-6 lists of binary digits, with target $\{-1, 0, 1\}$

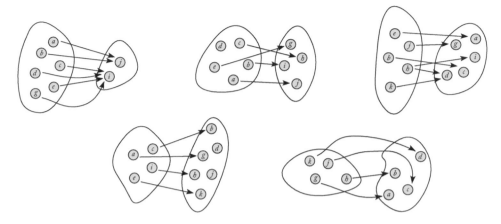

Figure 3.1. Exactly three of these are functions. Which three?

and gipo rule

$$f(s) = \begin{cases} -1 & \text{if } s \text{ ends in } (\ldots, 1, 1), \\ 0 & \text{if } s \text{ ends in } (\ldots, 0), \\ 1 & \text{if } s \text{ ends in } (\ldots, 1). \end{cases}$$

The element $s_0 = (0, 1, 0, 1, 1, 1)$ has $f(s_0) = -1$ and $f(s_0) = 1$, so this gipo is not a function.

These are not the main attributes of the definition, but it's easy to get confused about them. That's why they've been highlighted.

There are two other properties that some functions (and gipos) have and some do not. Zoomed-in bits of functions that do not have these properties are shown in Figure 3.2.

Figure 3.2. Pieces of gipos are shown. Any gipo containing the bit on the left cannot be one-to-one, while any gipo containing the bit on the right cannot be onto.

Definition 3.2.2. A function is *one-to-one*, also called *injective* or 1–1 or *into*, if whenever $f(a_1) = f(a_2)$, then $a_1 = a_2$.

Notice that the condition for injectivity is the converse of well-definedness (that defines a function). Three of the gipos shown in Figure 3.1 are one-to-one; can you discern which three?

Definition 3.2.3. A function is *onto*, also called *surjective*, if for every $b \in B$, there exists some $a \in A$ such that $f(a) = b$.

Notice that if a function is onto, then its target and range are equal. Four of the gipos shown in Figure 3.1 are onto; can you tell which four?

Definition 3.2.4. A function that is both one-to-one and onto is known as a *bijection*.

Here is another way to think of these definitions: in a surjection, every element of the target space is mapped to *at least* once; in an injection, every element of the target space is mapped to *at most* once; and in a bijection, every element of the target space is mapped to *exactly* once.

It's hard to know what terminology is best to use for functions. The only other term for bijection is *one-to-one correspondence*, but that could be confused with a function being one-to-one. And using bijection suggests using surjection and injection for parallelness. However, surjection and injection are hard to remember. Surjective is the same as onto, which seems to be a better term because *onto* refers to the function mapping *onto* every element of the target. Injective is the same as *into*, which is not frequently used. That's a pity because it refers to a copy of the domain landing *in* the target. (Every element of the domain lands on exactly one corresponding element of the target, so one can recover the domain by going backwards along the arrows.)

Example 3.2.5 (of a function that is one-to-one but not onto). Let $f : \mathbb{Z} \to \mathbb{Z}$ be defined by $f(k) = 2k$. Notice that f is not onto because $3 \in \mathbb{Z}$ is not the image of any $k \in \mathbb{Z}$. (If there were such a k, then $k = \frac{3}{2} \notin \mathbb{Z}$.) On the other hand, f is one-to-one, and we'll prove it: Let $f(k_1) = f(k_2)$. Using the definition of f, this becomes $2k_1 = 2k_2$, and dividing through by 2 we obtain $k_1 = k_2$ as desired.

Example 3.2.6 (of a function that is onto but not one-to-one). Let $g : \mathbb{N} \times \mathbb{Z}_2 \to \mathbb{W}$ be defined by $g(n,d) = n \cdot d$. Notice that for any $n \in \mathbb{N}$, the element $(n, 0)$ maps to 0. Therefore, g is not at all one-to-one. However, it is onto: any element of $\mathbb{W}$

is either 0 or an element of $\mathbb{N}$. We just saw that plenty of elements of $\mathbb{N} \times \mathbb{Z}_2$ map to 0, so consider $n \in \mathbb{N}$. Then $g(n,1) = n$, so given any $w \in \mathbb{W}$, we have found an element (n,d) of $\mathbb{N} \times \mathbb{Z}_2$ such that $g(n,d) = w$.

Example 3.2.7 (of a bijection). We will modify Example 3.2.5 to produce a bijection. Let $f : \mathbb{Z} \to 2\mathbb{Z}$ be defined by $f(k) = 2k$. Our proof of injectivity still holds, so we just have to prove that this map is surjective. Consider any $z \in 2\mathbb{Z}$. Then let $k = \frac{z}{2}$, so that $f(k) = f(\frac{z}{2}) = 2\frac{z}{2} = z$. We're done!

Example 3.2.8 (of a bijection proof). Let $g : \mathbb{W} \times \mathbb{Z}_2 \to \mathbb{Z}$ be defined by

$$g((n,t)) = \begin{cases} -n - 1 & \text{when } t = 0, \\ n & \text{when } t = 1. \end{cases}$$

We will show that g is a bijection.

First, we must show that g is injective.

Suppose $g((n_1,t_1)) = g((n_2,t_2))$. Then we have one of the following four cases, depending on the values of t_1, t_2:

1. $n_1 = n_2$ (in which case we're done).

2. $-n_1 - 1 = -n_2 - 1$, so that $n_1 = n_2$ (in which case we're done).

3. $-n_1 - 1 = n_2$, which is a contradiction because either n_1 or n_2 must be negative and there are no negative numbers in $\mathbb{W}$; thus, this case can't happen.

4. $n_1 = -n_2 - 1$, which cannot happen for exactly the same reasons.

We conclude that g is one-to-one.

Second, we must show that g is surjective.

Consider $z \in \mathbb{Z}$. If $z < 0$, then $-z \in \mathbb{N}$ so that $-z - 1 \in \mathbb{W}$ and $g((-z-1,0)) = -(-z-1) - 1 = z$. If $z = 0$, then $g((0,1)) = 0$. If $z > 0$, then $g((z,1)) = z$. Thus g is onto.

This all leads to a way of linking functions, sets, and counting.

> Fact 1. If there is an injective function from A to B, then $|A| \leq |B|$.

and

> Fact 2. If there is a surjective function from A to B, then $|A| \geq |B|$.

You may recall the Fact given in Section 1.5; this can be restated as

> Fact 3.   If there is a bijective function from A to B, then $|A| = |B|$.

While we're at it, let's state the pigeonhole principle in terms of functions. It is the contrapositive of Fact 1. First, though, you might want to visualize the situation: think of the left blobs of Figure 3.1 as being pigeons who fly along the arrows to their right-blob holes.

Hey.   If $|A| > |B|$, there is no injective function from A to B.

Wait! How is that the pigeonhole principle? Let's try again.

Hey Hey.   If $|A| > |B|$, there is no injective function from A to B and so every function from A to B must send at least two elements of A to a single element of B.

Hmm. Closer, but still not very clear.

> Hey Hey Hey.   If $|A| > |B|$, there is no injective function from A to B and so every function from A to B must send at least two elements of A to a single element of B. Let A represent pigeons and B represent pigeonholes, and now we see that any function of pigeons to holes must place at least two pigeons in some hole. Yeah.

We did not place any restrictions on the sizes of A and B in the above discussion. If we examine only finite sets, we have an interesting theorem with an informal proof (because the formal proof is unenlightening).

> **Theorem 3.2.9.** *Let A, B be finite sets and let f be a function $f : A \to B$. If $|A| = |B|$, then f is one-to-one $\iff f$ is onto.*

Proof: As with any if-and-only-if proof, two subproofs are needed.

($\Rightarrow$) Suppose f is one-to-one. Then there are at least as many elements in the range of f as there are in the domain (and no more, because f is a function). Therefore $|Range(f)| = |A|$, but also $|A| = |B|$ and thus $|Range(f)| = |B|$. This means that the range of f fills up the target space B, so f is onto.

($\Leftarrow$) Suppose f is onto. We will proceed by contradiction, so also suppose that f is not one-to-one. In other words, suppose at least two elements of A map to a single element of B. But $|A| = |B|$, so there are the same number of elements in f's domain and f's range (which is also its target because f is onto). That means some element of A isn't in f's domain… so f isn't a function after all… or some element of A maps to two elements of B… and in this case f isn't a function, either. Those are both contradictions to the conditions of the theorem, and so our assumption that f is not one-to-one must have been wrong. Thus f actually *is* one-to-one. $\square$

A version of Theorem 3.2.9 also holds for infinite sets; see Bonus Section 15.8 for more.

Check Yourself

The sum principle reveals that what seem like three problems are in truth eight.

1. Here are some gipos that have domain $\mathbb{N}$. For each gipo, determine whether it is a function, whether the target space could be $\mathbb{N}$, and whether it is one-to-one.

   (a) $f(n) = \frac{n}{3} + 1$.

   (b) $f(n) = n$.

   (c) $f(n) = n - 1$.

   (d) $f(n) = n^2 - 1$.

2. Here are some functions that have domain $\mathbb{Z}$ and target space $\mathbb{W}$. For each function, determine whether it is one-to-one or onto.

   (a) $f(k) = 0$.

   (b) $f(k) = |\lfloor \frac{k}{2} \rfloor|$. (The notation $\lfloor x \rfloor$ is known as the *floor* function, as it returns the integer equal to or just less than the input. Thus, $\lfloor \frac{k}{2} \rfloor$ returns $\frac{k}{2}$ if k is even and $\frac{k-1}{2}$ if k is odd.) (Oh, and there is a matching *ceiling* function, which returns the integer equal to or just greater than the input.)

   (c) $f(k) = k^2 + 2$.

   For those functions that are not onto, what is the range? Are any of the functions bijections?

3. **Challenge:** Write out proofs for Problems 1 and 2: that is, prove that the relevant gipos are well defined, one-to-one, and onto, and for those that are not, give counterexamples.

3.3   Try This! Play with Functions and Graphs

There are three subsections of playing; try to spread your playtime equally among them!

3.3.1   Play with Functions

Functions are related to counting, so let's count functions.

1. List all the functions from $\{a,b\}$ to $\{c,d,e\}$. Here's a start on that list:

$$f_1(a) = c, \quad f_1(b) = d.$$
$$f_2(a) = d, \quad f_2(b) = d.$$

How many functions are in your list?

2. How many functions are there from $\{a,b,c\}$ to $\{d,e\}$? Try to complete this computation by reasoning rather than by listing.

3. Without making lists or drawing pictures … how many functions are there from a two-element set to a ten-element set?

4. … from a ten-element set to a two-element set?

5. Generalize. That is, how many functions are there from an m-element set to a q-element set?

3.3.2   Play with Graphs

We need to have a couple of definitions before we can dive into exploration.

Definition 3.3.1. A *graph* is a set of dots (drawn as • or ◦) called *vertices* and a set of *edges* (drawn in any line- or curve-like way) that represent pairs of vertices. Thus, $G = (V, E)$, where V (or $V(G)$) is the vertex set and E (or $E(G)$) is the edge set. Elements of E are $e = \{v_1, v_2\}$ where $v_1, v_2 \in V$. (Sometimes we abbreviate to $v_1 v_2$.) The order of v_1 and v_2 does not matter (for now; see page 84 later). Two vertices joined by an edge are *adjacent* and that edge is *incident* to each of those vertices. The vertices adjacent to a vertex v are called v's *neighbors*. A few examples of graphs are shown in Figure 3.3.

Note.   The word *vertices* is plural. The singular form is *vertex*. It is criminal to leave the "s" off the plural form and use it as singular… so criminal, in fact, that we cannot even type the offending "word" here. Just don't do it.

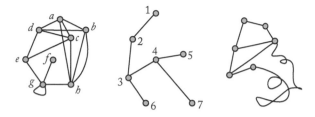

Figure 3.3. Three graphs.

In Figure 3.3, two of the graphs have vertices labeled (arbitrarily) and one does not. Vertex g has a *loop*, or an edge $\{g, g\}$, as well as a *multiple edge* connecting vertices b and h twice with *multiplicity* two. (In the case of multiple edges, our usual set-theoretic notation fails us, as $\{b, h\}$ and $\{b, h\}$ should describe the same edge... yet there are physically two different edges. One way around this is to mark the copies of the edge as $\{b, h\}_1$ and $\{b, h\}_2$.) The *degree* of a vertex is the number of edges that emanate from it; so, f has degree 1, 2 has degree 2, b has degree 4, and g has degree 5 (because both ends of the loop are incident to g). Now let's explore.

1. Draw your own graph with

   🐤 at least ten vertices,
   🐤 an edge with multiplicity three,
   🐤 at least three vertices that are all adjacent to each other, and
   🐤 a vertex with five neighbors.

   Draw this same graph again, but make sure that your second drawing has a different number of edge crossings than your first drawing.

2. Determine the degree of each vertex in the graph you just drew. Add up the numbers you get. How does this compare to the number of edges? Do the same with the unlabeled graph of Figure 3.3.

3. You now have four examples to work with: conjecture a relationship between the sum of the degrees of a graph (with a finite number of vertices) and the number of edges of that graph. Next, prove that your conjecture is correct.

4. Count the number of vertices of odd degree in each of the four graphs (including the one you created). For each graph, is the number even or odd? Make a conjecture about the number of vertices of odd degree a graph has. Can you prove it?

Figure 3.4. Five vertices hanging out (left), and six vertices chillin' (right).

3.3.3   A Dot Game

Let's play a game! You'll need two people, something to write on, and writing implements of two colors. Each person writes in only one color. Start with either five or six dots, no edges, as shown in Figure 3.4. Players alternate moves—a move consists of drawing an edge (we don't allow multiple edges or loops in this game)—and the goal of the game is to force the other player to complete a triangle in hir color. So, for example, the player who draws teal edges wants to force the player who draws purple edges to complete a purple triangle. The game ends when one player wins, or when all possible edges have been drawn (by the way, that's called a *complete* graph).

Play a few games so you can get the hang of how to strategize. Then try to answer these questions (assuming that neither player makes mistakes in play):

- Does the game always have a winner and a loser? Or, can it sometimes end in a draw?

- How does the game on five vertices differ from the game on six vertices?

- What if you start with fewer vertices? More?

Make some conjectures as to what is going on here, and try to prove them.

If you want to play the dot game by yourself later, you can do so at http://www.dbai.tuwien.ac.at/proj/ramsey/.

(This space inserted to encourage you not to look at the hints below until you've genuinely thought about what's going on with the dot game!)

Hints. If you have six dots, after the game is played, there are five edges coming out of each dot. Those edges come in two colors. How do they have to be split up between the colors? And what happens if you try to avoid making triangles in each color, starting from those edges? (How is the situation different if you have five dots?)

3.4   Functions and Counting

Hey! You! Don't read any further unless you have worked through the problems in Section 3.3.1. I mean it!

You probably discovered in Section 3.3.1 that the number of functions from an m-element set to a q-element set is q^m. We can denote a function by an ordered m-tuple, with the jth component corresponding to the image of the jth domain element. There are q choices for the image of each of the m domain elements, so a total of $q \cdot q \cdot \cdots \cdot q = q^m$ possible functions. (If this reminds you of the product principle from Chapter 1, you are correct and astute!)

Now imagine for a moment that we want to count only the injective (or one-to-one, or 1–1) functions. This is a significant restriction because we are not allowed to use any target space element more than once. So, we have q choices for the image of the first domain element, but only $q-1$ choices for the image of the second domain element. If there are m domain elements, then there are $q \cdot (q-1) \cdot \cdots \cdot (q-(m-1))$ possible functions. Notice that if $m > q$, we have a problem! Then there are no 1–1 functions (by the pigeonhole principle). This is secretly a preview of one type of counting addressed in Chapters 6 and 7. Counting surjections is more difficult; you may investigate it in Problem 24 at the end of Chapter 6.

There is also a meta-relationship between counting and functions: bijections are frequently used to reframe counting problems. Instead of directly counting elements of a set S, we create a bijection to a set T that we find conceptually easier to manage (and therefore count). We might, for example, count grey-spotted ducks by finding a bijection between ducks and set elements such that grey-spotted ducks are in correspondence with those elements that are divisible by 3. Then, we would count set elements that are divisible by 3 to obtain the number of grey-spotted ducks. We will use this technique in Chapters 6–9.

3.5   Graphs: Definitions and Examples

Hey! You! Don't read any further unless you have worked through the problems in Section 3.3.2. I mean it!

There are lots and lots of terms to use when discussing graphs—we saw *vertex* and *edge* above. Here are more. Imagine walking on a graph and you'll believe that a *walk* is encoded by a list of vertices alternating with edges. Both ends of

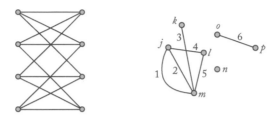

Figure 3.5. Two more graphs.

the list must be vertices; you can't just end a stroll mid-edge. A walk where no vertices repeat is a *path*, and a *cycle* is a walk whose only repetition is the first/last vertex. The *length* of a path or cycle is the number of edges it has. For example, in Figure 3.5 one walk of length 3 is l-4-j-1-m-2-j; it is not a path, but k-3-m-5-l is a path and j-2-m-1-j is a cycle of length 2. The *distance* between two vertices of a graph is the length of the shortest path between those vertices. A graph that is nothing but a path is called P_n; it has n vertices and length $n-1$. A graph with n vertices that is nothing but a cycle is called C_n. Some cycle and path graphs are shown in Figure 3.6. Any two vertices in a *connected* graph can be joined by some walk. A graph with no cycles is a *forest*, and a connected graph with no cycles is a *tree* (Really! Find one in Figure 3.3), and a *leaf* is a vertex of degree 1 (such as k in Figure 3.5).

Despite the large amount of terminology (it's all in the index, and listed in Section 3.11, for when you need to look it up later), we are now just touching on the basics of graphs, which will be the focus of Chapters 10–13. Later we have an entire chapter (Chapter 10) devoted to trees, and we will investigate special sorts of walks in Chapter 12.

A first result about graphs, which you have likely already discovered from Section 3.3.2, is known as the *handshaking lemma*. Wait, we'd better say that more officially.

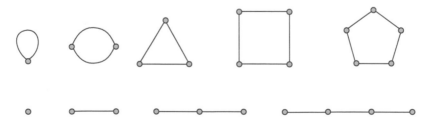

Figure 3.6. Cycles C_1, C_2, C_3, C_4, and C_5 (top); paths P_1, P_2, P_3, and P_4 (bottom).

> **Lemma 3.5.1 (the handshaking lemma).** *Because each edge is incident to two vertices, the sum of the degrees of the vertices of a graph must be twice the number of edges (and thus a multiple of two (i.e., even)).*

To visualize this, imagine that an edge in the graph is like two vertices shaking hands. Each edge has been counted twice, once for the vertex on each end/hand. This technique—overcounting and then tracking how much we overcounted—will be studied in more depth in Chapter 6 and used regularly thereafter.

It follows from Lemma 3.5.1 that the number of vertices of odd degree must be even. If not (notice the signal that we're about to do proof by contradiction), the number of vertices of odd degree is odd. Then, the sum of the degrees is $(\text{odd} + \cdots + \text{odd}) + (\text{even} + \cdots + \text{even})$, which simplifies to $\text{odd} + \text{even}$ because the sum of an odd number of odd numbers is odd. Of course, $\text{odd} + \text{even} = \text{odd}$... but that's a contradiction, because the sum of the degrees should be even.

Here is another basic term. A *simple* graph has no loops or multiple edges. Many people and textbooks mean that a graph (with no adjectives in front of the word) is simple unless otherwise specified. Around here, we like multiple edges and loops and so we generally allow them, though one challenge for you, the reader, is to figure out when loops or multiple edges cause problems but we, the author, haven't mentioned it. (For example, we are about to restrict ourselves to simple graphs for a while, but we won't say so.) This is good practice for being in the rest of the world....

A simple caution. Sometimes "graph" means "simple graph" and sometimes it doesn't. This depends on who is speaking/writing and on the situation, so be aware that you may have to figure out whether the presence (or absence) of loops and/or multiple edges makes any difference.

When you are trying to make a conjecture about graphs in general, it is a good idea to check your conjecture on several common classes of graphs, including trees, cycles, paths, and the graphs we are about to introduce in the next four paragraphs.

The *complete* graph with n vertices, called K_n (you know, K for *komplete*) has every possible edge; that is, every vertex is adjacent to every other vertex. Examples of complete graphs are shown in Figure 3.7.

Figure 3.7. Complete graphs K_1, K_2, K_3, K_4, and K_5.

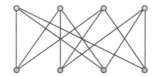 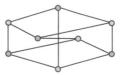

Figure 3.8. Two different drawings of the same bipartite graph.

There are two ways to think of a *bipartite* graph. One is to start with two piles of vertices V_1 and V_2, called *parts*, then draw some edges between the parts, but no edges within either part. This is a constructive way to produce a bipartite graph, but not every graph that is bipartite looks that way. Given a graph, it is bipartite if the vertices can be split into two parts so that neither part has internal edges. One common way of pointing out which vertices correspond to which part is coloring the vertices—V_1 gets one color and V_2 gets the other color. Both perspectives are illustrated in Figure 3.8.

A *complete bipartite* graph has all possible edges, so every vertex in V_1 is adjacent to every vertex in V_2. If $|V_1| = m$ and $|V_2| = n$, then we denote the corresponding complete bipartite graph by $K_{m,n}$. We can extend these notions to tripartite graphs (with three parts V_1, V_2, V_3) and multipartite graphs (with n parts $V_1, V_2, \ldots, V_n$). Examples are given in Figure 3.9.

Every vertex of K_n has degree $n - 1$. Every vertex of a cycle C_n has degree 2. Such graphs, where every vertex has the same degree, are called *regular*. Thus, K_n is $(n - 1)$-regular and C_n is 2-regular. The *degree sequence* of a graph is a list of the degrees of the vertices in increasing order. For example, the degree sequence of the first graph shown in Figure 3.3 is $(1, 3, 4, 4, 4, 4, 5, 5)$, and the degree sequence of C_n is $(2, 2, 2, \ldots, 2)$. The best graph of all is the *Petersen graph*, which

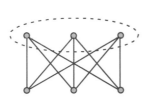 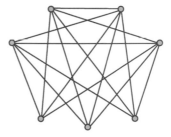

Figure 3.9. The complete bipartite graph $K_{3,3}$ and the complete tripartite graph $K_{2,2,3}$. Indicated by a dotted oval is part of this complete bipartite graph.

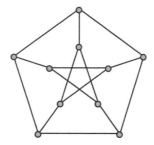

Figure 3.10. Your friend and mine, the Petersen graph.

is 3-regular, pictured in Figure 3.10, and either an excellent example of or a quick counterexample to almost every theorem and conjecture in graph theory. It is your friend.

Check Yourself ──────────────────────────────

Try these ten quickies.

1. Find the degree sequences of the graphs in Figure 3.5.

2. Look through the graphs pictured so far; identify one that is simple and one that is not simple.

3. For each graph in Figures 3.3 and 3.5, decide whether or not the graph is connected. Is any of the graphs a tree? A forest?

4. Find the longest possible path in the middle graph of Figure 3.3 and in the left-hand graph of Figure 3.5.

5. What is the largest cycle in any graph shown in Figures 3.3 and 3.5? How about the smallest?

6. There is at least one bipartite graph pictured in Section 3.3. Identify one; is it complete?

7. Draw K_7, C_8, and P_{10}.

8. Draw two 2-regular graphs on ten vertices, one of which is connected and one of which has two components.

9. What is the length of a smallest cycle in the Petersen graph?

10. Draw a bipartite graph with nine vertices.

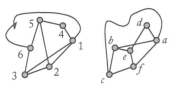
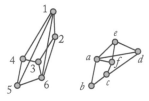

Figure 3.11. Are these two drawings of the same graph?

Figure 3.12. How about these?

3.6   Isomorphisms

We need a notion of when two graphs are the same. This will generalize to other types of mathematical objects as well, but we will begin with graphs. Should the graphs in Figure 3.11 be considered the same? How about in Figure 3.12?

The information encoded by a graph is only which vertices are adjacent. There is nothing in the definition of a graph that says how long or short or curly or straight the edges should be, whether or not they cross, what colors they might be, etc. So whatever definition of "sameness" we have, it should definitely include adjacency and exclude color.

Definition 3.6.1. Two graphs G, H are *isomorphic* if there exists a bijection $\varphi : V(G) \to V(H)$ such that $\{v_1, v_2\}$ is an edge in G if and only if $\{\varphi(v_1), \varphi(v_2)\}$ is an edge in H.

Basically, Definition 3.6.1 says that two graphs are isomorphic if two vertices are adjacent in one graph exactly when the corresponding vertices form an edge in the other graph. And even more basically, it says that two graphs are isomorphic if the vertices can be labeled in such a way that one graph can be redrawn to look exactly like the other.

Example 3.6.2 (of graph isomorphism). Consider the two graphs in Figure 3.13. Dragging vertex b to the left of vertex a (or, alternatively, dragging vertex 5 to the right of vertex 6) demonstrates that these are two different drawings of the same graph.

Figure 3.13. Each of these graphs is isomorphic to P_4.

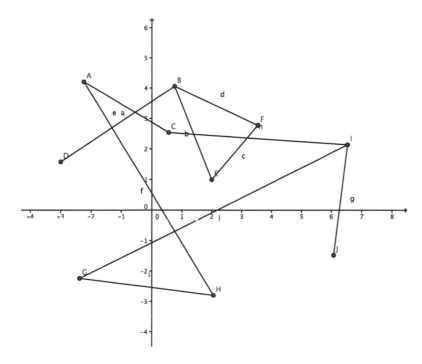

Figure 3.14. An arbitrary graph drawn in GeoGebra.

What one wants, in order to be able to determine whether two graphs are iso-morphic, is to be able to draw a graph and then drag the vertices around and have the edges follow, to see if it can be made to look like another graph. Hurray! Technology exists to enable this! The free software GeoGebra can be used to draw and fiddle with graphs. Go to www.geogebra.org and choose GeoGebra Geometry or GeoGebra Classic. Click on the point icon—it stays selected—and place a pile of vertices on the screen. Then click on the segment icon, and draw in edges by clicking on their respective vertices. A sample result is shown in Figure 3.14. Finally, click on the pointer-arrow icon to enable the dragging of vertices all over the place. Go to town with this! It's fun.

The notion of graph isomorphism is a specific example of the more general notion of isomorphism between two mathematical objects A and B. For our definition, we will need new notation: $\star_A$ represents an operation defined on A and $\star_B$ represents an operation defined on B. For example, $\star_{\mathbb{N}}$ might be addition in $\mathbb{N}$ and $\star_{\mathbb{Z}}$ might be addition in $\mathbb{Z}$. (The notation could also indicate multiplication in either of those cases.) When A and B are graphs, $\star_A$ and $\star_B$ represent adjacency of vertices, so that $a_1 \star_A a_2$ can return either *true* or *false*.

Figure 3.15. The two nonisomorphic (left) and three distinct (right) subgraphs of K_2. (See Section 3.7.1 for the definition of *subgraph*.)

Definition 3.6.3. Let $\varphi : A \to B$ be a function; φ is an *isomorphism* if it is a bijection and preserves every operation defined on A, that is, if $\varphi(a_1 \star_A a_2) = \varphi(a_1) \star_B \varphi(a_2)$. If there is an isomorphism mapping A to B, we say A and B are *isomorphic* and write $A \cong B$.

Example 3.6.4. The two graphs in Figure 3.13 of Example 3.6.2 are isomorphic because the function $a \mapsto 6, b \mapsto 5, c \mapsto 7, d \mapsto 8$ is an isomorphism. It is one-to-one and onto, and the edges $\{a,c\}, \{a,d\}, \{b,d\}$ map to the edges $\{6,7\}, \{6,8\}, \{5,8\}$; no other pairs of vertices correspond to edges in either graph.

Isomorphisms preserve just about any property you can think of. For example, two isomorphic graphs have the same number of vertices and the same number of edges; either they both have triangles or neither has any triangles.

The opposite of *isomorphic* is *nonisomorphic*, and this is not the same as *distinct*. For example, two copies of the same graph are distinct and also isomorphic; see Figure 3.15.

You may be wondering about the word "isomorphism." The particle *iso-* means "same" and the root *-morph-* means "shape." Throughout mathematics, isomorphisms are used to determine when two objects that seem different are secretly the same in some shape-like way.

Check Yourself

Test your understanding with these three brief problems.

1. Pick a graph from Figures 3.3 and 3.5 and draw it so that it looks different but is, in fact, the same graph.

2. List all nonisomorphic subgraphs of C_4. (See Section 3.7.1 for the definition of *subgraph*.)

3. Label the vertices of the graphs in Figure 3.8 and define a function between them that shows the graphs are isomorphic. (A GeoGebra file of Figure 3.8 is available for your playing pleasure at http://www.toroidalsnark.net/dmwdlinksfiles.html.)

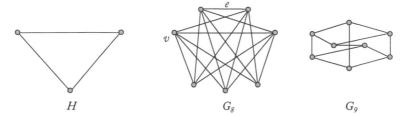

Figure 3.16. The triangle H is a subgraph of $G_8 = K_{2,2,3}$ but not of G_9. Note that the smallest cycle in G_9 has length 4.

3.7 Graphs: Operations and Uses

Just like sets can have subsets and we can take unions and complements of sets, we can define similar structures and operations on graphs.

3.7.1 Sets and Graphs Have Some Things in Common

A *subgraph* H of a graph G is a graph such that $V(H) \subset V(G)$ and $E(H) \subset E(G)$. (That wasn't so bad, was it? See Figure 3.16 for an example.) While technically the empty graph (no vertices) is a subgraph of every graph, in practice we ignore it. For graphs G_1 and G_2 (with disjoint vertex sets), the *graph union* $G_1 \cup G_2$ is another graph G_3 with $V(G_3) = V(G_1) \cup V(G_2)$ and $E(G_3) = E(G_1) \cup E(G_2)$. (That was probably not a surprise.) Notice that the union of two connected graphs will not be a connected graph—essentially, taking the union of two graphs consists of drawing them close to each other. Each individual connected piece of the union (or of any graph) is called a *component*.

Just as we could remove a subset from a set (see page 30), so too we can remove a subgraph from a graph. The graph $G \setminus e$ (or $G - e$) is G but with the edge e removed and e's vertices left intact. The graph $G \setminus v$ (or $G - v$), on the other hand, is G but with the vertex v and all its incident edges removed. This extends to a subgraph H of G, so that $G \setminus H$ (or $G - H$) removes H and all edges incident to any vertex in H. Examples of these three removals are given in Figure 3.17.

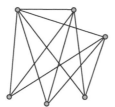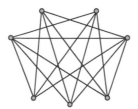

Figure 3.17. From left to right, $K_{2,2,3} \setminus v$, $K_{2,2,3} \setminus e$, and $K_{2,2,3} \setminus H$. (These refer to the same v, e, and H as labeled in Figure 3.16.)

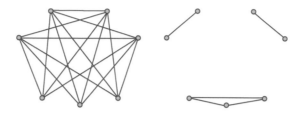

Figure 3.18. At left, $K_{2,2,3}$, and at right, $\overline{K_{2,2,3}}$.

We also have a notion of *graph complement*. This is not quite analogous to the notion of set complement: First, G and its complement $\overline{G}$ have the same set of vertices. Then, the complement of G is always taken relative to K_n, in the sense that overlaying G and $\overline{G}$ produces K_n. To form $\overline{G}$, we remove the edges of G from K_n, so that $\overline{G}$ has exactly the edges of K_n that G itself does not have. Figure 3.18 shows an example. For an interactive example, see http://www.mathcove.net/ petersen/lessons/get-lesson?les=37; there, you draw some vertices/edges and the graph complement is produced in real time.

3.7.2   How Are Graphs Useful?

Graphs are stand-ins for networks of many kinds: roads, cell-phone towers, friendships, circuitry, neurons, species, etc. For example, the vertices might represent cities, and the edges might represent roads between them. In such cases there are often numbers marked on the edges, called *weights*, that indicate the distance between the relevant cities when traveling along the given road. Or, vertices may represent cell-phone towers, with an edge present whenever two towers are able to communicate with each other. One can find graphs of the internet backbone, with vertices representing major servers and edges representing physical cables. Vertices could represent neurons in the brain, with edges representing which axons talk to which dendrites. Evolutionary biologists use graphs to indicate which species are evolutionary descendants of which others. Chemists use graphs to encode protein interactions. Teams in an informal Ultimate Frisbee tournament can be represented by vertices, with edges between teams that play each other. The edges can be directed after the tournament is over to indicate which team won each game, and from the resulting directed graph one can sometimes deduce who won the tournament. (In a *directed* graph, we write an edge as an ordered pair $e = (v_1, v_2)$ and draw it as an arrow $\rightarrow$.) So far, these applications are simply of graphs as models for other situations; mathematical theory is used in varying

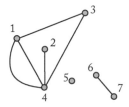

Figure 3.19. A graph to be stored.

amounts depending on the application, and we will see more applications of graph theory in later chapters (when we know some theory to apply).

As you might expect, graphs are not stored in computers as the pictures we draw. There are many ways of encoding graphs for computer use, but we will only indicate three here. First, we can store a graph as a pair of lists, where the first list is of the vertices and the second is of the edges as pairs of vertices. In this way, the graph in Figure 3.19 could be stored as

$$\{\{1,2,3,4,5,6,7\},\{\{1,3\},\{3,4\},\{2,4\},\{1,4\},\{1,4\},\{6,7\}\}\}.$$

Second, we can store a graph as an *adjacency matrix*, where each column and row corresponds to a vertex and each entry is the number of edges between the column vertex and row vertex. Using the same numbering as in our vertex/edge list, the graph in Figure 3.19 would have adjacency matrix

$$\begin{pmatrix} 0 & 0 & 1 & 2 & 0 & 0 & 0 \\ 0 & 0 & 0 & 1 & 0 & 0 & 0 \\ 1 & 0 & 0 & 1 & 0 & 0 & 0 \\ 2 & 1 & 1 & 0 & 0 & 0 & 0 \\ 0 & 0 & 0 & 0 & 0 & 0 & 0 \\ 0 & 0 & 0 & 0 & 0 & 0 & 1 \\ 0 & 0 & 0 & 0 & 0 & 1 & 0 \end{pmatrix}.$$

(For more examples, see http://www.mathcove.net/petersen/lessons/get-lesson?les=8; as you draw the graph vertex by vertex and edge by edge, this applet produces the corresponding adjacency matrix.) Finally, we could store a graph as a list of lists, where each list consists of a vertex and all of its neighbors. In this case, the graph in Figure 3.19 would be stored as

$$\{\{1,3,4,4\},\{2,4\},\{3,1,4\},\{4,1,1,2,3\},\{5\},\{6,7\},\{7,6\}\}.$$

Here are a few more quickies.

1. Draw $P_2 \cup C_3$.

2. What are $K_5 \setminus v$, $K_5 \setminus e$, and $\overline{K_5}$? (Note that the symmetry of K_5 means that it doesn't matter which vertex is chosen to be v or which edge is chosen to be e.)

3. Choose one of the graphs pictured in this chapter (other than the one in Figure 3.19) and encode it using vertex/edge lists, as an adjacency matrix, and using vertex/adjacency lists.

3.8 Try This! More Graph Problems

These problems are about graph structure and isomorphism. GeoGebra files for Figures 3.21, 3.22, 3.23, and 3.24 are available for your use at http://www.toroidalsnark.net/dmwdlinksfiles.html.

1. If you stick a vertex in the middle of an $(n-1)$-vertex cycle C_{n-1} (where $n-1$ is at least three) and connect it to all vertices on the cycle, you obtain the *wheel* graph, denoted W_n. (It has that extra vertex, see.) A few wheels are shown in Figure 3.20. (Note that W_4 is also K_4 but differently drawn than in Figure 3.7.) Let $n \geq 4$. Find and prove a formula for the number of edges of the wheel W_n.

2. Are the two graphs shown in Figure 3.21 isomorphic? If so, exhibit the isomorphism. If not, find a property that should be preserved by isomorphism for which the two graphs differ.

3. Let G be a graph with v vertices and e edges. In terms of v and e, how many edges does $\overline{G}$ have?

Figure 3.20. Wheels W_4, W_5, and W_6.

Figure 3.21. Two potentially isomorphic graphs.

Figure 3.22. Two potentially nonisomorphic graphs.

4. Prove that $C_5 \cong \overline{C_5}$. Can any other cycle graph be isomorphic to its complement? Justify your answer with example(s) or proof.

5. Are the two graphs shown in Figure 3.22 isomorphic? If so, exhibit the isomorphism. If not, find a property that should be preserved by isomorphism for which the two graphs differ.

6. Draw the seven nonisomorphic subgraphs of K_3 and the 17 distinct subgraphs of K_3.

7. Are the three graphs shown in Figure 3.23 isomorphic? (Are any two of them isomorphic?) If so, exhibit the isomorphism. If not, find a property that should be preserved by isomorphism for which the graphs differ.

8. Are the two graphs shown in Figure 3.24 isomorphic? If so, exhibit the isomorphism. If not, find a property that should be preserved by isomorphism for which the two graphs differ.

  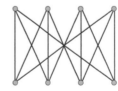

Figure 3.23. Three unruly graphs.

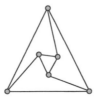

Figure 3.24. Two pointy graphs.

Figure 3.25. This 2-edge-colored K_5 has no monochromatic triangle.

3.9   Ramseyness

Hey! You! Don't read any further unless you have played the game in Section 3.3.3 and thought about the associated problems a lot. I mean it!

Hopefully you have discovered by coloring the edges of some complete graphs that if you start with five dots, you cannot force a win or loss and may have a draw, but that if you start with six dots, one player can always force the other to complete a triangle in hir color. Let's prove it! First, we will exhibit a K_5 with a draw; the colors in Figure 3.25 are black and teal. Now consider a single vertex (any vertex will do) of K_6. It has degree 5, so one of the two colors (let's say teal) must eventually be used on at least three of those edges. (If only one or two of the edges are teal, then the other four or three must be black and we could swap colors globally.) This situation is shown at left in Figure 3.26. Now assume the teal player makes no teal triangle, as otherwise we have a monochromatic triangle. Then each of the three pairs of teal edges must eventually be connected by a black edge (see Figure 3.26). But, ah! Those three black edges form a triangle. So if there is no teal triangle, there must be a black triangle; therefore, one player will always be forced to make a monochromatic triangle. (Here is the underlying structure of this proof: We want to show that A or B is true, where A, B represent *player must make monochromatic triangle*. Then if A is true the statement holds, so we prove that if A is not true, B must be true.)

Figure 3.26. At least three edges from a vertex are teal; then we need to avoid a teal triangle.

Instead of thinking of this as a game, we can rephrase the situation: Any K_6 that has edges colored using exactly two colors must have a monochromatic triangle. However, a 2-edge-colored K_5 need not have a monochromatic triangle. Similarly, neither a 2-edge-colored K_4 nor a 2-edge-colored K_3 necessarily has a monochromatic triangle. Therefore, six is the smallest number n such that a 2-edge-colored K_n must have a monochromatic triangle. In shorthand, we say $R(3,3) = 6$.

What?!? What wackiness is this? It's Ramsey wackiness! The *Ramsey number* $R(k,m)$ is the smallest number n such that a 2-edge-colored K_n must have either a K_k of one color or a K_m of the other color. In the case of $R(3,3)$, we must have a K_3 in one color or the other, which we've already referred to as a monochromatic triangle. Now read the fourth sentence of this paragraph aloud, twice. It's a darned complicated definition.

Relative to the infinitely many possibilities for k and m, not many Ramsey numbers are known. (There are upper and lower bounds for many of them.) Moreover, the idea of Ramsey numbers can be extended to using more than two colors for the edges of K_n; these are called multicolor Ramsey numbers. They are part of an area of graph theory research (unsurprisingly) called *Ramsey theory*, which is part of a larger area known as *extremal graph theory*.

Check Yourself ————————————————————————

Please do both of these problems.

1. What is $R(2,2)$?

2. Given three particular numbers k, m, N, what are the two ways you could show that $R(k,m) \neq N$?

3.10   Where to Go from Here

Did you find this chapter challenging? Exciting? Overwhelming? Fascinating? You might feel more mathematically grounded if you reread Section 2 of the student preface.

We are going to discuss certain aspects of graph theory in detail in Chapters 10–13. But there is so much more! Graph theory is a large subdiscipline of mathematics and is, in fact, one of the areas of mathematics in which the most papers are published. Graphs can be applied anywhere that networks are useful: for example, one modern research project revealed power relationships between committees of the US Congress; another tracked genes involved in a particular cancer.

You can take an undergraduate upper-level graph theory course at some institutions. Sometimes graph theory is half of the content in a combinatorics course (Yes, that doesn't make sense—why not call it Advanced Discrete Mathematics in that case? But it's often hard to change a course title), so ask instructors in your locality about what is offered that will give you more graph theory. Graph theory also plays a big role in computer science courses on networks and algorithms, and the study of network flows is central in the optimization area of operations research. Every computer algebra system (such as Maple, *Mathematica*, or Sage) has its own way of entering/displaying/manipulating graphs; and, there are many pieces of specialized graph theory software that have their own ways of storing graphs. These are interesting both from a mathematical point of view of investigating graphs and from a computer science point of view of understanding storing and manipulating the structure of a graph.

Should you wish to learn more about graph theory in general, an excellent place to start is *Introduction to Graph Theory* (which used to have the better title *Dots and Lines*) by Richard J. Trudeau [24]. Another lovely book, though more advanced, is *Introduction to Graph Theory* by Gary Chartrand and Ping Zhang [7]. *Introduction to Graph Theory* by Robin Wilson [25] is also readable. (Do you notice a theme in these book titles?) Probably the definitive reference on graph theory is Doug West's *Introduction to Graph Theory*.

To learn more about Ramsey theory and other parts of extremal graph theory, it is necessary to first understand general graph theory; at that point, you may look for books with titles like *Ramsey Theory* and *Extremal Graph Theory*. If you are interested in similar graph games, check out Chapter 20 in *Pearls of Discrete Mathematics* by Martin Erickson.

There are two different directions that you can take with functions. They are studied throughout higher mathematics in linear algebra, abstract algebra, topology, and real analysis courses. The algebraic classes focus on functions that preserve operations (as we introduced here), and the analytic classes focus on continuous functions. Seriously, you can't go wrong in learning more in both directions! Because different aspects of functions are emphasized in different contexts, there is no general source or text to suggest for reading more about functions as a whole. Within this very text, we will address some function properties on infinite sets in Chapter 15.

Credit where credit is due: Section 3.3.2 was inspired by [3]; Section 3.3.3 was inspired by Josh Greene and Ari Turner. The exposition of Section 3.12 was inspired by [4]. Finally, the problems in Section 3.8 were adapted from Richard Trudeau's *Dots and Lines* [24].

3.11 Chapter 3 Definitions

function: We call $f : A \to B$ a function when, given any element a of the set A as input, the function f outputs a unique element $f(a) = b \in B$.

map: As a noun, *map* is a synonym for *function*; as a verb, it expresses the action of a function, as in "f maps a to b."

well defined: The property of a function that if $a_1 = a_2$, then $f(a_1) = f(a_2)$.

domain: The set A from which function inputs are taken.

target: The set B from which outputs are selected. Also called the *target space*.

image: The element $f(a)$ is called the image of the element a.

range: All the elements of the target space that are mapped to by the function; that is, for $f : A \to B$, the range of f is $Range(f) = \{f(a) \mid a \in A\}$.

gipo: A thing that *given* *input*, *produces* *output*, but is not necessarily well defined. (Notice that every function is a gipo, but not every gipo is a function.)

one-to-one: Whenever $f(a_1) = f(a_2)$, then $a_1 = a_2$. Every element of the target space is mapped to *at most* once. Also denoted 1–1.

injective: One-to-one.

into: One-to-one.

onto: For every $b \in B$, there exists some $a \in A$ such that $f(a) = b$. Every element of the target space is mapped to *at least* once.

surjective: Onto.

bijection: A function that is both one-to-one and onto. Every element of the target space is mapped to *exactly* once.

one-to-one correspondence: A bijection.

floor function: The floor function returns the integer equal to or just less than the input.

ceiling function: The ceiling function returns the integer equal to or just greater than the input.

vertex: A dot, usually drawn as • or ◦, that can represent some object in a set of items.

vertices: Plural of *vertex*.

edge: A pair of vertices $e = \{v_1, v_2\}$ (sometimes abbreviated as $v_1 v_2$) that is usually represented by a line or curve between the dots representing v_1 and v_2.

graph: A pair $G = (V, E)$, where V is a set of vertices and E is a set of pairs of vertices.

adjacent: Two vertices joined by an edge are adjacent.

incident: An edge is incident to each of its endpoint vertices.

neighbor: Any vertex adjacent to a vertex v is a neighbor of v.

loop: An edge joining a vertex to itself.

multiple edge: More than one edge joining the same two vertices.

multiplicity: The number of edges in a multiple edge.

degree: The number of edges that emanate from a vertex.

degree sequence: A list of the degrees of the vertices in increasing order.

walk: A list of vertices alternating with edges, with both the start and end of the list vertices (not edges).

path: A walk where no vertices repeat.

cycle: A walk whose only repetition is the first/last vertex.

length: The number of edges of a path or cycle.

distance: The length of the shortest path between two vertices.

connected: A graph in which any two vertices are joined by some walk.

forest: A graph with no cycles.

tree: A connected graph with no cycles.

leaf: A vertex of degree 1.

simple graph: A graph that has no loops or multiple edges.

complete graph: A graph where every vertex is adjacent to every other vertex.

bipartite graph: A graph whose vertices can be separated into two piles, called *parts*, with edges between the parts, and no edges within either part.

complete bipartite graph: A bipartite graph with all possible edges; that is, if the parts are V_1, V_2, then every vertex in V_1 is adjacent to every vertex in V_2.

wheel graph: If you stick a vertex in the middle of an $(n-1)$-vertex cycle C_{n-1} (where $n-1$ is at least three) and connect it to all vertices on the cycle, you obtain the wheel graph, denoted W_n.

regular graph: A graph where all vertices have the same degree.

Petersen graph:

The best and most awesome graph, frequently a counterexample, named after Peter Christian Julius Petersen (1839–1910), who did more work outside of graph theory than in it and who was not the first to use the Petersen graph (apparently Kempe was; see page 427). Don't spell it Peterson or Pedersen or Petersön or Petersøn.

isomorphic graphs: Two graphs G, H are isomorphic if there exists a bijection $\varphi : V(G) \to V(H)$ such that $\{v_1, v_2\}$ is an edge in G if and only if $\{\varphi(v_1), \varphi(v_2)\}$ is an edge in H.

isomorphism: A gipo $\varphi : A \to B$ that is well defined, one-to-one, onto, and preserves every operation defined on A (that is, if $\varphi(a_1 \star_A a_2) = \varphi(a_1) \star_B \varphi(a_2)$). In other words, an isomorphism is an operation-preserving bijection.

subgraph: A subgraph H of a graph G is a graph such that $V(H) \subset V(G)$ and $E(H) \subset E(G)$.

graph union: For graphs G_1 and G_2 (with disjoint vertex sets), the graph $G_1 \cup G_2$ is another graph G_3 (not connected) with $V(G_3) = V(G_1) \cup V(G_2)$ and $E(G_3) = E(G_1) \cup E(G_2)$.

graph component: An individual connected piece of a graph.

graph complement: Let G have n vertices; to form $\overline{G}$, we remove the edges of G from K_n, so that $\overline{G}$ has exactly the edges of K_n that G itself does not have.

weights: Numbers marked on the edges (or vertices) of a graph to indicate information such as distance or traffic capacity or population.

adjacency matrix: A matrix representing a graph, where each column and row corresponds to a vertex and each entry is the number of edges between the column vertex and row vertex.

Ramsey number: The Ramsey number $R(k,m)$ is the smallest number n such that a 2-edge-colored K_n must have either a K_k of one color or a K_m of the other color.

3.12 Bonus: Party Tricks

This is a classic problem: Mei-Ting and Ri Zhao had a dinner party and invited four couples. Before sitting down at the table set for ten, there were formal introductions and people who did not know each other shook hands in greeting. After indicating that everyone should sit, Mei-Ting announced, "I have just noticed that no two of you shook hands the same number of times." Dear reader, how many times did Ri Zhao shake hands?

Pause. Any clue how to solve this problem? Think about it for at least one minute before proceeding.

Solution. To solve this problem, we will start by modeling the situation as a graph. Each of the ten dinner guests is a vertex, and two vertices are adjacent if (and only if) the two dinner guests shook hands.

Pause. What are the possible degrees for the vertices in this graph? Think about it for at least 30 seconds before proceeding.

Solution. For a given guest, there are nine possible other people with whom to shake hands. (No one needs to shake hir own hand.) And the guests arrived as couples, so no one needed to shake hands with hir partner; therefore, the maximum degree of a vertex is 8. Someone might have known everyone already, so it is also possible to have the minimum vertex degree of 0. Thus, the nine possibilities are $0, 1, 2, 3, 4, 5, 6, 7, 8$. We know there are nine people who Mei-Ting believes shook different numbers of hands. So, ignoring Mei-Ting's vertex, there are nine vertices with different degrees.

Consider the vertex of degree 8 and call the associated person Eight (and likewise with the other vertices). There is only one other person with whom Eight did not shake hands, so that person must be hir partner. We also know that the vertex of degree 8 is not adjacent to the vertex of degree 0, so therefore Zero must be the only person who did not shake hands with Eight. If we continue our reasoning in this fashion, we discover that Seven must be the partner of One because Seven didn't shake hands with two others; Zero is one of those two and One shook hands with Eight, so One didn't shake hands with Seven and must be Seven's partner. Keeping on, we see that Six must be partnered with Two, and Five must be partnered with Three, and Four must be partnered with... hmm... oh, it must be Mei-Ting. So Four must be Ri Zhao, as we had left Mei-Ting out of it all.

Now Mei-Ting and Ri Zhao have a less formal party. They simply invite some friends so that there are ten people at their house; some hang out in the house itself and others chat in the garden. Mei-Ting remarks that each person is friends with at least five other people at the party.

> Short activity:
>
> 1. Model this situation as a graph. What do the vertices represent? Fill in the blank: two vertices are adjacent if and only if _____.
>
> 2. What does Mei-Ting's observation tell you about the degrees of the vertices in the graph?
>
> 3. True or false (and explain): If there is always someone inside and always someone outside, then someone in the garden has a friend in the house.
>
> 4. Generalize the party to having $2n$ guests, each of whom is friends with at least n other guests. Describe your graphical model in this general party case; if there is always someone inside and always someone outside, then is it true that someone in the garden has a friend in the house? Explain.

3.13   Bonus 2: Counting with the Characteristic Function

The following seemingly silly function is ubiquitous in mathematics. Given a set S and a subset A, the *characteristic function* $\chi_A : S \rightarrow \{0, 1\}$ is defined as

$$\chi_A(s) = \left\{ \begin{array}{ll} 1 & \text{if } s \in A, \\ 0 & \text{if } s \notin A. \end{array} \right.$$

Pause. Here is a quickie exercise to make sure you understand this definition. Let $S = \{0.5, 1, 1.5, 2, 2.5, 3\}$ and let $A = \{0.5, 2, 2.5\}$; evaluate $\chi_A(0.5)$, $\chi_A(1)$, $\chi_A(1.5)$, and $\chi_A(3)$.

How can we use the characteristic function to count? Check *this* out: Consider the power set of $\{1, 2, \ldots, n\}$ (see page 30), and name it $\mathscr{P}_n$. Also, consider the set of all functions from $\{1, 2, \ldots, n\}$ to $\{0, 1\}$ and call that set $\mathscr{F}_n$.

Pause. What the heck? The power set is one thing, but the set of functions from one set to another? Wack. List four different elements of $\mathscr{F}_n$. Then compute $|\mathscr{F}_n|$. If you have trouble with that last bit, go to Section 3.3.1 and look again at the problems there.

Now, we are going to do something wackier still. Define a function $F : \mathscr{P}_n \to \mathscr{F}_n$ by $F(A) = \chi_A$. This makes sense because F takes a subset as input and returns a function as output.

> **Short activity:**
>
> 1. How does $|\mathscr{F}_n|$ compare to $|\mathscr{P}_n|$?
>
> 2. Forget the previous question. Instead, prove that F is a bijection; recall that this means you have to show that F is one-to-one and onto.
>
> 3. Explain how this gives yet another proof of Theorem 1.5.2.

This is an early example of a combinatorial theme: instead of counting something directly, exhibit a bijection f and count f(something).

3.14 Bonus Check-Yourself Problems

Solutions to these problems appear starting on page 598. Those solutions that model a formal write-up (such as one might hand in for homework) are to Problems 4, 6, and 9.

1. Let $S = \{s_1, s_2, \ldots, s_n\}$. How many functions are there with domain $\mathbb{Z}_3$ and target S? Of those functions, how many are one-to-one? How many are onto?

2. Draw all connected 3-regular graphs with four vertices.

3. Are the two graphs in Figure 3.27 isomorphic? Justify your response.

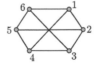 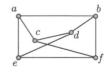

Figure 3.27. Two potentially isomorphic graphs.

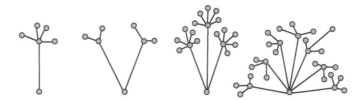

Figure 3.28. Some finger-finger graphs: $F_{1,4}$, $F_{2,2}$, $F_{3,5}$, and $F_{7,3}$.

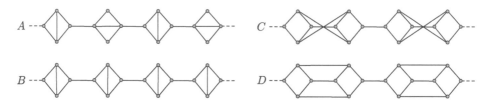

Figure 3.29. Who is who here?

4. Is the function $f : \mathbb{Z} \to \mathbb{Z}$ defined by $f(n) = \lfloor \sin(n) \rfloor$ a one-to-one function? Prove or disprove.

5. Is it possible to draw a graph with six vertices of degrees 2, 2, 3, 3, 4, and 4? If so, draw one. If not, explain why not.

6. A *finger-finger* graph is denoted by $F_{m,n}$ and has m fingers, from each of which grows n fingers; see Figure 3.28. Conjecture and prove formulas for the number of vertices and the number of edges of a finger-finger graph.

7. What can you say about the number of vertices of a 3-regular graph?

8. The following statement is true: *any cycle C_n with $n \geq k$ has complement $\overline{C_n}$ containing a triangle.* Determine k and prove the statement.

9. Consider the map $g : (\mathbb{N} \times \mathbb{N}) \to \mathbb{N}$ defined by $g((a,b)) = ab$. Is this one-to-one? Onto? Give proofs.

10. Shown in Figure 3.29 are four infinite graphs in pairs A,B and C,D. One of these pairs is isomorphic and the other nonisomorphic. Which is which? Justify your response.

3.15   Problems about Graphs and Functions

Some of the graph theory problems may seem at first like busywork rather than problems that enrich your cognition. However, a lot of gruntwork is necessary in order to develop good intuition about graphs. It's easy to be misled into thinking that the apparent simplicity of graphs indicates that they are straightforward to understand.

1. Try these three minis:

   (a) Draw the union of K_4 and C_3.

   (b) How many vertices and how many edges does the Petersen graph have?

   (c) Draw $\overline{W_6}$.

2. How many simple 3-regular graphs are there with five vertices? Prove that you have found them all.

3. Let $f : \{2,4,6,8,10\} \rightarrow \{1,3,5,7,9\}$ be an onto function. Prove that f is one-to-one.

4. Give an example of a graph that is 4-regular but neither complete nor complete bipartite. (While you're at it, give examples of 4-regular complete and complete bipartite graphs.)

5. Prove that $f : \mathbb{W} \rightarrow \mathbb{Z}$ defined by $f(k) = \lfloor \frac{k+1}{2} \rfloor (-1)^k$ is a bijection.

6. Find all ...

   (a) ... cycles that are also complete graphs.

   (b) ... cycles that are also wheels.

   (c) ... wheels that are also complete graphs.

   (d) ... cycles that are also paths.

   (e) ... paths that are also complete graphs.

   In each case, explain why your list is complete.

7. Let $S = \{s_1, s_2, \ldots, s_n\}$. How many functions are there with domain S and target $\mathbb{Z}_2$? Of those functions, how many are one-to-one? How many are onto?

8. Draw all nonisomorphic simple graphs with four vertices. (There are 11.) Verify that the handshaking lemma holds for each graph.

9. Do it again for graphs with five vertices. How many graphs did you find?

10. Examine Figure 3.11. Either show that the two graphs are isomorphic (by giving a bijection between their vertex labels that preserves adjacency) or explain why they are not isomorphic. Repeat this process with the two graphs in Figure 3.12. (GeoGebra files of Figures 3.11 and 3.12 are available for your playing pleasure at http://www.toroidalsnark.net/dmwdlinksfiles.html.)

11. Consider the Cartesian product $A \times B$, where A, B are finite nonempty sets, each with cardinality greater than 1. There are two functions with domain $A \times B$, called *projections*, with mapping rules $p_1(a,b) = a$ and $p_2(a,b) = b$. What is the target space of p_1? Of p_2? Are either of p_1, p_2 one-to-one? Onto?

12. Let $A = \{0,1,2,3,4\} \times \{0,1,2\}$, let $B = \{n \mid n$ is a positive factor of $144\}$, and let $f : A \rightarrow B$ with $f(a_1, a_2) = 2^{a_1} \cdot 3^{a_2}$. Is f one-to-one? Onto?

13. Perhaps keeping pigeons in mind, show that if a simple graph has at least two vertices, then two of its vertices must have the same degree.

14. Draw a graph with degree sequence $(1,1,2,2)$. Now draw one with degree sequence $(1,1,1,1,1,1)$. Can you find more graphs with these degree sequences?

15. Think of at least two different proofs that K_n has $\frac{n(n-1)}{2}$ edges.

16. Consider a cube, and make a graph from it by assigning a vertex to each corner of the cube and an edge to each edge of the cube.

(a) What are all the possible distances between two distinct vertices of the cube?

(b) How many different length-3 paths go from one corner of the cube to the opposite corner? Why?

(c) What is the average distance between any two distinct vertices of the cube? Explain.

17. The complete bipartite graph $K_{m,n}$ …

(a) … has $m+n$ vertices. Prove it.

(b) … has $m \cdot n$ edges. Prove it.

18. How many vertices and edges does the complete tripartite graph $K_{m,n,p}$ have? Prove your conjecture.

19. To what graph is $K_5 \setminus K_3$ isomorphic? How about $K_6 \setminus K_3$? (Any conjectures?)

20. The *star* graph on n vertices has one vertex adjacent to all other vertices (and no other adjacencies). Conjecture and prove a formula for the number of edges of the star graph on n vertices.

21. Two labeled infinite graphs are shown in Figure 3.30.

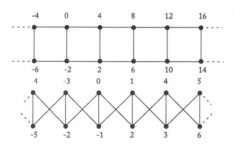

Figure 3.30. Two possibly isomorphic infinite graphs.

Show that they are isomorphic by defining a gipo between them and verifying

that the gipo is an isomorphism, or show that they are not isomorphic by finding a property that holds for one but not the other. (Note that if you want to show that the graphs are isomorphic, it will not be enough to just give a relabeling—that would only take care of finitely many vertices. You would need to give a rule for relabeling and show that this rule satisfies the properties of an isomorphism.)

22. Notice that many of the previous problems involve counting. Where did you use the sum principle? The product principle?

23. Some of the previous problems asked for proofs. What proof techniques did you use? Contradiction? Direct proof?

24. Encode K_5 as vertex/edge lists, as vertex/adjacency lists, and as an adjacency matrix. Do the same with P_5. How do the different storage methods compare for these two graphs? Does this suggest any general guidelines to you?

25. We showed that $R(3,3) = 6$ in Section 3.9. Certainly, it would be awesome if $R(4,4)$ were easy to guess— show that $R(4,4) \neq 8$. (It turns out that $R(4,4) = 18$.) You might find it helpful to experiment with different 2-colorings of the edges of various complete graphs at http://demonstrations.wolfram.com/ GraphsAndTheirComplements/.

26. Check out the graph G shown in Figure 3.31:

(a) Draw $G \setminus v_4$ (a.k.a. $G - v_4$).

(b) Now draw $G \setminus e$ (a.k.a. $G - e$).

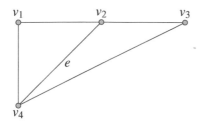

Figure 3.31. I'm G!

27. Consider $g : \{0,9,8,7,6,5\} \to \{4,3, 2,1\}$. Is it possible for g to be one to one? Onto?

28. Draw a connected graph with six vertices such that at least two vertices have degree 3 and at least one vertex has degree 1.

29. Delete things!
   (a) To what graph is $K_n \setminus v$ isomorphic? Explain.
   (b) To what graph is $C_n \setminus e$ isomorphic? Explain.

30. Let $A = \{2,4,6,8,10,12,14,16,18, 20\}$ and let $B = \{-1,0,1\}$.
   (a) How many functions can be defined $f : A \to B$? How many of those are one-to-one?
   (b) How many functions can be defined $f : B \to A$? How many of those are one-to-one?

31. Consider a function f defined on the set of finite graphs $\mathcal{G}$ by $f(G) = |V(G)|$. What is the range of f?

32. What is the length of the shortest walk you can take on the Petersen graph? What is the length of the longest path in the Petersen graph? Is K_3 a subgraph of the Petersen graph? Explain.

33. Let $X = \{1,2,3,4,5\}$ and let $Y = \{1,2, 3,4,5,6\}$.

(a) How many one-to-one functions are there from X to Y?
(b) How many one-to-one functions $f : Y \to X$ are there?
(c) How many injective functions are there from $X \to X$?

34. True or false: any two graphs with the same degree sequence are isomorphic. If true, provide proof; if false, provide a counterexample.

35. How many edges does a forest with n trees and v vertices (and v_i vertices in the ith tree) have?

36. Floors:
   (a) Let a function from the real numbers to the integers be defined by $f(x) = \lfloor x \rfloor$. Is f one-to-one? Is f onto? Explain why or why not in each case.
   (b) Let $f : \mathbb{Z} \to \mathbb{Z}$ be defined by $f(x) = \lfloor x \rfloor$. Is f one-to-one? Is f onto? Explain why or why not in each case.

37. Ceilings:
   (a) Let a function from the real numbers to the integers be defined by $c(x) = \lceil x - 2 \rceil$. Is c one-to-one? Is c onto? Explain why or why not in each case.
   (b) Let $c : \mathbb{Z} \to \mathbb{Z}$ be defined by $c(x) = \lceil x + 2 \rceil$. Is c a bijection?

38. Consider the map $f : \mathbb{W} \to \mathbb{N}$ defined by $f(x) = 2x + 1$. For every $n \in \mathbb{N}$, there exists some $w \in \mathbb{W}$ (let $w = n - 1$). Does this mean that f is onto? Explain.

39. Consider the *zero map* $z : A \to 0$, defined by $z(a) = 0$. Under what conditions is z one-to-one? Onto? A bijection?

40. Consider the Cartesian product $A \times B$, where A, B are finite nonempty sets, each with cardinality greater than 1. There are two functions called *inclusions*, with mapping rules $i_1(a) =$ (a, b_0) and $i_2(b) = (a_0, b)$. What is the domain of i_1? Of i_2? How about the target space of i_1 and i_2? Are either of i_1, i_2 one-to-one? Onto?

3.16   Instructor Notes

It is tough for students to absorb all the material in this chapter well over a single week of class, so you may wish to tailor your emphasis or expectations accordingly.

Assign the students to read Sections 3.1 and 3.2 and do the Check Yourself problems as preparation for the week. Start the first class of the week by having them work on the problems in Section 3.3. Collectively, these comprise a full class—so ask the students to work for a limited amount of time on each subsection, perhaps 10 minutes for Section 3.3.1 (students have a tendency to lollygag here) and 15 minutes for Section 3.3.2. Then, have students describe their results and take questions over the reading. Be sure to leave 10–15 minutes in class to describe the Section 3.3.3 dot game, have two students play it publicly at the front of the room, let the students loose to explore, and compile conjectures from the class.

As preparation for the next class meeting, assign students to think more about the dot game, read Sections 3.4–3.7, and do the associated Check Yourself problems. They often have trouble with the categorical notion of isomorphism, so it can be productive to begin class with a short lecture on isomorphisms in general that specializes to graph isomorphisms. While asking for questions over the reading, ask the students also what progress they have made in thinking about the dot game. It is likely that a student will produce a draw for the five-dot game, and possible that a student will contribute reasoning for why a win/loss is forced for the six-dot game. Remaining time can be used to work on problems from Section 3.8. Assign the students to read Section 3.9 as preparation for the third class.

A great warmup for the third class of the week is to interactively prove that every tree (with more than one vertex) has a leaf. (Besides, later in the course students will need this result.) Start by writing this theorem on the board and asking the students what it means. (If your classroom has a window, some students may puzzledly point outdoors; they need to recall what trees and leaves are in a graph-theoretic context.) Encourage students to contribute ideas for the proof and for what proof technique to use. They will often give the idea that one should walk along the graph until one runs into a leaf but be unable to formalize this into a proof by contradiction; help them realize that they cannot guarantee that they will run into a leaf without using the theorem they are trying to prove. Once they are able to conceptualize that an inability to find a leaf corresponds to walking along some cycle, they are usually able to close the deal on the proof.

It is also a good idea to review the proof that a win/loss is forced for the six-dot game if this did not arise in the previous class. Remaining class time would be used excellently by students working on and presenting solutions to the problems in Section 3.8. Use of GeoGebra will speed up their work on the visual problems—the author has GeoGebra files premade for these exercises, available at http://www.toroidalsnark.net/dmwdlinksfiles.html.

Finally, be aware that students may find the problems in Section 3.15 significantly harder than those in Chapters 1 and 2 because they use new terminology in conjunction with practicing earlier proof techniques. In addition, students are likely to still be processing concepts from earlier chapters. They will continue to work with these ideas over the semester. Some less challenging, but still proof-oriented, problems are presented in Section TI 2 for the purposes of supplementation.

Chapter 4 🐤🐤🐤🐤

Induction

4.1 Introduction and Summary

This proof technique is more complicated than the previous techniques, so it merits a chapter of its own. But there's only so much to say about mathematical induction, and then you just have to practice this proof technique for yourself. The basic idea is to reduce a theorem to a smaller case of the same theorem, and then to a smaller case, and so on to a small case you can deal with manually. Example 4.2.4 is particularly important, first because it shows why one must reduce to a previously known case (rather than building up from a known case), and second because it is a result we will use regularly in Chapters 10–13.

Unlike the previous topics addressed in this book, there are no everyday applications of induction. It instead applies to most of mathematics and computer sciences—as a proof technique, it arises frequently in each field at all levels.

4.2 Induction

We will begin with an example of a proof by induction using only basic arithmetic, then describe induction in general, and then give an extended example of a proof by induction about sets.

Example 4.2.1. Do you believe that if $n > 0$, then $2^n > n$? Let's check to see whether this statement holds for a few values of n.

When $n = 1$, we see that $2^1 = 2$, and $2 > 1$.

When $n = 2$, we see that $2^2 = 4$, and $4 > 2$.

When $n = 423$, we see that $2^{423} = 14{,}134{,}776{,}518{,}227{,}074{,}636{,}666{,}380{,}005{,}943{,}348{,}126{,}619{,}871{,}175{,}004{,}951{,}664{,}972{,}849{,}610{,}340{,}958{,}208$ and that is *certainly* greater than 423.

Yes, the statement *if $n > 0$, then $2^n > n$* seems to be true. We have checked it for small values of n, which are known as *base cases*.

To prove the statement, we will first suppose that when n is less than or equal to some indeterminate k (but greater than 0), $2^n > n$. This is called the *inductive hypothesis*.

If we can show that this assumption allows us to prove that $2^{k+1} > k+1$, then we will have proved that $2^n > n$ for *all $n > 0$*. This process is known as the *inductive step*. We will complete it as follows.

Consider 2^{k+1}. This can be rewritten as $2^{k+1} = 2 \cdot 2^k$. We note that $k \le k$, so the inductive hypothesis holds and we can use the fact that $2^k > k$. Multiplying this inequality by 2, we have $2 \cdot 2^k > 2 \cdot k$. We can rewrite again so that $2 \cdot k = k+k$, and as long as $k > 1$, we have $k+k > k+1$. Stringing this work together, we now see that when $k > 1, 2^{k+1} > k+1$, which is what we wanted to prove. (Well, except that we wanted $k > 0$. That's okay, because we checked the case for $n = 1$ separately.)

Notice what we did in the inductive step: The inductive hypothesis covers values of the index up to k, so we look at the index-$(k+1)$ version of the statement. We manipulate the index-$(k+1)$ statement so that we have found the index-k statement within it. Then, we apply the base case and the inductive hypothesis to show that the index-$(k+1)$ version of the statement is true. (We will often use index values lower than k, such as $k-1$ or $k-2$, as well.)

So what is going on with inductive proof? First, we want to prove some statement that has a variable n in it, and that variable takes values in $\mathbb{N}$. That's pretty important—induction only works when the statement you're trying to prove is indexed by the natural numbers. (We'll see why in a page or two.) Then, a proof by induction has three parts.

How to do a proof by induction:

- ⚓ Base case: Check to make sure that whatever you want to prove holds for small natural numbers, like 1, 2, or 3.

- ⚓ Inductive hypothesis: Assume that whatever you want to prove is true, as long as the variable in the statement is smaller than or equal to k; here, k is a specific (but unknown) value.

- ⚓ Inductive step: Consider the statement with $k+1$ as the variable. Use your knowledge that the statement is true when the variable is less than or equal to k in order to show that it's still true for $k+1$. (That is, use the base case(s) and inductive hypothesis.)

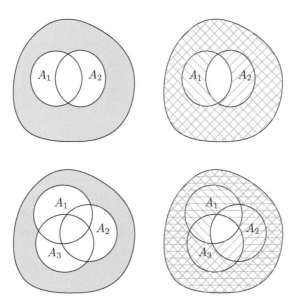

Figure 4.1. The left diagrams show $\overline{(A_1 \cup A_2 \cup \cdots \cup A_n)}$ for $n = 2,3$, whereas the right diagrams show $\overline{A_1} \cap \overline{A_2} \cap \cdots \cap \overline{A_n}$ for $n = 2,3$.

The inductive step shows that given any upper limit on n, that upper limit can be increased by one. Thus we can increase the possible values of n repeatedly to see that our statement is true for all $n \in \mathbb{N}$.

We turn now to a statement that uses sets rather than numbers. You have already proven the following theorem in the special case of just two sets—it is one of DeMorgan's laws—and here we extend the law to n sets.

Theorem 4.2.2 (DeMorgan's Laws for n sets). *Let $A_1, A_2, \ldots, A_n$ be n sets. Then for any $n \in \mathbb{N}$,*

$$\overline{(A_1 \cup A_2 \cup \cdots \cup A_n)} = \overline{A_1} \cap \overline{A_2} \cap \cdots \cap \overline{A_n}$$

and

$$\overline{(A_1 \cap A_2 \cap \cdots \cap A_n)} = \overline{A_1} \cup \overline{A_2} \cup \cdots \cup \overline{A_n}.$$

Proof: We will prove the first of DeMorgan's Laws here, and leave the other as an exercise (Problem 8 in Section 4.12).

(Base case) We begin by checking $n = 2$ and $n = 3$, using Figure 4.1. You can see that the teal-shaded areas on the left are the same as the multiply-hatched areas on the right. Unfortunately, Venn diagrams become increasingly difficult to draw and interpret as the number of sets increases, and there is no reasonable way to use Venn diagrams for an indeterminate number of sets. We will need to use symbols in order to generalize.

In symbols, our base cases are $\overline{(A_1 \cup A_2)} = \overline{A_1} \cap \overline{A_2}$ and $\overline{(A_1 \cup A_2 \cup A_3)} = \overline{A_1} \cap \overline{A_2} \cap \overline{A_3}$.

(Inductive hypothesis) As long as $n \le k$, $\overline{(A_1 \cup A_2 \cup \cdots \cup A_n)} = \overline{A_1} \cap \overline{A_2} \cap \ldots \cap \overline{A_n}$.

(Inductive step) Consider $\overline{(A_1 \cup A_2 \cup \cdots \cup A_{k+1})}$. We can also express this as $\overline{(A_1 \cup A_2 \cup \cdots \cup A_k \cup A_{k+1})}$, and we can add some parentheses to write $\overline{((A_1 \cup A_2 \cup \cdots \cup A_k) \cup A_{k+1})}$. If we rename the combined set in the new parentheses as $(A_1 \cup A_2 \cup \cdots \cup A_k) = B$, then $\overline{(A_1 \cup A_2 \cup \cdots \cup A_{k+1})}$ becomes $\overline{(B \cup A_{k+1})}$. That's just two sets, so our Venn-diagram-proved base case tells us that $\overline{(B \cup A_{k+1})} = \overline{B} \cap \overline{A_{k+1}}$. Let us resubstitute for B in order to obtain $\overline{(A_1 \cup A_2 \cup \cdots \cup A_k)} \cap \overline{A_{k+1}}$. We can now use the inductive hypothesis on $\overline{(A_1 \cup A_2 \cup \cdots \cup A_k)}$ because there are only k sets involved in that expression. Therefore, $\overline{(A_1 \cup A_2 \cup \cdots \cup A_k)} \cap \overline{A_{k+1}} = (\overline{A_1} \cap \overline{A_2} \cap \cdots \cap \overline{A_k}) \cap \overline{A_{k+1}}$. Linking all of our statements together, we have that $\overline{(A_1 \cup A_2 \cup \cdots \cup A_{k+1})} = \overline{A_1} \cap \overline{A_2} \cap \cdots \cap \overline{A_k} \cap \overline{A_{k+1}}$.

Because we have shown that $\overline{(A_1 \cup A_2 \cup \cdots \cup A_n)} = \overline{A_1} \cap \overline{A_2} \cap \cdots \cap \overline{A_n}$ for $n \le k$ implies that $\overline{(A_1 \cup A_2 \cup \cdots \cup A_{k+1})} = \overline{A_1} \cap \overline{A_2} \cap \cdots \cap \overline{A_k} \cap \overline{A_{k+1}}$, we know that $\overline{(A_1 \cup A_2 \cup \cdots \cup A_n)} = \overline{A_1} \cap \overline{A_2} \cap \cdots \cap \overline{A_n}$ for all n (including those larger than k). □

It's time for another example.

Example 4.2.3. We have the sets $\{1\}, \{1,2\}, \{1,2,3\}, \ldots \{1,2,3,\ldots,k\}, \ldots$ and so on. We want to show that the set of the first n natural numbers has 2^n subsets. (Yes, we already know this as a special case of Theorem 1.5.2. But it's always good to have more than one proof of a theorem!)

(Base case) The subsets of $\{1\}$ are $\{\}$ and $\{1\}$. There are two of them, which is 2^1. Excellent.

(We're done, but if you feel insecure, note that the subsets of $\{1,2\}$ are $\{\}$, $\{1\}$, $\{2\}$, $\{1,2\}$, and there are four of them, or 2^2.)

(Inductive hypothesis) We assume that a set containing the first n counting numbers has 2^n subsets, as long as $n \le k$.

(Inductive step) Examine the set $\{1,2,\ldots,k+1\}$. We would like to show that it has 2^{k+1} subsets. We need to understand how the subsets of $\{1,2,\ldots,k+1\}$ are related to the subsets of $\{1,2,\ldots,k\}$. (If there's no relationship, we can't use induction.) Every subset of $\{1,2,\ldots,k+1\}$ has one of two properties: either (a) $k+1$ is not an

Figure 4.2. The tree with one vertex (left) and every single tree with two vertices, all one of them (right).

element or (b) $k+1$ *is* an element. All the subsets with property (a) are also subsets of $\{1,2,\ldots,k\}$, and so we know there are 2^k of them by the inductive hypothesis. If we take each of the property (a) subsets and stick the element $k+1$ in, we get all the subsets with property (b). So there are the same number of them, and there are 2^k subsets with property (b). In total, we have $2^k + 2^k = 2 \cdot 2^k = 2^{k+1}$ subsets, and we're done.

Here's the important example promised in Section 4.1.

Example 4.2.4. We will prove using induction that any tree with n vertices has $n-1$ edges. (Recall that a tree is a connected graph with no cycles. It looks kind of like a tree in the sense that it branches.)

(Base case) We exhibit every tree with $n = 1, 2$ vertices in Figure 4.2. Notice that a tree with one vertex has zero edges and that a tree with two vertices has one edge, in accordance with the desired result.

(Inductive hypothesis) For any $n \le k$, we assume that any tree with n vertices has $n-1$ edges.

(Inductive step) Consider some tree, any tree really, with $k+1$ vertices. We would like to show that it has k edges. Close your eyes, stick out your hand, and grab an edge of the tree; hang the edge over a nail, and you'll have something like Figure 4.3. Let's call the tree T and the chosen edge e so that we can remove it by considering $T \setminus e$. Notice that $T \setminus e = S_1 \cup S_2$, where S_1 and S_2 are the two subtrees shown in Figure 4.3. Also notice that S_1 and S_2 are trees themselves, and

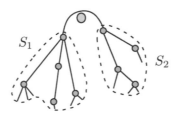

Figure 4.3. Choosing an edge of a tree separates it into two subtrees.

each has fewer than $k+1$ vertices, so that $|V(S_1)| \leq k$ and $|V(S_2)| \leq k$. Hey! That means we can apply the inductive hypothesis to S_1 and S_2! Doing so tells us that $|E(S_1)| = |V(S_1)| - 1$ and $|E(S_2)| = |V(S_2)| - 1$. Now consider $S_1 \cup S_2$. By the sum principle, $|V(S_1 \cup S_2)| = |V(S_1)| + |V(S_2)| = k+1$. Likewise, $|E(S_1 \cup S_2)| = |E(S_1)| + |E(S_2)| = (|V(S_1)| - 1) + (|V(S_2)| - 1) = (k+1) - 2 = k - 1$. The relationship between $S_1 \cup S_2$ and T is that T has one more edge... so put it back, showing that $|E(T)| = |E(S_1 \cup S_2)| + 1 = (k-1) + 1 = k$. Awesome. We're done.

Often, induction is presented as like climbing a ladder. A base case is like getting onto the ladder, near the bottom. The inductive hypothesis is like assuming you can get to the kth rung. Crucially, the inductive step tells you how to climb from the kth rung to the $(k+1)$st rung. After all, what do you need to know in order to climb a ladder? You need to know how to get on and how to get from one rung to the next. That'll take you as high as you need to go. The fact that you know how to get from rung k to rung $k+1$ means you can insert any values you like for k and you can get to (let's say) rung 10 by going rung 1 to rung 2, rung 2 to rung 3, rung 3 to rung 4, rung 4 to rung 5, rung 5 to rung 6, rung 6 to rung 7, rung 7 to rung 8, rung 8 to rung 9, and rung 9 to rung 10.

There's only one problem with this analogy. (Okay, there might be more than one problem. But there's one big problem.) It only works for simple things, like Example 4.2.3 above. What happens if there's more than one way to get from rung k to rung $k+1$? Or worse, suppose there are lots of different interpretations of the statement to be proved that all are described by the positive integer k? An example here would be all the internal computer networks that have five hubs and some number of channels connecting them. There are lots of ways to take a network with five hubs and create one with six hubs... way too many ways....

The correction to the ladder analogy is to think of the ladder as branching at every rung. And it doesn't just branch into two ladder paths, but a lot of ladder paths. And it happens at every rung, so that there are bunches of rungs at each level of the ladder. It's hard to imagine what that would look like, let alone how one would decide how to climb such a ladder. (A simplified version is shown in Figure 4.4.) Luckily, there's a way out of this problem.

When we do proofs by induction, we don't build up from the k case to the $k+1$ case. What we do in practice is start with the $k+1$ case and find a k case from which it came. So instead of climbing up from a k-level rung to a $(k+1)$-level rung, we take some $(k+1)$-level rung and look just below it to see the k-level rung (or a $(k-1)$-level rung, etc.). That's much easier. And then, in order to climb to a particular spot on our multi-branching ladder, we mark that spot, look downwards

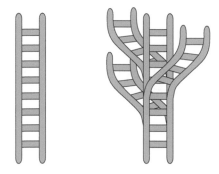

Figure 4.4. An ordinary ladder (left) and a simplified branching ladder (right).

to see what sequence of rungs leads to it, and use that sequence to climb to our particular spot. So an induction proof, in ladder terms, looks like this:

(Base case) Make sure that we can get onto the ladder.

(Induction hypothesis) Assume that if we happen to be on a k-level rung, we know how we got there.

(Induction step) From a $(k+1)$-level rung, figure out how to get back to a k-level (or lower) rung, so we'll know where we are.

This tells us why induction works for statements indexed by $\mathbb{N}$ only, not for those indexed by $\mathbb{Z}$ or other sets. Induction tells us how to climb down and up a ladder, but we can only climb down to a base case—not to arbitrary negative values of the index.

Finally, a note on writing: Every mathematician has slightly different preferences about how proofs by induction should be written. As a beginning proof writer, you should follow the format given here until you can consistently write correct proofs by induction. Then you can loosen up a bit, make your exposition more brief, and allow your own voice to come through.

4.2.1 Summation Notation

It can get pretty tiring to write out long sums, so we use compact notation for them. (We bring it up now because many proofs about sums use induction.) For example, we turn $1+1+1+1+1+1+1+1+1$ into $\sum_{j=1}^{9} 1$. This can be read as, "Add, from 1 to 9, the number 1." Of course, the result is 9. It is more common to involve variables, as with $\sum_{j=1}^{4}(j-2) = -1+0+1+2$ or with $\sum_{j=3}^{7}(-1)^{j+5} = 1-1+1-1+1$.

Summation notation always has three parts:

1. The summation sign $\sum$.

2. The function to be summed (above, this was 1 in the first example and $(j-2)$ in the second example); the function's variable is indicated in a subscript to the summation sign (in the examples above, the variable was j).

3. The index set, which can be a particular range of numbers (such as 3 to 8) or an indefinite range of numbers (such as 2 to n) or a set of elements (such as the vertices of a graph). If the index set is a set of elements, it is indicated in a subscript of the summation sign; if it is a range of numbers, the smallest number is indicated in a subscript and the largest number is indicated in a superscript of the summation sign.

The in-a-paragraph compact $\sum_{j=m}^{n}$ is usually written by hand as $\displaystyle\sum_{j=m}^{n}$.

Here are a few other instances of summation notation.

Example 4.2.5. Let f be a function whose domain includes $\mathbb{N}$. Then

$$\sum_{j=1}^{7} f(j) = f(1) + f(2) + f(3) + f(4) + f(5) + f(6) + f(7).$$

The sum of the first n natural numbers, except for 1, is

$$\sum_{j=2}^{n} j = 2 + 3 + 4 + \cdots + n.$$

The sum of the degrees of the vertices of K_3 (a triangle) is

$$2 + 2 + 2 = \sum_{v \in V(K_3)} deg(v).$$

The sum of the squares of the integers is

$$\sum_{k \in \mathbb{Z}} k^2 = \cdots + 36 + 25 + 16 + 9 + 4 + 1 + 0 + 1 + 4 + 9 + 16 + 25 + 36 + \cdots.$$

And, here is a proof by induction using summation notation.

Example 4.2.6. Let us prove that $\sum_{j=1}^{n} 2j = n + n^2$.

(Base case) When $n = 1$, we have $\sum_{j=1}^{1} 2j = 2 \cdot 1$ and $1 + 1^2 = 2$, so the statement is true. It's a little bit weird to compute sums without having any actual addition, so let's check for $n = 2$ as well: $\sum_{j=1}^{2} 2j = 2 + 4 = 6$, and $2 + 2^2 = 6$. That's better.

(Inductive hypothesis) We assume that when $n \leq k$, $\sum_{j=1}^{n} 2j = n + n^2$.

(Inductive step) Consider $\sum_{j=1}^{k+1} 2j$. (What are we trying to show? Plug $k + 1$ in for n in our inductive hypothesis to see what we seek.) We will separate out the last term by writing $\sum_{j=1}^{k+1} 2j = \sum_{j=1}^{k} 2j + 2(k+1)$. Now we can use the inductive hypothesis on $\sum_{j=1}^{k} 2j$ and substitute, getting $\sum_{j=1}^{k} 2j + 2(k+1) = k + k^2 + 2(k+1)$. Algebra shows that $k + k^2 + 2(k+1) = k^2 + 3k + 2 = k + 1 + k^2 + 2k + 1$, which factors to $(k+1) + (k+1)^2$, and putting it all together we see that $\sum_{j=1}^{k+1} 2j = (k+1) + (k+1)^2$—the result we desired.

4.2.2 Induction Types and Styles

You may notice that other books make a distinction between plain old induction and strong induction. It turns out that they are logically equivalent, so we will use whichever form of induction we please and just call it "induction."

Another thing: while in this book we do our proofs by induction full out, in more advanced texts and mathematics papers it is rare that the inductive hypothesis will be explicitly stated and sometimes the base case is not mentioned. It is expected that the reader can state the inductive hypothesis for hirself. Certainly the writer has checked appropriate base cases, but it is also expected that the reader will check base cases for hirself if ze has any doubt.

Check Yourself ——————————————————————

These simpler problems will prepare you for the challenges that lie ahead.

1. If the statement you want to prove is made in terms of n, should your inductive step be done using n or using k (or some other variable)?

2. Prove by induction that the path graph P_n has $n - 1$ edges.

3. Write $2! + 4! + 6! + 8! + 10!$ in summation notation. (Knowing what $2!, 4!, 6!$, etc. means is not necessary for completing this problem.)

4. Write $\sum_{j=0}^{6} \frac{3j-1}{2}$ out in full.

5. How is $\sum_{j=1}^{5} j^2 - j$ related to $\sum_{j=1}^{4} j^2 - j$? Try writing $\sum_{j=1}^{5} j^2 - j$ in terms of $\sum_{j=1}^{4} j^2 - j$. More generally, how is $\sum_{j=1}^{5} q(j)$ related to $\sum_{j=1}^{4} q(j)$? And even more generally, how is $\sum_{j=1}^{k+1} q(j)$ related to $\sum_{j=1}^{k} q(j)$?

Figure 4.5. Some mysterious dots.

4.3 Try This! Induction

It's time to practice induction for yourself (and with others). These problems vary substantially in difficulty—just so you're warned.

1. The *star* graph on n vertices has one vertex adjacent to all other vertices (and no other edges). Show, using induction, that the star graph on n vertices has $n - 1$ edges.

2. This problem is about adding the odd numbers; consider in particular $1 + 3 + 5 + 7 + \cdots + (2n - 1)$.

   (a) Write the above expression in summation notation.

   (b) What does Figure 4.5 have to do with this problem?

   (c) Find a formula for the sum you rewrote in part (a).

   (d) Now prove by induction that your formula is correct.

3. Prove that $\sum_{v \in V(G)} deg(v) = 2|E(G)|$, using induction. (In case you did not recognize this expression, it is the handshaking lemma!)

4. Show that if a letter requires postage of more than seven cents, one can apply exact postage using only three-cent and five-cent stamps. (Suggestion: use induction!)

5. Trees are the focus of this problem.

   (a) Draw a tree that has nine vertices, and label the vertices.

   (b) Redraw the tree so that it is clear that the tree is bipartite.

   (c) Mark the bipartitely drawn tree so that it is a tree with eight vertices connected to a tree with one vertex.

   (d) Mark (another copy of) the bipartitely drawn tree so that it is a tree with five vertices connected to a tree with four vertices.

   (e) Show by induction that every tree is bipartite.

4.4 More Examples

Here are three final examples of induction, two numerical and one geometric.

Example 4.4.1. Let us show that if $n > 1$, then $3^n > 3n$.

(Base case) When $n = 2$, we see that $3^2 = 9$, and $9 > 6 = 3 \cdot 2$.

(Inductive hypothesis) For any $2 < n \le k$, $3^n > 3n$.

(Inductive step) Consider the case $n = k + 1$. We may rewrite 3^{k+1} as $3 \cdot 3^k$. By the inductive hypothesis, $3^k > 3k$, so $3 \cdot 3^k > 3 \cdot 3k$. Now, $3 \cdot 3k = 9k = 3k + 6k = 3k + 3 + (6k - 3) > 3k + 3 = 3(k + 1)$. We know that $6k - 3 > 0$ because $k \ge 2$. Combining the above statements, we have $3^{k+1} > 3(k + 1)$ as desired.

Example 4.4.2. We will prove by induction that for $n \ge 2$, $2^n \le 2^{n+1} - 2^{n-1} - 2$.

(Base case) Because we are constrained to $n \ge 2$, we will examine the base case of $n = 2$. We know $2^2 = 4$ and $2^3 - 2^1 - 2 = 8 - 2 - 2 = 4$; because $4 \le 4$, all is well with the base case.

(Inductive hypothesis) For any $n \le k$, $2^n \le 2^{n+1} - 2^{n-1} - 2$.

(Inductive step) We would like to show that $2^{k+1} \le 2^{k+2} - 2^k - 2$. Consider the left-hand side of the statement. We know that $2^{k+1} = 2 \cdot 2^k$. The inductive hypothesis applies to that 2^k, so we have $2^{k+1} = 2 \cdot 2^k \le 2 \cdot (2^{k+1} - 2^{k-1} - 2)$. Multiplying out gives the expression $2^{k+2} - 2^k - 4$. Now... that's not far off from what we want to prove! Check this out—we know that $-4 \le -4 + 2 = -2$. Therefore, $2^{k+1} \le 2^{k+2} - 2^k - 4 < 2^{k+2} - 2^k - 2$ and we're done.

The trick used in Example 4.4.2, seen regularly in mathematics, is that if $a < b$ then $c + a < c + b$. For that matter, it is also true that if $a \le b$ and $d \le e$, then $a + d \le b + e$.

Example 4.4.3. Start with a circle and choose any $n \ge 3$ points on it. Now join consecutive points with line segments to form a polygon with n sides (see Figure 4.6).

Figure 4.6. Forming a polygon inside a circle using $n = 8$ points on the circle.

Figure 4.7. An indeterminate number $k + 1$ of points are chosen on a circle, as indicated by the dotted lines (left). Consider the polygon formed by k of them (right).

We claim that the sum of the interior angles of this polygon is $(n - 2) \cdot 180°$, no matter which n points are chosen on the circle. Induction says…

(Base case) Let $n = 3$. Then the polygon is a triangle, and every triangle has interior angle sum $180°$.

(Inductive hypothesis) A polygon formed by any $3 \leq n \leq k$ points on a circle has interior angle sum $(n - 2) \cdot 180°$.

(Inductive step) Pick any $k + 1$ points on a circle. We would like to show that the interior angle sum of the polygon formed by these points is $(k - 1) \cdot 180°$. To reduce to the case of k points so we can use the inductive hypothesis, pretend one of the $k + 1$ chosen points isn't there and look at the polygon formed by the k other points; see Figure 4.7. The inductive hypothesis applies to this polygon, so it has interior angle sum $(k - 2) \cdot 180°$. If we glue on the triangle formed by the pretend-it's-not-there point and the two points next to it on the circle, then we get our original polygon. The sum of the interior angles of the original polygon is the sum of the interior angles of the triangle ($180°$) plus the sum of the interior angles of the k-point polygon ($(k - 2) \cdot 180°$), so we have $180° + (k - 2) \cdot 180° = (k - 1) \cdot 180°$ as desired.

Check Yourself ⎯⎯⎯⎯⎯⎯⎯⎯⎯⎯⎯⎯⎯⎯⎯⎯⎯⎯⎯⎯⎯⎯⎯⎯⎯⎯⎯⎯⎯⎯⎯

Doing these exercises will assure that you understand the ideas behind these examples.

1. Use direct proof to show that $2^n \leq 2^{n+2} + 5$.

2. Show, by induction, that a polygon formed by n arbitrarily chosen points on a circle has exactly n edges.

Figure 4.8. A single duck is grey. Well, technically it is a Blue Swedish duck, but Blue Swedish ducks are slate grey.

4.5 The Best Inducktion Proof Ever

Hopefully you are starting to feel like you are getting the hang of induction. This proof is for the consideration of advanced induction studiers; it is subtle in its loveliness.

Theorem 4.5.1. *All ducks are grey.*

Proof: In order to prove the theorem by inducktion, we must restate it so that it has an index in the natural numbers. *All ducks are grey* is equivalent to saying *All ducks in a set of n ducks are the same color, and that color is grey.* We proceed by inducktion.

(Base case) Certainly it is true that a set with one duck has all ducks of the same color, and Figure 4.8 shows a grey duck.

(Inducktion hypothesis) Suppose that any set of n ducks is the same color (grey), as long as $n \leq k$.

(Inducktion step) Consider a set of $k+1$ ducks. We don't know what color they are, or even whether they are all the same color. Choose a duck arbitrarily and set it in the nearby water so it can swim about. This leaves us with k ducks. Aha! The inducktive hypothesis applies, so all of them are grey. Using a duck call, retrieve the swimming duck (of unknown color). Send one of the grey ducks to the water (in a different direction, so that there is no confusion between ducks.) Now we have $k-1$ grey ducks and one duck of unknown color, but together they are a set of k ducks and so the inducktive hypothesis holds—so all k of them are grey. Now recall the swimming grey duck, and see that all $k+1$ ducks are grey. Voilá! All ducks are grey. □

Because this is the best inducktion proof ever, it is worth considering carefully.

1. Go through the inducktive step of the proof for the case $n = 5$ ducks to see how the subsets of ducks interact.

2. Rewrite this proof for the statement *all owls are teal*, noting that whereas ducks swim about, owls fly and perch in trees.

3. Do you believe that all ducks are grey? Many students claim that they have seen white ducks, but Section 4.5 proves that all ducks are grey. (A "white" duck is very pale grey.) Remember, a correct proof compels assent—so either you believe a correct proof or you believe that the given proof is problematic. Try to find an error in the proof, or justify completely that all ducks are grey.

   Do *not* try looking this up (e.g., on the internet). That would spoil your fun! Instead, think through the details of the proof. Does the base case make sense? Is the inducktive hypothesis correctly stated? How does the inducktive step hold up under scrutiny?

4.6   Try This! More Problems about Induction

Just in case you have finished solving the problems in Section 4.3, here are a few 2^n-themed problems for you!

1. Compute 2, $2 + 2^2$, $2 + 2^2 + 2^3$, and $2 + 2^2 + 2^3 + 2^4$. Use your data to conjecture a simple formula for $\sum_{j=1}^{n} 2^j$. Now use induction to prove that your conjectured formula is correct.

2. Consider a $2^n \times 2^n$ grid with the upper-right-hand square missing; two are shown in Figure 4.9.

   (a) Can you always tile it with 3-square L-shaped tiles (no gaps and no overlaps, as shown in Figure 4.10)?

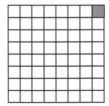

Figure 4.9. Example $2^2 \times 2^2$ and $2^3 \times 2^3$ grids, each with the upper-right corner missing.

Figure 4.10. How to tile a $2^1 \times 2^1$ grid with the upper-right corner missing with a three-square L-shaped tile.

   (b) If not, give a counterexample; if so, give a proof by induction.

  3. Prove that $2^n < n!$ for $n \geq 4$ by induction. (We write $n!$ as shorthand for $n \cdot (n-1) \cdot (n-2) \cdot \; \cdots \; \cdot 3 \cdot 2 \cdot 1$.)

4.7 Are They or Aren't They? Resolving Grey Ducks

Hey! You! Don't read this unless you have carefully read the proof given in Section 4.5. I mean it!

In the face of much clamor, we must regretfully admit that not all ducks are grey. So what is the flaw in the proof? Consider a pair of ducks.

🦆 If we try to consider the pair of ducks as a base case, we may have two ducks of the same color. Or we may have two ducks of different colors (see, for example, the duck heads in Figure 2.1 on page 29). **Lesson:** make sure to verify a nontrivial base case (even if you end up not using it in your final proof).

🦆 If we try to consider the pair of ducks as an instance of the inductive step, we set a duck afloat and are left with a single grey duck; then, we set the grey duck afloat and consider the remaining duck of an indeterminate color. And it's *still* of indeterminate color because there are no other ducks with which to use the inductive hypothesis! In set-theoretic terms, the problem is that when there are fewer than $n = 3$ ducks, the subsets of size $n - 1$ do not intersect. **Lesson:** make sure that the inductive step is not limited to certain values of n or k (unless these are restrictions placed on the theorem or resolved by checking sufficiently many base cases).

Check Yourself ────────────────────────────────────

  1. Prove that $3j^2 < 2j^3$. Be sure to use a base case of $j = 1$.

4.8   Where to Go from Here

Y'know, this would be a great point in the course to go back and reread Section 5 of the student preface, on tips for writing mathematics.

We have given a fairly comprehensive introduction to induction as a proof technique, but if you are interested, there *is* more. (There is *always* more…) You may want to learn about the seeming differences between strong and weak induction; many texts describe this, and Richard Hammack's *Book of Proof* is a good source. You may want to try dramatically more complicated induction proofs with more than one index—this uses double induction. Sadly, no good reference for double induction seems to be available. It appears that the most comprehensive source for proofs by induction is the recently published *Handbook of Mathematical Induction: Theory and Applications* by David S. Gunderson. At about 700 pages, it must go into more detail than it seems possible to desire.

If you enjoyed tiling the $2^n \times 2^n$ grid with L-shaped tiles, you will also enjoy http://www3.amherst.edu/~nstarr/puzzle.html, which has applets that allow you to tile 8×8 and $M \times N$ grids with L-shaped tiles.

Induction is used regularly by professional mathematicians and computer scientists in proofs, which is why so many undergraduate courses include induction as a topic. Of course, the statements used in research papers are much more sophisticated than those we use here, and the inductive steps are more subtle. We will see later (Chapter 8) that induction is intimately related to recursion, which is a common theme in computer science.

Credit where credit is due: Example 4.4.3 was inspired by [1]. Figure 4.14 was donated by Tom Hull. In Section 4.12, Problems 36 and 37 were donated by Karl Schaffer, and Problem 31 was inspired by work of Tom Leighton and Ronitt Rubinfeld.

4.9   Bonus: Small Crooks

The technique presented here is a very slick variant on proof by contradiction. Sometimes it is called *proof by smallest counterexample*, but it is also sometimes called a *minimal criminal* argument. (Yes, as you desire, Minimal Criminal is a band name: see http://www.minimal-criminal.com/. One of their songs is called "Graverobber from Outer Space." Rarely does one have this sort of satisfaction in mathematics.)

Here is the idea: You have some proposition that you hope is true, and it is indexed by $\mathbb{N}$ in some way (e.g., the number of vertices of a graph). You attempt a proof by contradiction by supposing the proposition is false. Therefore, there is a counterexample. And among all counterexamples, one must have the smallest index. Check that one out. Try to show that this counterexample implies the existence of an even-smaller-index counterexample. But that's a contradiction—you started with a counterexample that was smallest. Or, obtain a contradiction by showing that your smallest counterexample complies with the constraints of your proposition. So your proposition is true.

We start with a silly example.

Example 4.9.1. We claim that all natural numbers are interesting. Suppose not; then, there must be a smallest uninteresting natural number k. Ah, but k is interesting because it is the smallest uninteresting natural number! Thus, it is *not* the smallest uninteresting natural number. Contradiction.

A more serious example follows.

Example 4.9.2. Let us show that every natural number greater than 1 has a factorization into primes. Suppose not; then, there is some smallest natural number $k > 1$ that does not have a factorization into prime numbers. This k must not be prime, as otherwise it would be its own factorization into primes, and therefore $k = \ell \cdot m$ for some smaller natural numbers ℓ and m. Because $\ell, m < k$, they have factorizations into prime numbers. Thus $\ell \cdot m = k$ has a factorization into prime numbers, so k is not the smallest natural number that does not have a factorization into prime numbers—contradiction.

This proof uses a variant on the minimal criminal technique.

Example 4.9.3. We claim that every tree with at least one edge has at least two leaves. Given a tree T, look at the set of all paths in a tree and choose a path P of longest length. We claim that both ends of P are leaves. Suppose not; then, at least one end of P has degree 2 and so we can extend P to a longer path. This contradicts the longest-ness of P. (Okay, so maybe this should be called a *maximal* criminal argument ....)

Short activity:

1. Can you prove that a tree T with $|V(T)|$ vertices has $|V(T)| - 1$ edges using the minimal criminal technique?

2. Show that every connected graph has a walk that begins and ends at the same vertex and crosses every edge twice.

3. Consider a connected graph G where every vertex's degree is the average of the degrees of its neighbors. Prove that G must be regular.

4.10   Bonus 2: An Induction Song

By Induction

Max W. Chase, 2005

(sung to the tune of "Frère Jacques")

Take the base case,
Take the base case,
n is one,
n is one,
This is good to start with,
This is good to start with,
We're not done,
We're not done.

Now consider,
Now consider,
n less one,
n less one.
If we prove it for *n*,
If we prove it for *n*,
Then we're done,
Then we're done.

For all *n*,
For all *n*,
Now you see,
Now you see,
Our conjecture is true,
Our conjecture is true,
Q.E.D.,
Q.E.D.

4.11   Bonus Check-Yourself Problems

Seriously, do all of these problems by induction. That's what they're here for: induction practice. Solutions to these problems appear starting on page 600. Those solutions that model a formal write-up (such as one might hand in for homework) are to Problems 2, 7, and 8.

1. Prove that $\sum_{j=1}^{n} 3 + 5j = \frac{1}{2}(11n + 5n^2)$.

2. Prove that $n^4 < 3 \cdot 8^n$.

3. Show that every convex polygon can be decomposed into triangles.

4. Show by induction that $K_{m,n}$ has mn edges.

5. Prove that $\sum_{j=0}^{n}(j+1)(j-2)$
   $= \frac{1}{3}(n-3)(n+1)(n+2)$.

6. Prove $(2(n!))^2 < 2^{(n!)^2}$ for sufficiently large values of n.

7. Use induction to prove the sum principle for n finite sets.

8. Take a piece of paper and fold it—not necessarily in half, but definitely with a single straight crease somewhere in the paper. Fold the (still folded) paper again. In fact, fold it n times, wherever you like. Now unfold it completely. Prove by induction that you can always color the paper with two colors (teal and purple) so that no fold line has the same color on both sides.

9. For what values of n is $5^{n+2} < 6^n$? Prove it.

10. Prove that any natural number $n \geq 2$ can be written as the product of prime numbers.

4.12   Problems That Use Induction

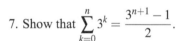

Recall from Section 4.6 that $n! = n \cdot (n-1) \cdot (n-2) \cdot \;\cdots\; \cdot 3 \cdot 2 \cdot 1$.

1. Prove that $\displaystyle\sum_{j=0}^{n-1} 3 = 3n$.

2. Prove that $25^n > 6^n$ using induction.

3. Show that $\displaystyle\sum_{j=-5}^{n-5} 4 = 4(n+1)$.

4. Compute $1,\ 1+2,\ 1+2+3$, and $1+2+3+4$. Draw these as dot diagrams (a row of one dot, with a row of two dots beneath, etc.). Use your data to conjecture a simple formula for $\sum_{i=1}^{n} i$. Now use induction to prove that your conjectured formula is correct.

5. Using induction, prove that $10n < n^2$ for $n \geq 11$.

6. Prove that any set with n elements has 2^n subsets, using induction. The proof in Example 4.2.3 for the subsets of $\{1, \ldots, k\}$ may inspire you.

7. Show that $\displaystyle\sum_{k=0}^{n} 3^k = \frac{3^{n+1} - 1}{2}$.

8. Prove the other of DeMorgan's laws for sets: Let $A_1, A_2, \ldots, A_n$ be n sets. Prove that for any $n \in \mathbb{N}$, $\overline{(A_1 \cap A_2 \cap \cdots \cap A_n)} = \overline{A_1} \cup \overline{A_2} \cup \cdots \cup \overline{A_n}$.

9. Show by induction that K_n has $\frac{n(n-1)}{2}$ edges.

10. Prove that
$$3\sum_{j=0}^{n-1} j(j-1) = n(n-1)(n-2).$$

11. Show that $((n+1)!)^n \leq 2! \cdot 4! \cdot \;\cdots\; \cdot (2n)!$.

12. Show using induction that for $n \in \mathbb{N}$,
$$\sum_{i=1}^{n} \frac{1}{i \cdot (i+1)} = \frac{n}{n+1}.$$

13. Prove that $n! < n^n$ as long as $n \geq 2$.

14. We know that $1 = 1$. It turns out that $2 + 3 + 4 = 1 + 8$ and that $5 + 6 + 7 + 8 + 9 = 8 + 27$ and that $10 + 11 + 12 + 13 + 14 + 15 + 16 = 27 + 64$. Does this generalize? Write out a general form for the pattern this follows, and then either give a counterexample or a proof.

15. Which two of the previous problems give different ways of counting the same quantity?

16. Use induction to prove the product principle for n finite sets.

17. Show that every tree is bipartite using induction. Why yes, that *is* one of the Try This! problems—it's worth writing up carefully.

18. Prove that $\sum_{j=1}^{n} j^3 = \left(\sum_{m=1}^{n} m \right)^2$.

19. Prove that $\sum_{j=1}^{n} j \cdot j! = (n+1)! - 1$.

20. Show that the sum of the interior angles of any n-gon (a polygon with n sides) is $\pi(n - 2)$. Notice that such polygons may be wildly irregular and even non-convex.

21. Write the equation $1 + 3 + 6 + 10 + \cdots + \frac{n(n+1)}{2} = \frac{n(n+1)(n+2)}{6}$ using summation notation. Prove that the equation is true for all positive integers n by using induction.

22. Suppose you have a 500-piece jigsaw puzzle showing the Ànec of Catalonia. To put it together, you must fit the pieces together. At any point, you either fit a new piece onto an existing chunk of puzzle or you fit two chunks together along a puzzle seam. Prove that no matter what order the pieces are placed in, there are exactly 499 piece/chunk fittings to solve the puzzle.

23. Show that a $2^n \times 2^n$ grid missing *any* square can be tiled with L-shaped tiles, as in the second problem of Section 4.6.

24. Draw three overlapping circles. Color the resulting regions using two colors, so that no two regions that share a curve get the same color. (This is known as 2-*coloring* the regions. Grey and white are popular colors for experimenters who use pencil on white paper.) Now draw two pairs of overlapping circles and a single circle overlapping none of the others; 2-color this configuration. Using the understanding gained from these experiments, prove that n circles drawn in the plane can be 2-colored, using induction.

25. **Challenge:** Analyze the proof you gave for the previous problem. Would it work for n overlapping squares? Triangles? What about for spheres in space?

26. Write the equation $(1 \cdot 3) + (2 \cdot 4) + (3 \cdot 5) + \cdots + n(n+2) = \frac{n(n+1)(2n+7)}{6}$ using summation notation. Now prove that the equation is true for all natural numbers $n \geq 1$ by using induction.

27. Let's dig into Problem 22 about the 500-piece jigsaw puzzle. Suppose the last step in solving the puzzle joined a chunk of 133 pieces with a chunk of 367 pieces. How many piece/chunk fittings did each of those two chunks require? What can you prove about an n-piece jigsaw puzzle? Do that proof.

28. Prove the second statement in Example 1.4.6, namely that $\sum_{i=1}^{n-1} i = \frac{n(n-1)}{2}$.

29. Prove that $n(n+1)(n+2)$ is a multiple of 3 for any natural n.

30. Prove that $n(n+1)(n+2)$ is a multiple of 6 for any natural n.

31. Think through the following proof: *We will show that any simple graph where every vertex has degree at least 1 is connected. As a base case, we have two vertices connected by a single edge. Now, suppose that for $n \leq k$, a simple graph with n vertices, each of which has degree at least 1, is connected. Consider a simple graph G with k vertices, each of which has degree at least 1. By the inductive hypothesis, it is connected. Add a vertex v to G so that we have G-with-v, which has $k+1$ vertices; in order that every vertex has degree at least 1, we also have to add an edge to v. But an edge in a simple graph must connect two vertices, so the other end of the edge must be incident to a vertex of G. Thus, G-with-v is connected.* What's wrong with this proof? It can't be right—consider the graph in Figure 4.11.

Figure 4.11. A graph that has all vertices of degree 1 but is not connected.

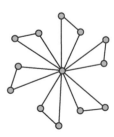

Figure 4.12. A 6-pin pinwheel graph.

32. Conjecture and prove by induction a formula for the number of edges of an n-pin pinwheel graph. (A 6-pin pinwheel graph is pictured in Figure 4.12.)

33. Conjecture and prove by induction a formula for the number of edges of an n-bubble bubblepath graph. (A 4-bubble bubblepath graph is pictured in Figure 4.13.) Note that both ends of the bubblepath are always bubbles.

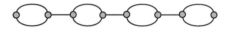

Figure 4.13. A 4-bubble bubblepath graph.

34. Here is a sketch of a flawed proof: *We will prove by induction that every graph with n vertices is bipartite. Our base case is two vertices connected by a single edge. For the inductive step, consider a generic $(k+1)$-vertex graph and remove a vertex. The inductive hypothesis applies to this k-vertex graph, so the result is bipartite. The deleted vertex is in the other part from its neighbors, so when we restore the vertex we see the original $(k+1)$-vertex graph is bipartite.* Where is the flaw?

35. Here is a sketch of a flawed proof: *We will prove by induction that every graph with n vertices and at least one leaf is bipartite. Our base case is two vertices connected by a single edge. For the inductive step, consider a generic $(k+1)$-vertex graph with a leaf and remove that leaf. The inductive hypothesis applies to this k-vertex graph, so the result is bipartite. The deleted vertex is in the other part from its only neighbor, so when we restore the vertex we see the*

original $(k+1)$-vertex graph that had at least one leaf is bipartite. Where is the flaw?

36. Let's play a pile game! Start with a pile of n small teal stones. Divide this into two (nonempty) piles of sizes j and $n - j$. Write down the product of the two pile sizes $j(n - j)$. Now repeat the process: at each step, divide a pile with more than one small teal stone into two smaller piles, and write down the product of the two pile sizes. When all that's left is a whole bunch of one-stone piles, add all the products you wrote down. Prove that no matter how the piles of stones were divided, the sum of the products is $\sum_{i=1}^{n-1} i$.

37. **Challenge:** Here's another pile game—again, start with a pile of n small teal stones. Again, divide this into two (nonempty) piles of sizes j and $n - j$. This time, write down the sum $\frac{1}{j} + \frac{1}{n-j}$. Now repeat the process: at each step, divide a pile with more than one small teal stone into two smaller piles, and write down the sum of the reciprocals of the two pile sizes. When all that's left is a whole bunch of one-stone piles, take the product of all the sums you wrote down. Let the result be denoted by $ps(n)$. Is $ps(n)$ well defined? If so, find a formula for $ps(n)$ and prove that it is correct. If not, find two sequences of pile divisions that give different results.

38. Draw n straight lines in the plane. Prove that the resulting regions can be colored using two colors, as in Figure 4.14, so that no two regions that share a line segment get the same color.

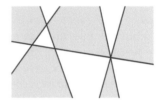

Figure 4.14. A 2-coloring of plane regions in teal and white.

39. Prove that $8^n - 1$ is a multiple of 7 for all $n \geq 1$.

40. Prove by induction that for $m > 1$, $K_{2,m} \setminus e$ has $2m - 1$ edges.

4.13   Instructor Notes

I don't know about your students, but mine always need a couple of class periods to deal with induction—even when they've seen it before in other classes. Assign Sections 4.1 and 4.2 as reading before the first class. The first two examples were chosen because they use scenarios with which students are familiar (counting, and one of DeMorgan's laws) and no new concepts (except, of course, for induction). Then begin with a general review of the inductive process, take questions over it, and give a simple sample proof by induction (doing one of the examples from the text is fine). If there is enough time, get students started on the problems in Section 4.3.

   If your students seem to be grasping induction well, leave about ten minutes of class to introduce your favorite version of the Section 4.5 proof that all ducks are grey. (Otherwise, leave this until the second class.) After this introduction, solicit student reaction. Here are some of the objections students will raise: "But I have a brown duck!" Response: You can't have a brown duck, because all ducks are grey. We just proved it. Objection: "There's some problem with the inductive step." Response: Oh, really? What is it? Students may claim that "it doesn't work" or some such, but what they identify as flaws are usually aspects of most induction proofs. That's one reason this is such a great example—it draws out the doubts students have about the structure of inductive proof. Often students will ask you to repeat the inductive step for clarity. (If you are running out of time, you can let them know that a version of this proof is in the book.)

Sometimes a student will notice the flaw in the proof quickly, that when $n = 2$ everything falls apart and so an additional base case should have been checked. Students can also correctly claim that the inductive step doesn't work if we have $k + 1 = 2$, because then there's no intersection between the two sets of k ducks. This is true and is equivalent to saying that the base case wasn't good enough.

Then assign Section 4.4 as reinforcing and enhancing reading for the second class. (If you presented the *all ducks are grey* proof at the end of the previous class, assign Section 4.5 as well.) Start the second class by asking for questions over the reading; then, either present your favorite version of the *all ducks are grey* proof or ask for student thoughts on it, and try to get the students to dig deeply into the workings of the "proof." Then have them work on problems from Section 4.3 in groups; be sure to leave ten or so minutes at the end of class to have students share their work publicly so everyone is on the same page.

If at the start of the third class no student has found the flaw in the *all ducks are grey* proof, walk the class through the proof again and point out the flaw. There will likely be a question or two. Use the remaining time to have students continue work on problems from Section 4.3 and, if there is time, work on problems from Section 4.6. Again, leave some time at the end of class to discuss the problems students have done and point out what they have learned from these problems. Assign Section 4.7 as reinforcing reading.

If you would like to foreshadow modular arithmetic while also placing students in groups for Try This! work, here's one way to do it: Number your students aloud and ask them to remember their numbers. With n students, choose a number k so that $\frac{k}{n}$ is close to the group size you like best. Then define $a \equiv b \pmod k$ and ask students to work with their classmates who have equivalent-to-them-mod k numbers. (This was suggested by David Cox.)

Proof clinic? Some instructors have found that their students benefited from devoting a review session or class period to basic proof writing after completing Chapters 1–4. Problems that might be useful for such an activity are given in Section TI.2. One way to conduct a classwide proof clinic is to post statements to be proven around the classroom and have students work in groups to write and share proofs publicly. (Excellent idea, Dana Rowland!) While students will still be processing induction, at this point they should be getting a better handle on other basic proof techniques so that they can be successful in practicing them in the context of discrete mathematics problems over the remainder of the term.

Chapter 5 🦆🦆🦆🦆🦆

Algorithms with Ciphers

5.1   Introduction and Summary

An algorithm is simply a list of instructions for completing a task. Algorithms form the core of computer science and are used regularly in all kinds of discrete mathematics. They even function as a proof technique—an algorithm can give a constructive proof of existence! The primary examples we will use in exploring algorithms are simple ciphers. Ciphers are ways to encode messages so that they are not easily read by people other than the intended recipient(s). In particular, we will investigate the shift cipher and the Vigenère cipher. Shift ciphers can be broken by hand, and the Vigenère with not much more work, but they are essentially the only ciphers that are understandable with the level of mathematics presented in this chapter. (Another interesting cipher is described in Section 16.10, but you'll need to digest most of Chapter 16 to understand it.) All of these ciphers use modular arithmetic, so our goals in this chapter are to learn some modular arithmetic, figure out how it is used in the shift and Vigenère ciphers, and understand what algorithms are used to encipher and decipher messages using these ciphers.

5.2   Algorithms

Before discussing algorithms, we'd better define them.

Definition 5.2.1. An *algorithm* is a finite list of unambiguous instructions to be performed on one or several inputs; some instructions may refer to others. If an algorithm produces an output and ends after executing a finite number of instructions, then we say that the algorithm *terminates*.

One of the best-known mathematical algorithms in the Unites States multiplies two multidigit integers.

Example 5.2.2. How to multiply two integers:

1. Label the two integers A and B with digits $b_1 b_2 \ldots b_n$.

2. Multiply A by b_n and name the result C_n.

3. Multiply A by b_{n-1} and write the result C_{n-1} under C_n but shifted so the last digit of C_{n-1} is directly under the second-to-last digit of C_n.

4. Repeat this process until A has been multiplied by b_1 and the result C_1 has been written, shifted to the left, under C_2.

5. Add the shifted results together to obtain D.

In Example 5.2.2, the inputs are A and B and the output is D. Because there are finitely many digits in each of A, B and finitely many instructions in the algorithm, the algorithm does terminate.

Our definition of algorithm differs from those in many texts; often, authors require an algorithm to terminate in order to truly be an algorithm. However, given that one of the most common questions asked about a proposed algorithm is, "Does that algorithm terminate?" we are convinced that in practice people consider non-terminating sets of instructions to be algorithms (just poor ones). Besides, if termination was required by definition, the answer to this common question would always be "yes" and no one would need to ask.

The hallmarks of an algorithm are clarity and precision. (Hey, those are hallmarks of proofs, too! Hmm….) There are lots of lists of instructions one could make that would not count as algorithms because they do not truly convey what is to be done.

Example 5.2.3. Consider this laundry nonalgorithm:

1. Put clothes in washer.

2. Turn washer on.

The inputs to the procedure should be clothes, and the desired output is clean clothes. The biggest problem here is that most washers have lots of settings, and some machines will not start the wash cycle after simply being turned on. But additionally, how many clothes are put in the washer—what if you have more clothes to wash than will fit? Then you can attempt to put them all in, but the washer may not turn on. (Notice also that the procedure does not address detergent.)

Here is another problematic example.

Example 5.2.4. How to (not necessarily) find the roots of a polynomial:

1. Set polynomial equal to zero.

2. Factor polynomial.

3. Read off the roots.

This list of instructions is ambiguous. What if the polynomial does not factor into linear terms? Then you cannot read the roots from the linear factors, so what do you do?

CVS corn removers (sublabel: for the removal of corns (yes, really)) have instructions for use. In summary, they say to put a salicylic-acid-impregnated sticky disc on the corn and cover it with a bandage. The penultimate instruction is, "After 48 hours, remove disk." However, what if you take a shower after 24 hours? Then the bandage will become wet and likely fall off. Should you replace the bandage? The instructions are not clear.

Then, there is a significant difference between clarity and precision for human interpretation and for machine interpretation. Consider this real-life algorithm, printed on bottles of Suave Naturals Shampoo.

To Use:

1. Massage through wet hair and scalp.
2. Rinse well.

A human knows what "Rinse well" means, but a computer would need a specific criterion that terminates rinsing (as in, "Rinse until the sulfate sensor reads below 0.001").

An algorithm is called *correct* if it does what it should do. The algorithm given in Example 5.2.2 is correct for multiplication but not correct for addition or division. There is a difference between an algorithm and its implementation (how an algorithm is made into executable code)—an algorithm may be correct while a poor implementation may not be. In practice, it is important to make sure both algorithms and their implementations are correct for all possible inputs.

Example 5.2.5 (of how to eat and not eat potato chips). Consider the following algorithm, with input a bag of potato chips.

1. Examine bag contents; if there are no chips, crumple bag and say "Curses!" Otherwise, proceed to step 2.

2. Pick up a chip.

3. Put it back in the bag.

4. If you are hungry, return to step 1. Otherwise, seal the bag.

This algorithm does not terminate when the executor starts hungry because no eating occurs nor does the bag's chip number change. We will try again.

1. Examine bag contents; if there are no chips, crumple bag and say "Curses!" Otherwise, proceed to step 2.

2. Pick up a chip.

3. Throw it away.

4. If you are hungry, return to step 1. Otherwise, seal the bag.

This algorithm terminates but does not do what we wish it to do, as (again) no eating occurs. We can do better.

1. Examine bag contents; if there are no chips, crumple bag and say "Curses!" Otherwise, proceed to step 2.

2. Pick up a chip.

3. Eat it.

4. If you are hungry, return to step 1. Otherwise, seal the bag.

Ah, yes. This is the expected algorithm for eating potato chips; it terminates *and* it is correct. Healthy eaters tend to use the following variant algorithm.

1. Examine bag contents; if there are no kale chips, crumple bag and say "Cruciferous!" Otherwise, proceed to step 2.

2. Pick up a kale chip.

3. Eat it.

4. If you are hungry, return to step 1. Otherwise, seal the bag.

5.2.1 Conditionals and Loops

A *conditional* is a statement within an algorithm that places conditions on an instruction. Example 5.2.5 uses a conditional in each algorithm. Here are three common styles of conditional:

❦ *if-then-else*, which usually takes the form "if (conditions), then (action set 1), else (action set 2)" and is read/understood as "If (conditions) are met, then do (action set 1); otherwise, if (conditions) are *not* met, then do (action set 2)";

❦ *until*, which takes the form "do (action set) until (conditions)" or "until (conditions), (action set)" and is read/understood as "Do (action set) until (conditions) are met and then go to the next instruction";

❦ *while*, which takes the form "do (action set) while (conditions)" or "while (conditions), (action set)" and is read/understood as "Do (action set) while (conditions) hold, and when (conditions) are no longer met, go to the next instruction."

The syntax used for each conditional varies from computer language to computer language, so we will simply use them English-wise here. Notice that algorithm conditionals extend the idea of implication (the logical conditional, see page 76) by giving the additional information of what to do when the *if* conditions are not met.

Example 5.2.6. Most medicines are labeled with conditionals as part of their algorithms for usage. They are not stated as conditionals, but that's what they are. Here's a sample dosage table:

for adults ages 12 and over	take 2 tablets
for children ages 6 to 12	take 1 tablet

This can be written as, "If you are age 12 or over, take 2 tablets; otherwise, if you are between ages 6 and 12, take 1 tablet; otherwise, seek the advice of a physician." In fact, this is a nested conditional, with one if-then-else within another.

Example 5.2.7. This algorithm sorts a single marble using conditionals.

1. Pick up a marble.
2. If it is red, place it in the left-hand pile. If it is green, place it in the right-hand pile. If it is neither red nor green, discard it.

Notice that the second instruction is one big ol' nested conditional, of the form if (red), then (left), else (if (green), then (right), else (if (neither) then (discard))).

That's a pretty useless algorithm, though. Who wants to sort a single marble? It would be much better to have an algorithm that sorts a pile of marbles. For this we need a *loop*, which gives an instruction to perform some set of actions more than once.

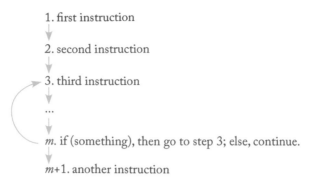

1. first instruction

2. second instruction

3. third instruction

...

m. if (something), then go to step 3; else, continue.

m+1. another instruction

Figure 5.1. A fake algorithm containing a loop.

Example 5.2.8. This algorithm sorts a bag of marbles using conditionals in a loop.

1. Pick up a marble.
2. If it is red, place it in the left-hand pile. If it is green, place it in the right-hand pile. If it is neither red nor green, discard it.
3. If marbles remain in the bag, go to step 1.

The third instruction has an implied "else" of being done with the task.

As you can see in Example 5.2.8, a loop earns its name because the instructions repeat, forming a string of instructions into a loop of instructions (see Figure 5.1). The process of repeating instructions is known as *iteration*. There is a danger inherent in loops: sometimes they go 'round and 'round forever, and this prevents an algorithm from terminating. So, it is important to consider how a loop can stop looping. Generally, this is accomplished by including a conditional in the loop that provides for an exit or by specifying the number of times the loop should be executed. Frequently, we wish to carry information from one iteration into the next iteration, such as the number of times a loop has been executed so far or a partial calculation. The usual way to do this is to set an iterating variable (e.g., k) with an initial value and then change the value of the variable during an iteration, e.g., by saying, "replace k with $k+1$." This means we should think of the variable k as a container that holds a value; the container is labeled "k," and when we change the value of the variable we stick the new value ($k+1$) in the container but leave the label on the container as it was ("k").

Example 5.2.9. We give three different algorithms that show how iterating variables are used.

1. Let $k = 1$ and let *value* $= 0$.

2. Replace k with $k + 1$.

3. If $k = 10$, output *value*; otherwise, go to step 2.

In this algorithm, the computer counts to ten in its head and just responds "0."

1. Let $k = 1$ and let *value* $= 0$.

2. Replace k with $k + 1$.

3. Output *value* and go to step 2.

This is a terrible algorithm. It does not terminate, so the computer counts to itself indefinitely and responds "0 0 0 0 0 0 ...."

1. Let $k = 1$.

2. Output k.

3. Replace k with $k + 1$.

4. If $k = 10$, output k, and stop; otherwise, go to step 2.

At least this algorithm is slightly more interesting; it counts to ten out loud.

Example 5.2.10 (of an algorithm including conditionals and iteration). Let us suppose that we have a large supply of Jelly Babies but a limited supply of aliens (n aliens, to be precise).

1. Let *alien* $= 1$.

2. Face the *alien*th alien.

3. Pick up a Jelly Baby. Say, "Would you like a Jelly Baby?" to the alien in front of you.

4. If the alien responds positively, then hand it the Jelly Baby; otherwise, if the alien responds negatively, then shrug and eat the Jelly Baby yourself; otherwise, if the alien is impassive, then shake your head and continue.

5. If *alien* $= n$, return to your companions. Otherwise, continue.

6. Replace *alien* with *alien* $+ 1$.

7. Go to step 2.

This algorithm includes a nested conditional; step 4 has the form *if A, then B, else (if C, then D, else (if E, then F))*. Note that the innermost if-then does not need an "else" because the only possibility left at that point is impassivity. (Well, okay, there might be some trouble in deciding whether an alien's response is positive, negative, or impassive, but those really are the only options.)

This algorithm also includes iteration. Steps 1, 5, and 6 collectively form a loop, by giving an initial value (step 1), iterating (step 6), and giving a condition on which we exit the loop (step 5).

The following examples are of practical mathematical interest, Example 5.2.11 because it is used by many computers and Example 5.2.12 because it has spawned an area of mathematics research.

Example 5.2.11 (of Russian-style multiplication). Believe it or not, this algorithm multiplies natural numbers! The input numbers are A and B. Recall from Chapter 3 that the floor function $\lfloor x \rfloor$ returns the greatest integer less than or equal to x.

1. Let $Astep = A$ and let $Bstep = B$.

2. Start a column by writing 0 at the top.

3. If $Astep$ is odd, then write $Bstep$ in the column; otherwise, continue.

4. Replace $Astep$ with $\lfloor \frac{Astep}{2} \rfloor$.

5. Replace $Bstep$ with $Bstep + Bstep$.

6. If $Astep = 1$, then write $Bstep$ in the column and go to step 7; otherwise, go to step 3.

7. Sum the column.

We explore how this algorithm produces the desired result (ordinary multiplication) in Problem 24 of Section 5.11.

Example 5.2.12 (the $3n + 1$ algorithm). Check *this* algorithm out. The input is some $n \in \mathbb{N}$.

1. If n is odd, replace n with $3n + 1$; otherwise, replace n with $n/2$.

2. If $n = 1$, output n; otherwise, go to step 1.

The Collatz conjecture states that the $3n + 1$ algorithm always terminates. However, as you might guess by the use of the word "conjecture," the truth of this statement is yet unknown. (It has been verified for all $n \leq 10^{18}$.) We explore the behavior of the $3n + 1$ algorithm in the Check Yourself problems at the end of this section.

5.2.2 Efficiency

An important issue, but one that we will not discuss in any depth here, is efficiency of algorithms. (For more detail, see Chapter 17.) In a practical sense, the amount of time an algorithm takes to run and the resources required to run it depend on the particular implementation of that algorithm. Thus, in mathematics and theoretical computer science, the efficiency of an algorithm is determined from its abstract description rather than from any implementation. For simple algorithms, it can be intuited whether an algorithm is efficient or inefficient.

Example 5.2.13. Both of these algorithms have as input $n \in \mathbb{N}$, and each performs the same task. Which algorithm is more efficient? The first algorithm ...

1. Let $k = 1$.

2. Let *sum* $= 0$.

3. Replace the value of *sum* with *sum* $+ k$.

4. If $k = n$, output *sum*; otherwise, replace the value of k with $k + 1$.

5. Go to step 3.

The second algorithm ...

1. Let $k = 1$.

2. Let *sum* $= 0$.

3. Replace the value of *sum* with *sum* $+ k$.

4. If $k < 2n$, replace the value of k with $k + 1$ and go to step 3; otherwise, continue to step 5.

5. Let *secondsum* $= 0$.

6. Let $m = n + 1$.

7. Replace the value of *secondsum* with *secondsum* $+ m$.

8. If $m < 2n$, replace the value of m with $m + 1$ and go to step 7; otherwise, continue to step 9.

9. Output *sum* $-$ *secondsum*.

Each algorithm computes $\sum_{j=1}^{n} j$; the first does so directly, whereas the second sums the first $2n$ natural numbers and subtracts $\sum_{j=n+1}^{2n} j$. The second algorithm does additional and unnecessary work, so the first algorithm is more efficient.

5.2.3 Algorithms and Existence Proofs

Back in Section 1.5 we described existence proofs—proofs that show something exists—in the context of the generalized pigeonhole principle. These were non-constructive existence proofs because they showed that items existed while not producing any examples of the desired items. Algorithms have outputs, and if the output of an algorithm is an object, then that algorithm functions as a constructive proof that the desired object exists; it gives us instructions for how to find or construct the object.

Example 5.2.14. We will now prove that there are n^2 functions from $\mathbb{Z}_2 \to \{1, \ldots, n\}$. The following algorithm provides a list of these functions.

1. Let $k = \ell = m = 1$.

2. Write $(m)\, f(0) = k, f(1) = \ell$ on a list.

3. If $\ell \geq n$, then continue; otherwise, replace ℓ with $\ell + 1$, replace m with $m + 1$, and go to step 2.

4. If $k \geq n$, then stop; otherwise, let $\ell = 1$, replace k with $k + 1$, replace m with $m + 1$, and go to step 2.

Notice that the algorithm does not create any maps that are not well defined, so each item in the generated list is a function. It also exhausts all possibilities for the image of 0 and for the image of 1, so every function from $\mathbb{Z}_2 \to \{1, \ldots, n\}$ is on the list. Moreover, because the algorithm has a nested loop in which each variable (k, ℓ) ranges from 1 to n, the list contains n^2 items. This completes the proof.

We will use algorithms to create constructive existence proofs in Chapters 10 and 12.

Check Yourself

There are only three of these, so please do them all.

1. Try performing the $3n + 1$ algorithm given in Example 5.2.12 for $n = 3, n = 4, n = 7$, $n = 8$, and $n = 13$. How many iterations are required for each of these numbers? Do any of the sequences generated appear within any of the others (and if so, which)?

2. Translate the instruction *replace t with $t/2$ while t is even* into plain English.

3. What does this list of instructions do? Comment on whether it forms an algorithm, and if so, whether it terminates and/or is correct.

 1. Let $n = 2$.
 2. Replace n with $n + 4$.
 3. If n is even, go to step 2; otherwise, continue.
 4. Output n.

5.3 Modular Arithmetic (and Equivalence Relations)

Modular arithmetic is essential for discrete mathematics, as you will soon see.

Definition 5.3.1. Let $a, b \in \mathbb{Z}$ and $n \in \mathbb{N}$. The expression $a \equiv b \pmod{n}$ means that when a is divided by n, it leaves the same remainder as when b is divided by n. This condition is equivalent to $a - b$ having remainder zero when divided by n. If $(a - b)/n \in \mathbb{Z}$, then we say that n *divides* $(a - b)$ and write this as $n|(a - b)$. Similarly, $n|(a - b)$ also means $(a - b) = kn$ for some $k \in \mathbb{Z}$. (This last version is the most useful in writing proofs.) We verbalize $a \equiv b \pmod{n}$ by saying that a and b are *congruent modulo n* or by saying that *a is congruent to b modulo n*. The set of different remainders obtainable by dividing integers by n is called the set of *integers modulo n*.

Notice the distinction between the similar symbols $/$ and $|$; the former means "divided by" and indicates an action, whereas the latter means "divides" and gives a description of an expression. That is, $/$ is used when the result of division is wanted, and $|$ is used to give information about the expression following it.

Example 5.3.2 (of some modular arithmetic calculations). One of our favorite sets is $\mathbb{Z}_2 = \{0, 1\}$. This is the integers modulo 2 (and so we unconsciously use modular arithmetic all the time in discrete mathematics!). Every even number is congruent to 0 (mod 2), and every odd number is congruent to 1 (mod 2). For example, $91,305,743,890 \equiv 0 \pmod{2}$, $4,589 \equiv 1 \pmod{2}$, and $547,392 \equiv 0 \pmod{2}$. In conversation, we might say, "What's 4,378 (mod 2)?" (pronounced "What's 4,378 mod 2?"), with the answer, "Oh, it's zero."

In our daily lives, we usually compute time modulo 12. For example, if at 10 a.m. a friend says, "See you at 2 p.m.," you know that's four hours away because $2 - 10 = -8 \equiv 4 \pmod{12}$.

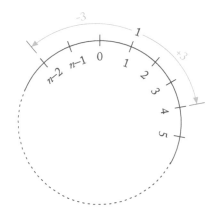

Figure 5.2. Modular arithmetic envisioned as arithmetic on a circle.

You can think of modular arithmetic as regular arithmetic done on a circle with n marks instead of as done on an infinite number line. Counting and adding are done clockwise, and subtracting is done by counting counter-clockwise. This is directly analogous to our usage of analog clocks, as shown in Figure 5.2.

We are able to operate on both sides of the $\equiv$ sign in much (but not quite) the same way we operate on both sides of an equals sign. For example, we know that if $a = b$ then $b = a$, and if additionally $b = c$ then $a = c$. Similarly, if $a \equiv b$ (mod n), then $b \equiv a$ (mod n); and, if $a \equiv b$ (mod n) and $b \equiv c$ (mod n), then $a \equiv c$ (mod n). (These are called the *symmetric* and *transitive* properties, respectively.) Witness the following result.

Theorem 5.3.3. *Let $a, b, c \in \mathbb{Z}$ and $n \in \mathbb{N}$. If $a \equiv b$ (mod n), then $ac \equiv bc$ (mod n).*

Proof: Because $a \equiv b$ (mod n), we know that $n | (a - b)$. This means that $(a - b) = nk$ for some $k \in \mathbb{Z}$. We multiply through by c to get $(a - b)c = (ac - bc) = nkc$. Renaming kc as q, we see that $(ac - bc) = nq$ for some $q \in \mathbb{Z}$. Therefore, $n | (ac - bc)$ and $ac \equiv bc$ (mod n). □

Example 5.3.4 (of modular arithmetic properties exhibited on actual integers).
$20 \equiv 12$ (mod 8) and also $12 \equiv 20$ (mod 8). Each of 20 and 12 has a remainder of 4 when divided by 8. $20 \equiv 12$ (mod 8) and $12 \equiv 4$ (mod 8), so $20 \equiv 4$ (mod 8). $12 \equiv 4$ (mod 8), so $3 \cdot 12 \equiv 3 \cdot 4$ (mod 8) or $36 \equiv 12$ (mod 8); both 36 and 12 reduce to 4 (mod 8) as $36 = 4 \cdot 8 + 4$ and $12 = 8 + 4$. $5 \equiv 11$ (mod 3), so $4 \cdot 5 \equiv 4 \cdot 11$ (mod 3), or $20 \equiv 44$ (mod 3) as $20 = 6 \cdot 3 + 2$ and $44 = 14 \cdot 3 + 2$. $22 \equiv 4$ (mod 6) because 22 has a remainder of 4 when divided by 6.

Likewise, $40 \equiv 4 \pmod 6$, and $40 \equiv 22 \pmod 6$, ... and $40 \equiv 745{,}408 \pmod 6$ as well.

Theorem 5.3.3 lets us multiply at will. In contrast, the converse is not true—division doesn't always work.

Example 5.3.5. $18 \equiv 24 \pmod 6$, so $9 \cdot 2 \equiv 12 \cdot 2 \pmod 6$. However, $9 \not\equiv 12 \pmod 6$ because $9 \equiv 3 \pmod 6$ but $12 \equiv 0 \pmod 6$.
$36 \equiv 24 \pmod{12}$, but while $4 \cdot 9 \equiv 4 \cdot 6 \pmod{12}$, notice that $9 \not\equiv 6 \pmod{12}$. (For the reader interested in more modular arithmetic, note that $9 \equiv 6 \pmod 3$ ... what might be going on there?)

On the other hand, there are certain conditions under which division is possible. We will explore this in Problem 19 of Section 5.11.

Congruence modulo n is defined on $\mathbb{Z}$; it is a specific example of a general idea.

Definition 5.3.6. An *equivalence relation* defined on a set S must satisfy

- 🦆 the *symmetric* property (if $s_1 \sim s_2$, then $s_2 \sim s_1$),

- 🦆 the *reflexive* property ($s \sim s$), and

- 🦆 the *transitive* property (if $s_1 \sim s_2$ and $s_2 \sim s_3$, then $s_1 \sim s_3$).

Here, the symbol $\sim$ acts as the verb "is equivalent to," just as $\equiv$ represented "is congruent to." We noted just before Theorem 5.3.3 that congruence modulo n is symmetric and transitive; of course $a \equiv a \pmod n$, so congruence modulo n is also reflexive. Therefore, it is an equivalence relation.

Let us develop this idea further using the most discrete of all sets, $\mathbb{Z}_2$, also known as the integers modulo 2. Because all odd integers are congruent to 1 (mod 2), we can write the odd integers as the set $\{1 + 2k \mid k \in \mathbb{Z}\}$. We will refer to this set as $[1]$. Similarly, the even integers can be written as $\{0 + 2k \mid k \in \mathbb{Z}\} = [0]$. The notation $[a]$ means "all the elements equivalent to a using some equivalence relation," so in $\mathbb{Z}_2$ we could also refer to $[1]$ as $[3]$ (though this would be less evocative). In set notation, $[a] = \{s \in S \mid s \sim a\}$, and hopefully when you see $[a]$ you will know from context which equivalence relation is meant. The sets $[a]$ and $[b]$ are called *equivalence classes*.

For example, we could consider the equivalence relation of people who have the same favorite Doctor. (Who?) Certainly someone has one favorite doctor, and so is equivalent to hirself (reflexive); if I have the same favorite Doctor as you, then you have the same favorite Doctor as me (symmetric); if I have the same favorite Doctor as you, and you have the same favorite Doctor as Madame Vastra, then I have the same favorite Doctor as Madame Vastra (transitive). There are currently 13 Doctors, so 14 equivalence classes (we add one for people who dislike all of the Doctors). Here is another example.

Example 5.3.7. The integers modulo 4, or $\mathbb{Z}_4$, can be represented as follows:

$$\{\ldots,-8,-4,0,4,8,\ldots\} = \{0+4k \mid k \in \mathbb{Z}\} = \{z \in \mathbb{Z} \mid z \equiv 0 \pmod 4\}$$
$$= [0] = [12].$$
$$\{\ldots,-7,-3,1,5,9,\ldots\} = \{1+4k \mid k \in \mathbb{Z}\} = \{z \in \mathbb{Z} \mid z \equiv 1 \pmod 4\}$$
$$= [1] = [5].$$
$$\{\ldots,-6,-2,2,6,10,\ldots\} = \{2+4k \mid k \in \mathbb{Z}\} = \{z \in \mathbb{Z} \mid z \equiv 2 \pmod 4\}$$
$$= [2] = [18].$$
$$\{\ldots,-5,-1,3,7,11,\ldots\} = \{3+4k \mid k \in \mathbb{Z}\} = \{z \in \mathbb{Z} \mid z \equiv 3 \pmod 4\}$$
$$= [3] = [-9].$$

An equivalence relation splits a set up into distinct chunks. We have seen that $\mathbb{Z} = [1] \cup [0]$ because every number is odd or even. Additionally, $[1] \cap [0] = \emptyset$. These are the two criteria required for...

Definition 5.3.8. A *partition* of a set A is a set of subsets $A_1, A_2, \ldots, A_n$ such that $A_1 \cup A_2 \cup \cdots \cup A_n = A$ and $A_i \cap A_j = \emptyset$ for all $i \neq j$.

Figure 5.3 shows some sets of subsets that are and are not partitions. Notice that a set of subsets can fail to be a partition in two different ways: some subsets may overlap, or the union of the subsets may not be the entire set. In Example 5.3.7, the four equivalence classes $[0],[1],[2],[3]$ partition the set $\mathbb{Z}$, as do the four equivalence classes $[12],[5],[18],[-9]$ and the four equivalence classes $[563560],[867157],[-95814],[459551]$.

More generally, we have...

Theorem 5.3.9. *If a set S has an equivalence relation $\sim$, then the equivalence classes of $\sim$ partition S.*

Proof: In accordance with the definition, we must show that $\bigcup_{s \in S}[s] = S$ and that for $s_1, s_2 \in S$, we have $[s_1] = [s_2]$ xor $[s_1] \cap [s_2] = \emptyset$. First things first: for every $s \in S$, we can examine $[s]$, which certainly contains the element s. Thus,

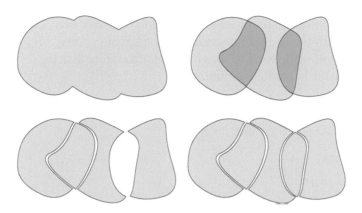

Figure 5.3. A set (upper left), with two nonpartitions (upper right and lower left) and a partition (lower right).

$S = \bigcup_{s \in S} s \subset \bigcup_{s \in S}[s]$. Also, $[s]$ only contains elements from S, so $\bigcup_{s \in S}[s] \subset S$. Therefore, $\bigcup_{s \in S}[s] = S$. Now notice that $[s_1] = [s_2]$ if and only if $s_1 \sim s_2$, because that is how $[s_1]$ and $[s_2]$ are defined; they are each composed from elements that are equivalent under $\sim$. We can conclude that if $[s_1] \neq [s_2]$, then $s_1 \not\sim s_2$. In this case, let's look at $[s_1] \cap [s_2]$. We hope it has no elements. But, suppose it does have some element s_3. (Notice: proof by contradiction coming up!) Then $s_3 \in [s_1]$ and $s_3 \in [s_2]$, so $s_3 \sim s_1$ and $s_3 \sim s_2$. By the symmetric property of equivalence relations, we also know $s_1 \sim s_3$, and that allows us to apply the transitive property of equivalence relations to see that $s_1 \sim s_2$. Hey! That contradicts our assumption that $s_1 \not\sim s_2$! Therefore, $[s_1] \cap [s_2] = \emptyset$ and we are done. $\square$

We can do arithmetic on equivalence classes modulo n in exactly the way we can do arithmetic on integers modulo n.

Example 5.3.10. Suppose $a \equiv b \pmod{n}$ and $c \equiv d \pmod{n}$. By substituting appropriately in each expression, $a + c \equiv b + c \equiv b + d \pmod{n}$. We could restate the situation: Suppose $[a] = [b]$ and $[c] = [d]$. Then it is true that $[a+c] = [b+c] = [b+d]$. A more concrete version of this same example is that $2 \equiv 5 \pmod{3}$ and $11 \equiv 8 \pmod{3}$, so $2 + 11 \equiv 5 + 11 \equiv 5 + 8 \equiv 1 \pmod{3}$. We could also say that modulo 3, $[2] = [5]$ and $[11] = [8]$, so $[2+11] = [5+11] = [5+8] = [1]$.

What's interesting about the arithmetic shown in Example 5.3.10 is that equivalence classes are *sets*, but in this context they act like *numbers*. It turns out that $[a] + [b] = [a + b]$ and also that $[a] \cdot [b] = [ab]$. The proofs of these facts require mathematical sophistication of a level beyond this text. For more mathematics of this sort, please investigate your nearest abstract algebra course. Most of the time in this text, we will work using $\equiv$ (mod n) notation rather than the $[a]$ notation of equivalence classes in modular arithmetic.

Check Yourself

These problems take very little time. So do them!

1. True or false:

    (a) $2 \equiv 10$ (mod 12).
    (b) $2 \equiv -10$ (mod 12).
    (c) $22 \equiv 10$ (mod 12).
    (d) $-2 \equiv 10$ (mod 12).

2. What is the set $[2]$ if we are working modulo 3?

3. Show that $=$ is an equivalence relation.

4. Is $\{1,2\},\{2,3\},\{3,4\},\{4,5,6\}$ a partition of $\{1,2,3,4,5,6\}$?

5. Create a partition of $\{1,4,2,7,9,14,89,246\}$.

6. **Challenge:** We know that $=$ has the property that if $a = b$, then $ac = bc$; Theorem 5.3.3 says that this property also holds for $\equiv$ (mod n). Think of another property that holds for $=$ in ordinary arithmetic, and test to see whether that property holds for $\equiv$ (mod n).

5.4   Cryptography: Some Ciphers

The field of cryptography is concerned with making communication secure from anyone other than the sender and the recipient. We will study some elementary methods of *encryption* (the process of taking messages and converting them to forms that are not directly readable) and *decryption* (converting the received text to readable messages) that use modular arithmetic. These *substitution ciphers* encrypt via letter-by-letter substitutions. We refer to a readable message as *plaintext*, a message encrypted with a known cipher as *ciphertext*, and a communication we cannot read as *wacktext*. (Say it three or four times: wacktext, wacktext, wacktext, wacktext.)

To use modular arithmetic in encryption/decryption, we first need to convert messages into numbers. For the moment, we will ignore case (capital vs. small letters) and punctuation. The simplest way to do the conversion is as follows.

Letter	a	b	c	d	e	f	g	h	i	j	k	l	m
Number	0	1	2	3	4	5	6	7	8	9	10	11	12
Letter	n	o	p	q	r	s	t	u	v	w	x	y	z
Number	13	14	15	16	17	18	19	20	21	22	23	24	25

Using this translation, "duck" becomes 3 20 2 10.

Now, you might think it would be simpler to convert a to 1, b to 2, and so on to z becoming 26. However, that is problematic for use with modular arithmetic, so it turns out to be less simple. (We will see why shortly.)

5.4.1 Shift Ciphers

A *shift cipher* encrypts by shifting each number by some fixed amount. The classical example (literally!) of a shift cipher is the *Caesar cipher*; it uses shift 1. (Well, rumor has it that Augustus Caesar used a shift of 1. Apparently Julius Caesar preferred a shift of 3, again according to internet rumor.)

Example 5.4.1 (of Caesar ciphers and ducks). The message *duck* converts to 3 20 2 10. We shift each number by 1, obtaining $(3+1)\,(20+1)\,(2+1)\,(10+1)$, or 4 21 3 11. This converts to the encrypted message *evdl*. Now, suppose we receive the message *ifo*. To decrypt this, we first convert it to 8 5 14; then, we shift each number by -1, obtaining $(8-1)\,(5-1)\,(14-1)$, or 7 4 13. This, in turn, converts to *hen*.

Now we will see why modular arithmetic is necessary by considering the message *zebra*; it converts to 25 4 1 17 0. This then encrypts to 26 5 2 18 1. Wait a minute! How do we convert this back to letters? We have nothing in our table that corresponds to the number 26... so we must do our shifting modulo 26. In other words, we compute $(25+1 \pmod{26})\,(4+1 \pmod{26})\,(1+1 \pmod{26})$ $(17+1 \pmod{26})\,(0+1 \pmod{26})$ so that our ciphertext is 0 5 2 18 1, or *afcsb*.

Similarly, on receipt of the ciphertext *tobaaz*, we convert to 19 14 1 0 0 25 and then shift by $-1 \pmod{26}$ as follows: $(19-1 \pmod{26})\,(14-1 \pmod{26})$ $(1-1 \pmod{26})\,(0-1 \pmod{26})\,(0-1 \pmod{26})\,(25-1 \pmod{26})$ becomes 18 13 0 -1 -1 24. Uh-oh. We have no letter that corresponds to -1. However, because the underlying set is $\mathbb{Z}_{26}$, we know that every integer is in an equivalence class that can be represented by a number from 0 to 25. We note that $-1 \equiv 25 \pmod{26}$, and now our message is 18 13 0 25 25 24, or *snazzy*.

Example 5.4.1 shows why the use of modular arithmetic prohibits converting a to 1, b to 2, and so on; in order to have 26 letters, we need to operate in $\mathbb{Z}_{26}$ by computing modulo 26, and one of those numbers is certainly 0.

A different example of a shift cipher is *ROT13*, popularized with the advent of Usenet (this is a pre-World Wide Web system, kiddos). It was beloved for sending encrypted punchlines to jokes, and the favored cipher because encryption and decryption are the same operation: add 13 (mod 26). (Notice that $-13 \equiv 13$ (mod 26).)

Example 5.4.2 (of ROT13 in use). Question: Why did the duck cross the road? Answer: *Gb yrnq ure qhpxyvatf gb gur Choyvp Tneqra.*

First we convert the ciphertext to numerals, to wit: 6 1 24 17 13 16 20 17 4 16 7 15 23 24 21 0 19 5 6 1 6 20 17 2 7 14 24 21 15 19 13 4 16 17 0. Then we decrypt by adding 13 (mod 26) to each letter. Here are the first few: $6+13 \equiv 19$ (mod 26); $1+13 \equiv 14$ (mod 26); $24+13 \equiv 37 \equiv 11$ (mod 26); $17+13 \equiv 30 \equiv 4$ (mod 26). (Conveniently, this also corresponds to simply switching numerical rows in the number/letter conversion table.) In full, the decrypted text is 19 14 11 4 0 3 7 4 17 3 20 2 10 11 8 13 6 18 19 14 19 7 4 15 20 1 11 8 2 6 0 17 3 4 13.

You are invited to convert this to appropriate letters in order to read the riddle's answer.

Shift ciphers are not very secure because a frequency analysis (determining which letters are used most frequently) allows a person who has intercepted the ciphertext to guess that it is a shift cipher and by how many positions the text is shifted.

5.4.2   The Vigenère Cipher

Slightly more secure is the *Vigenère cipher*. Instead of shifting each letter by a fixed amount, the Vigenère cipher shifts each letter by an amount determined by alignment with a *key word*.

Example 5.4.3. We will encrypt the message *cake is delicious*. As usual, we first convert the message to numbers: 2 0 10 4 8 18 3 4 11 8 2 8 14 20 18.

Let the key word be *duck*, or 3 20 2 10. We repeat the key word over the length of the message, so that the first letter of the message will be shifted by 3, the second letter shifted by 20, the third letter shifted by 2, the fourth letter shifted by 10, the fifth letter shifted by 3, the sixth letter shifted by 20, and so forth and so on. Writing the key in a row below the plaintext message, we have

2	0	10	4	8	18	3	4	11	8	2	8	14	20	18
3	20	2	10	3	20	2	10	3	20	2	10	3	20	2

Now, we add modulo 26 to obtain

   5   20   12   14   11   12   5   14   14   2   4   18   17   14   20,

or *fumo lm foocesrou*.

   Now, suppose we have received the message *cyev kj xayx vir voq*, and we know
that the key word is *quack*. We first convert to numbers and lay out the key word
as we did before…

2	24	4	21	10	9	23	0	24	23	21	8	17	21	14	16
16	20	0	2	10	16	20	0	2	10	16	20	0	2	10	16

… but this time we *subtract* the repeated key word from the message, working
number by number and modulo 26. This gives us

   12   4   4   19   0   19   3   0   22   13   5   14   17   19   4   0,

and you can convert this to letters to see the secret message.

   While the repeated use of a key word is what most people mean when referring
to a Vigenère cipher, it is not exactly what Blaise de Vigenère (1523–1596) himself
proposed. What Vigenère did was start encrypting with a single copy of a key
word, and then, instead of repeating the key word, he started using the letters of
the plaintext itself, or of the ciphertext as it was generated from the key word (and
then from itself). This sounds as though it would produce undecryptable ciphertext
(i.e., wacktext), but not so: one can decrypt as many letters as are in the key word,
and thereby obtain the next "key" letters to use in the decryption.

Example 5.4.4. This time, we will encrypt the message *cake is delicious*, again
using the key word *duck*, but using the original version of the Vigenère cipher.
Example 5.4.3 gives the conversion of the message and key word to numbers:

2	0	10	4	8	18	3	4	11	8	2	8	14	20	18
3	20	2	10	2	0	10	4	8	18	3	4	11	8	2

   Notice that the second row of numbers begins with the converted *duck* and is
followed by the text of the message. We then add modulo 26 to obtain

   5   20   12   14   10   18   13   8   19   0   5   12   25   2   20,

or *fumo ks nitafmzcu*.

Decrypting is slightly more complicated in this instance. Suppose we have been sent the message *watk ygawums vxiznt bcddgg bus lal vvl* and that we know the key word is *egg*. We will just decrypt the first three words, and we will start by converting them to numbers:

22 0 19 10 24 6 0 22 20 12 18 21 23 8 25 13 19

Now, we write *egg* (as 4 6 6) underneath the first three numbers because it is the key word. But that's all we can do—the remainder of the "key word" has to come from the message itself.

22	0	19	10	24	6	0	22	20	12	18	21	23	8	25	13	19
4	6	6														

We subtract the key word numbers from the message, modulo 26, to obtain

22	0	19	10	24	6	0	22	20	12	18	21	23	8	25	13	19
4	6	6														
18	20	13														

Hmm… 18 20 13 is *sun*. In any case, we write our three plaintext numbers under the next three ciphertext numbers:

22	0	19	10	24	6	0	22	20	12	18	21	23	8	25	13	19
4	6	6	18	20	13											
18	20	13														

Then we subtract modulo 26 to obtain three more plaintext numbers,

22	0	19	10	24	6	0	22	20	12	18	21	23	8	25	13	19
4	6	6	18	20	13											
18	20	13	18	4	19											

which convert to *set*. Hmm. We continue in this fashion, writing our new plaintext numbers underneath the next ciphertext numbers

22	0	19	10	24	6	0	22	20	12	18	21	23	8	25	13	19
4	6	6	18	20	13	18	4	19								
18	20	13	18	4	19											

and subtracting modulo 26 to obtain more plaintext numbers.

22	0	19	10	24	6	0	22	20	12	18	21	23	8	25	13	19
4	6	6	18	20	13	18	4	19								
18	20	13	18	4	19	8	18	1								

We leave the remainder of the translation to you.

Back in the days when people wrote messages by hand (when was that? Oh, that still happened in the 1980s, but here we're talking about the 1550s), a shift table was devised to make manual Vigenère encryption/decryption faster. The top row of this shift table is the usual alphabet, as is the left column. The second row and column are shifted-by-one copies of the alphabet, beginning with B and ending with A. The third row and column are shifted-by-two copies of the alphabet, beginning with C and ending with B. The remainder of the shift table is filled in the same fashion. This way, when encrypting manually one can choose a plaintext letter from the top row and a key word letter from the left column; the letter underneath the plaintext letter and in the row selected by the key word letter is the ciphertext letter. To decrypt manually, one would find the row corresponding to the key word letter and find the ciphertext letter in that row; then, one would zip upwards to find the corresponding plaintext letter in the top row.

5.4.3 Decryption and the Real World

Substitution ciphers are only the tip of the iceberg when it comes to encryption, and a message encrypted using a substitution cipher can easily be read by someone who intercepts it. It is easy to obfuscate by breaking a message into blocks of uniform length and leaving the ciphertext in numerical form. Also, using an alphabet of a length other than 26 (by allowing some punctuation or paying attention to case) makes decryption more difficult. And in reality there is an additional layer of security beyond any individual cipher—there are so many encryption methods that only an expert can recognize which type has been used on a given message. Still, no method described here is a match for an even mildly experienced computer hacker.

The sciences of creating and cracking encryption schemes are incredibly important in commerce, finance, and national security. As a result, there are academic, governmental, and industrial research jobs specifically tasked with the hardware, software, and theory of cryptography. On the creation side, unbreakable encryption is needed for banks to communicate internally and externally so that account transactions cannot be altered or fabricated, and for businesses to accept credit card information over the internet without leaving customers vulnerable to identity theft (a common method for this is described in Section 16.10), and for governments to communicate with intelligence personnel (spies!) in the field. The flip side of that last need is the desire of governments to crack encryption schemes used by terrorists so that their nefarious plans can be uncovered. During times of war, it has been (and is, and will be) crucial to intercept and understand enemy communications about troop movements and bombing targets. The twin necessities of inventing and breaking new methods of encryption will always be in tension.

Check Yourself

Do a representative sampling of these problems to be sure you understand how to use the various ciphers presented in this section.

1. Encrypt the message *lemon drops* using a Caesar cipher.

2. Decrypt the message *pvaanzba ohaf* using ROT13.

3. Encrypt the message *quilt blocks* using a shift cipher with shift 7.

4. Decrypt the message *bdpja lxxtrnb*, which was encrypted using a shift of 9.

5. Encrypt the message *lions tigers and bears oh my* using a Vigenère cipher and key word *zoo*.

6. Decrypt the message *wwrfw aiw wowl*, which was encrypted with a standard Vigenère cipher using key word *ears*.

5.5   Try This! Encryptoequivalent Modulalgorithmic Problems

Please have fun with these problems.

1. Consider the symbols ab...xyzAB...XYZ.

   (a) Let $\alpha \sim_1 \beta$ if the symbols α and β represent the same letter. Is $\sim_1$ an equivalence relation? If so, what are the equivalence classes?

   (b) Let $\alpha \sim_2 \beta$ if the symbols α and β are the same case (upper or lower). Is $\sim_2$ an equivalence relation? If so, what are the equivalence classes?

   (c) Now assign $a \mapsto 0$, $b \mapsto 1$, ..., $z \mapsto 25$, $A \mapsto 26$, ..., $Z \mapsto 51$. Notice this converts our symbols to elements of $\mathbb{Z}_{52}$. If we apply $\sim_1$, what are the corresponding equivalence classes in $\mathbb{Z}_{52}$? What happens if we instead apply $\sim_2$?

   (d) **Challenge:** Examine the equivalence classes of $\mathbb{Z}_{52}$ under each of $\sim_1$ and $\sim_2$. Do the classes themselves correspond to other familiar sets?

2. Before proceeding, review the formal definition of congruence modulo n.

   (a) Prove (using the definition of congruence mod n) that if $a \equiv b \pmod{n}$ and $c \equiv d \pmod{n}$, then $ac \equiv bd \pmod{n}$.

   (b) Prove that if $a \equiv b \pmod{n}$, then $a^k \equiv b^k \pmod{n}$.

   (c) Is $92^{57} \equiv 6^{57} \pmod{43}$? Explain.

3. Describe three different algorithms you use in everyday life. Write them as lists of instructions in human-readable form. What changes would need to be made for these instructions to be specific enough for a machine to follow?

4. Create a plaintext message and encrypt it using a shift cipher. Copy the ciphertext onto a separate piece of paper and note the amount of the shift you used. Trade ciphertexts with a partner and decrypt the ciphertext you receive.

5. Write an algorithm for encrypting with the Caesar cipher that is precise enough for a computer to follow.

6. Create a plaintext message and encrypt it using a Vigenère cipher, either the standard sort or Vigenère's original cipher. Copy the ciphertext onto a separate piece of paper and note the key word you used. Trade ciphertexts with a partner—but do not disclose which variety of Vigenère cipher you used—and decrypt the ciphertext you receive. Try to figure out whether your partner used original or standard Vigenère. How might you decide?

5.6 Where to Go from Here

The study of algorithms leads more deeply into computer science and also more deeply into mathematics. We will use algorithms in Chapter 8 to solve certain recurrence relations, and in the context of graph theory in Chapters 10 and 12. In computer science, there are entire courses on algorithm design and implementation and how to determine the efficiency of algorithms. The analysis of algorithms is highly mathematical, so we address it in Chapter 17.

The Collatz conjecture (Example 5.2.12) is one of the easiest-to-state open problems in mathematics. It has generated a lot of research. For a summary of the history, current state of knowledge, and related generalizations of the Collatz conjecture, see http://mathworld.wolfram.com/CollatzProblem.html. For information on how the Collatz conjecture has been verified for all $n \leq 10^{18}$ (and perhaps larger

n by the time this textbook is printed!), check out the website "Computational Verification of the 3x+1 Conjecture" at http://sweet.ua.pt/tos/3x+1.html.

Modular arithmetic is but the beginning of a branch of mathematics called *elementary number theory*, which is a subbranch of number theory (of course). If you want to learn some number theory, read Chapter 16; if you want to learn even more number theory, consult the references given there.

To automate encryption/decryption for shift ciphers, use http://www.dcode.fr/shift-cipher, and to automate encryption/decryption for the standard Vigenère cipher, use http://www.dcode.fr/vigenere-cipher.

The original work in which Blaise De Vigenère suggested the ciphers that bear his name is *Traicté des chiffres* from 1586. An interesting analysis of this work is contained in "Blaise De Vigenère and the 'Chiffre Carre,'" by Charles J. Mendelsohn in the *Proceedings of the American Philosophical Society*, Vol. 82, No. 2 (1940).

You will need to learn a large chunk of number theory in order to study practical cryptography. A start on this is given in Chapter 16. In Section 16.10 we give an introduction to the RSA algorithm for encryption, which is used widely on secure websites.

It is worth noting that cryptography research includes designing algorithms, implementing algorithms in software, and designing hardware with specific cryptographic functions. This means that there are many directions in which a study of cryptography can proceed—there are both mathematics courses and computer science courses in cryptography, as well as graduate programs and industrial workshops.

The concept of equivalence classes can be seen in the context of abstract algebra, where it leads to the idea of quotient objects. To learn more about abstract algebra and quotients, see *Visual Group Theory* by Nathan Carter, as well as *Contemporary Abstract Algebra* by Joe Gallian.

Credit where credit is due: Some of the commentary on algorithms in this chapter was inspired by [1] and [5]. Example 5.4.2 was inspired by [19]. Section 5.9 was inspired by [4]. The Doctors on page 139 are those from the long-running television show *Doctor Who*. My colleagues at the Centre for Textiles and Conflict Studies provided inspiration for Bonus Check-Yourself Problems 2 and 4. Bonus Check-Yourself Problem 10 includes a quote from Jane Austen. Bonus Check-Yourself Problem 1 was suggested by Tom Hull. In Section 5.11, Problem 2 refers to the *Cake Wrecks* blog (see www.cakewrecks.com) and Problem 15 refers to the *SuperFriends* cartoon from the 1980s (in which the Wonder Twins took shapes of an animal and water, respectively). Regarding Problem 40 in Section 5.11, Susan is a resident of *Sesame Street*, and the message is, of course, a line from that excellent show's theme song.

5.7 Chapter 5 Definitions

algorithm: A finite list of unambiguous instructions to be performed on one or several inputs; some instructions may refer to others.

terminate: An algorithm terminates when it produces an output and ends after executing a finite number of instructions.

correct algorithm: An algorithm that does what it should do.

conditional: A statement within an algorithm that places conditions on an instruction.

if-then-else: A conditional that usually takes the form "if (conditions), then (action set 1), else (action set 2)" and is read/understood as "If (conditions) are met, then do (action set 1); otherwise, if (conditions) are *not* met, then do (action set 2)."

until: A conditional that takes the form "do (action set) until (conditions)" or "until (conditions), (action set)" and is read/understood as "Do (action set) until (conditions) are met and then go to the next instruction."

while: A conditional that takes the form "do (action set) while (conditions)" or "while (conditions), (action set)" and is read/understood as "Do (action set) while (conditions) hold, and when (conditions) are no longer met, go to the next instruction."

loop: An instruction to perform some set of actions more than once. (The instructions repeat, forming a string of instructions into a loop of instructions.)

iteration: The process of repeating instructions.

cruciferous: An adjective used to describe vegetables from the family Cruciferae (a.k.a. Brassicaceae). The many cruciferous vegetables include kale, cabbage, broccoli, arugula, turnips, and wasabi.

existence proof: A proof that shows that something exists.

constructive proof: A proof that produces an example of a desired object.

divides: Short for "divides evenly."

congruent modulo n: Two integers a and b are congruent modulo n when $(a - b) = kn$ for some $k \in \mathbb{Z}$.

integers modulo n: The set of different remainders obtainable by dividing integers by n.

symmetric property: This holds if when $s_1 \sim s_2$, then $s_2 \sim s_1$ for all $s_1, s_2 \in S$.

reflexive property: This holds if $s \sim s$ for all $s \in S$.

transitive property: This holds if when $s_1 \sim s_2$ and $s_2 \sim s_3$, then $s_1 \sim s_3$ for all $s_1, s_2, s_3 \in S$.

equivalence relation: An operation $\sim$ defined on a set S that satisfies the symmetric property, the reflexive property, and the transitive property.

equivalence class: All the elements equivalent to a using some equivalence relation, i.e., $[a] = \{s \in S \mid s \sim a\}$.

partition: A set of subsets $A_1, A_2, \ldots, A_n$ of a set A such that $A_1 \cup A_2 \cup \cdots \cup A_n = A$ and $A_i \cap A_j = \emptyset$ for all $i \neq j$.

encryption: The process of taking messages and converting them to forms that are not directly readable.

decryption: The process of converting received text to readable messages.

substitution cipher: A cipher that encrypts via letter-by-letter substitutions.

plaintext: A readable message.

ciphertext: A message encrypted with a known cipher.

wacktext: A communication we cannot read.

shift cipher: A cipher that encrypts by shifting each number by some fixed amount.

Caesar cipher: A shift cipher that shifts by 1.

ROT13: A shift cipher that shifts by 13 (and therefore encryption and decryption both proceed by adding 13 (mod 26)).

key word: A set of letters that provides the information needed to decrypt a cipher.

Vigenère cipher: A cipher that shifts each letter by an amount determined by alignment with a key word. Named after Blaise de Vigenère (1523–1596), who was a diplomat. In modern standard usage, a Vigenère uses the key word repeatedly to decrypt an entire message; originally, Vigenère himself used just one copy of the key word and then used the plaintext (or ciphertext) as it was generated for subsequent decryption (or encryption).

5.8   Bonus: Algorithms for Searching Graphs

Suppose we need to look at every vertex in a graph. There are tons of reasons (mathematical, computer scientific, etc.) why we might want to do this, but here is one practical example: You maintain a website with many pages. Each page links to some of the other pages and also has some external links. Once each year, you need to check every link to make sure none of them are broken. Now, you could think of the link structure as a graph, where every hyperlink represents a vertex and two hypertext-vertices are adjacent if clicking on one leads to a page containing the other. (This is a directed graph, though one can always hit the back button to travel to a prior vertex.) An external hyperlink is considered to be a vertex of degree 1 as it does not point to any page on your website. What you need is an algorithm that visits every vertex in the graph, so that you can automate the link checking. In this section, we will give you *two* algorithms that will do the job.

How to search a connected graph, depth-first:

1. Examine the first vertex of the graph that you come across.

2. Mark the vertex as seen.

3. If this vertex has a neighbor that has not yet been seen, examine the unseen neighbor and go to step 2; otherwise, continue.

4. If the current vertex is also the first vertex examined, then be done; otherwise, return to the previous vertex and go to step 3.

In our example of checking all the links on a website, this is equivalent to clicking on the first link you see until you reach a page on which the first link has been clicked, and then clicking on the next link on the page. When there are no links left to click on the current page, you hit the back button until you get to a page with an unclicked link. If you are back at the starting page and all links have been clicked, then you've clicked all possible links. This assumes, of course, that every page on the website is linked from some other page on the website so that each graph vertex is reachable. (Both algorithms need to be modified for nonconnected graphs or general digraphs; the vertices must be labeled and ordered to assure that all are reached.)

How to search a connected graph, breadth-first:

1. Examine the first vertex of the graph that you come across.

2. Mark the vertex as seen.

3. Examine all unseen neighbors of the vertex and mark them as seen.

4. For each of the neighbors considered in the previous step, execute step 3.

5. If no unseen neighbors were identified, be done; otherwise, go to step 4.

This is equivalent to checking each link on the first page, then following the first internal link on the page and checking all the links on the page you reach, then returning to the first page, following the second internal link on the page and checking all the links on the page you reach, etc. This process checks all the links that are one click away from the first page. Next, it checks all the links that are two clicks away from the first page, then all the links that are three clicks away from the first page, and so on.

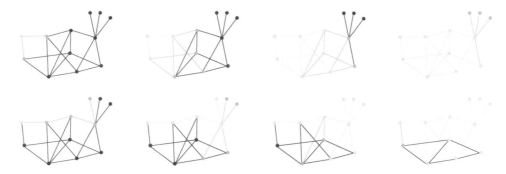

Figure 5.4. A breadth-first search (across the top row) and a depth-first search (across the lower row) are both performed on a particular graph, starting with the upper-left vertex.

Let us contrast depth-first and breadth-first searches by seeing how they proceed on a particular graph; this is shown in Figure 5.4. For another example, see http://demonstrations.wolfram.com/GraphSearchingBreadthFirstAndDepthFirst/ where you can use a slider to control how many steps of the search have been done.

> Practice.   Redraw the graph in Figure 5.4 twice. Choose a different starting vertex than the upper-left vertex, and perform a depth-first and a breadth-first search using the new starting vertex. Now, draw a completely new graph, never before seen by you, and perform a depth-first and a breadth-first search on that graph.

5.9   Bonus 2: Pigeons and Divisibility

First, here's a statement of the generalized pigeonhole principle in terms of partitions (see Definition 5.3.8): Suppose a set S has more than pq elements. If we partition S into p parts, then at least one of the parts has more than q elements.

Now... holy cow.   Look at the sequence $4, 44, 444, 4444, 44444, 444444,$ $4444444, 44444444, \ldots$. Do you believe that one of the first 63 elements of this sequence is divisible by 63?

Well, if not, too bad. We're going to prove that it's true, using proof by contradiction and the pigeonhole principle.

Suppose not. That is, suppose that none of the first 63 elements of $4, 44, 444,$ $4444, 44444, 444444, 4444444, 44444444, \ldots$ is divisible by 63. Take those 63 elements and find their remainders after division by 63. Cast in the language of modular arithmetic, compute $4 \pmod{63}, 44 \pmod{63}, 444 \pmod{63}, 4444$

(mod 63), etc. These will be numbers in the 1 to 62 range. None of them will be 0 because none of the sequence elements are divisible by 63. Now, there are 63 remainders and 62 numbers, so at least two of them have to be the same by the pigeonhole principle.

What? What's that you're saying? You don't know how that helps? Right. Hang on a minute. First, notice that you just proved a lemma.

> **Lemma 5.9.1.** *At least two of the first* 63 *elements of* 4, 44, 444, 4444, 44444, 444444, 4444444, 44444444, ... *have the same remainder on division by* 63.

Hey, don't knock it—even if that didn't help prove the statement that one of the first 63 elements of 4, 44, 444, 4444, 44444, 444444, 4444444, 44444444, ... is equivalent to 0 (mod 63), it would be pretty cool.

But anyway. Go back to that remainder list. Use it to find two sequence elements that have the same remainder r after division by 63. Call them a_1 and a_2. Now, $a_1 = 63p + r$ and $a_2 = 63q + r$. So $a_1 - a_2 = 63p + r - (63q + r) = 63(p - q)$, and that means that $a_1 - a_2 \equiv 0$ (mod 63).

I know what you're thinking—this still doesn't help, because yeah, we have a number divisible by 63, but it's not one of the elements of the sequence! Elements a_1 and a_2 are both strings of 4s (though they have different lengths), and so $|a_1 - a_2|$ is a bunch of 4s followed by a bunch of 0s. Still not helpful, but we're almost there. Notice that if you chop off all of the 0s, you get a bunch of 4s— and it's fewer than sixty-three 4s, because the longer of a_1 and a_2 is no more than 63 digits in length. That means the remaining bunch of 4s is one of the first 63 elements of 4, 44, 444, 4444, 44444, 444444, 4444444, 44444444, ... Now for the denouement: 63 has no factors in common with 10, so because integer factorization is unique, $a_1 - a_2$ is divisible by 63 and also by 10, and therefore chopping off the ending 0s leaves a number that is divisible by 63. Ha! We're done.

(There's a related problem, on which numbers divide numbers like 111, 1111, 11111, etc., that you may enjoy exploring at http://demonstrations.wolfram.com/ThePigeonholePrincipleRepunits/.)

Pause. What was special about the digit 4 and the number 63 in the proof given above? Think on this for at least 30 seconds. Your goal is to see how this proof, and therefore the lemma and theorem, can be generalized. Spend at least two minutes contemplating how to generalize this mathematical situation—and bookmark/close the book if your eyes are likely to wander further.

What would a generalization of our original statement look like? We would consider the sequence $k, kk, kkk, kkkk, kkkkk, kkkkkk, \ldots$, and desire to show that one of the first m elements is divisible by m. In order to generalize our statement, we need to determine what restrictions our proof placed on k and m. So let's go through the proof step by step and see what we find.

We first proved the lemma that at least two of the first m elements of $k, kk, kkk,$ $kkkk, kkkkk, kkkkkk, \ldots$ have the same remainder on division by m. There are at most $m - 1$ different values of $k \pmod{m}, kk \pmod{m}, \ldots$ among the first m remainders. Thus, by the pigeonhole principle, two of these remainders must be the same. What are the crucial parts of this argument? Well, m has to be larger than $m - 1$ in order to apply the pigeonhole principle, and $m > m - 1$ is always true no matter what m is. The value of k is irrelevant here—in fact, it doesn't matter what bunch of m numbers we pick. This means we have in fact now shown that *at least two of the first m elements of any integer sequence have the same remainder on division by m.*

The next step is to produce a number that is $0 \pmod{m}$. We do this by finding two sequence elements that have the same remainder r after division by m. If we call them $a_1 = mp + r$ and $a_2 = mq + r$, we find that $a_1 - a_2 = m(p - q)$ so that $a_1 - a_2 \equiv 0 \pmod{m}$. This time, k doesn't even get a mention, and the rest is arithmetic, so we still have no restriction on m. (At this point, you might reasonably wonder whether we will end up with any restrictions at all. Keep reading to find out!)

We then observed that $|a_1 - a_2|$ is a bunch of ks followed by a bunch of 0s. Aha! This does at least restrict us to a sequence of the form $k, kk, kkk, kkkk, kkkkk,$ $kkkkkk, \ldots$, as otherwise we have no idea what the digits of $|a_1 - a_2|$ will be. Then we note that the bunch of ks is one of the first m elements of our sequence. This is true because the length of $|a_1 - a_2|$ is less than m, which is true because each of a_1 and a_2 has length no more than m. No restrictions here. But the denouement gives a restriction: m has to have no common factors with 10 so that chopping off all those 0s leaves a number divisible by m.

Now, there have been no restrictions on k so far. But let's look a tiny bit closer. We were sort of assuming that k was a digit, meaning that $1 \leq k \leq 9$. But does that have to be true? For example, what if $k = 0$? Well, we get a silly statement and silly result. (You may decide for yourself whether we should allow $k = 0$ or not.) And what if $k > 9$? The only place in the proof that k comes into play is in showing that (bunch of ks) $-$ (other bunch of ks) $= k$s followed by 0s. And that's true no matter how many digits k has.

Conclusion: For any $m \in \mathbb{N}$ divisible by neither 2 nor 5, and any $k \in \mathbb{N}$, at least one of the first m elements of the sequence $k, kk, kkk, kkkk, kkkkk,$ $kkkkkk, \ldots$, is divisible by m.

Related problem. Consider a prime p other than 2 or 5. Show that if you consider $p, p^2, p^3, \ldots$, one of these must have last digit 1. In fact, one must have last digits 01; why? And in fact, one must have last digits 000000000000000001; why?

5.10 Bonus Check-Yourself Problems

Solutions to these problems appear starting on page 603. Those solutions that model a formal write-up (such as one might hand in for homework) are to Problems 3 and 6.

1. Find the smallest nonnegative integer x that satisfies the equation $3(x + 7) \equiv 4(9 - x) + 1 \pmod 5$.

2. Encrypt this message from a supportive shark using a shift-by-10 cipher: YOU ARE SUPER GREAT AND FACES ARE HIGH IN PROTEIN

3. Prove, using only the definition of congruence modulo n, that if $a \equiv b \pmod n$, then $a + c \equiv b + c \pmod n$.

4. While you are distraught over your latest discrete math exam, a passerby shoves a scrap of paper into your hand that reads *xvghdibhvivozz 21*. You suspect that this could be a shift cipher. What does the message say?

5. Here is an algorithm:

    1. Get a pot, a cover, a stove, and an egg.
    2. Put the egg in the pot.
    3. Fill the pot with enough water to cover the egg.
    4. Turn a burner to high heat.
    5. Set the pot on the burner.
    6. Put on a hat.
    7. Wait until the water boils.
    8. Wait for 3 minutes.
    9. Remove the pot from the heat and add a cover.
    10. Wait for 10 minutes.
    11. Crack the shell of the egg.
    12. Drain the water, replace with cold water, and let stand for 3 minutes.
    13. Put away the egg.

    What are the inputs? What are the outputs? Does the algorithm terminate? What does the algorithm do? Are there any problems with this algorithm?

6. Let $a \sim b$ exactly when ab^2 is even. Is $\sim$ an equivalence relation?

7. Write an algorithm that lists the first 10 negative multiples of 9.

8. Encrypt *the foam shark visor is intended only for children* using the original Vigenère cipher with key word *pickles*.

9. Let $\sim$ be defined so that $a \sim b$ exactly when $b - a \geq 2$. Is this an equivalence relation? If so, list the equivalence classes. If not, which of the three properties (reflexive, symmetric, transitive) does not hold?

10. Decrypt *xx ut e kcyrp nvavximtsfl ixoegwwpbggn* using a Vigenère cipher and the key word *pemberley*. Is this an original or a standard Vigenère cipher?

Figure 5.5. A sanitary cheese preserver (left) and the inscription on its lid (right).

5.11 Problems about Algorithms, Modular Arithmetic, and Ciphers

1. Consider the inscription shown in Figure 5.5. What are these instructions designed to do? Are they an algorithm?

2. Encrypt the message *naked mohawk baby carrot jockeys* using a shift cipher with shift 24.

3. Finish decrypting the message given in Example 5.4.4.

4. Let $\sim$ be defined so that $a \sim b$ exactly when $a \cdot b$ is divisible by 3. Is this an equivalence relation? If so, list the equivalence classes. If not, which of the three properties (reflexive, symmetric, transitive) does not hold?

5. List the equivalence classes of $\mathbb{Z}_n$ in both equivalence-class notation and set notation, and verify that they partition $\mathbb{Z}$.

6. Prove that $a \equiv b \pmod{n}$ if and only if $n|(a - b)$; that is, check that the condition given in Definition 5.3.1 is correct.

7. These instructions were found on an actual chopstick wrapper:

   1. Tuck under thumb and hold firmly.
   2. Add second chopstick, hold it as you hold a pencil.
   3. Hold first chopstick in original position, move the second one up and down, now you can pick up anything.

   What are the inputs for this algorithm? Is the algorithm correct? Comment on the algorithm's clarity.

8. Decrypt the message *fqenjxpno fv rtnffrz fvug*, which was encrypted using ROT13.

9. Prove, using only the definition of congruence modulo n, that if $a \equiv b$ (mod n) and $b \equiv c$ (mod n), then $a \equiv c$ (mod n).

10. What does this list of instructions do? Comment on whether it is an algorithm and whether it terminates.

    1. Let $n = 3$.

    2. Replace n with $n + 4$.

    3. If n is even, go to step 2.

11. Decrypt the message *kqfnjzykwj qpmljd uki xaegd aaf dua*, which was encrypted using the original Vigenère cipher with key word *fish*.

12. The input for this algorithm is $n \in \mathbb{N}$. What does the algorithm do?

    1. Let $k = 1$.

    2. Let *result* $= 1$.

    3. Replace *result* with $k \cdot$ *result*.

    4. If $k < n$, replace k with $k + 1$ and go to step 3.

    5. Output *result*.

13. What kind of proof is used in proving Theorem 5.3.3?

14. Let $\sim$ be defined so that $a \sim b$ exactly when $a + b$ is even. Is this an equivalence relation? If so, list the equivalence classes. If not, which of the three properties (reflexive, symmetric, transitive) does not hold?

15. Encrypt the message *iron bars procrastinate rhymes with twin powers activate* using the standard Vigenère cipher with key word *silly*.

16. Is $[2] = [123]$ modulo 6? Explain.

17. Write an algorithm that counts to 18 by twos.

18. Prove, using only the definition of congruence modulo n, that if $a \equiv b$ (mod n) and $c \equiv d$ (mod n), then $a + c \equiv b + d$ (mod n).

19. Our goal in this problem is to determine when the converse of Theorem 5.3.3 holds and when it does not, namely, when does $ac \equiv bc$ (mod n) imply that $a \equiv b$ (mod n)?

    (a) Let us recall our counterexample: $18 \equiv 24$ (mod 6), but $9 \not\equiv 12$ (mod 6). In fact, $18 \equiv 24 \equiv 0$ (mod 6). Find another example in which $ac \equiv bc \equiv 0$ (mod n) and $a \not\equiv b$ (mod n). (Try not to have $n = 6$.)

    (b) In your example, was n even? If so, find another example in which n is odd.

    (c) Make a conjecture: under what conditions does the converse of Theorem 5.3.3 hold?

    (d) **Challenge:** Perhaps there is something special about zero... or perhaps not. Use the definition of congruence modulo n to figure out whether there are a, b, c, n such that $ac \equiv bc$ (mod n) and $ac \not\equiv 0$ (mod n) and $a \not\equiv b$ (mod n).

20. Is set containment $\subset$ an equivalence relation?

21. Decrypt the message *q kiv pih kpmmhjczomz*, which was encrypted using a cipher that shifted by eight letters.

22. Decrypt the message *dtrn utzaknr dbbv utzaknr T lbt'b zoam po gabbd hvlwoec*, which was encrypted using a Vigenère cipher with key word *lady*. Which type of Vigenère cipher was used?

23. The company Charmed, I'm Sure makes bracelets. Each bracelet has four

charms, Apple, Banana, Cherry, and Fig (or $\{A, B, C, F\}$ for short). The way these bracelets are made is by sending a line of charms into a machine, where they get attached to circular chains. There are 24 different orders in which a set of 4 charms might get fed into the machine. (Why?) Once the bracelets are complete, some of these 24 orders look the same. If we consider the equivalence relation $b_1 \sim b_2$ when bracelets b_1 and b_2 look the same, how does this partition the set of 24 charm orderings?

24. **Challenge:** Understand the algorithm for Russian multiplication given in Example 5.2.11.

(a) Execute the algorithm using $A = 12$ and $B = 7$.

(b) Now choose two different values for A, B and execute the algorithm again.

(c) When do you write down *Bstep* and when not?

(d) When you sum the column, what multiple of B (how many copies of B) do you obtain?

(e) Does this have anything to do with binary numbers?

(f) How exactly does this algorithm give the same result as usual multiplication?

25. **Challenge:** Write actual code in an actual programming language that...

(a) ... encrypts a message using a shift cipher of 12 letters.

(b) ... decrypts a message encrypted using a shift cipher of 6 letters.

(c) ... encrypts a message using the original Vigenère cipher.

(d) ... decrypts a message using the standard Vigenère cipher.

26. Let $\sim$ be defined so that $a \sim b$ exactly when $a^2 = b^2$. Is this an equivalence relation? If so, describe the equivalence classes. If not, which of the three properties (reflexive, symmetric, transitive) does not hold?

27. Is $\{3k + 1 \pmod{15} \mid k \in \mathbb{Z}\} \cup \{5k - 2 \pmod{15} \mid k \in \mathbb{Z}\} \cup \{6k + 2 \pmod{15} \mid k \in \mathbb{Z}\}$ a partition of $\mathbb{Z}_{15}$? Why or why not?

28. Create a partition of $\mathbb{Z}_{12}$ and three non-partitions of $\mathbb{Z}_{12}$, two that violate exactly one of the conditions to be a partition and one that violates both conditions.

29. Fill in these addition and multiplication tables modulo 5:

+	0	1	2	3	4
0					
1					
2					
3					
4					

×	0	1	2	3	4
0					
1					
2					
3					
4					

30. (a) Find the smallest nonnegative number x such that $x \equiv 107 \pmod{7}$.

(b) Compute the smallest x such that $x \equiv 483 \pmod{9}$.

(c) Calculate the smallest positive value of $8 + 8 \pmod{9}$.

(d) Is $10 \equiv 7 \pmod 3$?

(e) What about 49... is it $\equiv 17 \pmod{22}$?

31. Encrypt the message *grackles are shiny and like shiny things* using the Caesar cipher.

32. The input for this algorithm is $n \in \mathbb{N}$. What does the algorithm do?

    1. Let $k = 0$.
    2. Let *result* $= 1$.
    3. Replace *result* with $k \cdot result$.
    4. If $k < n$, replace k with $k+1$ and go to step 3.
    5. Output *result*.

33. Let $\sim$ be defined so that $a \sim b$ exactly when $a + 2b$ is even. Is this an equivalence relation? If so, list the equivalence classes. If not, which of the three properties (reflexive, symmetric, transitive) does not hold?

34. Decrypt the message *ctzt dhp gkt qvym pediyk dhp jvohibs*, which was encrypted using the standard Vigenère cipher with key word *kale*.

35. Consider the following, printed on boxes of Opti-Free RepleniSH.

    Directions for care of your lenses:
    To clean, recondition, disinfect and remove protein from your contact lenses:

1. Thoroughly rinse each side of the lens (5 seconds) with OPTI-FREE® RepleniSH® Multi-Purpose Disinfecting Solution.

2. Fill your lens case with fresh OPTI-FREE® RepleniSH® Multi-Purpose Disinfecting Solution.

3. Store lenses in the closed lens case overnight or at least 6 hours. After soaking, lenses are ready to wear.

What are the inputs for this algorithm? Does the algorithm terminate?

36. Write an algorithm that sums the first n squares.

37. Is $9 \equiv 3 \pmod 6$? What about $389 \equiv 87 \pmod{92}$?

38. Encrypt the message *it was a dark and stormy night* using the original Vigenère cipher with key word *weather*.

39. What is the smallest number greater than n that is $\equiv 1 \pmod n$? For a number k between 1 and n, what is the smallest number larger than k but also $\equiv k \pmod n$?

40. Encrypt the message *can you tell me how to get to Sesame Street* twice, once using the standard Vigenère cipher and once using the original Vigenère cipher, both with key word *susan*.

5.12 Instructor Notes

The material in this chapter is intentionally light so as to make room in a semester schedule for review and the giving of an exam. So, the class-time plan here is for about one-and-one-half classes. The fact that many students have more of a passing familiarity with algorithms and with modular arithmetic than with other topics in a discrete mathematics course is helpful.

Assign the students to read Sections 5.1–5.4 in preparation for the week and to do some of the Check Yourself problems. Twenty pages of reading seems like a lot, but it's very fast and students find most of it to be easy; in particular, they will zip right through the algorithm and cipher material. Really! You probably don't believe this, but it is true (and verified by people other than the author). Usually students do not feel that modular arithmetic presents any difficulty, but they do need some supervised practice with it, and equivalence relations/classes can present a challenge.

Begin by having the students open to Section 5.5. It works well to introduce the first problem to the students in the form of a tiny interactive lecture and then alternate between having them work in pairs and conducting a large group discussion. Similarly, the second problem can be introduced via interactive lecture and segue into work in pairs or groups. It also presents a good opportunity to ask for questions over the reading. To add a bit of dramatic flair to the first problem's introduction, show a video (http://www.youtube.com/watch?v=qTvhKZHAP8U) of Big Bird singing the "Abcdefghi..." song.

Problem 5 in Section 5.5 can be profitably introduced via interactive lecture as well because it allows a discussion of how modular arithmetic is used algorithmically in these substitution ciphers. If you want to use actual code or code-like pseudocode, python syntax is a good choice because students pick it up very quickly. If you plan to cover Chapter 17 later, you'll want to assign Problem 24 in Section 5.11 now.

Section 5.5 is probably too much for a single class period; a good warmup for a second class period is asking the class to describe the Vigenère cipher to you. This can easily transition into the remaining problems in the activity.

An excellent way to break students into groups for work in this chapter is to count them off from 1 to n (assuming you have n students), choose a $k \approx \frac{n}{4}$, and tell them to collect into groups where all members have numbers congruent $\mod k$. Then ask them to decide whether their group forms an equivalence class (or not). This gives students practice with modular arithmetic and equivalence classes while also allowing a bit of controlled chaos into the classroom.

Theme I Supplement

TI.1   Summary of Theme I Proof Techniques

Template for a direct proof:

1. Restate the theorem as *if (conditions) are true, then (conclusion) is true.*

2. On a scratch sheet, write *assume* or *suppose (conditions) are true.*

3. Take some notes on what it means for (conditions) to be true. See where they lead.

4. Attempt to argue in the direction of *(conclusion) is true.*

5. Repeat attempts until you are successful.

6. Write up the results on a clean sheet, as follows.

   - 🦆 **Theorem:** (State theorem here.)

   - 🦆 **Proof:** Suppose (conditions) are true.

   - 🦆 (Explain your reasoning in a logically airtight manner, so that no reader could question your statements.)

   - 🦆 Therefore, (conclusion) is true. (Draw a box or checkmark or write Q.E.D. to indicate that you're done.)

How to apply the pigeonhole principle:

1. Figure out what represents the pigeons.

2. Figure out what represents the pigeonholes.

3. Figure out how pigeons correspond to holes.

Template for a proof that $A \subset B$:

🐤 Let a be any element of A.

🐤 (Reasoning, statements.)

🐤 Therefore, $a \in B$, and so $A \subset B$.

Double-Inclusion.   To show that $A = B$, show first that $A \subset B$ and then show that $B \subset A$.

Biconditionals ($\Longleftrightarrow$s):

(a) First, write ($\Rightarrow$) to indicate you'll prove that P implies Q (and then do so).

(b) Then, write ($\Leftarrow$) to indicate you'll prove that Q implies P (and then do so).

Be sure to start a new paragraph for each implication.

Template for proving the contrapositive:

1. State the theorem as *if (conditions) are true, then (conclusion) is true.*

2. Restate the theorem in the equivalent form *if ¬(conclusion) is true, then ¬(conditions) is true.*

3. On a scratch sheet, write *assume* or *suppose ¬(conclusion) is true.*

4. Take some notes on what it means for ¬(conclusion) to be true. See where they lead.

5. Attempt to argue in the direction of ¬(conditions) is true.

6. Repeat attempts until you are successful.

7. Write up the results on a clean sheet, as follows:

- 🐤 **Theorem:** (State theorem here.)

- 🐤 **Proof:** Suppose ¬(conclusion) is true.

- 🐤 (Explain your reasoning in a logically airtight manner, so that no reader could question your statements.)

- 🐤 Therefore, ¬(conditions) is true, so our original theorem holds and we are done.

Template for a proof by contradiction:

1. Restate the theorem as *if (conditions) are true, then (conclusion) is true.*

2. On a scratch sheet, write *suppose not.* Then write out (conditions) and the negation of (conclusion).

3. Try to simplify the statement of ¬(conclusion) and see what this might mean.

4. Attempt to derive a contradiction of some kind—to one or more of (conditions) or to a commonly known mathematical truth.

5. Repeat attempts until you are successful.

6. Write up the results on a clean sheet, as follows:

- 🐤 **Theorem:** (State theorem here.)

- 🐤 **Proof:** Suppose not. That is, suppose (conditions) are true but (conclusion) is false.

- 🐤 (Translate this to a simpler statement if applicable. Derive a contradiction.)

- 🐤 Contradiction!

- 🐤 Therefore, (conclusion) is true. (Draw a box or checkmark or write Q.E.D. to indicate that you're done.)

How to do a proof by induction:

- 🐦 **Base case:** Check to make sure that whatever you want to prove holds for small natural numbers, like 1, 2, and 3.

- 🐦 **Inductive hypothesis:** Assume that whatever you want to prove is true, as long as the variable in the statement is smaller than or equal to k; here, k is a specific (but unknown) value.

- 🐦 **Inductive step:** Consider the statement with $k+1$ as the variable. Use your knowledge that the statement is true when the variable is less than or equal to k in order to show that it's still true for $k+1$. (That is, use the base case(s) and inductive hypothesis.)

TI.2   Potential Practice Proof Problems

Here is a panoply of plain practice proof problems. Many of those pertaining to set theory were provided by David Cox.

TI.2.1   Problems Pertaining to Chapter 1

1. Prove that the sum of an odd number and an even number is odd.

2. Prove, or find a counterexample: for k any integer and n any odd number, kn is odd.

3. Prove that for finite sets A, B, C, the number of elements of $A \times B \times C$ is $|A| \cdot |B| \cdot |C|$.

4. Prove, or find a counterexample: if m is odd, then $4m - 3$ is odd.

5. Prove that if $n + 6$ is even, then n is even.

6. Prove that the binary representation of an even number ends in 0.

7. Prove that if a is even and b is odd, then $ab + 1$ is odd.

TI.2.2   Problems Pertaining to Chapter 2

8. Prove that $A \subseteq B$ if and only if $A \cup B = B$.

9. Let A, B, C be sets; show that if $A \subseteq B$, then $A \cap C \subseteq B \cap C$.

10. Show that for sets A, B, if $A \cup B \subseteq A \cap B$, then $A = B$.

11. Prove that $A \subseteq B$ implies that $A \times C \subseteq B \times C$.

12. For sets A, B, C, show that $A \times (B \cap C) = (A \times B) \cap (A \times C)$.

13. Prove that if $m^2 - 2m$ is even, then m is even.

14. Show that $(A \setminus B) \cap (B \setminus A) = \emptyset$.

15. Show that if $z \in \mathbb{Z}$ and $z^2 \mid z$, then $z \in \{-1, 0, 1\}$.

16. Consider the proposition *if n is even, then $n^2 + n$ is even*.

    (a) Prove it.
    (b) State the converse.
    (c) If the converse is true, prove it; if it is false, give a counterexample.

17. Using element notation, prove that if $A \cap B = A$, then $\overline{A} \cup B = U$.

18. Prove, using contradiction or the contrapositive, that if the average age of four children is ten years old, then at least one child is at least ten years old.

19. Prove that if a natural number n is even, then $n + 2$ is even...

    (a) ... using a direct proof.
    (b) ... by proving the contrapositive.
    (c) ... using proof by contradiction.

TI.2.3 Problems Pertaining to Chapter 3

20. Prove that K_4 is isomorphic to W_4.

21. Prove that $f : \mathbb{Z} \to \mathbb{Z}$ defined by $f(z) = z + 3$ is a well-defined bijection.

22. Prove that $f : \mathbb{Z} \times \mathbb{Z} \to \mathbb{Z} \times \mathbb{Z}$ defined by $f((a, b)) = (-b, a)$ is a well-defined bijection.

23. Prove that there is no graph with degree sequence $(1, 1, 2, 3, 4, 4, 5, 7)$.

24. Prove that the cycle graph C_n has n edges.

25. Is it possible for a simple graph with 6 vertices to have 42 edges? Explain.

26. Let $f : \mathbb{N} \times \mathbb{N} \to \mathbb{N}$, with $f(a,b) = a^2 + b^2$. Decide (with proof) whether f is one-to-one, onto, both, or neither.

27. Let $f : \mathbb{N} \times \mathbb{N} \to \mathbb{N}$, with $f(a,b) = a + b$. Decide (with proof) whether f is one-to-one, onto, both, or neither.

TI.2.4   Problems Pertaining to Chapter 4

28. Prove, in two different ways, that when $n > 1$, $2n > n + 1$.

29. Prove that $\displaystyle\sum_{j=1}^{n} (2j + 3) = n^2 + 4n$.

30. Prove that $\displaystyle\sum_{j=1}^{n} (3j^2 + 1) = \frac{2n^3 + 3n^2 + 3n}{2}$.

31. Prove that $12^n > 3^n$ for all $n \in \mathbb{N}$.

32. Show that for $n \in \mathbb{N}$, $\displaystyle\sum_{i=1}^{n} \frac{2}{(i+1)(i+2)} = \frac{n}{n+2}$.

TI.2.5   Problems Pertaining to Chapter 5

33. Let $\sim$ be an equivalence relation on a set S. Using the definition of equivalence class (see page 139), prove that $[a] = [b]$ if and only if $a \sim b$.

34. Prove that if $a \equiv b \pmod{n}$ and $c \equiv d \pmod{n}$, then $ac \equiv bd \pmod{n}$ without applying the definition of congruence $\bmod\, n$.

35. Prove that $3 | x \iff 3 | x^2$.

36. Let $\sim$ be defined so that $a \sim b$ exactly when $a + b$ is odd. For each of the three equivalence relation properties (reflexive, symmetric, transitive), prove that the property holds or demonstrate why the property does not hold. Is this an equivalence relation? If so, list the equivalence classes.

37. Let $\sim$ be defined so that $a \sim b$ exactly when $|a| = |b|$. Is this an equivalence relation? If so, list the equivalence classes. If not, which of the three properties (reflexive, symmetric, transitive) does not hold?

38. Let $\sim$ be defined so that $a \sim b$ exactly when $a \leq b + 2$. Is this an equivalence relation? If so, list the equivalence classes. If not, which of the three properties (reflexive, symmetric, transitive) does not hold?

TI.3   Problems on the Theme of the Basics

These problems could be used for studying for (or writing!) in-class or take-home exams, or just for more enrichment. They are not given in any particular order. In fact, they have been intentionally mixed up so that they are *not* in chapter order, so that the solver cannot use the ordering of the problems as a clue in solving them.

1. In a 1922 arsenic poisoning, at least 20 of the > 50 victims worked in a 12-story building. Prove that at least two of the victims worked on the same floor.

2. Determine whether the graphs in Figure TI.1 are isomorphic: exhibit an isomorphism or find a property for which the two graphs differ. (A GeoGebra file for Figure TI.1 is available for your use at http://www.toroidalsnark.net/dmwdlinksfiles.html.)

Figure TI.1. Two suspicious graphs.

3. Prove that $\sum_{j=1}^{n} j^2 = \dfrac{n(n+1)(2n+1)}{6}$.

4. Find the smallest number of vertices needed to draw a graph with

   - ❦ an edge with multiplicity 3,
   - ❦ at least two edges crossing,
   - ❦ at least three vertices that are all adjacent to each other, and
   - ❦ a vertex with five neighbors.

   Now draw that graph.

5. Let n be an integer. Is it always true that the difference between two consecutive cubes is never even? Explain.

6. Give a counterexample to each of the following statements... unless you think the statement is true, in which case give a one-line justification.
   - (a) If $n^2 = 4$, then $n^3 = 8$.
   - (b) If $\sin(x) = 0$, then $\cos(x) = 1$.
   - (c) If $\cos(x) = 0$, then $\sin(x) = 1$.
   - (d) If $x^3 = x$, then $x^2 = 1$.

7. Find $|\mathbb{Z}_2 \times \mathbb{Z}_3|$.

8. **Challenge:** Cast your memory all the way back to Example 1.5.5. How far does that example generalize? Consider the statement *given any length-k list of m-digit numbers, two subsets have the same sum.* Find constraints on k, m for this statement to be true.

9. Write the converse of *if a graph G has 30 vertices, then G is not blue.*

10. Consider a function on the real numbers defined by $f(1) = q$ and $f(a+b) = f(a) \cdot f(b)$ for all real numbers a, b.
    - (a) Prove by induction that $f(n) = q^n$ for all $n \in \mathbb{N}$.
    - (b) Show that if $q \neq 0$, then $f(0) = 1$.

11. Prove that a connected simple graph with ten vertices must have two vertices of the same degree.

12. For which n is K_4 a subgraph of K_n? Explain.

13. How many possible passwords have 6–12 characters, where the characters must be alphanumeric but case does not matter?

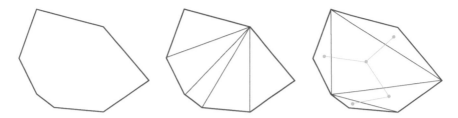

Figure TI.2. An arbitrary heptagon with two different decompositions into triangles and a graph associated to one of them.

14. **Challenge:** Try to figure out what the Cartesian product of two graphs should be, without looking it up. If you do eventually search for information on graph products, please forgive the mathematical community for using such terrible and inconsistent notation. (The author has been unable to excuse it.)

15. Show that any 11 numbers selected from $\{3,\ldots,30\}$ must include two that have a common divisor (other than 1).

16. Show that for $n \in \mathbb{N}$,
$$\sum_{i=1}^{n} \frac{3}{(i+2)(i+3)} = \frac{n}{n+3}.$$

17. Give three examples of functions $f : \mathbb{Z} \to \mathbb{Z}$, where one is one-to-one but not onto, the second is onto but not one-to-one, and the third is neither one-to-one nor onto.

18. Consider any convex polygon with at least four sides and decompose it into triangles by connecting vertices. (Do not let chords cross or create new vertices.) See Figure TI.2 for an example of this process.

    (a) Show that no matter how you decompose the polygon into triangles, at least two of the triangles have two sides (each) in common with the original polygon.

(b) For any decomposition of a convex polygon into triangles, create a graph as follows. Place a vertex in each triangle and join two vertices when their two surrounding triangles share an edge.

    (i) What kind of graph is this? (You may justify your answer.)

    (ii) Does every graph of this type arise from some convex polygon? Explain.

    (iii) What aspect of the graph corresponds to a triangle that has two edges in common with the original polygon?

    (iv) What theorem about graphs did you prove in the previous part of this problem?

19. Prove or give a counterexample: Every multiple of 6 that is ≥ 12 is the sum of two consecutive primes. (For example, $30 = 13 + 17$.)

20. Show that for a fixed $r \in \mathbb{N}$ and any $n \in \mathbb{N}$, $\displaystyle\sum_{i=1}^{n} \frac{r}{(i+r-1)(i+r)} = \frac{n}{n+r}.$

21. Consider $a, b \in \mathbb{Z}$ and let $a \sim b$ if $a \equiv b$ (mod 3) or if $a \equiv b$ (mod 5). Is this an equivalence relation? Explain which properties of equivalence relations hold

and which don't. If this *is* an equivalence relation, then list the equivalence classes.

22. Suppose all of G's vertices have degree 3. Prove that G can be decomposed into copies of $K_{1,3}$ (a.k.a. *claws*), with vertices of different copies possibly overlapping, if and only if G is bipartite.

23. You get on an elevator. There are nine people already in the elevator, and six floor-indicator buttons are lit. What is a reasonable conclusion and why? (Strange but true: If instead there are six people already in the elevator, and nine floor-indicator buttons are lit, what is a reasonable conclusion and why?)

24. Write each of the following statements using formal logic notation.

    (a) For every integer n, $2n \neq 9$.

    (b) There exists a triangle T that is equilateral and has perimeter 10.

    (c) Every circle has an integer diameter or an integer area.

    (d) Every two natural numbers have an integer between them.

25. How many different programs to do a single task can you write if there are six available algorithms for the task and each has been coded in four different ways?

26. Give examples of sets A, B such that $|A| = 8$, $|B| = 6$, and $|A \setminus B| = 5$.

27. Prove that $\sum_{j=1}^{n} 2j - 3 = (n-1)^2 - 1$.

28. Show that if you choose k integers arbitrarily, then at least two of them will have the same remainder on division by $k - 1$. (What principle did you use in solving this problem?)

29. Let b represent the statement *the cat likes to eat broccoli*, let p represent the statement *the cat likes to play*, and let s represent the statement *the cat likes to sleep*. Write each of the following sentences using formal logic notation.

    (a) The cat likes to eat broccoli but does not like to play.

    (b) The cat likes to sleep and eat broccoli, or ze likes to play.

    (c) The cat does not like to eat broccoli, but ze likes to play or to sleep.

30. Encrypt the text *rubber baby buggy bumpers* using a shift-by-15 cipher.

31. Prove that $\sum_{j=1}^{n} 2^j = 2^{n+1} - 2$.

32. Decrypt the message *O lkfz me hkaj atd sy negrz ot tne jatck frour*, which was encrypted using a Vigenère cipher with key word *gaga*. Which type of Vigenère cipher was used?

33. What does this list of instructions do? Comment on whether it is an algorithm and whether it terminates.

    1. Let $n = 2$.
    2. Replace n with $n + 3$.
    3. If n is even, go to step 1; otherwise, go to step 2.
    4. Output n.

34. Let $C = \{d, e, \{d, e\}, f\}$. List the elements in each of the following sets (or write $\emptyset$ if appropriate). Be careful with your notation.

    (a) $C \setminus \{d, e\}$.

    (b) $C \setminus \{\{d, e\}\}$.

35. *Ittw etaiz mbmit* (8).

36. A tap-dancing duck makes sequences of sounds with its feet. You count the number of taps between pauses. Show that if

you listen long enough, say, for ten tap sequences, then two of the numbers of taps have the same value mod 9.

37. How many natural numbers $k < 100$ have the property $k \equiv 3 \pmod{7}$?

38. Let t represent the statement *Amali is tall*, let d represent the statement *Amali is dark*, and let b represent the statement *Amali is beautiful*. Write an English sentence equivalent to each of the following formal-logic expressions.

    (a) $(t \vee d \vee b) \wedge \neg(t \wedge d \wedge b)$.

    (b) $\neg t \wedge \neg d$.

    (c) $d \wedge \neg(t \wedge \neg d)$.

    (d) $(t \wedge d) \vee (\neg t \wedge d)$.

39. For the Cartesian product $\mathbb{Z} \times \mathbb{Z} = \mathbb{Z}^2$, let $(a,b) \sim (c,d)$ if $a - c = b - d$. Is $\sim$ an equivalence relation?

40. Prove that if $A \subseteq B$, then $\overline{B} \subseteq \overline{A}$. Do this once using Venn diagrams and once using element notation.

41. Let $A = \{-2,-1,0,1,2\}$ and let $B = \{q,r,s,t\}$.

    (a) How many functions $f : A \to B$ can be defined? Explain.

    (b) How many *one-to-one* functions $f : A \to B$ can be defined? Explain.

    (c) Suppose that $f(0) = r, f(1) = q$, $f(2) = t$. How many functions $f : A \to B$ satisfy these conditions?

    (d) Suppose again that $f(0) = r$, $f(1) = q$, $f(2) = t$. How many *onto* functions $f : A \to B$ satisfy the conditions?

42. Let $A = \{2k \mid k \in \mathbb{Z}\}$, $B = \{3k + 1 \mid k \in \mathbb{Z}\}$, and $C = \{6k + 5 \mid k \in \mathbb{Z}\}$. Show that $\{A,B,C\}$ is *not* a partition of $\mathbb{Z}$.

43. Prove that if $A \subseteq B$, then $A \cup (B \setminus A) = B$. Use element notation in your proof.

44. Encipher *graph theory is the bomb* using a Vigenère cipher with key word *bat*.

45. Which one of the following statements is true (and why)?

    (a) $-3 \equiv 21 \pmod{12}$.

    (b) $-3 \equiv 15 \pmod{12}$.

46. Let $f : \mathbb{N} \times \mathbb{N} \to \mathbb{N}$, with $f(a,b) = 2^a \cdot 3^b$. Decide (with proof) whether f is one-to-one, onto, both, or neither.

47. Prove that if $a \equiv b \pmod{n}$ and $c \equiv d \pmod{n}$, then $a - c \equiv b - d \pmod{n}$.

48. The *distance* between vertices x and y of a graph G, denoted $d(x,y)$, is the length of the shortest path joining x and y. If x,y,z are vertices of G, is it always true that $d(x,y) + d(y,z) \geq d(x,z)$? Give a proof or a counterexample.

49. The coop next door contains a flock of chickens and a few ducks. Each duck is basically either brown, white, or grey. Every chicken in the flock is red xor black. How many different color pairs of birds are there if one is a duck and the other is a chicken? Explain *very* briefly.

50. Draw Venn diagrams representing the sets $(A \setminus B) \cup (B \setminus C)$, $(A \setminus B) \cup (B \cap C)$, $(A \cup C) \setminus (A \setminus B)$, $(A \cup C) \setminus (A \cap B)$, and $(A \cup B) \setminus (B \setminus C)$. Do any two (or more) of these represent the same set?

Credit where credit is due: Problem 18 was donated by David Cox. The message in Problem 30 is a traditional tongue twister. For more on the situation of Problem 1, see *The Poisoner's Handbook* by Deborah Blum.

Part II

Theme: Combinatorics

Chapter 6 🐦🐦🐦🐦🐦🐦

Binomial Coefficients and Pascal's Triangle

6.1   Introduction and Summary

This is the start of our four-chapter-long focus on combinatorics, and it is jam-packed with (related-to-each-other) ideas! Combinatorics is the science of counting. We will begin by considering the number of ways to choose some objects from a larger pile of objects. This will lead us to investigate the links between choosing objects, Pascal's triangle, and powers of $(x+y)$. (At first glance, these don't seem to have anything to do with each other... surprise!) We will also see how these ideas tie in to the factorial function $(n!)$ and how factorials relate to arranging objects in different orders. As before, every exploration is followed by reinforcing material in a subsequent section.

This chapter also introduces two counting techniques, both of which were foreshadowed in earlier chapters. Careful overcounting is exactly what it sounds like—we overcount carefully so that we can compensate appropriately and determine an exact count. Combinatorial proof is the process of counting the same thing two different ways, where one of the ways represents the amount we desire to calculate (but can't yet) and the other way is something we already know how to calculate. Collaborating on challenging problems will help you practice these techniques, so such problems are provided.

The bonus sections will be of particular interest to those who like algorithms—and to those who like playing games. Go look at them (after you've worked through Sections 6.2–6.5)!

6.2   You Have a Choice

Let us look back to Problem 5 in Section 1.2: Some people heading to a party stop by an ice-cream store to buy quarts of ice cream. The store has five flavors of ice cream. How many orders of three quarts could they make? What if the three flavors have to be different? What if no one will agree to order squirrel ice cream?

You may not remember how you solved this problem (it was a while ago!), so here is one way to approach the first question. The three quarts could be all the same flavor; there are five different ways that could happen. Or, two of the quarts might be one flavor, and one might be a different flavor; in this case, we need to know how many ways there are to choose two out of the five flavors. If we number the flavors $1, 2, 3, 4, 5$, then the different possible flavor pairs are $(1,2), (1,3), (1,4), (1,5), (2,3), (2,4), (2,5), (3,4), (3,5), (4,5)$: there are ten of them. (We do need to count each one twice because ordering one quart of squirrel ice cream and two quarts of licorice ice cream is different from ordering one quart of licorice ice cream and two quarts of squirrel ice cream.) The remaining possibility is that the three flavors could be different from each other; here we need to know how many ways there are to choose three out of the five flavors. The possible flavor triples are $(1,2,3), (1,2,4), (1,2,5), (1,3,4), (1,3,5), (1,4,5), (2,3,4), (2,3,5), (2,4,5), (3,4,5)$: there are ten of them. So, in total, there are $5 + 10 \cdot 2 + 10 = 35$ possible three-quart orders.

Surely there is a faster way to know how many choices there are than to list them each time. Indeed, we will learn two different ways to determine this number (but in a later section; keep your eyes out). And, there must be a shorter way to say, "the number of ways there are to choose two out of the five flavors" than to use all those words. Indeed, there is; read on to the next paragraph.

Definition 6.2.1. The symbol $\binom{n}{k}$ is pronounced "n choose k," will be referred to as a *choice number*, and means the number of ways one can choose k things from a pile of n different things. The order in which the k things are chosen does not matter. Expressions using choice numbers are said to be written using *choice notation*.

For nonexample, the number of ways of choosing two teal balls from a pile of three teal balls is just one, if the balls are truly identical. On the other hand, for example, if the three teal balls are different somehow (perhaps because they are numbered), then there are three different ways to choose two of them. (Borifying this by placing it in a set-theoretical context, $\binom{n}{k}$ is also the number of k-element subsets of an n-element set.)

For now, let us imagine that we have a bag of n sugar-numbers. These are wrapped maple sugar candies in the shapes of numbers (see Figure 6.1), and each sugar-number in our bag is different. Our model for choice is that you reach into the bag and grab a handful of sugar-numbers. (Public Service Apology: we're

Figure 6.1. The sugar-number 2.

sorry if this has made you crave maple sugar candy.) If there are n sugar-numbers in the bag, and you grab a handful of k sugar-numbers, then there are $\binom{n}{k}$ ways for that grab to occur. Borific side note: there is a one-to-one correspondence between a handful of k sugar-numbers and the k-element subset of $\{1, \ldots, n\}$ with those k numbers in it.

We will now derive an important relation that can be expressed using choice notation. Moreover, we will derive this relation using *combinatorial proof*, a characteristic approach to solving combinatorics problems by counting some quantity in two different ways. Section 6.8 provides more on combinatorial proof.

Start with a bag of n sugar-numbers, numbered 1 through n. Consider the set H of all possible handfuls of k sugar-numbers. Each of them either (a) contains the sugar-number n xor (b) does not contain the sugar-number n. Call the first set N and call the second set D (for "Does not contain n"). Now, $H = N \cup D$. We know that $|H| = \binom{n}{k}$, and we are about to find ways to determine $|N|$ and $|D|$. Then we will have two ways to express $|H|$, one as $\binom{n}{k}$ and one as $|N| + |D|$. (Notice that we can use the sum principle here because N and D are disjoint.)

First, think about N. Every handful of k sugar-numbers in N contains the sugar-number n. If we take the n out of a handful, we have a corresponding handful of $k - 1$ sugar-numbers. Because there can't be an n in there, the handful might as well have been taken from a bag with the first $n - 1$ sugar-numbers in it. Thus, we have a one-to-one correspondence between N and the set of all possible handfuls of $k - 1$ sugar-numbers drawn from a bag of $n - 1$ sugar-numbers. (See Figure 6.2 for a visualization.) Call that latter set M (for "k Minus one"). We know how to express $|M|$ in other terms—it's $\binom{n-1}{k-1}$. And, because of the one-to-one correspondence, we know that $|M| = |N|$.

Now, think about D. Every handful of k sugar-numbers in D does *not* contain the sugar-number n. So, those k sugar-numbers might as well have been taken from a bag with only the first $n - 1$ sugar-numbers in it. Thus, we have a one-to-one correspondence between D and the set of all possible handfuls of k sugar-numbers drawn from a bag of $n - 1$ sugar-numbers. Call that latter set Y (for "Yuck, can't

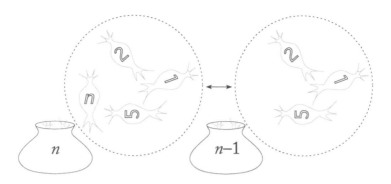

Figure 6.2. An element of N in correspondence with one of M.

think of a name"). Again, we know how to express $|Y|$ in other terms—it's $\binom{n-1}{k}$. And again, our one-to-one correspondence tells us that $|Y| = |D|$.

Denouement.   Putting this all together, we have that $\binom{n}{k} = |H| = |N| + |D| = |M| + |Y| = \binom{n-1}{k-1} + \binom{n-1}{k}$. This fact, that $\binom{n}{k} = \binom{n-1}{k-1} + \binom{n-1}{k}$, is the most basic of choice notation identities. Notice that it re-expresses $\binom{n}{k}$ in similar notation but with smaller numbers (or indices); this type of equation is called a *recursion* and will be the focus of Chapter 8.

Check Yourself ——————————————————————————————

Do all of these problems—they're worth it.

1. Write the solutions to the questions that begin this section in choice notation.

2. Compute $\binom{4}{2}$ using the basic choice notation identity (and a little bit of exhaustive listing).

3. Compute $\binom{5}{2}$ using the basic choice notation identity and the previous problem.

6.3   Try This! Investigate a Triangle

Our investigations begin by switching gears away from choice notation. Maybe sometime in your past you've seen this creature:

$$
\begin{array}{ccccccccccccc}
 & & & & & & 1 & & & & & & \\
 & & & & & 1 & & 1 & & & & & \\
 & & & & 1 & & 2 & & 1 & & & & \\
 & & & 1 & & 3 & & 3 & & 1 & & & \\
 & & 1 & & 4 & & 6 & & 4 & & 1 & & \\
 & 1 & & 5 & & 10 & & 10 & & 5 & & 1 & \\
1 & & 6 & & 15 & & 20 & & 15 & & 6 & & 1 \\
\end{array}
$$

It's created by starting out with the 1 at the top, and then in each successive row, each entry is the sum of the two numbers diagonally left and diagonally right above it. (Those 1s on the edges? They're created by adding the numbers 1 and 0, with 0 represented above by " ".)

1. Write out the next two or three rows of this triangular array, to verify that you understand it.

2. Do you see any connection between the triangular array above and the choice notation identity $\binom{n}{k} = \binom{n-1}{k-1} + \binom{n-1}{k}$? (If so, what is the connection?) Is there a way to connect numbers n and k to the triangular array? (If so, what is it?)

3. Use the choice notation identity $\binom{n}{k} = \binom{n-1}{k-1} + \binom{n-1}{k}$ to figure out how many different Sushi Samplers (three kinds of rolls) can be ordered from a 12-roll sushi menu (salmon skin, avocado, california, yellowtail/scallion, dragon, futo maki, alaska, kanpyo, eel/avocado, shrimp tempura, spicy tuna, cucumber). Or, use the triangular array to determine how many different quadruple-chunk cookie recipes can be made from the eight common inclusions milk chocolate chunk, white chocolate chunk, dried cranberry, peanut butter chip, raisin, butterscotch chip, macadamia nut, and dark chocolate chunk. And why not switch? Use the identity to count cookie recipes and use the array to count Sushi Samplers. (How different are these methods?)

4. You have probably noticed some left-right symmetry in the triangular array. (If not, go notice it now.) Using choice notation, express this symmetry for a particular pair of entries in the triangular array. Pick three more symmetric pairs and use choice notation to express their symmetry. Can you now express the symmetry of the entire array using choice notation? Please attempt to do so. Using sugar-numbers, prove that this symmetry makes sense. (By the way, this is another example of combinatorial proof.)

5. Let n be odd. What is the relationship between the number of subsets of $\{1, 2, \ldots, n\}$ of odd size and the number of subsets of $\{1, 2, \ldots, n\}$ of even size? The previous problem may be of assistance.

6. How does the total number of subsets of $\{1, 2, \ldots, n\}$ relate to the triangular array? (What is that number again, anyway?)

6.4   Pascal's Triangle

Hey! You! Don't read this unless you have worked through the problems in Section 6.3. I mean it!

The triangular array introduced in Section 6.3 above is called *Pascal's triangle* (after Blaise Pascal from the 1600s, even though it was known at least 500 years earlier). The rows are indexed by n, and the entries are indexed by k, as follows. We consider the top 1 to be 0 over in row 0 (that is, the initial 1 is at $k = n = 0$) and count n downwards and k to the right of the left edge of the triangle. In this way, the generation of the triangle by adding pairs of entries in the previous row corresponds to the choice notation identity $\binom{n}{k} = \binom{n-1}{k-1} + \binom{n-1}{k}$. So the zeroth row is secretly $\binom{0}{0}$, the first row is secretly $\binom{1}{0}$ $\binom{1}{1}$, the second row is secretly $\binom{2}{0}$ $\binom{2}{1}$ $\binom{2}{2}$, etc. More generally, $\binom{n}{k}$ is the kth number in the nth row of the triangular array. One great utility of Pascal's triangle is that it's pretty fast to generate, so for small values of n (like $n \leq 10$) one can quickly look up a given $\binom{n}{k}$ in the triangle. This is much faster and easier than repeatedly applying $\binom{n}{k} = \binom{n-1}{k-1} + \binom{n-1}{k}$.

From Pascal's Triangle we can see that

$$\binom{n}{k} = \binom{n}{n-k}$$

because the triangle has symmetry over the vertical midline. We could prove this combinatorially by noting that if we choose k candies (such as sugar-numbers) from a bag of n candies, there are $n - k$ candies remaining. So there are the same number of ways of choosing k candies as there are of choosing $n - k$ candies from a bag containing n candies. (Do not confuse this with choosing candles from a bag. If lit, they are dangerous.)

Additionally, we can use the correspondence between $\binom{n}{k}$ and the number of size-k subsets of $\{1, 2, \ldots, n\}$ to gain some insight into the latter, and vice versa. The number of subsets of $\{1, 2, \ldots, n\}$ is 2^n, and that's also the sum of the numbers across a row of Pascal's triangle—each entry in the row counts the number of subsets of a particular size, and every subset-size is represented. Now notice that in odd-n rows of Pascal's triangle, the set of odd-k entries is equal to the set of even-k entries because $n-k$ is even when k is odd; therefore, there are the same number of odd-sized subsets of $\{1,2,\ldots,n\}$ as there are even-sized subsets of $\{1,2,\ldots,n\}$. In fact, this tells us that the number of odd-sized (and of even-sized) subsets of $\{1,2,\ldots,n\}$ must be 2^{n-1}.

Pascal's triangle can also be written as a right triangle with n increasing downwards and k increasing to the right.

1	0	0	0	0	0	0	0	0	0	...
1	1	0	0	0	0	0	0	0	0	...
1	2	1	0	0	0	0	0	0	0	...
1	3	3	1	0	0	0	0	0	0	...
1	4	6	4	1	0	0	0	0	0	...
1	5	10	10	5	1	0	0	0	0	...
1	6	15	20	15	6	1	0	0	0	...
1	7	21	35	35	21	7	1	0	0	...
1	8	28	56	70	56	28	8	1	0	...
:	:	:	:	:	:	:	:	:	:	⋱

Instead of taking the sum of the two entries diagonally above the desired entry, we take the sum of the entry directly above and the entry diagonally left above. We can extend this to a rectangle by filling in 0s in the remaining entries—after all, there are no ways of choosing more than n items from a pile of n items, i.e., when $k > n$ for $\binom{n}{k}$.

Check Yourself ─────────────────────────────────

These are extra-fast.

1. Quickly compute $\binom{8}{3}$.

2. Quickly compute $\binom{7}{5}$.

3. How many ways are there to choose two ducks out of a raft of nine ducks?

4. How many ways are there for two of the author's three cats to sit on her bed?

5. Invent a question to which the answer is $\binom{13}{4}$.

6.5   Overcounting Carefully and Reordering at Will

Let us take a cue from *Math Curse* by Jon Sczieska (rhymes with "Fresca") and Lane Smith. (If you have not read this book, go directly to the nearest public library, without passing GO or collecting $200, and get it. If you are unsure as to whether your local university library has this seminal tome, check http://www. worldcat.org/title/math-curse/oclc/32589625 and enter your zip code.) There are 24 students in the narrator's class. She counts the desks in the room as divided in four rows, eight rows, three rows, and two rows; she counts the students by twos; she wonders how many fingers and ears and tongues are in the class.

Super-quickie questions:

1. How many of the 24 desks are in each row if there are four rows? … eight rows? … three rows? … two rows?

2. How many pairs of students does the narrator count?

3. If there are 216 fingers in the class, how many fingers does each student have?

4. If there are 47 ears in the class, how many ears does each student have (and who is one of the students)?

Let's face it: you were just doing division. And that's basically all you were doing. But just as when you were adding and multiplying in Chapter 1, the basic operation is but a paltry reflection of the larger cognition in which you were engaged. That is, it doesn't look like you were doing much, but in reality you did something pretty complicated. Namely, you counted indirectly—you noticed that the students had been overcounted by a certain factor and divided by that factor.

Another perspective on this phenomenon is to cast it in the language of functions. (If you feel that this borifies matters, then read it quickly and think not of it again; use the concept but not the language.) In the case of counting desks by rows, we are secretly mapping the set of desks onto the set of rows. If there are the same number of desks in each row (let's say k), then we have a k-to-one and onto function, or a k-to-one correspondence; see Figure 6.3. (This is also known as a *many-to-one correspondence* and is related to the generalized pigeonhole principle—see page 18.)

> Fact.   Just as a one-to-one correspondence is a bijection and implies that two sets have the same size, a k-to-one correspondence indicates that the size of the domain is k times the size of the target space.

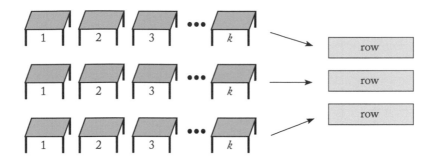

Figure 6.3. A k-to-one correspondence between desks and rows.

Example 6.5.1 (of careful overcounting). Consider a polygon with p sides. We would like to know how many vertices the polygon has. Each side of the polygon touches two vertices, so there are a total of $2p$ vertices. However, each vertex touches two sides, and that means we have overcounted by a factor of two. Thus the polygon has $\frac{2p}{2} = p$ vertices.

Example 6.5.2. Consider a graph G in which each vertex has degree r. Let $|V|$ be the number of vertices in G and let $|E|$ be the number of edges in G. We will use overcounting to prove that $|E| = \frac{r|V|}{2}$. (This is a special case of the handshaking lemma proved in Chapter 3.)

Because each vertex has degree r and there are $|V|$ vertices, the total degree of G is $r|V|$. An edge has two ends, each of which is counted once in the total degree of G, so the total degree is also $2|E|$. (More formally, there is a two-to-one correspondence between vertex/edge incidences of G and the set E.) This means that $r|V|$ counts each edge twice, so $|E| = \frac{r|V|}{2}$. (By the way, this is also a combinatorial proof because we counted the same quantity two different ways.)

Here is the general structure underlying Example 6.5.2; it provides a template you can apply to other problems. Suppose there are a things, and each of the a things has b aspects. By the product principle, there are $a \cdot b$ variants in total. But there might be a different way to think about these ab variants. It may be that the number ab can be written as $c \cdot d$, where this represents c things that each have d aspects. (This would give a d-to-one correspondence between the set of ab variants and the set of c things.) In trying to solve a problem, we would probably know the value of d and be seeking the value of c. To complete our translation, in Example 6.5.2 we have $a = |V|$, $b = r$, $c = |E|$, and $d = 2$.

Let us do a more complicated example to see how a k-to-one correspondence can be noticed and then used.

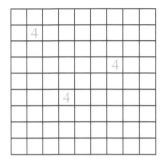 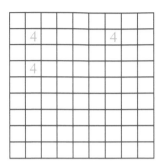

Figure 6.4. (Left) This may be a start on a solution. (Right) Well, *that's* not going to work....

Example 6.5.3. In a Sudoku board, a player cannot have two of the same number appear in the same row or the same column. How many different ways can we place three 4s on a 9×9 grid so that no two of them share a row or a column? (For this example, we will ignore the fact that on a Sudoku board, a player also cannot have two of the same number appear in the same 3×3 block.) Figure 6.4 shows a possibility and an impossibility.

There are nine possibilities for the row of the first 4, and also nine possibilities for the column of the first 4. We can't reuse that row or that column, so there are only eight possibilities for the row and for the column of the second 4, and neither of those two columns nor rows may be reused, so there are only seven possibilities for the row and for the column of the third 4.

However, for each of the nine row choices for the first 4, we could use any of the nine columns. So there are 9^2 ways to place the first 4. For each of those, there are (by similar reasoning) 8^2 ways to place the second 4, and then for each of those there are 7^2 ways of placing the third 4, for a total of $(9 \cdot 8 \cdot 7)^2$ ways of placing the three 4s.

Uh, but wait. We kind of overcounted there. If we place the first 4 in row 1 and column 2, and the second 4 in row 3, column 4, that gives the same result as placing the first 4 in row 3, column 4, and the second 4 in row 1 and column 2. (Notice that if we'd wanted to place a 4, a 5, and a 6, or perhaps 4s of three different colors, there would have been no overcounting and our initial answer would have been correct.) So now what we need to figure out is (1) whether we are overcounting uniformly—that is, do we have a k-to-one correspondence?—and (2) what the heck k is, if so.

Often when we (intentionally) overcount, we will rely on some symmetry to help us create a k-to-one correspondence. In the case of a Sudoku board, every row looks like every other row; for this purpose, they're all the same. This is true for the columns as well. (And, as we just discovered, all the 4s are identical.) This symmetry tells us that we have some sort of uniform overcounting. So what is k in this case? For a given placement of three 4s, there are six ways we could have chosen it: let the placement be denoted by (a, b, c), where a represents the location of the first 4, b the location of the second 4, and c the location of the third 4. If we were to place the second 4 in location c and the third in location b, that'd be the same configuration of 4s; all possible permutations of (a, b, c) give this same result, and there are six of them. (Those six permutations are (a, b, c), (a, c, b), (b, a, c), (b, c, a), (c, a, b), and (c, b, a).)

So we have not $(9 \cdot 8 \cdot 7)^2$ ways of placing three 4s in a Sudoku board, but $\frac{(9 \cdot 8 \cdot 7)^2}{6} = \frac{254,016}{6} = 42,336\ldots$ that's still a lot.

The key to figuring out Example 6.5.3 was determining that we were over-counting and by what factor k. However, there was a subproblem we glossed over: we had to know that there were six possible orderings of the letters a, b, c. Let's generalize this—we might be in trouble if we had to know how many orderings there were of $a, b, c, d, e, f, g, h, i, j, k, l, m, n, o, p$. First we will do an example. Luckily we will need nothing more sophisticated than the product principle.

Example 6.5.4. Suppose you have five different citrus fruits (blood orange, clementine, kumquat, pink grapefruit, and uglifruit) and you can't decide in what order to eat them. How many possibilities are there? There are five choices for the first-eaten fruit. Then, for each of those first-fruit choices, four options remain for the second-eaten fruit; for each of those choices, three choices remain for the third-eaten fruit; for each of those choices, there are two options for the fourth-eaten fruit, and there is then only one fruit left to eat. By the product principle, there are $5 \cdot 4 \cdot 3 \cdot 2 \cdot 1$ orders in which to eat the five citrus fruits. (A word of advice: don't store the uglifruit in the refrigerator.)

In general, suppose we have n items and we want to know how many ways there are to order them, also referred to as the number of *permutations* of the set. There are n choices for which item appears first in the ordering, then $n - 1$ items remaining to choose for the second spot in the ordering, then $n - 2$ items remaining to choose for the third spot in the ordering, and so forth. By the product principle, we have a total of $n \cdot (n - 1) \cdot (n - 2) \cdot \cdots \cdot 3 \cdot 2 \cdot 1$ ways to order n items. We abbreviate this as $n!$. *Wait*, you say, *why are you shouting numbers at me?* Oops. Let's have a…

Definition 6.5.5. The notation $n!$ is pronounced n *factorial* and means $n \cdot (n - 1) \cdot (n - 2) \cdot \cdots \cdot 3 \cdot 2 \cdot 1$. (It would be confusing if we pronounced it "En!!" excitedly, tempting though that might be.) We define $0! = 1$ because there is exactly one way to rearrange nothing.

Let us make a quick comparison between permutations of a set and subsets of that same set.

Example 6.5.6. Does a set with n elements have more subsets or more ways of being ordered? Check out the data:

$|S|$	1	2	3	4	5	6	n
subsets	2	4	8	16	32	64	2^n
permutations	1	2	6	24	120	720	$n!$

At first, it looks like the subsets are in the lead. But then suddenly at $n = 4$, the permutations overtake them and go dashing ahead! To see that this trend continues and that permutations will win the race for even moderately large n, notice that when increasing n by 1, the number of subsets is multiplied by 2, whereas the number of permutations is multiplied by $n + 1$.

Check Yourself ─────────────────────────────────

Try to do at least two of these problems.

1. You receive a shipment of 36 legs for stools to go with the stock of mass-manufactured stool seats you already have. How many stools can you complete?

2. Suppose we wanted to place all nine different numbers onto a Sudoku board without reusing rows or columns—how many ways would there be to do it?

3. On the other hand, what if we wanted to place nine 4s onto a Sudoku board without reusing rows or columns? (Again, we will ignore the fact that on a Sudoku board, a player also cannot have two of the same number appear in the same 3×3 block.) How many different ways would there be to make that placement?

4. How many orderings *are* there of $a, b, c, d, e, f, g, h, i, j, k, l, m, n, o, p$?

6.6   Try This! Play with Powers and Permutations

Enjoy these problems; the first three are quite fundamental.

1. Let us return to sugar-numbers for a moment.

   (a) Pull a sugar-number out of the bag of n sugar-numbers. How many ways are there to do this?

(b) Now pull another sugar-number out of the bag and put it next to the first sugar-number. How many ways are there to do this?

(c) Keep pulling sugar-numbers out of the bag until you have a line of k sugar-numbers. How many ways are there to produce this line?

(d) Basically, you have chosen k sugar-numbers from a bag of n sugar-numbers. But your resulting number of ways to do this is not the same as $\binom{n}{k}$. (If you don't believe this, check for a few values of n and k.) What information do you need to take into consideration? Do so.

(e) Rewrite your expression so that it only uses factorials. Now you have a handy formula for computing $\binom{n}{k}$! (No, not "n-choose-k-factorial." This is the interjection sort of exclamation point.)

2. Go back, back into your memory, far back to the days of high school algebra. You will certainly recall that $(x+y)^2$ expands to $x^2 + 2xy + y^2$. Expand $(x+y)^3$ by hand. Wait—please follow this algorithm:

   1. Let $j = 3$.
   2. Expand $(x+y)^j$.
   3. If you see something familiar from earlier in this chapter, go to step 4. If not, replace j with $j+1$ and go to step 2.
   4. Make a conjecture about the expansion of $(x+y)^n$ for any $n \in \mathbb{N}$.

3. Prove the conjecture you made in the previous problem by choosing xs.

4. Can you write your (now proven) conjecture in summation notation?

5. Using some sort of fact you've developed somewhere recently, compute $\sum_{j=0}^{6} 4^j \binom{6}{j}$.

6. Write $\binom{n}{0} - \binom{n}{1} + \binom{n}{2} - \cdots \pm \binom{n}{n}$ in summation notation. Then evaluate it, somehow.

6.7 Binomial Basics

Hey! You! Don't read this unless you have worked through the problems in Section 6.6. I mean it!

A *binomial* is a polynomial with exactly two terms, such as $3a - 2b^5$. Consider the simple binomial $(x+y)$. We can rewrite $(x+y)^n$ as $(x+y) \cdot (x+y) \cdot (x+y) \cdot \cdots \cdot (x+y)$, where there are n copies of $(x+y)$ in that product. If we expand the binomial power $(x+y)^n$ into a polynomial, we know what all of the variable parts of the terms are—they're $x^n, x^{n-1}y, x^{n-2}y^2, \ldots, x^2y^{n-2}, xy^{n-1}, y^n$. But we don't know what the coefficients are. Well, we know they should be called *binomial coefficients* because they are coefficients of a binomial expansion, but that's not yet helpful. Notice that each term has variables whose degrees total to n. (Yes, literally go back a few sentences and notice it actively.) We will have one occurrence of a given term in the full expansion for every way there is of forming it, meaning, for example, that there will be only one copy of x^n because we can only form x^n by multiplying together all n copies of x in $(x+y) \cdot (x+y) \cdot (x+y) \cdot \cdots \cdot (x+y)$. We can form $x^{n-1}y$ by multiplying together all but one of the xs and the remaining y, and we can figure out the number of different $x^{n-1}y$s we can have by counting the number of ways of choosing that single y. Oh, hey! That's $\binom{n}{1}$, and in fact, the number of ways of forming x^n was $\binom{n}{0}$. Similarly, the number of ways to form $x^{n-k}y^k$ is $\binom{n}{k}$ because we choose k copies of y from the n copies of $(x+y)$ (or, equivalently, because we choose $n-k$ copies of x from the n copies of $(x+y)$). Yup.

Conclusion 1: **Another name for $\binom{n}{k}$ is binomial coefficient.**

Conclusion 2: **We have proven a theorem.**

Theorem 6.7.1 (the binomial theorem).

$$(x+y)^n = \sum_{k=0}^{n} \binom{n}{k} x^{n-k}y^k = \binom{n}{0}x^n + \binom{n}{1}x^{n-1}y + \cdots + \binom{n}{n-1}xy^{n-1} + \binom{n}{n}y^n.$$

This theorem has lots of cool consequences. My favorite is this one: the sum across row n of Pascal's Triangle is 2^n. I know, you're like, *Whaaaa?* but it's true. (Well, you won't be so surprised if you finished the last problem in Section 6.3. But this is a different approach, so pay attention in any case.)

Mini-activity (do it right now!):

1. Add up all the numbers in the third row of Pascal's triangle. Do it again with the fifth row and with the sixth row. (Now do you believe that the sum will be 2^n?)

2. Look at the binomial theorem and see if you can use it to prove that the sum across row n of Pascal's triangle is 2^n. Try for at least two minutes.

3. Cover your eyes and wait 30 seconds.

4. Hop up and down on your left foot (if you have one; if not, improvise).

5. Let $x = y = 1$ in the binomial theorem and see what happens.

6. Explain why this gives yet another proof that a finite set of size n has 2^n subsets.

7. Go tell someone you know from high school about these interconnections because it's all so cool.

Another consequence of the binomial theorem is that instead of multiplying out binomials until the end of time, you can be sleek and use Pascal's triangle instead. Here is an example.

Example 6.7.2. We will expand $(3a - 2b^5)^5$ the easy way. First, let $x = 3a$ and let $y = -2b^5$. Then, read off the coefficients of $(x + y)^5$ from Pascal's triangle $(1, 5, 10, 10, 5, 1)$ and slap them in front of the appropriate monomials in the expansion to get $x^5 + 5x^4y + 10x^3y^2 + 10x^2y^3 + 5xy^4 + y^5$. Then plug in for x and y to get $(3a)^5 + 5(3a)^4(-2b^5) + 10(3a)^3(-2b^5)^2 + 10(3a)^2(-2b^5)^3 + 5(3a)(-2b^5)^4 + (-2b^5)^5$. Finally, expand out the coefficients to get $3^5a^5 - (10 \cdot 3^4)a^4b^5 + (10 \cdot 27 \cdot 4)a^3b^{10} - 720a^2b^{15} + (15 \cdot 16)ab^{20} - 32b^{25}$. To do this the *easiest* way, first be confident and then skip some of the intermediate steps.

You may have seen the following basic fact about binomial coefficients before; people often learn it as a definition, when in fact it's a computational tool that needs to be justified.

A formula for computing binomial coefficients. $\quad \dbinom{n}{k} = \dfrac{n!}{k!(n-k)!}.$

Problem 1 in Section 6.6 leads to one proof of this fact, and Problem 25 in Section 6.15 gives another. It is one of the three most popular ways of verifying *binomial identities* (equations involving binomial coefficients); the other two are induction (See? It's everywhere!) and combinatorial proof.

Check Yourself

Make sure you understand how to do binomial computations by attempting these problems.

1. Find the coefficient of x^5y^5 in $(x + y)^{10}$.

2. Find the coefficient of the monomial containing c^3 in $(5b^2 - 4c)^4$.

3. Compute, by hand, a numerical value for $\binom{36}{32}$.

4. **Challenge:** Create a binomial $(x + y)^n$ (with n greater than one) with neither x nor y a constant, such that when expanded it will have a constant term.

6.8   Combinatorial Proof

Here is the long- and multiply-promised introduction to combinatorial proof. (After all the buildup earlier in the chapter, this may seem anticlimactic.) Because a combinatorial proof proceeds by determining the size of one set in two different ways, it is sometimes also called *bijective proof*. A bijection is formed from the one-to-one correspondence between the two descriptions of the set. Essentially, combinatorial proof works by looking at a set from multiple perspectives and gaining new information from each perspective.

We have already seen three combinatorial proofs in this chapter:

- 🦆 In the proof that $\binom{n}{k} = \binom{n-1}{k-1} + \binom{n-1}{k}$ in Section 6.2, we viewed the set of ways to choose k objects from n objects also as the union of two disjoint subsets; then, we figured out how to express the sizes of those subsets in new ways.

- 🦆 The proof that $\binom{n}{k} = \binom{n}{n-k}$ in Sections 6.3 and 6.4 counted the number of ways to choose k objects from n objects and the number of ways to choose all but k objects (that is, to choose $n - k$ objects) from a set of n objects. Handily, these are the same number.

- 🦆 When we proved a special case of the handshaking lemma (for regular graphs) in Example 6.5.2, we double-counted the number of edges twice, once by counting the total number of ends of edges and once by counting the edges touching each vertex.

Notice that in each case, we began with an equation. We then interpreted each side of the equation in a different way. The challenge in combinatorial proof is figuring out a good way to interpret a mathematical expression! We will discuss this in the context of Example 6.8.1. Our goal will be to decide what we are counting if one side of the equation is the number of ways some choosing can happen and then answer the question, "Why does the other side of the equation count the same thing (but in a different way)?"

Example 6.8.1. Consider the identity $\sum_{k=0}^{n} \binom{n}{k}\binom{3n}{n-k} = \binom{4n}{n}$. We will prove its validity by interpreting each side of the equation as a different way of counting *something*. The right-hand side of the equation is somewhat straightforward; $\binom{4n}{n}$ is by definition the number of ways to choose n objects from a set of $4n$ objects, so that will be our *something* in this proof. In keeping with our previous themes, we will imagine that we have a bag of $4n$ delicious Ahiru hard candy drops. On the left-hand side of the equation appear choice notations with n and $3n$ in the top slots.

These total to $4n$, so perhaps we can partition the bag of Ahiru drops into two flavors of drops. Let us say that our bag contains n chrysanthemum drops and $3n$ lemon drops. Now we will reinterpret the equation. On the right-hand side of the equation, we are choosing any n candies from the bag, regardless of flavor. On the left-hand side, things are a bit more complicated. Imagine that within the bag we have separated the candies into two piles, one for the chrysanthemum drops and one for the lemon drops. Then we choose k drops from the chrysanthemum pile and the remaining $n - k$ drops from the lemon pile. However, the number of ways we can do that depends on what k is, and k could take any value from 0 to n. Therefore, we need to add up all the possibilities. If we only take lemon drops, then there are $\binom{3n}{n-0}$ ways to choose them; because $\binom{n}{0} = 1$, this is the same as $\binom{n}{0}\binom{3n}{n-0}$. If we take one chrysanthemum drop and $n - 1$ lemon drops, there are $\binom{n}{1}$ ways to grab the chrysanthemum drop, and for each of those possibilities, there are $\binom{3n}{n-1}$ ways to choose the lemon drops, so in total we have $\binom{n}{1}\binom{3n}{n-1}$ ways to grab a chrysanthemum drop and $n - 1$ lemon drops. Analogously, we have $\binom{n}{k}\binom{3n}{n-k}$ ways to choose k chrysanthemum drops and $n - k$ lemon drops. Summing, we obtain $\sum_{k=0}^{n} \binom{n}{k}\binom{3n}{n-k}$. This completes our combinatorial proof.

Check Yourself

These may seem challenging but are useful practice before attempting full combinatorial proofs.

1. Let $m \le n$. What might $\binom{n}{3}\binom{n}{m-3}$ be counting?

2. What might $\binom{n}{6}\binom{6}{k}$ be counting?

3. What might $\binom{n}{k}2^k$ be counting?

4. **Challenge:** Create a situation similar to the lemon-and-chrysanthemum-drops situation, and use this to write down a new binomial identity.

6.9   Try This! Pancakes and Proofs

You will especially want to collaborate on the last three of these problems.

1. Check out the octahedron in Figure 6.5.

   (a) How many vertices does it have?

   (b) How many edges are there touching each vertex?

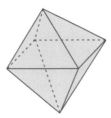

Figure 6.5. I am an octahedron because I am solid and have eight faces.

    (c) Using your previous two answers, compute how many edges there are, total.

2. At the Local Order-place of Pancakes (LOOP), all you can do is order pancakes in stacks of three. In fact, you must specify which pancake is on top, which is in the middle, and which is on the bottom. The waitstaff refer to this as "stack order." There are six pancake flavors (blueberry, chocolate chip, apple cinnamon, pecan, buttermilk, and banana).

    (a) How many possible pancake orders are there (specifying flavor and stack order)?

    (b) How many different pancake orders involve three different flavors of pancakes?

    (c) How many different pancake orders involve one banana, one chocolate chip, and one pecan pancake?

    (d) How many three-different-flavor pancake orders provide subtly different nutrition? (How many are nutritively the same?)

3. Recall the complete graph K_n. (See page 77 for images in Figure 3.7.)

    (a) How many edges does K_n have? Suggestion: use overcounting or use induction.

    (b) What property holds for any pair of vertices in K_n?

    (c) Use your previous two answers to construct a combinatorial proof of a formula for $\binom{n}{2}$.

4. Use a combinatorial proof to show that $\binom{w}{p}\binom{p}{m} = \binom{w}{m}\binom{w-m}{p-m}$.

5. Use a combinatorial proof to show that $\displaystyle\sum_{k=0}^{n}\left(\binom{n}{k}\right)^{2}=\binom{2n}{n}$.

6. Use a combinatorial proof to show that $\displaystyle\binom{n+m}{r}=\sum_{i=0}^{r}\binom{n}{i}\binom{m}{r-i}$.

6.10 Where to Go from Here

If you haven't looked at Section 4 of the student preface, on problem-solving prompts, lately, this would be a good time to do so; you've just started working on serious discrete mathematics problems, and there are more to come.

Combinatorics is used extensively in computer science: all counting techniques are important for computing the efficiency of algorithms. Section 6.12 gives an elementary example of such a computation that involves binomial coefficients.

Binomial coefficients are used throughout combinatorics; we will need them in the rest of the book, particularly in Chapters 7 and 9. To learn more, see Chapter 5 of [10]. (It has lots and lots of binomial identities—in fact, it has a top ten list of them!) This is in fact an excellent book for self-studying combinatorics more generally. Reference [4] is more advanced, but quite well written and gives introductions to many combinatorial subfields in which research is active. While they are not a central focus of combinatorics research, there are still open questions about binomial coefficients; for example, no one knows the prime factorization of $\binom{2n}{n}$.

To learn more about combinatorial proofs, read *Proofs That Really Count: The Art of Combinatorial Proof* by Art Benjamin and Jennifer Quinn. Combinatorial proof is highly desired as it usually produces the most elegant explanations, so it is regularly used in research papers.

Credit where credit is due: The uglifruit tip in Example 6.5.4 came from Prof. George Piranian (1914–2009) of the University of Michigan. An emeritus professor at the time, he regularly appeared (seemingly from nowhere) to give the author luncheon advice when she was in graduate school. Section 6.12 was inspired by Section 17.2.1 in [4]. Section 6.13 was inspired by Section 5 of Chapter 3 in [1]. Some problems in Sections 6.3 and 6.6 were inspired by [3]. Example 6.5.6 was inspired by [15] and [16]. In Section 6.15, Problem 33 was inspired by Sam Oshins's character who was inspired in turn by this book (see page 148), and Problem 13 arose from the excellent survival ballroom dance class the author took from Dr. Kathleen Kerr many years ago. Everyone should learn both leading and following!

6.11   Chapter 6 Definitions

choice number: A number of the form $\binom{n}{k}$ (that is, n choose k).

combinatorial proof: A characteristic approach to solving combinatorics problems by counting some quantity in two different ways.   Combinatorial proof works by looking at a set from multiple perspectives and gaining new information from each perspective.

Pascal's triangle: A triangular array of numbers in which the kth entry in the nth row is $\binom{n}{k}$. The top row is a single "1" and this is counted as the 0th entry of the 0th row.

permutation: An ordering of items.

factorial: An operation on a natural number n, denoted $n!$, that returns $n \cdot (n-1) \cdot (n-2) \cdot \cdots \cdot 3 \cdot 2 \cdot 1$.

binomial: A polynomial with exactly two terms, such as $2x - y$.

binomial coefficient: A coefficient in a binomial expansion, such as the coefficient "2" in $w^2 + 2wr + r = (w+r)^2$.

binomial identity: An equation involving binomial coefficients.

bijective proof: A proof that shows the equality of the sizes of two sets by demonstrating a one-to-one correspondence between two descriptions of the sets.

6.12   Bonus: Sorting Bubbles in Order of Size

Imagine that we have a bunch of bubbles, all of different sizes. (We can measure size by the radius or the diameter of the bubbles.) We would like to order the bubbles from smallest to largest. Here is one way to do it:

1. Look around for the smallest bubble. Put it at the start of a line. (Okay, yes, in reality it might pop, but assume the bubbles are made of something like bubblegum so they can be moved around.)

2. Find the next-smallest bubble. Put it next in the line.

3. Repeat until all the bubbles are in a line.

   That's problematic, though. It seems like an algorithm, but it's not something a computer could deal with. And worse, there's all kinds of steps encapsulated in the command, "Find the smallest bubble." How exactly do we do that? (Isn't that kind of equivalent to the original problem?) Let's try again.

1. Pick up the first two bubbles. Put the smaller one to the left of the larger one.

2. Pick up another bubble. If it is larger than the last bubble in line, put it to the right of that bubble. If it is smaller than the last bubble in line, put it to the left of that bubble.

3. Repeat until all the bubbles are in a line. But wait! They might not be in order!

4. Go back to the left end of the line. Compare the first two bubbles in line. Put the smaller one to the left of the larger one.

5. Check the next bubble in line against the previous bubble. If the smaller isn't to the left of the larger, switch them.

6. Repeat until the end of the line. But wait! The bubbles *still* might not be in order!

7. If any switches were made along the line, go back to the left end of the line and do comparisons again. If no switches were made along the line, be done because the bubbles are in order according to size.

Hey, we might as well assume that the bubbles are in a line to begin with. After all, our arbitrary picking up of bubbles creates an order to the bubbles, which is like lining them up. That simplifies the algorithm a little bit.

1. Start at the left of the bubble line. Compare the first two bubbles. If the larger is to the left of the smaller, switch them.

2. Move one bubble down the line. Compare it to the previous bubble. If the larger is to the left of the smaller, switch them.

3. Repeat until the end of the line. If any switches were made along the line, go back to the left end of the line and do comparisons again. If no switches were made along the line, be done because the bubbles are in order according to size.

This algorithm is commonly called *bubble sort*. (No, really. It is.) To see an example in action, check out https://www.cs.usfca.edu/~galles/visualization/ComparisonSort.html (and choose Bubble Sort from the menu). Bubble sort is not used in practice because it is not very efficient, but encoding it is a standard computer science exercise—if you are a student of computer science, you might try writing it out in pseudocode.

What we would like to do is figure out how many switches we have to make to order the bubbles, *in a worst-case scenario*—depending on the initial order of the bubbles, we might go through the line many times or just a few times. This is one measure of the efficiency of the algorithm. (For more on algorithmic efficiency, see Chapter 17.)

Notice first that no matter where the largest bubble starts, when we finish the first pass through the line the largest bubble will be on the right end of the line. At worst, the largest bubble is at the start of the line, so with n bubbles, we will have to do $n - 1$ switches.

But hey! If we know the largest bubble ends up on the right, we don't have to look at it again on subsequent passes through the line! So that means that on the next pass through, if the second-largest bubble is at the start of the line,

1. we will have to do $n - 2$ switches, and

2. it will end up in the second position from the right.

Hmm. Does anything seem familiar here? There are two things that might come to mind:

1. Each time we go down the line, in the worst-case scenario we will have to do one less switch.

2. Each time we go down the line, we're sort of reducing the problem to the worst-case scenario for a shorter line.

Hmmmmmmm.

Let's leave that line of thinking for now and just say that this is related to induction (see Chapter 4) and recursion (see Chapter 8).

Okay, so in a worst-case scenario, we do

$n - 1$ switches the first time through the line,

$n - 2$ switches the second time through the line,

$n - 3$ switches the third time through the line,

$n - 4$ switches the fourth time through the line, $\ldots$

$n - n$ switches the nth time through the line $\ldots$

Hey wait, if we make zero switches, we're not truly going through the line. So that last line should read

one switch the $(n - 1)$st time through the line.

So our total number of switches is

$$(n-1)+(n-2)+\cdots+1 = 1+\cdots+(n-2)+(n-1) = \sum_{i=1}^{n-1} i = \frac{n(n-1)}{2} = \binom{n}{2}.$$

Huh! A binomial coefficient pops up after all of that!

> Activities:
>
> 1. Perform the bubble sort algorithm on a line of bubbles with sizes 2.3, 6.4, 1.2, 7.9, 5.5, 3.4, 2.9, 5.7.
>
> 2. Given bubbles of sizes 1 through 9, place them in an order that would take the bubble sort algorithm the largest number of switches to complete.
>
> 3. Write pseudocode for the bubble sort algorithm.
>
> 4. Implement the bubble sort algorithm in your favorite programming language.

6.13   Bonus 2: Mastermind

Once upon a time, there was a two-player game called *Mastermind.* It was popular in the 1970s and 1980s; one can still obtain a physical copy but these days few people play it with pieces rather than on phones or tablets. One end of the board has a bin of pegs in six possible colors (red, yellow, green, blue, black, and white). The board has a hidden area at the other end, behind which one player places four colored pegs in a row; color repetition is allowed. The bulk of the board consists of rows of four holes into which the second player places a guess sequence of colored pegs. The goal of the game for the second player is to discover the hidden peg sequence before running out of rows. Most mathematically inclined second players can win, so truly the goal is to guess the sequence in the smallest number of moves possible, and usually the players agree that a win will only count if it is achieved in eight or fewer guesses. On one side of each row are four tiny holes into which tiny marker pegs are placed. After the second player places guess pegs into a row, the first player places one black marker peg per correctly colored-and-placed guess peg and one white marker peg per correctly colored-but-not-placed guess peg. Modern (computer) versions of the game have options to use more colors (often eight or nine) and/or more guess peg holes (usually five).

For our discussion, we will call the six guess peg colors **R**, **Y**, **G**, **B**, **K**, and **W** and the two marker peg colors **k** and **w**. With this nomenclature, we give an example.

Example 6.13.1. Suppose you and I are playing. I hide the peg sequence **RYKR**. You guess **GGRR**, so I respond **kw** because one of your reds is in a correct position and one is in an incorrect position, but neither of your greens appears in my sequence. You then guess **BBYY**, and I respond **w** because neither of your yellows is in a correct position.

For maximal enjoyment of this topic, you may wish to play a few games of *Mastermind* on a computer or mobile device in order to familiarize yourself with the thinking of a mastermind. Note how many guesses it takes each time so you can compare with our later results.

Now for some mathematics. If the second player seriously wants to win, ze needs to definitively determine what the peg sequence is. There is some huge number of possibilities for the possible guess sequences, and ze needs to narrow this huge number down to one so that ze is certain of hir final answer. How large is that huge starting number?

Pause. Try to think of how large this number will be. Don't read further until you've thought for at least 30 seconds. (Maybe close the book so you're not tempted to look further for the answer.)

Solution. Let's use the product principle. There are four pegs in the sequence, and each can be one of six possible colors, so we see that there are $6 \cdot 6 \cdot 6 \cdot 6 = 1,296$ possible sequences for the first player to choose. How will we get this cut down to one?

6.13.1   One Strategy for Playing

We will focus on the strategy employed by the second player, as this is where binomial coefficients and permutations and algorithms arise. Usually when one plays games, one employs intuitive strategies of what will help one win. However, in games such as *Mastermind* where one's number of moves is limited, developing efficient strategies and following them (sometimes disregarding intuition) is helpful. (This is akin to basing policy decisions or medical diagnoses on the results of studies rather than on gut feelings or anecdotal evidence. As a mathematics student, you should learn how to act in accordance with logical thinking, even if you do not always choose to act this way.) We will attempt to use mathematics to develop a good strategy for *Mastermind*, and to make sure it is an easily usable strategy, we will write it as an algorithm.

A naive approach is to start with six guesses: **RRRR, YYYY, GGGG, BBBB, KKKK,** and **WWWW.** The only responses player 1 can give are black markers (why?), and between the six guesses there will be exactly four black markers (why?), so these guesses tell us which colors are present in the hidden peg sequence and how many times each occurs.

Pause. If we know which colors are present in the hidden peg sequence and how many times each occurs, what is the largest number of possibilities left for the sequence? Don't read further until you've thought for at least 30 seconds. (Maybe close the book so you're not tempted to look further for the answer.)

Solution. This time we will use permutations. The most possibilities are present when the four colors are different. Those four colors might have been placed in any order, so there are 4! = 24 different possibilities that remain. We have used six guesses to cut down the number of possibilities from 1,296 to 24, or by a factor of 54. That's pretty good bang for the buck.

If by chance we have a repeated color, the number of possibilities remaining is even smaller. There are lots of cases here, represented without loss of generality using examples from **R, Y,** and **G**: we could have something like **RRYG** or **RRYY** or **RRRY** or **RRRR.** We also know which type of hidden peg sequence we have from our original six guesses. Let's figure out how many possibilities there are if our original six guesses reveal that we have two **R**s, one **Y**, and one **G**.

Pause. Write out all possible peg sequences using two **R**s, one **Y**, and one **G**. Do you see any patterns that would allow you to count these efficiently instead of by brute force? Think on this for at least 30 seconds.

Solution. We'll approach this in two ways. First, let's pretend the two reds are different by labeling them **Ra** and **Ro**. Then we have 24 different orderings of **RaRoYG**; however, **Ra** is the same color as **Ro** so we counted each possibility twice (e.g., **RaRoYG** and **RoRaYG** both correspond to **RRYG**). Thus, we need to divide by 2 and we obtain 12 possibilities.

On the other hand, we could think of the holes into which the pegs are placed. Two of those holes will be filled with **R**s, and there are $\binom{4}{2}$ ways to place them. This leaves two blank holes that could be filled with **Y** on the left or **G** on the left (so two possibilities), and the product principle says we get a total of $\binom{4}{2} \cdot 2 = 6 \cdot 2 = 12$ possibilities.

We will leave the counting of the possibilities of the other cases as an exercise. Still, the way we treat any one of the cases (including the one with four different colors) can be seen as an example of an overall algorithm:

1. Let the set of possible hidden peg sequences under consideration be called H.

2. Begin with all possibilities, so that $|H| = 1{,}296$.

3. Guess one of the possibilities in H and get feedback from the other player.

4. Using this knowledge, revise H by removing all hidden peg sequences that are not possible sequences for this game.

5. If $|H| \neq 1$, go to step 3.

Our goal is to create a version of this algorithm that loops through step 3 the fewest times possible.

6.13.2   Mini-Project

1. As a warm-up, use the following guesses and feedback information to determine the hidden peg sequence.

Guess	Response
RYRY	*none*
GBGB	*none*
KWKW	**bbb**
KKWK	**ww**

2. Let's make sure we're fully warm—determine the hidden peg sequence from the following guesses and feedback.

Guess	Response
KKWW	*none*
KBKB	*none*
GGYY	**bb**
GRGR	**bw**
RYRY	**bbww**

3. Why is there no hidden peg sequence that could correspond to these guesses and feedback? Explain.

Guess	Response
GKGK	w
WYWY	b
BRBR	w
GWBR	bwww

4. Same here: why is there no hidden peg sequence that could correspond to these guesses and feedback? Explain.

Guess	Response
RYGG	bbww
GYRG	bwww
YRGG	wwww

5. How many different feedback sequences could player 1 give? Explain.

6. This exercise was promised earlier in this section: How many hidden peg sequences use…

   (a) … two **R**s and two **Y**s?

   (b) … three **R**s and one **Y**?

   (c) … four **R**s?

   Try to count efficiently.

7. Now we will start to develop a reasonable algorithm for beating *Mastermind*. Above, it was suggested that the six guesses **RRRR**, **YYYY**, **GGGG**, **BBBB**, **KKKK**, and **WWWW** would tell us which colors are present in the hidden peg sequence and how many times each occurs. Suppose we only use the first five of the guesses. How does that change the information we have?

8. After determining which colors are present in the hidden peg sequence, we know that there are at least two colors not present (because there are six possible colors and only four positions for pegs). Let *C* be a color that is present in the sequence, and let *N* be one of the colors that is not present. What information is gained from the guesses *CNNN*, *NCNN*, *NNCN*, and *NNNC*?

9. Using the ideas developed in the previous two questions, write an algorithm that takes no more than 11 guesses.

Still, we should be able to do better. If you've played *Mastermind* several times, you know that you often do better than 11 guesses without using mathematics. There is an excellent algorithm that uses no more than *five* guesses, introduced by Donald Knuth. Sadly, it is not quick to develop or explain, as Knuth generated it using a computer. However, the algorithm is of the same type as the overall algorithm given above, and the paper in which Knuth published his algorithm and proof that it works is relatively easy to read. The reference is Donald E. Knuth, "The Computer as Master Mind," *J. Recreational Mathematics*, Vol. 9, No. 1 (1976–77), pp. 1–6, and it can be downloaded from http://colorcode.laebisch.com/links/Donald.E.Knuth.pdf (at least as of April 2018).

6.14   Bonus Check-Yourself Problems

Solutions to these problems appear starting on page 606. Those solutions that model a formal write-up (such as one might hand in for homework) are to Problems 9 and 10.

1. Find a combinatorial proof for the identity $\sum_{k=0}^{n} k\binom{n}{k} = n2^{n-1}$.

2. Show that if n is even and k is odd, then $\binom{n}{k}$ is even.

3. Evaluate $\sum_{r=0}^{2m} 3^r 2^{2m-r} \binom{2m}{r}$.

4. The four students Ariel, Bingwen, Clarissa, and Dwayne have albums they need to listen to for a music appreciation class: *Duck Rock* (by Malcolm McLaren), *Duck Stab* (by The Residents), *Quack* (by Duck Sauce), and *This Time* (by Galapagos Duck).

   (a) How many ways are there to match the students with the albums?

   (b) The library has two listening rooms, each of which has two listening stations. How many ways are there to pair the students in the rooms?

   (c) Suppose the students have to sign up in advance, so they have to specify which listening station each student is using. Now how many ways are there for the students to be distributed into the rooms?

5. Give a combinatorial proof that $\binom{n}{4} = \frac{n!}{4!(n-4)!} = \frac{n(n-1)(n-2)(n-3)}{24}$.

6. At the art museum, you are decorating a round spinny top with stickers. However, this is an anti-creative art museum, so there are only four equally spaced spots on the spinny top that are designated for receiving stickers, and there are only two colors of sticker available—gray and grey. How many ways are there to "decorate" the spinny top? (There are quotation marks because it is hard to envision the spinny top as actually being decorated....)

7. There are 18 students gathering to work on making a campus duck pond. They need to work in groups of three on various tasks. How many ways are there for the students to form groups?

8. Conjecture and prove a binomial identity for $\sum_{i=0}^{n} \binom{i}{5}$.

9. Find the coefficient of $x^4 y^6$ in $(5x^2 - 3y^3)^4$.

10. Prove that $\binom{2n}{2} = 2\binom{n}{2} + n^2$.

6.15 Problems Binomially Combinatorial in Nature

1. These are super-quickies:

    (a) How many ways can you choose one insect from a collection of 5 butterflies and 11 ants?

    (b) How many ways can you choose one flower from a garden of two roses, six lilies, and eight birds of paradise?

    (c) How many ways can you choose two pieces of fruit from a bowl containing one persimmon and four mangoes?

    (d) How many ways can you choose nine spiders from ten spiders?

2. In Section 6.2, we proved that $\binom{n}{k} = \binom{n-1}{k} + \binom{n-1}{k-1}$. Adapt this proof to use any particular sugar-number (dye it pink) instead of the sugar-number n.

3. When a new session of Congress is in session, at the first formal meeting of the House Ways and Means Committee, every pair of Representatives must shake hands. (Indeed, that's made up.) The committee has 41 members. (That, on the other hand, is *not* made up.) Give two different expressions (with explanation) for the number of handshakes that occur at the first formal meeting.

4. Using the computational tool that $\binom{n}{k} = \frac{n!}{k!(n-k)!}$, find formulas for $\binom{n}{0}, \binom{n}{1}, \binom{n}{2}$, and $\binom{n}{3}$. Give combinatorial proofs of the first two of these formulas.

5. Prove that if p is prime, then $p \mid \binom{p}{j}$ for $0 < j < p$.

6. Prove that for $n \geq 2$, $2\binom{n}{2} + \binom{n}{1} = n^2$.

7. Use a combinatorial proof to show that $\binom{2n}{n}$ is even for $n > 0$.

8. Let n be even. What is the relationship between the number of subsets of $\{1, 2, \ldots, n\}$ of odd size and the number of subsets of $\{1, 2, \ldots, n\}$ of even size? Explain. Problem 5 of Section 6.3 may be of assistance.

9. Compute $\sum_{j=0}^{m} 3^j \binom{m}{j}$. The result should be somewhat familiar. Then, use the binomial theorem to verify the result.

10. The Massachusetts Megabucks Doubler lottery game lets you pay $1 in exchange for the privilege of filling out a ticket with six different numbers that range from 1 to 49. How many ways are there to fill out the ticket?

11. Compute $\sum_{k=0}^{n} \frac{1}{k+1}\binom{n}{k}$ for $n = 3, 4, 6, 8$. (A computer algebra system can make quick work of this task.) Using this information, conjecture a formula for this sum.

12. **Challenge:** Use induction to prove that $\sum_{k=0}^{n} k\binom{n}{k} = n2^{n-1}$.

13. In a ballroom dance class, participants are divided into couples for each drill session. One partner leads and the other follows for three minutes, and then the couple switches roles for the next three minutes.

    (a) Only four people show up on time. How many ways are there to pair them up?

    (b) If instead six people show up on time, how many ways are there to pair them up?

    (c) Assume all m people in the class arrive on time. (There are an even number of people in the class.) How many ways are there to pair them up?

14. Consider the previous problem, this time assuming that we specify which member of each couple leads first. How many ways are there to pair-and-specify the dancers?

15. Figure 6.6 shows a polyhedron with 60 vertices. (Note that at each vertex, two squares meet one triangle and one pentagon.) How many edges does it have? Explain, perhaps by overcounting carefully.

Figure 6.6. I am a Small Rhombicosidodecahedron. (Yes, "Small" is officially my first name. One of my relatives has the first name "Large.")

16. There is a very quick way to compute $\sum_{\ell=0}^{q} 2^{\ell}(-1)^{q-\ell}\binom{q}{\ell}$; do use it.

17. Find a combinatorial proof that $n\binom{n-1}{k} = (k+1)\binom{n}{k+1}$.

18. Find a combinatorial proof that $n\binom{n-1}{k-1} = k\binom{n}{k}$.

19. Find a combinatorial proof that $n(n-1)\binom{n-2}{k-2} = k(k-1)\binom{n}{k}$.

20. Calculate $\sum_{k=0}^{n}\binom{n}{2k}$ for $n = 2,3,4$. Use these data to find an identity of the form $\sum_{k=0}^{n}\binom{n}{2k} = ???$. Now prove that your identity holds.

21. Compute $\sum_{r=0}^{s} 64^{r}(-64)^{s-r}\binom{s}{r}$.

22. Show that $\sum_{i=1}^{n}\binom{i}{2} = \binom{n+1}{3}$.

23. Use Problems 6 and 22 to develop (and prove) an identity for $\sum_{j=1}^{m} j^2$. Be careful about the lower index for your sum.

24. **Challenge:** In Section 3.4, we counted the number of functions from an m-element set to an n-element set (and obtained n^m) and the number of injections from an m-element set to an n-element set with $m \leq n$ (and obtained $n \cdot (n-1) \cdots \cdot (n-(m-1))$, or $\binom{n}{m}m!$). Suppose we count *surjections* from an m-element set to an n-element set. How many are there?

25. Find a more direct combinatorial way to show that $\binom{n}{k} = \frac{n!}{k!(n-k)!}$ than in Problem 1 of Section 6.6. Here are some questions to guide you in your search.

    (a) First, we lay our n sugar-numbers out in a line, in some order (probably with no discernable pattern). How many ways are there to do this?

    (b) Separate the first k of the sugar-numbers from the remaining $n - k$. In how many ways could the first k sugar-numbers have been ordered?

    (c) In how many ways could the remaining $n - k$ sugar-numbers have been ordered?

    (d) Assemble the information you have gathered so far to produce a new explanation for how to compute $\binom{n}{k}$.

26. Revisit the octahedron from Figure 6.5.

    (a) Using the number of vertices, and the relationship between vertices and faces, count the number of faces.

(b) Using the number of edges, and the relationship between edges and faces, count the number of faces.

27. Let us construct a new Pascal-ish triangle. We start with a 1 at the top, and then in each successive row, each entry is the sum of the number diagonally left above and *twice* the number diagonally right above.

(a) Write out the first 5 rows of this Pascal-ish triangle.

(b) What is the sum of the nth row?

(c) Let the kth number in the nth row be denoted by $[n \triangle k]$. Write a binomial-like theorem for the Pascal-ish triangle, of the form $(stuff)^n = \sum_{k=0}^{n} [n \triangle k] x^{n-k} y^k$.

(d) Now use the binomial theorem to get a new expression of the form $(stuff)^n = \sum_{k=0}^{n} \binom{n}{k} things$. What formula do you now have for $[n \triangle k]$?

(e) **Challenge:** We know that $\binom{n}{k}$ counts the number of subsets of $\{1, \ldots, n\}$ of size k. Find a characterization for $[n \triangle k]$ in terms of subsets of $\{1, \ldots, n\}$.

28. Find the coefficient of $x^5 y^3$ in $(12x - 4y^3)^6$.

29. Return to the 60-vertex polyhedron in Figure 6.6, where at each vertex, two squares meet one triangle and one pentagon. How many faces of each size, and in total, does the polyhedron have?

30. The garden has room for eight rows of plants. Your seed bin contains 12 vegetable-seed packets and six flower-seed packets.

(a) How many ways are there to choose four kinds of vegetables and four kinds of flowers to plant?

(b) You decide you'd rather have six rows of vegetables bordered by two rows of flowers. Now how many ways are there to choose the seeds to go in your garden?

(c) In each of the situations (a) and (b), how many different ways are there to plant your garden?

31. Your duck-loving young cousin has pulled the books *Hey, Duck!*, *Just a Duck?*, and *Sleepover Duck!* by Carin Bramsen, *Giggle, Giggle, Quack* and *Duck for President* by Doreen Cronin, *Duck, Death and the Tulip* by Wolf Erlbruch, and *Duck on a Bike* by David Shannon off a shelf. How many ways are there to put four of the books back on the shelf?

32. Evaluate $\sum_{a=0}^{c} (-4)^a \binom{c}{a}$.

33. Intrepid spy Pvaanzba Ohaf looks under a door and sees 21 human foot-containing shoes. How many people are in the room?

34. **Challenge:** You are given a long sum $\cdots + 495 b^{18} + 4{,}455 b^{16} + \cdots$ and know this is a binomial expansion. What are the values of x, y, and n?

35. A hungry person needs a snack. This could be two mini granola bars (there are four different flavors and no one wants two of the same flavor for a snack) or a cracker pack (there are seven kinds of these) or a packet of M&M's (milk chocolate, dark chocolate, peanut, peanut butter, mint). How many ways can a snack be selected?

6.16   Instructor Notes

This is a very problem-oriented chapter; there are enough problems between Sections 6.3, 6.6, and 6.9 to use an entire week (or more!) of class time. As preparation for the first class meeting, assign students to read Sections 6.1 and 6.2 and attempt the Check Yourself problems. At the first class meeting, my suggestion is to warm up the students by reminding them of the definition of $\binom{n}{k}$ and computing a couple of small-value binomial coefficients, such as $\binom{4}{2}$ and $\binom{5}{4}$. Then have the students work on Section 6.3. For the second class meeting, have students read Sections 6.4 and 6.5 and attempt the Check Yourself problems. At that meeting, ask whether they have questions over the reading and then have the students work on Section 6.6. For the third class meeting, have students read Sections 6.7 and 6.8 and attempt the Check Yourself problems. Start the last class meeting of the week by reminding students that because $1 + 1 = 2$, the binomial theorem shows that the sum of the entries in the nth row of Pascal's triangle is 2^n. Then have the students work on Section 6.9. They are likely to find the combinatorial proof problems quite difficult; the goal is to have them start to understand combinatorial proof, not to master it. Be sure to leave time for discussion each day so the students achieve some closure on the main concepts. If some students whip through these problems, a nicely challenging supplement that is suitable for groupwork would be Problems 13 and 14 of Section 6.15.

Chapter 7 🦆🦆🦆🦆🦆🦆🦆

Balls and Boxes and PIE: Counting Techniques

7.1   Introduction and Summary

In this chapter, we will consider more advanced counting questions and techniques. At the start, we will examine a range of combinatorial questions commonly encountered and reframe all of these in the language of distributing balls into boxes. Later in the chapter, we will try to figure out how many ways there are to distribute balls into boxes. That sounds simple, but there are lots of variations: Are the balls identical, or are they labeled? What about the labeling of the boxes? Does it matter whether any box is empty? It turns out that addressing some of the possible scenarios is quite difficult and well beyond this text! We will focus on balls-and-boxes problems that can be answered using binomial coefficients or permutations in interesting ways. We will also work with the principle of inclusion-exclusion, or PIE for short. Back in the old days of Chapter 1, we used the sum principle to determine the size of a collection of disjoint sets; PIE reveals how to determine the size of a collection of sets that are *not* disjoint. One application of PIE is to Venn diagrams and how to figure out the number of objects represented by one of the regions of a Venn diagram given sufficient information about the other regions.

7.2   Combinatorial Problem Types

The most difficult part of solving a counting problem is determining what type of problem it is. There are certain words, phrases, and ideas commonly used in counting problems that can help determine the problem type, so we will outline those here. We begin with an example.

Example 7.2.1. Consider the question, "How many different seven-digit telephone numbers are there?" We may think of each digit of the phone number as a slot or

Figure 7.1. A line of empty, labeled boxes.

box into which we must place a number. The order of the digits in the telephone number does matter (as you've surely noticed if you've accidentally reversed a pair of digits while dialing), and so these boxes are distinguishable, or labeled. We may also think of the possible one-digit numbers as balls labeled with those ten different one-digit numbers. We have a large number of balls with each label, and we must place one labeled ball into each labeled box. Thus, we have now converted the question into, "How many ways are there to place one labeled ball, with ten possible labels, into each of seven labeled boxes?"

Notice that we did not discuss how to *solve* the problem in Example 7.2.1. In this section, we are primarily concerned with identifying which problem type we have encountered. Once the problem type has been identified, the solution becomes straightforward (after you have understood that solution type! These are detailed in Section 7.4). We will translate all problems into the same framework so that we have a single typing system; that universal framework is balls and boxes.

We will only consider balls-and-boxes problems that involve *labeled* boxes because unlabeled-box problems lead to concepts beyond this course (such as integer partitions and Stirling numbers). This also simplifies matters because every problem can be conceived as beginning with a row of empty boxes in a line (see Figure 7.1). After you see our list of problem types (below), you will surely think of other related problem types that could conceivably be encountered. Rest assured that those problems have solutions you don't yet want to confront.

> Question A. *How many ways are there to place k differently labeled balls, at most one per box, into n labeled boxes?*
>
> Notice that k must be less than or equal to n for this to be possible. See Figure 7.2 for a visualization of this question. The fact that the balls are labeled means that it matters which ball is placed in a given box. That is, the order in which they appear matters. This is sometimes abbreviated as *order matters* or *ordered*. The restriction that at most one ball may be placed in any box means that each ball label will only appear once. In other words, the ball labels do not repeat; this is referred to as a problem *without repetition*.

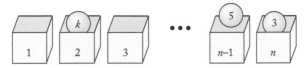

Figure 7.2. Shown are k differently labeled balls, placed (at most one per box) into n labeled boxes.

Example 7.2.2. In your extended family (nephews, nieces, cousins, …), there are 12 children between the ages of four and ten. You have knitted six differently colored pairs of mittens. How many ways are there to give out the mittens at the wintertime family reunion? Because the pairs of mittens are different colors, they are not identical, and therefore, it matters which child receives which pair of mittens. Additionally, the children are all different. The problem does not explicitly say that no child can receive more than one pair of mittens, but it would be reasonable to interpret the problem in this way (as there are not enough mittens to go around). Translating, we may consider each pair of mittens to be a labeled ball and each child to be a mitten-receiving box. Our question then becomes, "How many ways are there to place six labeled balls, at most one per box, into 12 labeled boxes?" One might wonder why we chose to consider mittens as balls and children as boxes, rather than the other way around. This is because we are not asking multiple children to share a pair of mittens, as would happen if we were placing all 12 balls (children) in only six boxes (pairs of mittens).

Question B. *How many ways are there to place k identical (unlabeled) balls, at most one per box, into n labeled boxes?*

(Again, k must be less than or equal to n for this to be possible.) See Figure 7.3 for a visualization of this question. Because the balls are not labeled, it does not matter which ball ends up in which box. In other words, the order in which they appear in the box line does not matter. This is often abbreviated as *order doesn't matter* or *unordered*.

This is a problem *without repetition*: each box label will only appear once because of the restriction that at most one ball may be placed in any box.

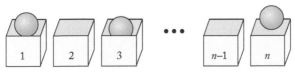

Figure 7.3. Shown are k unlabeled balls, placed (at most one per box) into n labeled boxes.

Figure 7.4. Balls with k different labels, placed (with exactly one per box) into n labeled boxes.

Example 7.2.3. You are asked to distribute fliers for the student dance company performance at your introductory salsa class. However, the publicity director only gave you 15 fliers and there are 40 students in your salsa class. How many ways are there to distribute the fliers? Here, the fliers are identical, and so unlabeled, but the students are all different (and so labeled). Therefore, if we are to make balls and boxes of this situation, the students must correspond to boxes (and that leaves the fliers as balls). Moreover, it would be silly to give multiple fliers to any person when we want to distribute publicity materials, so we will distribute at most one flier per student. Our question has become, "How many ways are there to place 15 identical balls, at most one per box, into 40 boxes?"

> Question C. *How many ways are there to place balls, exactly one per box, with k different possible labels, into n labeled boxes?*
>
> See Figure 7.4 for a visualization of this question. Here, *order matters* (the balls are *ordered*). The order in which the balls appear matters because the balls are labeled; it makes a difference which ball is placed in which box.

Example 7.2.4. In order for a computer system to be at all secure, there must be enough different possibilities for passwords that a hacker can't just try all of them and thus succeed in breaking into a given account. The Bank of Önd uses eight-character passwords and allows the use of letters and numbers. How many possible passwords are there? The order of the characters in a password certainly matters, so the eight positions are labeled; additionally, which character appears in each position matters, so the characters are labeled as well. In creating a password, a user may use a letter or number more than once in different positions. Thus, the 26 letters and 10 numbers cannot be boxes and instead must be balls. The positions naturally correspond to boxes. Our question is now, "How many ways are there to place one ball, with 36 different possible labels, into each of eight labeled boxes?"

We also addressed this problem type in Example 7.2.1.

Figure 7.5. Shown are k unlabeled balls placed into n labeled boxes.

Question D. *How many ways are there to place k unlabeled balls into n labeled boxes?*

Notice that some boxes may be empty, and others may have multiple balls. See Figure 7.5 for a visualization of this question. Because the balls are not labeled, *order doesn't matter* (the balls are *unordered*); it does not make a difference which specific ball ends up in which box, so the order in which the balls appear does not matter.

The fact that a box may contain more than one ball means that the use of a box as a ball destination can be repeated. Thus, this is sometimes referred to as *repetition allowed* or *with repetition*.

Example 7.2.5 (secretly three examples). Suppose an ogre is distributing 43 cupcakes to 12 baby mice. The ogre is, of course, cruel (almost by definition) and will sometimes deny cupcakes to some of the baby mice. (Yes, it's terrible. Try to make it through.) How many ways are there for the ogre to distribute the cupcakes? The baby mice are all different, and thus labeled. On the other hand, the cupcakes are all the same, and thus unlabeled. (Even if a human can tell the difference between delicately decorated cupcakes, an ogre will treat all tiresome cupcakes as the same.) Therefore, the cupcakes correspond to balls and the baby mice correspond to boxes. It only remains to determine what constraints (or lack thereof) there are on the number of balls per box. No constraints are given, and with 43 cupcakes for 12 baby mice, there are enough to go around, so our question has become, "How many ways are there to place 43 unlabeled balls into 12 labeled boxes?"

Consider a second question. How many ways are there to line up five grey ducks and two white ducks? We will later see that thinking of this problem in terms of balls and boxes may not be the fastest way to develop a solution, but it is worth seeing that the problem does fit into our framework. The initial difficulty is that it does not appear that either the grey ducks or the white ducks are labeled, and we know that every ball-and-boxes problem requires labeled boxes. Let us suppose that the grey ducks correspond to balls. There are three categories of grey duck in our duck-y lineup, namely those to the left of both white ducks, those between the two white ducks, and those to the right of the two white ducks—they will become our labeled boxes. That transforms our question into, "How many ways are there to place five unlabeled balls into three labeled boxes?" Now suppose instead that the white ducks correspond to balls. In our duck-y lineup, there are six categories of white duck, namely those to the left of all five grey ducks, those between the first two grey ducks, those between the second and third grey ducks, those between the third and fourth grey ducks, those between the fourth and fifth grey ducks, and those to the right of all five grey ducks. These six different categories of white duck can become our labeled boxes. This transforms our question into, "How many ways are there to place two unlabeled balls into six labeled boxes?" Probably you are quite concerned now, because we have created two *different* questions from our original query about duck lineups. As you will discover in Section 7.4.3, these two questions have the same answer—and in Problem 21 in Section 7.11, you will discover that for any similar situation, those two questions will *always* have the same answer.

Finally, consider a third question. How many nonnegative integer solutions are there to the equation $a + b + c = 5$? This is completely mysterious (there are no obvious candidates for balls *or* boxes) until we observe that $5 = 1 + 1 + 1 + 1 + 1$. Each of a, b, and c is a sum of some number of 1s (perhaps zero 1s), where there are five total 1s. Thus, we need to distribute the 1s into the variables—so the 1s are unlabeled balls. Moreover, the variables are different from each other; we usually consider $a = 1, b = 4, c = 0$ to be a different solution than $a = 4, b = 1, c = 0$. Thus, the variables can be viewed as labeled boxes. Now, our question has become, "How many ways are there to place five unlabeled balls into three labeled boxes?"

> Question E.   *How many ways are there to place k labeled balls into n labeled boxes, where k_j balls are placed into the jth box?*
>
> See Figure 7.6 for a visualization of this question. The balls are *unordered* (*order doesn't matter*). Because the balls are labeled, it matters which ball ends up in which box. However, the order in which balls appear in a given box does not matter.

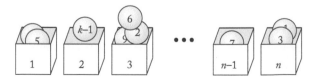

Figure 7.6. A total of k differently labeled balls, with k_j placed in the jth of n labeled boxes.

Example 7.2.6. At a party, there are 20 bite-sized pieces of fancily decorated strawberry cake and three fairies who would like some. After some fey nitternattering, the fairies decide that the first fairy will receive five bites of strawberry cake, the second fairy will receive eight bites of strawberry cake, and the third fairy will receive seven bites of strawberry cake. How many ways are there to distribute the fancily decorated strawberry cake bites among the fairies? Here the reader has to do a bit of deciphering. The fairies are certainly different from each other, and because they will receive cake, it seems likely that we could consider them boxes. This would make the cake pieces balls—but are they labeled or unlabeled? If they were unlabeled, then there would only be one way to distribute them. Thus, it seems they should be labeled. In fact, the wording of the problem supports this: When fancily decorated cake is cut, different pieces have differing amounts and patterns of frosting. Thus, our question is now, "How many ways are there to place 20 labeled balls into 3 labeled boxes, with those boxes receiving 5, 8, and 7 balls respectively?"

Interestingly, the answer to this question type is the same as the answer to the question, "How many ways are there to rearrange the letters of a given word?" but this question cannot be recast as a balls-and-boxes question in an intuitively helpful fashion.

Check Yourself

Rephrase the following situations as balls-and-boxes problems. Unless you have been instructed to consider only a few of the problems, please try to address all of them.

1. How many ways are there to give four snacks to six puppies, with no more than one snack going to each puppy?

2. How many ways are there to give four snacks to six puppies, with gluttony and cruelty allowed?

3. How many ways are there to deal five ace cards and four queen cards to nine card players?

4. How many ways are there to feed 12 spinach stems (of different lengths) to four ducks such that the grey duck gets five spinach stems, the white duck gets five spinach stems, and the pale-grey and black ducks get one spinach stem each?

5. How many ways are there to arrange 5 monster figurines and 12 angel figurines in a line on a shelf?

6. How many ways are there to distribute three chocolates (one white, one milk, and one dark) to four classmates, at most one chocolate per classmate?

7. How many ways are there to put a star sticker (they come in gold, silver, red, green, and blue) on every student's paper in your class?

8. How many ways are there to give two different catnip toys to five cats, such that no cat gets more than one toy?

7.3   Try This! Let's Have Some PIE

Think for a moment way back to the beginning, to Chapter 1, where we encountered the sum principle; it allowed us to calculate the size of two disjoint sets (said boringly). For example, the total number of pastilles, some of which are in the set of anise pastilles A and some of which are in the set of violet pastilles V, is $|A| + |V|$. But we are at a loss when there is overlap between our sets, as when we wish to know the total number of pastilles, but some of the pastilles are anise-violet flavor and we are only told $|A|$ and $|V|$—the sum would double-count the anise-violet pastilles. Let's experiment.

1. At Bakery Patitsa (in Bulgaria), there are 25 pies available: 10 contain blueberries, 15 contain peaches, and 5 are blueberry-peach pies. The only other offering at Bakery Patitsa is chocolate mousse pies. How many fruit pies are available? How many chocolate mousse pies are available?

2. Let's borify: If we have $A \subset Q$ and $B \subset Q$, and you are given the values of $|A|$, $|B|$, and $|A \cap B|$, then how can you determine $|A \cup B|$? How about $|Q \setminus (A \cup B)|$?

3. Back at the bakery, there has been a disaster. A contingent of crazy criminal clowns arrived and stole a bunch of the pies, taking some with them to eat, throwing others at each other's (and customers'!) faces, and smearing pie filling on the glass of the cases before leaving. While a counter clerk cleans the cases, the bakers rush to the market to get more ingredients and frantically bake more pies. The result is seven apple-containing pies, eight strawberry-containing pies, five rhubarb-containing pies, three pies containing both apple and strawberry, four pies containing both strawberry and rhubarb, and one apple-strawberry-rhubarb pie that is bought seconds after it comes out of the oven. (Later it was noticed that this last pie was also the one pie containing both apple and rhubarb.) How many total pies did the bakers bake?

4. What would a convenient formula be for $|A \cup B \cup C|$ in terms of sizes of individual sets and their intersections?

5. In an explosion of creativity, the bakers combine apples, blueberries, cherries, dates, figs, gooseberries, and huckleberries into all sorts of pies. Can you conjecture a formula to help determine the total number of pies baked, given the number of pies containing each subset of the filling fruits? If this proliferation of pies is too overwhelming to contemplate, first consider the situation of the trainee baker who is only allowed to use four fruit fillings in all possible combinations.

7.4 Combinatorial Problem Solutions and Strategies

In this section, we will outline solutions for the problem types introduced in Section 7.2 and introduce a few combinatorial strategies for approaching different types of problems. Most of our solutions proceed by combinatorial proof.

7.4.1 Strategy: Slots

For some combinatorial problems, the simplest way to approach a solution is to imagine filling some slots. By *slot*, we mean an area reserved in which to place something, visualized as ___. This harkens back to Chapter 1; see Figure 1.3 in particular for how slots were used with the product principle.

Example 7.4.1. We return to the situation of Example 7.2.1: How many different seven-digit telephone numbers are there? We may think of this as needing to fill the seven slots ___ ___ ___ ___ ___ ___ ___. Now, there are ten different digits that can be placed in each slot, which by the product principle tells us there are $10 \cdot 10 \cdot 10 \cdot 10 \cdot 10 \cdot 10 \cdot 10 = 10^7$ different seven-digit telephone numbers.

Example 7.4.2. The second question of Example 7.2.5 asks, "How many ways are there to line up five grey ducks and two white ducks?" Again we have seven slots ___ ___ ___ ___ ___ ___ ___, but this time we are not going to fill them one at a time. Instead, we will place the white ducks and then let the grey ducks walk to the remaining slots. There are $\binom{7}{2}$ ways to place the white ducks because we wish to choose two of the seven slots for them to stand in. That's all there is to it! Notice that we also could have placed the grey ducks first and let the white ducks walk to the remaining two slots, which would have resulted in $\binom{7}{5}$ ways; luckily, $\binom{7}{5} = \binom{7}{2}$. (Actually, no luck was involved.)

Example 7.4.3. Now for a new question. How many seven-digit telephone numbers have exactly three even entries? We still have seven slots, but we need to do some fancier counting than in the previous examples. There are $\binom{7}{3}$ ways to determine which three digits will be even. Then, for each of those ways, there are five ways to fill each even-digit slot (because there are five even digits, namely, 0, 2, 4, 6, and 8) and five (odd) ways to fill each of the remaining four slots. Thus, the product principle tells us that there are $\binom{7}{3} \cdot 5^3 \cdot 5^4$ telephone numbers with three even entries.

7.4.2   Strategy: Stars and Bars

By *stars*, we mean $\star \star \star \ldots \star$, and by *bars* we mean $| \, | \, \ldots \, |$. Here is one possible arrangement of four stars and two bars: $\star \, | \, \star \, \star \, | \, \star$. When we reframe a problem so that it is about arranging stars and bars, we call the discussion a *stars-and-bars argument*. Stars-and-bars arguments work well for problems that ask for two different types of items to be arranged in a line and for problems that ask for some objects to be distributed among some beings. Here are two examples.

Example 7.4.4. Suppose we have six teal owls and three orange cats. How many ways are there to arrange them in a line so no two cats sit next to each other?

First, we will line up the teal owls. A cat can sit to either side of any owl, so there are five places between owls, one place to the left of all the owls, and one place to the right of all the owls for a total of seven places a cat can sit (see Figure 7.7). We just have to choose three of those, so the total number of ways is $\binom{7}{3}$. Here, the owls were like stars, and the cats were like bars—there are $\binom{7}{3}$ ways to arrange six stars and three bars so that no two bars are adjacent.

Figure 7.7. Six owls, with the seven places near them in which a cat could sit.

Let's generalize. Suppose that instead we have s stars and b bars to arrange, and we do not want any two bars to be adjacent. Again, line up the stars; there are $s-1$ spots between the stars plus one spot to either side of the line, so a total of $s+1$ spots in which bars can be placed. We can put at most one bar in each spot, so we just need to choose b of the $s+1$ spots. This gives us $\binom{s+1}{b}$ ways to arrange the stars and bars so that no two bars are adjacent. Notice that if there are too many bars, the answer will appropriately be zero.

Example 7.4.5. Back in Example 7.2.5, we had an ogre distributing 43 cupcakes to 12 baby mice. One way the ogre can accomplish its task is to line up the cupcakes, hand some out to the first baby mouse, tell that baby mouse to shoo, hand some more cupcakes out to the second baby mouse, tell *that* baby mouse to shoo, etc. So the ogre is simply deciding when to shift giving cupcakes from one baby mouse to the next.

We will line up 43 stars (to reframe the problem and get rid of the ogre). Among them we must place 11 bars. Why 11, you ask? Well, let's return to cupcakes for a moment. We need to separate our line of cupcakes into 12 line segments, one for each baby mouse. So there are 11 separators—the cupcakes to the left of the first separator will be allocated to the first baby mouse, the cupcakes between the first and second separators will be allocated to the second baby mouse, and so on, until we see the cupcakes after the 11th separator given to the 12th (last) baby mouse.

Now, our question has become, "How many ways are there to arrange 43 stars and 11 bars?" That's not so bad. There are 54 slots in which a star or bar can be placed, so there are $\binom{54}{11} = \binom{54}{43}$ ways to arrange them.

Again generalizing, suppose we are given s stars and b bars to arrange. There are $\binom{s+b}{b} = \binom{s+b}{s}$ ways to arrange them, because there are $s+b$ spaces to fill with stars and bars and b of them must be chosen to be bars, or s of them must be chosen to be stars.

An Issue.   Sometimes problem solvers are confused about when to add one (as in the number of places cats can sit) and when to subtract one (as in the number of pauses between baby mice). There is no general rule as to how many slots or bars there will be for a given problem. You just have to figure it out every time. One of the goals of this chapter is to help you become familiar with the process of ferreting out the details of how to count in a given situation. Hopefully, the many examples given here will be of service.

7.4.3   Solutions to Problem Types

We now solve all the problems outlined in Section 7.2 (and one bonus problem type).

> Question A.   *How many ways are there to place k differently labeled balls, at most one per box, into n labeled boxes?*

Solution. If $k > n$, then there are zero ways. So let us assume $k \leq n$. Because the balls are labeled, it matters which ball lands in which box. Consider the first ball; there are n choices for which box to toss it in. Similarly, there are $(n-1)$ choices for the box that will hold the second ball, $(n-(j-1)) = (n-j+1)$ choices for the box that will hold the jth ball, and $(n-(k-1)) = (n-k+1)$ choices for the box that will hold the kth ball. Thus, there are $n(n-1)(n-2)\cdots(n-k+1) = \frac{n!}{(n-k)!}$ ways to place k differently labeled balls, at most one per box, into n labeled boxes. We could also solve this problem by noting that there are $\binom{n}{k}$ ways of picking the boxes that will receive balls and $k!$ ways to specify which label is on which ball, for a total of $\binom{n}{k}k! = \frac{n!}{(n-k)!}$ ways to place k differently labeled balls, at most one per box, into n labeled boxes.

For Example 7.2.2, there are $\binom{12}{6}6! = \frac{12!}{6!} = 665{,}280$ ways to distribute six different pairs of mittens to 12 young relatives.

> Question B.   *How many ways are there to place k identical (unlabeled) balls, at most one per box, into n labeled boxes?*

Solution. Again, if $k > n$, then there are zero ways. So let us assume $k \leq n$. We just need to choose which k of the n boxes will get balls, as the order of the balls does not matter. Thus, we have $\binom{n}{k}$ ways to place k identical balls, at most one per box, into n labeled boxes.

For Example 7.2.3, there are $\binom{40}{15} = 40{,}225{,}345{,}056$ ways to decide which of the 40 students in your salsa class will receive the 15 fliers about the student dance company performance.

> Question C.   *How many ways are there to place balls, exactly one per box, with k different labels, into n labeled boxes?*

Solution. For each of the n boxes, we have k choices as to the type of ball that will be placed in the box. Thus, by the product principle, we have k^n ways to place balls, exactly one per box, with k different labels, into n labeled boxes.

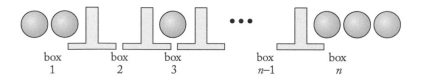

Figure 7.8. The balls from Figure 7.5 lined up between separators.

For Example 7.2.4, there are $36^8 = 2{,}821{,}109{,}907{,}456$ different eight-character alphanumeric passwords available for use by Iceland's Bank of Önd. Some quick googling suggests that an ordinary brute-force password-cracking program can whip through 100 million passwords per second, so it would take about 45 minutes to search through all possible Bank of Önd passwords. Of course, in practice the search would end when the user's password is encountered, and how fast that is depends on where the user's password falls in the ordering of potential passwords that the cracking program uses. Also, in practice there are more efficient ways to crack a password. (Perhaps the Bank of Önd should use longer passwords, or make them case-sensitive.)

> **Question D.** *How many ways are there to place k unlabeled balls into n labeled boxes?*

Solution. Notice first that there are no restrictions specifying how many or how few balls may land in a given box. Therefore there may be zero, one, or many balls in a box. We will line up the balls and place separators in the line to determine which balls land in which box. All the balls to the left of the first separator will be tossed in the first box; those balls between the first and second separators will land in the second box; and in general, the balls between the $(j-1)$st and jth separators will land in the jth box (see Figure 7.8). Those balls to the right of the last separator will be in the nth box, so there must be $n-1$ separators. (If you wish to think literally in terms of boxes, each separator corresponds to the right side of one box glued to the left side of the next box, and the whole assembly is placed under the line of balls and moved upwards to catch the balls in the boxes.)

With k balls and $n-1$ separators, we have a stars-and-bars situation: we want to know how many ways there are to line up k stars and $n-1$ bars. This becomes a situation with slots, in which we have $k+n-1$ slots and must place k stars and $n-1$ bars in those slots. (Two bars next to each other corresponds to an empty box.) There are $\binom{k+n-1}{k} = \binom{k+n-1}{n-1}$ ways to do this, and thus $\binom{k+n-1}{k} = \binom{k+n-1}{n-1}$ ways to place k unlabeled balls into n labeled boxes.

For Example 7.2.5, we have $\binom{43+12-1}{11} = 95,722,852,680$ ways to distribute 43 cupcakes to 12 baby mice. We also have $\binom{5+2}{2} = 28$ ways to arrange five grey ducks and two white ducks, which can be recast as $\binom{5+3-1}{3-1} = \binom{7}{2}$ ways if the grey ducks are the balls and the white ducks are the boxes, or $\binom{2+6-1}{2} = \binom{7}{2}$ ways if the white ducks are the balls and the grey ducks are the boxes. Finally, there are $\binom{7}{2} = 28$ nonnegative integer solutions to the equation $a+b+c = 5$. We will now give one more example for good measure.

Example 7.4.6. Suppose you wish to get a dozen tulip bulbs at the garden center. (They are on sale for $5/dozen.) There are 16 different kinds of tulip bulbs available. How many ways are there to fill your dozen bag? We place 12 slots in a line to represent our dozen tulip bulbs. We will place 15 dividers to mark the places in the line where our tulip bulbs switch between the 16 types. One such arrangement is

$$t_1 \; t_1 \mid \mid t_3 \; t_3 \; t_3 \; t_3 \; t_3 \mid t_4 \mid t_5 \mid t_6 \mid \mid t_8 \mid \mid \mid \mid \mid t_{13} \mid \mid \mid .$$

Thus, we have 27 spots from which we need to choose the locations of the tulips (or of the dividers). The total number of ways to fill the dozen bag is then $\binom{27}{12} = \binom{27}{15} = 17,383,860$, which is a lot. No wonder it takes so long to make decisions at the garden center!

> Question D′.   *How many ways are there to place k unlabeled balls into n labeled boxes, so that each box contains at least one ball?*
>
> This is a variation on the previous problem, and we must require that $k \geq n$ so that there are enough balls to go around.

Solution 1. Let us start by placing one ball into each box; there is only one way to do this, as the balls are unlabeled. Now we have $k - n$ unlabeled balls remaining to place into the n boxes. But this is exactly the same as Question D, so there are $\binom{(k-n)+n-1}{n-1} = \binom{k-1}{n-1} = \binom{k-1}{k-n}$ ways to place the remaining balls in the boxes.

Solution 2. We will convert this to a stars-and-bars problem. Let the k balls be represented by a line of k stars. We want to place bars into the line of stars in such a way that they separate the line into n clumps of stars. This means we need to place $n - 1$ bars as separators. How many places are there in which bars can be placed? There are $k - 1$ places between stars. (We cannot place a bar to the left or right of the star line, as it would mean there was an empty box. And we cannot place two bars into the same place, as that would also indicate an empty box.) That means we have $\binom{k-1}{n-1}$ ways to place k unlabeled balls into n boxes so that each box contains at least one ball.

Because this is a new problem type, it deserves some examples to go with it. We will use variations on Example 7.2.5.

Example 7.4.7 (secretly two examples). Let's talk about feeding folks cake again. This time we will feed identical miniature cupcakes to baby mice, but each (polite) baby mouse will pick up some cupcakes in turn. Baby mice are nice and, of course, will make sure that everyone gets at least one cupcake. Therefore, we just need to help the baby mice decide when one will stop munching and let the next one start. With c cupcakes, there are $c - 1$ different times that a baby mouse could stop eating, and if we are feeding b baby mice, there are $b - 1$ transitions between baby mice eating. Therefore, the number of ways the b baby mice can distribute their c cupcakes so that every baby mouse gets at least one cupcake is $\binom{c-1}{b-1}$.

Now let us suppose that we want to count *positive* integer solutions to the equation $a + b + c = 5$. We can put a 1 in each of the a, b, c variables, and then have two 1s left to distribute among the three variables. There are $\binom{2+3-1}{2} = 6$ ways to place two bars (that demarcate three variables) among two stars (the 1s), and so six positive integer solutions to the equation $a + b + c = 5$. Alternatively, we could choose two of the three variables to get the extra 1s and thus would have $\binom{3}{2} = 3$ ways of doing it, but we could also give both 1s to one variable and there are $\binom{3}{1} = 3$ ways of doing that. In total, there are again six positive integer solutions to the equation $a + b + c = 5$.

> Question E. *How many ways are there to place k labeled balls into n labeled boxes, where k_j balls are placed into the jth box?*

Solution 1. We know that $k_1 + k_2 + \cdots + k_n = k$ because there are k total balls with k_1 balls assigned to the first box, k_2 balls assigned to the second box, and k_n balls assigned to the nth box. Within a given box, the balls are unordered (because we could have reached a hand in and stirred them). So we can just choose k_1 of the k balls for the first box, k_2 of the remaining balls for the second box, and so forth. Notice that when we get to the nth box, we will have no choice because there will be exactly k_n balls left. Therefore, using the product principle, we see that there are $\binom{k}{k_1}\binom{k-k_1}{k_2}\binom{k-k_1-k_2}{k_3}\cdots\binom{k_n+k_{n-1}}{k_{n-1}}$ ways to place k labeled balls into n labeled boxes, where k_j balls are placed into the jth box. (Notice that $k - k_1 - k_2 - \cdots - k_{n-2} = k_n + k_{n-1}$.)

Solution 2. There are $k!$ ways to lay out all the balls in a line. We can then scoop up the first k_1 balls and put them in the first box, the next k_2 balls and put them in the second box, and so forth. However, this overcounts because the balls in a single box could be stirred into any order within the box. So we need to divide by $k_1!$ to account for the orderings within the first box all being equivalent, by $k_2!$ to account for the orderings within the second box all being equivalent, and so forth. Thus, there are $k!/(k_1!k_2!\ldots k_n!)$ ways to place k labeled balls into n labeled boxes, where k_j balls are placed into the jth box. Some symbolic manipulation will show that this solution is equivalent to the choice notation solution above. (See Problem 3 in Section 7.11.)

In Example 7.2.6, there are $\binom{20}{5}\binom{15}{8} = 99{,}768{,}240 = \frac{20!}{5!8!7!}$ ways to give 20 bites of fancily decorated strawberry cake to three fairies, so that the first of the fairies gets 5 bites, the second gets 8 bites, and the third gets 7 bites.

> Question F. *How many ways are there to rearrange the letters of a given word?*

Solution. In mathematics, a *word* is a (finite) sequence of letters from a given alphabet—not necessarily a word found in a dictionary. If a k-letter word has no repeated letters, there are $k!$ rearrangements of the letters, or *anagrams* of that word. If a k-letter word has some repeated letters, then $k!$ overcounts the number of anagrams because it treats two identical letters as being different. Thus, we need to divide by the number of orderings of each repeated letter. If there are n different letters in the word and the jth letter is repeated k_j times, then the number of anagrams of the word is $k!/(k_1!k_2!\cdots k_n!)$. (For most words, the denominator will contain a bunch of 1!s.) An example will make this clearer.

Example 7.4.8 (of anagramming three words). First we will count the anagrams of BOXES; there are 5! of them because all letters are different.

Now consider the word BALLS. Let us label the repeated letters so that BALLS becomes BAL_1L_2S. BAL_1L_2S has 5! anagrams, though we've overcounted because L_1 and L_2 could have appeared with L_1 first or L_2 first, so BALLS has $\frac{5!}{2!}$ anagrams.

Finally, notice that PARALLELOGRAM has $\frac{13!}{3!2!3!}$ anagrams because the three As could have been in any of 3! orders, the two Rs could have been in any of 2! orders, and the three Ls could have been in any of 3! orders.

Solutions summary. The grid in Table 7.1 presents our results in abbreviated form; each entry contains a question letter and the solution formula.

How many ways... ... n labeled boxes?	at most one per box	any number per box	exactly one per box
k labeled (ordered) balls	A: $\binom{n}{k}k! = n(n-1)$ $\cdots(n-k+1)$	E, F: (k_j balls unordered within box) $\frac{k!}{k_1!k_2!\ldots k_n!} = \binom{k}{k_1}$ $\binom{k-k_1}{k_2}\binom{k-k_1-k_2}{k_3}$ $\cdots\binom{k_n+k_{n-1}}{k_{n-1}}$	——
k unlabeled (unordered) balls	B: $\binom{n}{k}$	D, D′: $\binom{k+n-1}{k} = \binom{k+n-1}{n-1}$ and $\binom{k-1}{n-1} = \binom{k-1}{k-n}$	——
unlimited balls, k different labels (order matters)	——	——	C: k^n

Table 7.1. Solutions summary.

7.4.4 Denouement: Bijective Counting, Again

Just to call attention to what's going on underneath the hood in this section: we are using bijections, lots of them, to accomplish our counting. For example, the number of ways to choose a dozen duck eggs that come in five colors (white, brown, cream, green, and speckled) is $\binom{16}{4}$. How can we see the bijections that are involved? We will make a one-to-one correspondence between duck eggs and stars, and a one-to-one correspondence between the spaces between piles of like-colored eggs and bars (see Figure 7.9). Then we'll spread each pile of eggs out into a line and discover that we have a bijection between choices of a dozen duck eggs and star-and-bar sequences of length 16 with exactly 12 stars and 4 bars. There are certainly $\binom{16}{4}$ such sequences because we can make a one-to-one correspondence

Figure 7.9. Eggs correspond to stars and spaces to bars.

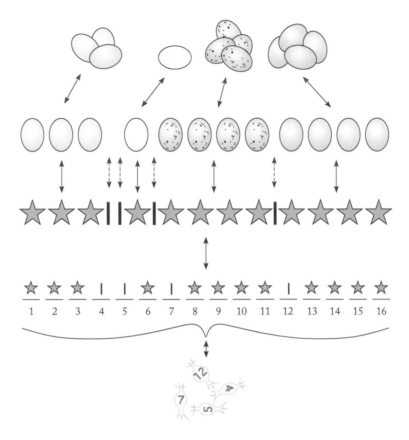

Figure 7.10. A choice of a dozen eggs corresponds to a choice of sugar-numbers.

between 16-slot sequences and a bag of 16 sugar-numbers, and choosing which slots to fill with bars (or which slots to fill with stars) is the same as choosing 4 sugar-numbers (or 12 sugar-numbers) from the bag. A sample choice of eggs with this sequence of bijections is shown in Figure 7.10.

While we're discussing bijections, this seems like an appropriate moment to mention the connection between counting functions of certain types and counting the number of ways to place balls into boxes. In Section 3.3.1 you counted the number of functions from an m-element set to a q-element set; this is the same as counting the number of ways to assign one of the q elements of the target to each of the m elements of the domain, which is in turn the same as counting the number of ways to place balls with q labels into m labeled boxes. This is Question C, with solution q^m. Section 3.4 also addresses counting the number of injective functions from an m-element set to a q-element set. Here, we cannot use any target-space element more than once, and some elements of the target space may end up unused; however, all of the elements of the domain are used. Therefore the target-space elements must represent boxes rather than balls. Thus, this situation is like placing m labeled balls into q labeled boxes so that each box gets at most one ball. This is Question A, with solution $\binom{q}{m}m! = q \cdot (q-1) \cdot \cdots \cdot (q-(m-1))$.

Check Yourself

Solve the following problems (which, yes, are the same ones posed in the Check Yourself list for Section 7.2).

1. How many ways are there to give four snacks to six puppies, with no more than one snack going to each puppy?

2. How many ways are there to give four snacks to six puppies, with gluttony and cruelty allowed?

3. How many ways are there to deal five aces and four queens to nine card players?

4. How many ways are there to feed 12 spinach stems (of different lengths) to four ducks such that the grey duck gets five spinach stems, the white duck gets five spinach stems, and the pale-grey and black ducks get one spinach stem each?

5. How many ways are there to arrange 5 monster figurines and 12 angel figurines in a line on a shelf?

6. How many ways are there to distribute three chocolates (one white, one milk, and one dark) to four classmates, at most one chocolate per classmate?

7. How many ways are there to put a star sticker (they come in gold, silver, red, green, and blue) on every student's paper in your class?

8. How many ways are there to give two different catnip toys to five cats, such that no cat gets more than one toy?

7.5 Let's Explain Our PIE!

Hey! You! Don't read this unless you have worked through the problems in Section 7.3. I mean it!

So, let's check out what we've learned (and then figure out why the theme is PIE rather than, say, pie). We wanted to understand how to find the size of a union of

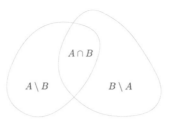

Figure 7.11. The intersection of two sets can be partitioned into three disjoint subsets.

sets when these sets overlap, so that we can avoid accidental overcounting and only overcount intentionally. Check out the handy Venn diagram in Figure 7.11: notice that there are three disjoint regions, $A \setminus B$, $B \setminus A$, and $A \cap B$. Certainly, $|A \cup B| = |A \setminus B| + |B \setminus A| + |A \cap B|$. But generally, this is not the information we're given— after all, if that were the case, we could simply have used the sum principle. If we look at this from an overcounting perspective, we see that $|A| + |B|$ has counted the elements of $|A \cap B|$ exactly twice. Thus, $|A \cup B| = |A| + |B| - |A \cap B|$, as you discovered yourself in Section 7.3.

Example 7.5.1. A map is labeled with numbers 1, 2, 3, 4, 5 along the left edge and letters A, B, C, D, E, F along the bottom. Locations on the map are indicated by their sectors, e.g., A-4 or D-2. How many ways are there to select two sectors that both have odd numbers *or* both have vowels? The total number of sectors is 30, and $\frac{3}{5}$ of these (18) are labeled with odd numbers; $\frac{2}{6}$ of these (10) are labeled with vowels. So to get a pair with odd numbers there are $\binom{18}{2}$ ways, and to get a pair with vowels there are $\binom{10}{2}$ ways. Still, we have overcounted the pairs with both odd numbers and vowels, so we need to know how many such pairs there are. There are three odd numbers and two vowels, so six sectors are labeled with both odd numbers and vowels, and thus there are $\binom{6}{2}$ pairs of such sectors. Therefore, there are $\binom{18}{2} + \binom{10}{2} - \binom{6}{2} = 183$ pairs of sectors that both have odd numbers or both have vowels.

Figure 7.12 shows a Venn diagram for sets A, B, C. We can extend our reasoning from two sets to three and obtain the formula $|A \cup B \cup C| = |A| + |B| + |C| - |A \cap B| - |A \cap C| - |B \cap C| + |A \cap B \cap C|$.

Example 7.5.2. Thirty-five cats are surveyed. Twenty like catnip, 25 like tuna, and 23 like to sleep in the sun; 10 like all three; 15 like catnip and tuna; and 17 like tuna and sleeping in the sun. How many like catnip and sleeping in the sun?

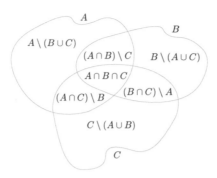

Figure 7.12. All disjoint subsets of three intersecting sets are labeled.

Notice first that while this is a PIE situation, we are not trying to calculate $|C \cup T \cup S|$. Instead, we desire $|C \cap S|$. We could use the three-set PIE formula, plug in the numbers we have, and solve for $|C \cap S|$, but here is a visual approach. Highlighted at left in Figure 7.13, we see $C \cap S$. From the given information, we can fill in $|C \cap T \cap S| = 10$, from which we can deduce that $|(C \cap T) \setminus S| = 15 - 10 = 5$ and $|(T \cap S) \setminus C| = 17 - 10 = 7$ (see the center diagrams of Figure 7.13). Then we are able to see that $|T \setminus (S \cup C)| = 25 - (10 + 5 + 7) = 3$. Now in Figure 7.14, we will focus on the part of the Venn diagram left blank in Figure 7.13 and start by noting that $|(C \cup S) \setminus T| = |C \cup S \cup T| - |T| = 35 - 25 = 10$. We also see that $|C \setminus T| = 20 - 15 = 5$ and that $|S \setminus T| = 23 - 17 = 6$. Because $10 = 5 + 6 - 1$, we deduce that $|(C \cap S) \setminus T| = 1$, and therefore, $|C \cap S| = 10 + 1 = 11$.

Notice how this goes: to describe the size of a union of sets, we *include* the individual set sizes, but this overcounts pairwise intersections, so we *exclude* them (by subtracting their sizes). But in turn, this undercounts triple intersections, so we *include* them again by adding their sizes. If there are fourfold intersections, we will have overcounted them, so we would then *exclude* (by subtraction) their

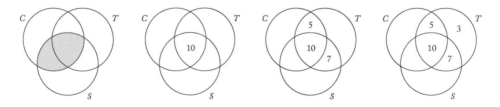

Figure 7.13. Venn diagrams showing an approach to determining $|C \cap S|$.

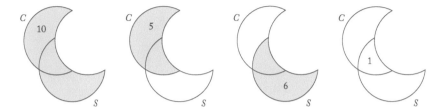

Figure 7.14. Venn diagrams for $(C \cup S) \setminus T$.

sizes. This is the origin of the descriptor *inclusion-exclusion* for this technique; while one might call it "the inclusion-exclusion principle" to parallel the sum and product principles, that doesn't lend itself to a happy acronym like PIE (really, IEP isn't as fun).

Example 7.5.3 (of apple PIE). A Cortland apple bushel has 35 apples, of which 20 are large, 25 are green, and 23 have stems. Ten of the apples are large, green, and have stems; 15 are large and green; 17 are green and have stems. How many of the Cortland apples are large and have stems?

The apple bushel has $|L \cup G \cup S| = 35$ apples, and we want to know $|L \cap S|$. We know from the PIE formula that $35 = (20 + 25 + 23) - (15 + 17 + |L \cap S|) + 10 = 68 - 32 - |L \cap S| + 10 = 46 - |L \cap S|$, so $|L \cap S| = 46 - 35 = 11$.

> **Summary of PIE for two or three sets.** We have $|A \cup B| = |A| + |B| - |A \cap B|$, and $|A \cup B \cup C| = |A| + |B| + |C| - |A \cap B| - |A \cap C| - |B \cap C| + |A \cap B \cap C|$.

It is not super-difficult to see how to extend our formulas to n sets in concept— just take an alternating sum of many-fold intersections—but figuring out notation with which to express this idea is a nightmare. Here is one version:

> **Theorem 7.5.4 (general PIE).** *Let $A_1, A_2, \ldots, A_n$ be subsets of a finite set B. Then*
>
> $$\left| \bigcup_{i=1}^{n} A_i \right| = \sum_{j=1}^{n} (-1)^{j-1} \left(\sum_{\text{all } \binom{n}{j} \text{ intersections involving } j \text{ sets}} \left| \bigcap A_i \right| \right).$$

And you knew we would prove that this is a correct formula, right?

Proof: We would like to show that each element of $\bigcup_{i=1}^{n} A_i$ is counted exactly once by that monstrous double sum on the right-hand side of the equation. To that end, consider $a \in A_1 \cup A_2 \cup \cdots \cup A_n$ and let the number of A_i in which a is contained be k. The items being added up on the right-hand side are all sizes of j-fold intersections of the A_i, and a is in such a j-fold intersection if and only if a is an element of all j sets in the intersection (duh, but important). The number of different j-fold intersections containing a is $\binom{k}{j}$. So let's count: there are k "intersections" containing a once, $\binom{k}{2}$ intersections containing a twice, $\binom{k}{3}$ intersections containing a three times, and so forth, so that a is counted a total of $k - \binom{k}{2} + \binom{k}{3} - \cdots + (-1)^{k-1}\binom{k}{k}$ times. Now notice that this looks eerily familiar: if we have $x = 1, y = -1$ in the binomial theorem (and change our index variable letters from n, k to k, j), we get $(1-1)^k = \sum_{j=0}^{k} \binom{k}{j}(1)^{k-j}(-1)^j$, or $0 = \binom{k}{0} - k + \binom{k}{2} - \binom{k}{3} + \cdots + (-1)^k \binom{k}{k}$, which can be rewritten as $k - \binom{k}{2} + \binom{k}{3} - \cdots + (-1)^{k-1}\binom{k}{k} = \binom{k}{0} = 1$, as desired. $\square$

To uncrunch our heads, let's do another example.

Example 7.5.5. Consider all possible permutations of the numbers $0, 1, 2, 3, 4, 5, 6, 7, 8, 9$. How many of them have the substrings 53, 02, or 28? The permutation **4536028**179 has all three, but 8042957316 has none. First, let's see how many permutations have the substring 53. Because all digits of our permutation are distinct, we can think of 53 as a single digit within a permutation with eight other numbers $0, 1, 2, 4, 6, 7, 8, 9$; thus, there are 9! different permutations containing 53. Likewise, there are 9! permutations containing 02 and 9! permutations containing 28. However... there is overlap. We can think of 53 and 02 as single digits within permutations also containing $1, 4, 6, 7, 8, 9$, for a total of 8! permutations, and likewise with 53 and 28. The only way that 02 and 28 can appear together in a permutation is if they overlap as 028, so we can think of that as a single digit in a permutation with numbers $1, 3, 4, 5, 6, 7, 9$, and we again (slightly surprisingly) have a total of 8! permutations. Using this same technique, we find that there are 7! permutations containing 53, 028 (02 and 28), and $1, 4, 6, 7, 9$. Putting it all together, we have $3 \cdot 9! - 3 \cdot 8! + 7! = 972{,}720$.

Check Yourself

Please do at least the first three of these problems.

1. Write out Theorem 7.5.4 for four sets A_1, A_2, A_3, A_4.

2. How many permutations of the numbers $0, 1, 2, 3, 4$ have substrings 03 or 21?

3. How many permutations of the numbers $5, 6, 7, 8, 9$ have the substrings 59 or 85?

4. **Challenge:** Write and solve your own three-set PIE problem (but not involving pies).

7.6 Try This! What Are the Balls and What Are the Boxes? And Do You Want Some PIE?

It is useful to discuss these problems with others, to practice explaining why each problem is of the appropriate type.

1. (This problem comes from a real-life situation faced by the author.) If the Smith College fencing team offers eight kinds of fudge and you want to order $2\frac{1}{2}$ pounds, and the fudge comes in $\frac{1}{4}$-pound boxes, how many different orders could you make?

2. How many anagrams are there (including those that aren't dictionary words) of PIES?

3. How many anagrams are there (including those that aren't dictionary words) of TELEPHONE?

4. Suppose that N and M are sets in the universe U. Express $|\overline{N} \cup \overline{M}|$ in terms of the cardinalities of N, U, M and combinations of these sets. (Did you use symbols or did you use pictures in your expression?)

5. How many ways are there to arrange 15 bluejays and 3 white cats to sit in a line? What if the white cats may not sit next to each other?

6. There are 184 students taking classes in French. Of these, 112 are taking Intermediate French, 84 are taking French Literature, and 46 are taking French Cultural History. There are 66 students taking Intermediate French and French Literature, 37 students taking Intermediate French and French Cultural History, and 30 taking French Cultural History and French Literature. If 45 students take none of these three French classes, how many take all three (Intermediate French, French Literature, and French Cultural History)?

7. How many seven-digit telephone numbers are there with all digits different? Answer the same question including area codes (so for ten-digit telephone numbers). What if we include country codes (each is two or three digits long) as well?

8. A mathematically inclined club is forming a recruitment committee with five members. They have calculated that there are 8,568 ways to form this committee. Two of the club members are named Joaquín and Ana. The club

calculates that 1,820 of the possible committees would have Joaquín on them but not Ana, 1,820 would have Ana but not Joaquín, and 560 would have both Joaquín and Ana.

(a) How many potential committees have either Joaquín or Ana?

(b) How many potential committees have neither Joaquín nor Ana?

(c) One semester, Joaquín and Ana carpool to meetings, so they insist that if either one of them is on the committee, then both should be on the committee. How many potential committees meet this condition?

(d) Later, Joaquín and Ana have a fight and refuse to work together. Joaquín says ze won't be on the committee if Ana is on it, and Ana says that ze won't be on the committee if Joaquín is on it. How many of the potential committees meet this condition?

7.7 Where to Go from Here

As previously mentioned, there are balls-and-boxes questions whose solutions are too advanced for this text. Among other topics, they lead to integer partitions and Stirling numbers—research is active on both of those topics. Sometimes the collective variety of balls-and-boxes questions is known as *the Twelvefold Way*. For a first introduction to these ideas and many more related combinatorial concepts, check out [18], which is eminently readable (but be prepared for challenging problems!). The gold standard for combinatorial reference is the hardcore and awesome *Enumerative Combinatorics, Volume I* by Richard Stanley.

If you enjoyed PIE (and who doesn't?), you may want to study *derangements* (rearrangements in which no item ends up in its starting place). Learn how to use PIE in this way in [18], [10], or *Enumerative Combinatorics, Volume I*. That last gives extensions of PIE to linear algebra, permutations, and more.

As with Chapter 6, the likeliest courses to introduce ideas related to those in this chapter are courses in combinatorics.

Credit where credit is due: Example 7.2.5 was inspired by the *Babymouse* books by Jennifer and Matthew Holm. Go read them. Example 7.2.6 was inspired by Terry Moore's comic character Kixie. Example 7.5.5 was inspired by [17]. Bonus Check-Yourself Problem 3 was inspired by an *Art of Problem Solving* student whose username is ninjataco. In Section 7.11, Problem 34 was inspired by Sam Oshins's character who was inspired in turn by this book (see page 148); Jayke Bouche in Problem 28 is a friend of the author. Some problems in this chapter were adapted from [8], [1], [11], and [18].

7.8   Chapter 7 Definitions

order matters: "Item 1, then item 2" is different from "item 2, then item 1."

ordered: Order matters.

order doesn't matter: "Item 1, then item 2" is the same as "item 2, then item 1."

unordered: Order doesn't matter.

without repetition: Once an item has been used, it cannot be used again; for example, in forming a number with distinct digits.

repetition allowed: Once an item has been used, it can be used again; for example, in forming a number with possibly repeated digits.

stars-and-bars: A problem reframed so that it is about arranging stars and bars, and then solved.

word: A (finite) sequence of letters from a given alphabet—not necessarily a word found in a dictionary.

anagram: A rearrangement of the letters of a word.

inclusion-exclusion: A process of careful over- and undercounting in which we *include* individual set sizes by adding them (but this overcounts pairwise intersections) and *exclude* sizes of pairwise intersections by subtracting them (but this undercounts triple intersections) and include sizes of triple intersections by adding them... and so on.

7.9   Bonus: Linear and Integer Programming

Notice.   The material in this section is not particularly related to the rest of the chapter. It gives an introduction to a mathematical subfield that is not usually a part of the undergraduate curriculum. Linear and integer programming are ways to set up and solve a large class of practical problems, and they are part of the larger field of operations research (also called industrial engineering or management science and including the subfields of combinatorial optimization and discrete optimization). The material is placed here because this is the first point in the text where the reader has enough familiarity with discrete mathematics to appreciate these techniques. We will give introductions to other types of operations research problems later in the text, namely network flows (Bonus Section 12.10), minimum-weight spanning trees (Sections 10.3 and 10.4), and the Traveling Salesperson Problem and shortest path calculations (Section 12.4). In this section, we will discuss how to set up linear and integer programming problems, and in Bonus Section 10.11, we will introduce an approach to solving integer programming problems.

If you have taken calculus, you may recall that the word "optimization" is used in that context for the process of finding maximum or minimum values of a continuous function. More generally (and literally), *optimization* is the process of finding the optimal (best) solution to a problem.

Example 7.9.1. In assembling a fruit basket for your Great-Aunt Mildred, you want to spend the least amount of money possible. However, you know her taste— she adores mangoes, so you must include some; on the other hand, she despises apples, so you'd better not put any of *them* in there. There are 10 types of fruit from which to choose (oranges, grapes, mangoes, pineapple, kiwi, persimmons, bananas, apples, pears, and plums) and you may select 12 pieces of fruit to put in the basket. What is the best selection?

We will begin by defining some variables and some constants.

x_1 is the number of oranges selected and c_1 is the cost for an orange.

x_2 is the number of bunches of grapes selected and c_2 is the cost for a bunch of grapes.

x_3 is the number of mangoes selected and c_3 is the cost for a mango.

x_4 is the number of pineapples selected and c_4 is the cost for a pineapple.

x_5 is the number of kiwi fruits selected and c_5 is the cost for a kiwi fruit.

x_6 is the number of persimmons selected and c_6 is the cost for a persimmon.

x_7 is the number of bunches of bananas selected and c_7 is the cost for a bunch of bananas.

x_8 is the number of apples selected and c_8 is the cost for an apple.

x_9 is the number of pears selected and c_9 is the cost for a pear.

x_{10} is the number of plums selected and c_{10} is the cost for a plum.

Now, the total cost of the fruit basket will be $x_1c_1 + x_2c_2 + \cdots + x_{10}c_{10} = \sum_{j=1}^{10} x_j c_j$. We would like to select fruit so as to make this expression have minimum possible value. But we must also use the other information we have. In order to include at least one mango, we know $x_3 > 0$, and in order to avoid apples, we know $x_8 = 0$. Selecting 12 pieces of fruit means that $x_1 + x_2 + \cdots + x_{10} = \sum_{j=1}^{10} x_j = 12$. There are some hidden pieces of information, as well; in particular, we cannot have negative numbers of pieces of fruit, so $x_j \geq 0$ for all j.

In summary, we have converted the Great-Aunt Mildred's Fruit Basket Problem into a new problem, namely,

$$\text{minimize the quantity} \quad \sum_{j=1}^{10} x_j c_j,$$

subject to the constraints $\quad x_j$ is an integer, $\quad x_3 > 0,$

$$x_j \geq 0 \text{ for all } j, \quad x_8 = 0,$$

$$\sum_{j=1}^{10} x_j = 12.$$

A solution to this problem will give values for the x_j.

Now, we will not discuss how to *solve* problems such as that of Great-Aunt Mildred's Fruit Basket because that would entail an entire course. We will talk about how solvable they are, but first we should be clearer in describing these problems. For a problem to be formulable for linear or integer programming, it must have certain properties.

- ✤ There must be a function to be maximized or minimized. This is often termed the *objective function* or *cost function*.

- ✤ The variables involved must take on real values or integer values.

- ✤ The objective function must not involve any powers or products of variables; it must be *linear*.

- ✤ There must be restrictions placed on the variables, usually termed *constraints*.

- ✤ The constraints must be writable as inequalities or equalities, for which one side of the equation/inequality is a constant and the other is a sum of variables, perhaps with coefficients added. (These are also linear.)

A linear objective function together with linear constraints is a *linear program*. In Example 7.9.1, our variables had to be whole numbers. Thus, that objective function and its constraints together constitute an *integer program*. It is also possible for a problem to involve some variables that must be integers and others that need not be. Such setups are called *mixed integer-linear programs*. The science of creating linear programs and solving them is known as *linear programming*. Three acronyms you may run across are LP, IP, and MILP (can you discern what they stand for?); they can be used to refer to individual problems (as in, "Oh, that's an MILP") or more general studies (as in, "This technique is used in IP").

Example 7.9.2 (of linear and integer versions of a problem). Before leaving the house in the morning, you like to have a smoothie. You keep bananas, milk, oranges, frozen blueberries, protein powder, soy milk, frozen strawberries, and honey on hand to make your smoothies. Each of these ingredients has a cost c_j and will be added to a smoothie in some quantity x_j. You know the nutritional profile (calories, protein, carbohydrates, fats, vitamins) of each ingredient. There is some minimum number of calories you need for breakfast, and you want to meet the recommended amounts of certain vitamins and other nutrients. You may also want to have no more than a certain amount of calories in your smoothie. Probably you do not want multiple types of milk in a single smoothie. This gives a set of constraints. Your objective may be to minimize the total cost to make a smoothie. Or, perhaps you want to maximize nutritive value. This situation forms a linear program because you can cut up the fruit and measure any amount of liquid or powder desired. Sometimes you are in a rush to leave the house, so all you have time for is to grab some pieces of fruit and a protein bar or a small milk carton. However, you have the same constraints and the same objective. This situation forms an integer program because you don't have time to cut up fruit or put milk or parts of protein bars in separate containers; you can only grab unit amounts of these foods.

Example 7.9.3 (of real life IP problems). Here are three practical problems that are usually set up as integer programs:

1. A community needs firehouses. There are several proposed sites available. Where should firehouses be built so that (a) nowhere in the community is more than two miles from a firehouse and (b) the cost is as low as possible?

2. A team of explorers is heading out into the wild for a week. Each explorer needs food, clothing, and equipment but can only carry 10 kg in hir backpack. What should the team pack?

3. A corporation wants to invest a maximum of $9,000 in one of five different projects. Which one will likely be most profitable?

It is known how to solve any linear program quickly. To understand how the most-used algorithm works, you will need to know linear algebra or some convex geometry. On the other hand, there is no known quick way to solve an integer program. We will encounter one solution technique in Section 10.8 and apply it in Section 10.11. This technique uses the fact that every integer program has an associated linear program (found by removing the constraints that specify the variables must be integers), along with the fact that the associated linear problem can be solved quickly. Even with this advantage, it is only slightly better than using brute force.

What is meant by using brute force as a solution? Basically, one makes a list of all possible values for each variable and tries out every combination of those values. Some combinations will violate a constraint or several constraints; among those combinations of values that honor all constraints, one must check to see which combination gives the best output for the objective function.

Realistic problems usually have large numbers of variables. Putting these problems into standard forms for LP/IP introduces many more variables. For example, inequalities are made into equalities by the introduction of additional variables (called *slack* variables), and integer variables are converted into binary variables. In this way, a simple problem that at first glance seems to have fewer than 10 variables can quickly become a problem with more than 100 variables. Therefore, a brute-force approach will involve checking a very large number of combinations; in the case of a binary integer program with n variables, that will be 2^n combinations. Any improvement that can be made over using brute force— even if the improvement is only for certain types of integer programs—is valuable, and finding such improvements is an active area of mathematics research. (It is a good question whether such research falls under pure or applied mathematics; perhaps it is both!)

Hopefully, this introduction has convinced you that LP/IP/MILP is super-cool, mega-interesting, and seriously important… or at least cool enough, interesting enough, or important enough to learn more about it. Find ways to study it in your future.

Questions for LP/IP practice:

1. Set up the linear program described in Example 7.9.2: list and name the variables and constants, write down an objective function, and create the constraints indicated by the example as well as any hidden constraints you can discover.

2. Set up the integer program described in Example 7.9.2: list and name the variables and constants, write down an objective function, and create the constraints indicated by the example as well as any hidden constraints you can discover.

3. In your work with Example 7.9.2, you created some inequalities. For each inequality, introduce additional variables to make that inequality into an equation. How many additional variables do you need?

4. Using Example 7.9.1, rewrite each variable as a sum of binary variables. You will find it helpful to remember that there can be no more than 12 pieces of any one fruit.

7.10   Bonus Check-Yourself Problems

Solutions to these problems appear starting on page 608. Those solutions that model a formal write-up (such as one might hand in for homework) are to Problems 4 and 6.

1. Around Halloween, one can find bags of minipacks of SweeTarts. There are three SweeTarts in each pack, and the available color-flavors are orange, pink, purple, and blue.

   (a) How many different kinds of three-SweeTart minipacks are there?

   (b) Actually, if you open a pack reasonably (instead of ripping it completely apart), you get only one SweeTart out to eat at a time. How many different experiences of three-SweeTart minipacks are there?

2. In a 300-home neighborhood of Batamji, there are four different kinds of trees (magnolias, cypress, willow, and river birch). Forty homes have just cypress trees; 32 homes have just willow trees; 9 homes have just river birch. Seventy homes have magnolia and willow; 47 homes have magnolia and cypress; 40 homes have cypress and river birch; 61 homes have magnolia and river birch; 44 homes have cypress and willow; 56 homes have willow and river birch. Twelve homes have magnolias, cypress, willow, and river birch; 38 homes have magnolias, cypress, and willow; 19 homes have magnolias, willow, and river birch; 28 homes have magnolias, cypress, and river birch; 29 homes have cypress, willow, and river birch. How many homes have just magnolia trees?

3. A hungry ninja is making tacos with the following ingredients: beans, guacamole, cheese, tomatoes, scallions, salsa, and lettuce. How many ways can the ninja assemble tacos for different meals (breakfast, snack, lunch, tea, dinner), the first of which has three fillings, the next two of which have four fillings, and the final two of which have five fillings?

4. The Edgy Ruck company uses length-10 serial numbers that mix letters (except Y) and numbers. How many serial numbers are there that have a 7 in the fourth slot and a consonant in the eighth slot, or have a letter in the fifth slot and a vowel in the ninth slot?

5. All that is left of your Hello Kitty Jelly Belly sampler is the 12 Very Cherry flavored Jelly Bellies (because you *hate*

that flavor) and you have four friends who volunteer to eat them for you. How many ways are there to hand out the Jelly Bellies?

6. You've made a pile of eight cute notes for your best friend to find. Ze has 12 folders, one for each of hir classes and activities. How many ways are there to tuck the notes into folders? (Of course, you will not put more than one note in a folder. That would be excessive.)

7. The computer print-out says it all: Your first student needs three Learning Modules inserted, your second student needs five Learning Modules inserted, and your third student needs 54 Learning Modules inserted from the bank of 62 new government-approved-topic Learning Modules. But wait... The computer print-out doesn't say which

Learning Modules should go to which student. How many ways can you assign Learning Module topics to students?

8. Your spiky little plant has once again outgrown its pot, and when you split off all the small bits into different pots, you discover you have 23 spiky plant-spawn. You've promised eight people they can have baby spiky plants, but really you want to get rid of *all* of the spiky plant-spawn so they don't take over your house. How many ways are there to distribute the 23 baby spiky plants to the eight people?

9. How many anagrams are there of the word ENUMERATE?

10. How many ways are there to list the 50 U.S. states so that no two states beginning with "A" are next to each other?

7.11   Problems about Balls, Boxes, and PIEs

In this section, a *word* is a (finite) sequence of letters from a given alphabet, not necessarily a word found in a dictionary.

1. How many words of length w can be made using an alphabet with ℓ letters?

2. How many telephone numbers have no 0 in the prefix (the first three numbers before the hyphen)?

3. True or false: if $A \cap B \cap C = \emptyset$, then the sum principle applies so $|A \cup B \cup C| = |A| + |B| + |C|$.

4. How many words of length w can be made using a set of alphabets with a_j letters for the jth letter in a word?

5. You bring a bag of 12 snacks to an animal shelter and discover that there are 18 animals there. To be fair, you don't

want to give more than one snack to any individual animal. How many ways are there of distributing snacks?

6. How many seven-digit telephone numbers have an odd number of even numbers?

7. Consider the alphabet A, B, C, D, E, F.

   (a) How many four-letter words contain the subword ACE?

   (b) How many four-letter words don't begin with F or don't end in E?

   (c) How many five-letter words contain the subword CAB?

(d) How many four-letter words begin with C or end in two vowels?

8. Consider the alphabet A, B, C, D, E, F and make words without repetition of letters allowed.

   (a) How many six-letter words are there?

   (b) How many words begin with D or E?

   (c) How many words end in B or A?

   (d) How many words begin with D or E and end in B or A?

   (e) How many have first letter neither D nor E and last letter neither B nor A?

9. At the chocolate store, you decide to get a 20-chocolate box for your beloved. There are creams (maple, vanilla, orange, lemon, chocolate) and caramels (milk chocolate, dark chocolate, walnut) and coated nuts (milk chocolate peanuts, dark chocolate peanuts, milk chocolate almonds, dark chocolate almonds, white chocolate almonds), as well as cherry cordials, white chocolate truffles, solid chocolate pieces (in both milk and dark chocolate), and yogurt- and chocolate-coated pretzels. How many ways are there to fill the box?

10. **Multinomial mini-project:** The following problems introduce multinomial coefficients and the multinomial theorem. A *multinomial coefficient* is denoted by $\binom{k}{k_1,k_2,\ldots,k_n}$ and counts the number of ways, given a pile of k things, of choosing n mini-piles of sizes $k_1, k_2, \ldots, k_n$ (where $k_1 + k_2 + \cdots + k_n = k$).

(a) Show that multinomial coefficients give the solution to the number of ways of placing k labeled balls into n labeled boxes, such that the jth box holds k_j balls. (This needs to be a combinatorial proof: it then allows you to use $\binom{k}{k_1}\binom{k-k_1}{k_2}\binom{k-k_1-k_2}{k_3}\cdots\binom{k_n+k_{n-1}}{k_{n-1}}$ in calculating multinomial coefficients.)

(b) Show twice, once by symbolic manipulation and once by combinatorial proof, that $\binom{k}{k_1,k_2} = \binom{k}{k_1}$. Thus, multinomial coefficients for $n = 2$ agree with binomial coefficients.

(c) Show that $\frac{k!}{k_1!k_2!\ldots k_n!} = \binom{k}{k_1}\binom{k-k_1}{k_2}\binom{k-k_1-k_2}{k_3}\cdots\binom{k_n+k_{n-1}}{k_{n-1}}$.

(d) Show that multinomial coefficients give the solution to the number of ways of anagramming a word.

(e) Calculate $\binom{10}{3,2,5}$ and $\binom{12}{4,4,2,1}$.

(f) Show that it doesn't matter in what order the k_j appear in a multinomial coefficient. For example, $\binom{12}{4,4,2,1} = \binom{12}{4,1,2,4}$ and $\binom{10}{3,2,5} = \binom{10}{5,3,2}$.

(g) Show that $\binom{k}{k_1,k_2,k_3} = \binom{k-1}{k_1-1,k_2,k_3} + \binom{k-1}{k_1,k_2-1,k_3} + \binom{k-1}{k_1,k_2,k_3-1}$. In fact, do this both symbolically and using combinatorial proof. Thus, trinomial coefficients are generalizations of binomial coefficients.

(h) **Challenge:** Write down a general multinomial identity of that form.

(i) Prove the multinomial theorem shown in Figure 7.15. (Examining your proof of the binomial theorem

$$(x_1 + x_2 + \cdots + x_n)^k = \sum_{k_1+k_2+\cdots+k_n=k} \binom{k}{k_1, k_2, \ldots, k_n} x_1^{k_1} x_2^{k_2} \cdots x_n^{k_n}$$

Figure 7.15. The wider-than-a-column multinomial theorem.

will probably help.) Thus, multinomial coefficients are generalizations of binomial coefficients.

11. A common scheme for motor vehicle license plates is to require that the first three characters be letters and the last three characters be numbers. How many vehicles could a state have using such a scheme? In 2009, there were 10,699,846 valid vehicle registrations in New York state. Is the license plate scheme described in this problem appropriate for New York state?

12. A task force working to preserve the nose of Saint Tabascus is forming a chemistry committee with five members. They have calculated that there are 65,780 ways to form this committee. Of these, 24,480 have exactly one woman, 22,848 have exactly two women, 8,568 have exactly three women, 1,260 have exactly four women, and 56 have exactly five women.

   (a) How many committees have at least three women?

   (b) How many committees have no women?

   (c) How many committees have at most two women?

   (d) **Challenge:** How many members does the Preserve the Nose of Saint Tabascus Task Force have? And how many of those are women?

13. There are 8 strands of embroidery floss in a basket (all of different colors) and 11 different small pillows that could be embellished (with at most one strand of floss each). How many ways are there to embellish the pillows?

14. How many anagrams are there of the word SUPERCALIFRAGILISTICEX-PIALIDOCIOUS?

15. One college sent another a report saying that 119 students took Calculus I in a Fall semester. The report notes that during the next term, 96 of these students took Calculus II, 53 of them took Discrete Mathematics, and 39 of them took Physics II. The report also says that 38 of the students took both Calculus II and Discrete Mathematics, 31 of the students took both Discrete Mathematics and Physics II, 32 of the students took both Calculus II and Physics II, and 22 of the students took all three courses. We examine the report and sense an error is present. Why?

16. There are 20 plants in your garden but only enough fertilizer spray to boost 8 of them. How many ways are there to fertilize the plants?

17. How many different seven-digit phone numbers begin with 231- and contain at least one 9?

18. At a meeting of h hobgoblins and t trolls, it is decided that the collection of n chipmunk snax will be distributed in some way. Every hobgoblin must get at least one chipmunk but trolls are not required to receive chipmunks, yummy though they are. How many ways can the chipmunks be fed to the hobgoblins and trolls?

19. How many positive integer solutions are there to the equation $y_1 + y_2 + y_3 + y_4 + y_5 + y_6 = 13$?

20. Let's count banana splits. These are ice-cream treats that have three scoops of ice cream (two or three of the scoops could be the same flavor), three toppings (two or three of the toppings could

be the same flavor), whipped cream (always), a choice of nuts or no nuts, and a choice of a cherry or no cherry, all placed atop two banana halves. If there are 18 different flavors of ice cream and 5 choices of toppings, how many different banana split orders are possible? Note that people do care which toppings end up on which scoops, so the positions of the scoops should be labeled.

21. Suppose we want to arrange p things of one type and q things of another type into a line.

    (a) Argue that the p things can act as unlabeled balls and the q things can form separators between labeled boxes. How many ways are there to line up all the things?

    (b) Argue that the q things can act as unlabeled balls and the p things can form separators between labeled boxes. How many ways are there to line up all the things?

    (c) Aside from the fact that we are counting the same situation in two different ways, why do we get the same answer? (That is, give a symbolic explanation.)

22. When we kindly distributed 43 cupcakes to 12 baby mice, every baby mouse was guaranteed a single cupcake at least. What if a beneficent person of some sort guaranteed at least *two* cupcakes per baby mouse—how many ways would there be to feed the cupcakes to the baby mice?

23. Let us say that two words are equivalent if they are anagrams of each other. How many six-letter words are equivalent to BLEFLA? (Note that the words do not have to be sensical.) Is this an equivalence relation on the set of all words?

24. A person is murdered at twilight near the outskirts of a town of population 5,000, in view of several witnesses. The dastardly murderer has been described as male, between the ages of 16 and 35, with glasses, and under 6 feet tall. The police search Department of Motor Vehicles records for the city in an effort to narrow down the identity of the criminal.

Let F indicate that the driver is female, let O indicate that the driver is over the age of 35, let E indicate that the driver is not required to wear corrective lenses, and let T indicate that the driver is over 6 feet tall.

The police obtain the data given in Table 7.2. In a discussion of these data, the

characteristics	F	O	E	T
residents with characteristics	2,400	3,600	1,200	800
characteristics	F, O	F, E	F, T	O, E
residents with characteristics	1,200	900	700	800
characteristics	O, T	E, T	F, O, E	F, O, T
residents with characteristics	700	700	800	500
characteristics		F, E, T	O, E, T	F, O, E, T
residents with characteristics		500	400	200

Table 7.2. Police data from DMV records.

police chief says the murderer doesn't reside in the city. A detective responds by saying that the DMV records are incorrect. Why does each crime-fighting official make the claim that ze does?

25. How many passwords can be constructed that have between 6 and 12 characters and must use at least one letter and at least one number (and no special characters), but are not case sensitive?

26. How many passwords can be constructed that have between 8 and 12 characters and must use at least one upper-case letter, at least one lower-case letter, and at least one number (and no special characters)? If we assume that a pretty good hacker can attempt 1 billion passwords per second, how long would it take to crack a password made with this scheme?

27. According to PetHelpful.com, the top 10 duck snacks (in descending order) are algae, strawberries, mealworms, dandelions/clover, scrambled eggs, crickets, kale, feeder fish, earthworms, and marigolds. You have acquired a sample box of duck-snack packs, with one sample pack of each of these top 10 duck snacks. How many ways are there to distribute one pack to each of your 6 ducks? What if you just got the ducks, and not only are they all grey (of course), but you don't know them well enough to tell them apart?

28. Jayke has a bag of 54 super-fancy candies (different flavors and wrappers and...) to give out to people waiting in line. Ze gives two candies to the first person, and then because the next person has been waiting longer, gives hir

an additional candy, and indeed gives out one more piece of candy to each person in line than ze gave to the previous person. How many ways are there for Jayke to give out the candies?

29. There are seven flavors of sorbet available at Tebros, namely blackberry-lime, meyer lemon, watermelon, blueberry, pink grapefruit champagne, dandelion-lilac, and durian.

    (a) How many ways are there to order a three-scoop cone at Tebros?

    (b) How many ways are there to order a five-scoop bowl at Tebros?

    (c) How many ways are there to order two different hand-packed pints of sorbet to take home?

    (d) How many different anagrams are there of TEBROS?

    (e) How many different anagrams are there of SCOOP?

30. How many ways can 11 chemistry textbooks and 6 post-modern novels be arranged on a bookshelf?

31. A narwhal, a unicorn, and a rhinoceros go into a bar. How many ways can they order Italian sodas (there are 18 flavors available) from the jackalope bartender?

32. At Half Do– – –n–t, one buys—you guessed it!—half dougnuts. On any given day there are eight flavors available.

    (a) How many ways are there to pick three different flavors?

    (b) How many ways are there to order three half doughnuts?

    (c) If there are a dozen half doughnuts in the *Chocolate or Apple!* case, of which six are chocolate apple and eight have apple, then how many are plain chocolate and how many are plain apple half doughnuts?

33. Today you go to Half Do–––n–t with your BFF.

    (a) Your BFF wants to split a four-half-doughnut order with you. Ze will pick two different half doughnuts, and you will pick two different half doughnuts. How many ways are there to choose the order (noting which half doughnuts belong to which person)?

    (b) What if your BFF insists that there be no overlap between your half doughnut flavors and hir half doughnut flavors? How many ways are there to choose the order (noting which half doughnuts belong to

which person)? How many four-half-doughnut orders are possible here?

34. International individual of mystery Pvaanzba Ohaf is faced with a corridor of 13 doors.

    (a) How many ways can Pvaanzba Ohaf select four doors on which to paint numbers $(1, 2, 3, 4)$ in invisible ink?

    (b) How many ways can Pvaanzba Ohaf paint three doors puce, five doors pink, two doors periwinkle, and three doors peach?

    (c) How many anagrams are there of PVAANZBA OHAF?

    (d) How many anagrams of PVAANZBA OHAF contain exactly three As?

35. How many anagrams are there of DISCRETE MATHEMATICS WITH DUCKS?

7.12   Instructor Notes

Most students will find solving counting problems to be confusing. And indeed, this topic *is* confusing! It is included to give students a taste of the complexity involved in combinatorics and to let them experience counting problems that are a step above the nigh-trivial straightforward $\binom{n}{k}$-type problems. The goal is not to have students master counting problems, but to have students acquire enough experience that they can work through counting problems using the text as an aid.

To that end, here is one way to approach this material. Ask students to read Sections 7.1 and 7.2 before the first class of the week. Then, split this class between an interactive lecture and letting the students loose to work on Section 7.3. The interactive lecture can give balls-and-boxes sample problems and discuss how to solve them (thus foreshadowing Section 7.4). I, and others, have done this successfully by listing the problems on different sections of the board and then asking the students for ideas on how to solve them. They are often creative (and wrong) in their attempts, and this provides good clarification between problem types—it's easy for students to misread problem statements! After the "easy" types have been addressed, the instructor then takes over to present the stars-and-bars argument.

Students should then be asked to read Section 7.4, to deepen and reinforce their understanding, for the following class. (Section 7.5 could be assigned reading after the first or second class meeting.) Spend the bulk of the next two class meetings on having the students do problems from Section 7.6 in groups—but of course, take time to ask them if they have questions over the reading or Check Yourself problems. They will argue back and forth about how to determine the type of each problem and then how to get the solution to make sense; it takes quite a while, but these are productive arguments. About five to ten minutes before the end of such a class meeting, survey the students to see whether any problems have been solved by most groups and then discuss those problems.

A fun showmanship opportunity for starting a class based primarily on doing problems is to show the Flash movie *The Donuts*, available at https://youtu.be/2bhKHg8xX4M—ask students how it relates to the class, and it is likely that someone will identify this as an opportunity for inclusion/exclusion analysis.

There are a couple of side effects of combining many problem types into one chapter. One is that students are likely to confuse problem types, either by misreading problems (or leaving out crucial words) or misinterpreting problems. Another is that some students will invent somewhat laborious counting arguments for individual problems. Be aware that you may need to wade through such proposed solutions. A common error is that a student will separate counting into two stages, using a variable for how the stages are to be split, and then not realize that ze must sum over that variable to obtain all possibilities. While these issues are annoying to deal with, an unnecessarily complicated solution does have the bonus of revealing a combinatorial identity (when the resulting expression is equated with a simpler solution).

Chapter 8 🐥🐥🐥🐥🐥🐥🐥🐥

Recurrences

8.1 Introduction and Summary

Our introduction to recursion is the Fibonacci sequence, which is $1, 1, 2, 3, 5, 8,$ $13, \ldots$ and is defined by $F_n = F_{n-1} + F_{n-2}$. There are lots of interesting identities one can prove about the Fibonacci sequence (just as we saw there are for binomial coefficients). Many of these can be proven by induction—as foreshadowed in Chapter 4, this proof technique is pervasive.

We will then study integer sequences in general, with a focus on recursively defined sequences. For example, the sequence $1, 2, 4, 8, \ldots$ is recursively defined by $a_n = 2a_{n-1}$. This sequence is also defined by the formula $a_n = 2^n$. A general question is how to find formulas (in terms of n rather than a_{n-1}, etc.) for recurrences; this is important because it's faster to compute using a formula than using a recurrence. Once a formula is found, we need to prove that it is correct. Induction is the proof method of choice, as the recursive definition of a sequence usually provides the induction step in the proof.

In this text, we will only learn the very basics of finding formulas for sequences and recurrences. The simplest sequences are arithmetic and geometric sequences; they have linear and exponential formulas, respectively. A formula for an arithmetic sequence can be found using first differences, and a formula for a generalized geometric sequence can be found using a characteristic equation. We will explain how and why these solution techniques work. As frosting on the cake, we will show how to find a closed form for the Fibonacci sequence!

8.2 Fibonacci Numbers and Identities

You are well aware that $1 + 1 = 2$ (one hopes). Similarly, you must know that $1 + 2 = 3$ and that $2 + 3 = 5$ and also that $3 + 5 = 8$. But you may not have detected the pattern in these three statements—they look like statements about addition, but secretly they are a way of generating a sequence of numbers. Let's take some notes.

The numbers that appear above are first $1, 1$, and these are added together to produce 2 (duh). This gives us $1, 1, 2$ so far. Then we add the last two numbers in our short sequence together and append the result to the end, yielding $1, 1, 2, 3$. If we continue, we have $1, 1, 2, 3, 5, 8, 13, \ldots$. (You might enjoy generating some more of this sequence yourself.) If we name the nth one of these numbers F_n, then the general rule via which we extend our sequence is $F_n = F_{n-1} + F_{n-2}$. Notice that in order to begin, we need a first and second number so that we have enough information to generate a third number.

This, in case you haven't encountered it before, is the famous *Fibonacci sequence*, named after some guy from the Middle Ages who liked to talk about rabbits. (That's only sort of true, and also only sort of untrue. See the "MacTutor History of Mathematics Archive" biography at http://www-groups.dcs.st-and.ac. uk/~history/Biographies/Fibonacci.html for more and more interesting information.) We may consider $F_n = F_{n-1} + F_{n-2}$ to be the defining characteristic of the Fibonacci sequence.

There are all sorts of interesting facts about Fibonacci numbers. For example, let's try adding them up as follows:

$$1 = 1.$$
$$1 + 1 = 2.$$
$$1 + 1 + 2 = 4.$$
$$1 + 1 + 2 + 3 = 7.$$
$$1 + 1 + 2 + 3 + 5 = 12.$$
$$1 + 1 + 2 + 3 + 5 + 8 = 20.$$
$$1 + 1 + 2 + 3 + 5 + 8 + 13 = 33.$$

Perhaps you don't yet see a pattern. But if we add 1 to both sides of each of these statements, you will:

$$1 + 1 = 1 + 1 = 2.$$
$$1 + 1 + 1 = 2 + 1 = 3.$$
$$1 + 1 + 2 + 1 = 4 + 1 = 5.$$
$$1 + 1 + 2 + 3 + 1 = 7 + 1 = 8.$$
$$1 + 1 + 2 + 3 + 5 + 1 = 12 + 1 = 13.$$
$$1 + 1 + 2 + 3 + 5 + 8 + 1 = 20 + 1 = 21.$$
$$1 + 1 + 2 + 3 + 5 + 8 + 13 + 1 = 33 + 1 = 34.$$

Aha! We get the Fibonacci sequence back (minus the first few terms)! In general, it appears that $F_1 + F_2 + \cdots + F_n + 1 = F_{n+2}$. And you know what comes next—we need to prove this statement. Our old friend induction will be of use.

Theorem 8.2.1. *For $n \in \mathbb{N}$, $1 + \sum_{j=1}^{n} F_j = F_{n+2}$.*

Proof: We begin with a base case. However, we have already done a bunch of base cases just above (in creating the statement of the theorem), so we can safely dispense with that. Our inductive hypothesis should be that as long as $n \leq k$, $1 + \sum_{j=1}^{n} F_j = F_{n+2}$. So we look at $1 + \sum_{j=1}^{k+1} F_j$ and recognize that $1 + \sum_{j=1}^{k+1} F_j = 1 + \sum_{j=1}^{k} F_j + F_{k+1}$. This allows us to use our inductive hypothesis, so by substituting in, we have $1 + \sum_{j=1}^{k} F_j + F_{k+1} = F_{k+2} + F_{k+1}$. But, aha! The rule for how Fibonacci numbers are generated, $F_n = F_{n-1} + F_{n-2}$, can be rewritten by substituting $k + 2$ for n, so that it says $F_{k+2} + F_{k+1} = F_{k+3}$. Notice that this rule is exactly what we need to complete the inductive step. Putting it all together, we have $1 + \sum_{j=1}^{k+1} F_j = F_{k+3}$, and hey, that's what we wanted to prove, so our inductive step is complete. We. Are. Done. $\square$

Here's another wild property of the Fibonacci numbers: take three Fibonacci numbers in a row from the sequence, multiply the two outer Fibonacci numbers together, and you'll be 1 off from the square of the middle Fibonacci number. What?!? Check it out!

$$2, 3, 5 \rightarrow 2 \cdot 5 = 10 \text{ and } 3^2 = 9 = 10 - 1.$$
$$1, 2, 3 \rightarrow 1 \cdot 3 = 3 \text{ and } 2^2 = 4 = 3 + 1.$$
$$3, 5, 8 \rightarrow 3 \cdot 8 = 24 \text{ and } 5^2 = 25 = 24 + 1.$$
$$21, 34, 55 \rightarrow 21 \cdot 55 = 1,155 \text{ and } 34^2 = 1,156 = 1,155 + 1.$$

Holy cow. Here is the general theorem.

Theorem 8.2.2. *For $n > 1, n \in \mathbb{N}$, $F_{n-1}F_{n+1} = (F_n)^2 + (-1)^n$.*

Proof: Again, induction will come to our aid. Here's a base case (in addition to our experimentation above, for good measure): Considering $1, 1, 2$, we see that $1 \cdot 2 = 2$ and $1^2 + 1 = 2$. This is lovely, but we need to make sure the sign (addition versus subtraction) works out! Therefore, we must look at the power to which -1 is raised. Here, the first number is $1 = F_1$, so this tells us $n - 1 = 1$ so that

$n = 2$. This gives $1^2 + (-1)^2 = 1 + 1 = 2$, as desired. Now for the inductive hypothesis; it should say that for any $n \leq k$, $F_{n-1}F_{n+1} = (F_n)^2 + (-1)^n$. So, let us look at F_kF_{k+2} as this is the left-hand side of the statement for the case $n = k+1$. Again we will use the fact that $F_n = F_{n-1} + F_{n-2}$ as the crucial ingredient for the inductive step; here, we will note that substituting $k+2$ for n gives us $F_{k+2} = F_{k+1} + F_k$. Substituting into F_kF_{k+2} produces $F_kF_{k+2} = F_k(F_{k+1} + F_k) = F_kF_{k+1} + (F_k)^2$. Our inductive hypothesis says that $(F_k)^2 = F_{k-1}F_{k+1} - (-1)^k$, so $F_kF_{k+1} + (F_k)^2 = F_kF_{k+1} + F_{k-1}F_{k+1} - (-1)^k$. Factoring out an F_{k+1}, we have $F_{k+1}(F_k + F_{k-1}) - (-1)^k$, and using the definition of the Fibonacci sequence again, this equals $F_{k+1}(F_{k+1}) - (-1)^k$. That's almost what we desire—we want to obtain the statement that $F_kF_{k+2} = (F_{k+1})^2 + (-1)^{k+1}$. But aha, $-(-1)^k = (-1)(-1)^k = (-1)^{k+1}$, so we can substitute and achieve our desired conclusion. □

As you will soon discover, there are many more such identities—more than you can shake a stick at. (You can, in fact, shake a stick at a multitude of items at once. See http://www.gocomics.com/theargylesweater/2012/03/22.) You are encouraged to experiment—really, play around—with the Fibonacci numbers and see what you can discover.

Check Yourself ─────────────────────────────────

These won't take long—do them all.

1. List the third, sixth, and seventh Fibonacci numbers.

2. Write out the ten Fibonacci numbers after 55.

3. Find a formula for the sum of the first n odd-index Fibonacci numbers (F_1, F_3, etc.).

4. Find a formula for the sum of the first n even-index Fibonacci numbers (F_2, F_4, etc.).

8.3   Recurrences and Integer Sequences and Induction

The Fibonacci sequence has three properties that we have been exploiting:

🦆 It is an integer sequence.

🦆 It is defined by a recurrence.

🦆 Induction is useful in proving identities involving it.

We will deal with these in turn, though it turns out that only two of the properties matter—the third one follows logically from the other two.

Integer sequences are just what they sound like—they are integers listed in some order. Technically, an *integer sequence* is a function $s : \mathbb{N} \to \mathbb{Z}$ or $s : \mathbb{W} \to \mathbb{Z}$, but rarely do we think of a sequence in this manner. A generic integer sequence is denoted $a_1, a_2, a_3, \ldots, a_n, \ldots$ or $a_0, a_1, a_2, \ldots, a_n, \ldots$; in either case, a_n is called the nth term of the sequence. Usually, though not always, the integers count something. Often, we know the first few numbers of an integer sequence and would like to know what sort of rule produces the rest of the sequence. Such a rule could be a formula into which one plugs n and gets out the nth integer in the sequence; this is called a *closed form* or a *closed-form formula* for the sequence. Or, the rule could be a way to generate more terms of the sequence, knowing some of the previous terms; this is called a *recurrence* or a *recurrence relation*, and the process of using a recurrence is called *recursion*. Sometimes a closed form is called an *explicit formula*, in contrast to the implicit expression of a recurrence. More excitingly, we would also like to know what else is counted by our proposed integer sequence. (This can be an excellent source of ideas for combinatorial proof.)

The easiest way to find out what's going on with a particular sequence of numbers is to consult the Online Encyclopedia of Integer Sequences (OEIS). The OEIS is described at http://oeis.org/wiki/Welcome, and you can look up sequences as in a dictionary at http://oeis.org/. Once upon a time, a long, long time ago now, there was a small blue book called *A Handbook of Integer Sequences* (1973) that really *was* like a dictionary—first were sequences starting with 1, then those starting with 2, and so on. (This later became *The Encyclopedia of Integer Sequences* (1995).) The whole thing started when Neil J. A. Sloane was in graduate school and started collecting sequences. Sloane still does a lot of personal maintenance on the OEIS database.

Anyway, because the OEIS is an encyclopedia, the information is abbreviated. It gives many terms for each sequence and a recurrence or closed form if they are known, as well as different descriptions for what the sequence counts and references to literature that discusses the sequence. There are no explanations, just lists of facts, so it is still quite worthwhile to learn how to go between recurrences and closed forms and how to prove facts about sequences.

Furthermore, for any given initial set of terms of a sequence, there may be many, many different sequences that have those same initial terms! For example, $1, 1, 2, 3, \ldots$ leads to at least 3,776 different sequences, and even $1, 1, 2, 3, 5, \ldots$ leads to at least 672 different sequences. So we *must* understand the mathematical situation that leads to a sequence in order to know how it continues and what defines it.

Definition 8.3.1. A *recurrence* is a statement of the form $a_n = $ (some stuff, some of which involves $a_{\text{smaller than } n}$).

Example 8.3.2 (of recurrences). The recurrence $a_n = 2a_{n-2} + 1$ governs the sequence $1, 0, 3, 1, 7, 3, 15, 7, \ldots$, as well as the sequence $2, 4, 5, 9, 11, 19, 23, \ldots$. The recurrence $a_n = 3na_{n-1} - 6a_{n-2} + a_{n-5}$ governs the sequence $-1, 1, 0, -1, 3, 23,$ $466, 1{,}260, 983, \ldots$.

Just as the Fibonacci recurrence relation indicates that two initial values are needed to generate the rest of the sequence, we can determine how many initial values are required by any recurrence relation. Any term $a_{\text{smaller than } n}$ can be written as $a_{n-\ell}$ for some ℓ, and the largest such ℓ that appears in the recurrence relation is the number of initial values needed.

Example 8.3.3 (of generating terms from a recurrence). Consider the recurrence $a_1 = 1, a_2 = 1, a_n = -a_{n-1}a_{n-2} + n$. We already have a_1, a_2, so let's start with $n = 3$.

$a_3 = -a_2a_1 + 3 = -1 \cdot 1 + 3 = 2.$ Keep going. $a_7 = -4 \cdot 1 + 7 = 3.$
$a_4 = -a_3a_2 + 4 = -2 \cdot 1 + 4 = 2.$ $a_8 = -3 \cdot 4 + 8 = -4.$
$a_5 = -a_4a_3 + 5 = -2 \cdot 2 + 5 = 1.$ $a_9 = -(-4)3 + 9 = 21.$
$a_6 = -a_5a_4 + 6 = -1 \cdot 2 + 6 = 4.$ $a_{10} = -21(-4) + 10 = 94.$

Our first 10 terms are thus $1, 1, 2, 2, 1, 4, 3, -4, 21, 94$.

Example 8.3.4. The recurrence $a_n = a_{n-2} + a_{n-4}$ requires four initial values. If we let those be $1, 0, 0, 1$, we obtain the sequence $1, 0, 0, 1, 1, 1, 1, 2, 2, 3, \ldots$.

Notice that recurrence relations are great in that they define sequences entirely. But they are terrible in that if you want to know the 1,000th term of the sequence, you first have to compute the 999 previous terms. And while that might not take a computer very long, it is a huge pain if those terms have to be recalculated every time you want a new value, and that is just a waste of processor time and memory. It is therefore quite useful to have a closed form in addition to a recurrence.

Whenever you encounter a recurrence and the word "proof" is floating in the air or your mind nearby, you should immediately have a flashing sign light up in your head that reads INDUCTION. Why is this? Recall that induction is effective for statements indexed by the natural numbers, and an integer sequence is indexed by the natural numbers. Also, the process of induction involves a base case or two and then a reduction from the $(k+1)$st case to earlier case(s) (the kth case or before). A recurrence exactly translates the nth term of a sequence into a formula involving earlier terms—rewriting, it can translate the $(k+1)$st term of a sequence into a formula involving earlier terms. Basically, a recurrence is the key to an inductive step—and in the case of proving that a closed-form formula is correct, it *is* the inductive step!

Example 8.3.5. Consider the sequence $1, 2, 4, 8, \ldots$, recursively defined by $a_n = 2a_{n-1}$. We suspect that a closed form for this sequence is $a_n = 2^n$. Let's prove it—by induction, of course.

(Base cases) As base cases, note that $2^1 = 2$ and $2^2 = 4$. Hmm. This must mean that the first element of the sequence is $a_0 = 1 = 2^0$.

(Inductive hypothesis) Our inductive hypothesis is that for $n \leq k$, $a_n = 2^n$.

(Inductive step) Now we will consider a_{k+1} in service of the inductive step. We immediately notice that the recursive definition of the sequence can be rewritten to read $a_{k+1} = 2a_{k+1-1} = 2a_k$. But, we know that $a_k = 2^k$ by our inductive hypothesis, so substitution gives us $a_{k+1} = 2 \cdot 2^k = 2^{k+1}$ as desired.

We can generalize this procedure to a mock induction proof for verifying that a closed form matches a recurrence.

How to prove that a closed form for a recurrence is correct. Suppose that we have a sequence with initial values $a_1, a_2, a_3, a_4, \ldots$, a recursive definition $a_n =$ (some stuff, some of which involves $a_{\text{smaller than } n}$), and a formula $a_n =$ (a function of n).

1. Decide to proceed by induction.

2. Check base cases by plugging $n = 1, 2, 3$ into (the given function of n); hopefully the result will be a_1, a_2, a_3.

3. State the inductive hypothesis: for $n \leq k$, $a_n =$ (the proposed function of n).

4. Consider the $(k+1)$st term, a_{k+1}.

5. Rewrite the recurrence by substituting $k + 1$ in for each copy of n, obtaining something like $a_{k+1} =$ (some stuff, some of which involves $a_{\text{smaller than } k+1}$).

6. For every $a_{\text{smaller than } k+1}$ in the rewritten recurrence, use the inductive hypothesis to substitute the formula for (the given function of smaller than $k + 1$).

7. Do some symbolic manipulations so that the rewritten recurrence becomes a function of $k+1$.

8. Notice that you now have $a_{k+1} = $ (the given function of $k+1$) and therefore you're done.

There is a sequence-y situation in which induction is not necessary. Suppose you know some sequence a_n, with its initial terms and recurrence relation and closed form, and you encounter a sequence b_n. You intuit (and then prove) that b_n has the same recurrence relation as a_n and then discover that b_n has the same initial terms as a_n. That means that $b_n = a_n$, so it has the same closed form!

A special case. If sequences a_n and b_n can be defined by the same recurrence relation and have the same initial terms, then $a_n = b_n$.

The remainder of this chapter focuses on finding closed forms for recurrences. We will build up our understanding from looking at recurrences from an intuitive standpoint to learning how to recognize some specific types of recurrences and solve for their associated closed forms.

Check Yourself

Sample enough of these problems that you are comfortable with using the ideas of this section.

1. Find a closed form for each of these sequences:

   (a) $1, 2, 3, 4, 5, \ldots$
   (b) $1, 2, 4, 8, 16, \ldots$
   (c) $1, 2, 6, 24, 120, \ldots$
   (d) $1, 4, 9, 16, 25, \ldots$
   (e) $1, -4, 16, -64, \ldots$

2. Write out the first five terms of each of the sequences defined by these closed-form formulas:

   (a) $a_n = n^2 - 3n + 4$.
   (b) $a_n = n - n^2 + 5$.
   (c) $a_n = 3F_n - 1$.
   (d) $a_n = 3\binom{n}{2}$.
   (e) $a_n = 2^n - 2$.

(f) $a_n = \binom{n+2}{3}$.

(g) $a_n = (n-2)^2$.

3. Give an example of a sequence that is not an integer sequence.

4. Find a sequence different from $a_n = 2^n$ that satisfies the recurrence $a_n = 2a_{n-1}$.

5. Find the recurrence that defines the sequence $1, 3, 6, 10, 15, \ldots$.

6. Find a description for the sequence $2, 3, 5, 7, 11, 13, \ldots$. Can you find a closed form or a recurrence that defines this sequence?

8.4 Try This! Sequences and Fibonacci Identities

Some of these problems will be done much more efficiently with collaboration. (That's one way of warning you of a challenge.) You will need internet access in order to complete Problem 4.

1. Write out the first several terms of the integer sequence defined by $a_1 = 1$, $a_2 = 2, a_n = a_{n-1} + 2a_{n-2}$.

   (a) How do things change if you begin with $a_1 = 2, a_2 = 1$?

   (b) What if you begin with $a_1 = a_2 = 1$?

2. For each of the following closed forms, write out the first several terms of the sequence (at least five) and use this to create a recurrence for the sequence.

   (a) $a_n = 3n + 1$.

   (b) $a_n = 2n + 7$.

   (c) $a_n = 3n^2 + n$.

   (d) $a_n = 2^{2n-1} - 1$.

3. For each of the following recurrence relations, write out the first several terms of the sequence (at least five) and use this to find a closed form for the sequence. Then prove that each conjectured closed form is correct. (Is there a flashing sign in your head that indicates a fruitful approach?)

   (a) $a_1 = 3; a_n = 2a_{n-1}$.

   (b) $a_1 = 4; a_n = 4 + a_{n-1}$.

   (c) $a_1 = 1; a_n = 4 + a_{n-1}$.

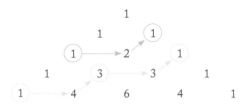

Figure 8.1. Marking certain entries in Pascal's triangle.

4. Choose one of the sequences you generated above, and input the first five terms into the Online Encyclopedia of Integer Sequences at http://www.oeis.org. How many different known sequences contain these terms? Find the specific sequence you were working with; how many different things does it count? (Can you figure out why it counts any of them?)

5. Write out several rows of Pascal's triangle. Circle the first 1 of any row. Go over one entry to the right and then one entry up and to the right, and circle that second entry (see Figure 8.1). Repeat this procedure until you run out of triangle. Then compute the sum of the entries you marked.

   Do this procedure for several rows. What numbers do you get? Make a conjecture, then formulate your conjecture as an identity involving binomial coefficients. Finally, prove your conjecture is correct.

6. Prove or disprove that $F_n + \sum_{j=1}^{k} F_{n+2j+1} = F_{n+2k+2}$.

7. Prove that $F_{n+1}^2 - F_{n+1}F_n - F_n^2 = (-1)^n$.

8. Prove a few things:

   (a) Show that $F_{n+3} = 2F_{n+1} + F_2 F_n$.

   (b) Show that $F_{n+4} = 3F_{n+1} + 2F_n$.

   (c) Show that $F_{n+5} = 5F_{n+1} + 3F_n$.

   (d) Now show that for a fixed k, $F_{n+k} = F_k F_{n+1} + F_{k-1} F_n$.

9. Prove that $F_{2n} = F_n F_{n+1} + F_{n-1} F_n$.

8.5 Naive Techniques for Finding Closed Forms and Recurrences

The least formulaic way to find a closed form associated with a given recurrence is to generate a lot of terms and stare at them until you have a conjecture for a possible closed form. Then try to prove it. This sort of naive attempt frequently does not work, so we will introduce progressively more sophisticated techniques for finding closed forms.

A method that is very bash-y (and only sometimes works) is to iterate the recurrence. (This process is known as *recursing*, as it performs recursion.)

Example 8.5.1. Let us start with $a_n = a_{n-1} + (n-3), a_1 = 1$. We will plug in the recurrence, but translated for a_{n-1}. First, note that $a_{n-1} = a_{(n-1)-1} + ((n-1) - 3)$ $= a_{n-2} + (n-4)$. So $a_n = a_{n-2} + (n-4) + (n-3)$. A few more iterations produces $a_n = a_{n-5} + (n-7) + (n-6) + (n-5) + (n-4) + (n-3)$. Eventually, we get the statement $a_n = a_{n-(n-2)} + (n-(n-2)-2) + \cdots + (n-4) + (n-3) = a_2 + 0 + \cdots + (n-4) + (n-3)$. We compute $a_2 = 0$, and then have $a_n = 0 + 0 + 1 + \cdots + (n-3) = \sum_{j=1}^{n-3} j$. Some experimentation shows that the sum is $\frac{(n-2)(n-3)}{2}$, so we conjecture that $a_n = \frac{(n-2)(n-3)}{2}$. We can prove this is correct using induction (but we will not).

If you try either of these techniques in an attempt to find a closed form for the Fibonacci sequence, or even on $a_n = a_{n-2} + 8n^2 + 3$, it won't work. For some sequences you may get lucky—you might reformulate a sequence so that it becomes clear that it has the same recurrence and initial terms as a sequence you know about. And you can look up a closed form for the Fibonacci sequence (but don't do it! We'll compute one later in Example 8.8.5).

Check Yourself ─────────────────────────────────────

Figure out closed forms for some of these recurrences and check (briefly) by induction that they're correct.

1. $a_0 = 1, a_n = 3a_{n-1}$.

2. $a_1 = 1, a_n = -2a_{n-1}$.

3. $a_0 = 1, a_n = 2a_{n-1} - 1$.

4. $a_1 = 0, a_2 = 1, a_n = a_{n-1}a_{n-2}$.

5. $a_1 = a_2 = 1, a_n = a_{n-1}a_{n-2}$.

8.6   Arithmetic Sequences and Finite Differences

Let's start with an example. Consider the recurrence

$$a_n = a_{n-1} + 3, a_0 = 2.$$

It produces the sequence

$$2, 5, 8, 11, 14, \ldots.$$

Examine the differences between consecutive terms, $a_{n+1} - a_n$.

$$
\begin{array}{ccccc}
2 & 5 & 8 & 11 & 14 \quad \ldots \\
 & 3 & 3 & 3 & 3
\end{array}
$$

Notice that all of the differences are 3. *All* of them. And, the closed form for this recurrence is $a_n = 2 + 3n$. (Check it—it works.)

A sequence with constant differences is called an *arithmetic sequence*. In fact, if $a_n = a_{n-1} + d, a_0 = c$, then the closed form is $a_n = c + dn$. We can see why by examining the differences between terms.

$$
\begin{array}{ccccc}
a_0 & a_1 & a_2 & a_3 & a_4 \quad \ldots \\
 & d & d & d & d
\end{array}
$$

With each successive term, we add another d, so by the nth term we've added dn. This approach generalizes to recurrence relations of the form $a_n = a_{n-1} + p_k(n), a_0 = c_0$, where $p_k(n)$ denotes a polynomial of degree k (i.e., the highest power is k) with input variable n.

Example 8.6.1. The recurrence $a_n = a_{n-1} - 10n + 8, a_0 = 2$ produces the sequence $2, 0, -12, -34, -66, \ldots$. Let us check the differences between consecutive terms as before.

$$
\begin{array}{ccccc}
2 & 0 & -12 & -34 & -66 \quad \ldots \\
 & -2 & -12 & -22 & -32
\end{array}
$$

Uh-oh. We didn't get constant differences. Well, let us forge ahead and take differences again in case that helps. (We already took *first* differences, and now we will take *second* differences.)

$$
\begin{array}{ccccc}
2 & 0 & -12 & -34 & -66 \quad \ldots \\
 & -2 & -12 & -22 & -32 \\
 & & -10 & -10 & -10
\end{array}
$$

Hey! Those are constant! Cool. Uh... so how do we use this to get a closed form? Well, we had a linear function before, and this time it will be quadratic. Why? Before, the closed form was linear because we were adding the same thing over and over. The first differences were constant. Now, the second differences are constant, so the first differences will be linear. Then, adding linear terms over and over produces a quadratic function. Therefore, the closed form will look something like $a_n = c + dn + fn^2$, and we need to determine c, d, and f.

When $n = 0$, we have $a_0 = 2 = 2 + d0 + f0^2$, so we know $c = 2$.

When $n = 1$, we have $a_1 = 0 = 2 + d + f$. That doesn't seem helpful yet, but if we combine it with $a_2 = -12 = 2 + 2d + 2^2 f = 2 + 2d + 4f$, we can make progress. We now have two linear equations and two unknowns, and using high-school algebra, we can solve for d and f.

The equation $0 = 2 + d + f$ is the same as $f = -d - 2$, and substituting that into $-12 = 2 + 2d + 4f$ gives us $-12 = 2 + 2d + 4(-d - 2) = 2 + 2d - 4d - 8 = -6 - 2d$ or $-6 = -2d$ so that $d = 3$.

Then $f = -3 - 2 = -5$, so our closed form is $a_n = 2 + 3n - 5n^2$.

As promised, this process generalizes to polynomials of higher degree! Then the kth differences of the corresponding sequence will be constant, and the closed form will be $a_n = c_0 + c_1 n + c_2 n^2 + \cdots + c_k n^k$. Looking at $a_0, a_1, a_2, \ldots, a_k$ will give us c_0 and k linear equations in k unknowns, and we can solve them by hand as done above (though solving them is faster if you know some linear algebra).

How to find a closed form for a recurrence relation of the form $a_n = a_{n-1} + p_k(n), a_0 = c_0$:

1. Check to see whether the sequence has constant differences. If so, then the closed form is $a_n = c + dn$.

2. The kth differences of the corresponding sequence will be constant, so figure out k. (It is one more than the highest power in the polynomial.)

3. Use $a_0, a_1, a_2, \ldots, a_k$ to get c_0 and k linear equations in k unknowns.

4. Solve these by hand, either using high-school algebra or linear algebra. Or feed them to a computer-algebra system.

5. The k unknowns are $c_1, \ldots, c_k$. The closed form is $a_n = c_0 + c_1 n + c_2 n^2 + \cdots + c_k n^k$.

Side note.   The calculus-savvy and adventurous reader can attempt to understand why taking successive finite differences should determine a closed form for some recurrences. Envision a discrete version of calculus—where taking differences is like taking a derivative discretely. Finding a closed form from kth differences is then like integrating k times (again, discretely). Yes, that's not really an explanation… but it is a hint towards investigations you might do in order to justify this process to yourself.

Finally, please notice that this solution technique is only useful for a very limited sort of recurrence relation ($a_0 = c_0, a_n = a_{n-1} + p_k(n)$). The recurrence has to have *one* lower term, with a coefficient of 1. And the extra stuff added on has to be a polynomial, not any other sort of function. But, that's the way things are—in general, it is *not* easy to find closed forms for recurrence relations.

Check Yourself

Do a representative sampling of these problems.

1. Can a closed form for the recurrence $a_0 = 2, a_n = 3a_{n-1} - 7$ be found using the techniques of this section? Why or why not?

2. Can a closed form for the recurrence $a_0 = 22, a_n = a_{n-1} + 9n^9 - 12n^7 + n^6 - 43n - 7$ be found using the techniques of this section? Why or why not?

3. Can a closed form for the recurrence $a_0 = 2, a_n = a_{n-1} + 3^{n-7}$ be found using the techniques of this section? Why or why not?

4. Can a closed form for the recurrence $a_0 = 2, a_1 = 22, a_n = a_{n-1} + a_{n-2} - 22$ be found using the techniques of this section? Why or why not?

5. Try to use kth differences to find a closed form for the recurrence $a_0 = 1, a_n = 2a_{n-1}$. What happens?

6. Try to use kth differences to find a closed form for the recurrence $a_0 = 0$, $a_n = a_{n-1} + (n+1)(-1)^{n+1}$. What happens?

7. Find a closed form for $a_0 = 8, a_n = a_{n-1} - 4$ and check that your formula is correct.

8. Find a closed form for $a_0 = 3, a_n = a_{n-1} + 2$ and check that your formula is correct.

9. Challenge: Create your own recurrence for which a closed form can be found using kth differences, and find that closed form.

8.7 Try This! Recurrence Exercises

The power of combined minds will produce the creativity necessary to solve these problems.

1. Variations on a recurring theme:

    (a) In the first Check Yourself problem on page 255, you found a closed form for the recurrence $a_0 = 1, a_n = 3a_{n-1}$. Recall that closed form.

    (b) Examine $a_0 = 1, a_1 = 3, a_n = 2a_{n-1} + 3a_{n-2}$. Generate some terms of the sequence, conjecture a closed form, and then prove that your closed form is correct.

    (c) Now examine $a_0 = 1, a_1 = -1, a_n = 2a_{n-1} + 3a_{n-2}$. Generate some terms of the sequence, conjecture a closed form, and then prove that your closed form is correct.

    (d) This time, look at $a_0 = a_1 = 2, a_n = 2a_{n-1} + 3a_{n-2}$, and conjecture what kind of closed form a_n might have. Now generate some terms of the sequence, conjecture a closed form, and then prove that your closed form is correct.

    (e) How do things change if we examine $a_0 = a_1 = 2, a_n = 2a_{n-1} + 3a_{n-2}$?

2. Consider the sequence given by $a_1 = 1, a_n = a_{n-1} + (2n - 3)$. Can a technique from this chapter be used to find a closed form? Write out the first several terms of the sequence. Can you intuit a closed form based on the terms of the sequence? (Hint: $\square$.) Find a closed form for this recurrence and use induction to prove that the closed form is correct.

3. For each of the following recurrence relations, first determine whether a closed form can be found using a technique introduced in this chapter or whether you must experiment and use your intuition to find a closed form. Then, find a closed form and show that it is correct (using, of course, induction).

    (a) $a_1 = 1, a_n = a_{n-1} + 2^n$.

    (b) $a_0 = 0, a_n = a_{n-1} + 2n$.

    (c) $a_0 = 4, a_n = a_{n-1} + 6$.

    (d) $a_1 = a_2 = 2, a_n = a_{n-1}a_{n-2}$.

8.8 Geometric Sequences and the Characteristic Equation

In this section, we will learn how to find closed forms for linear homogeneous recurrence relations with constant coefficients. (Try to say *that* three times fast!) Of course, you have no clue what those are—nor should you.

Definition 8.8.1. A *linear* recurrence relation is one where none of the a_{n-j} terms are raised to powers other than 1 and none are multiplied by each other.

For example, $a_n = a_{n-1} + n^2$ is linear, whereas $a_n = (a_{n-1})^2 - n$ is not, nor is $a_n = a_{n-3}a_{n-1}$.

Definition 8.8.2. A *homogeneous* recurrence relation is one that evaluates to 0 when 0 is plugged in to the a_j on the right-hand side.

For example, $a_n = a_{n-1} + 346a_{n-7} + 2a_{n-74}$ gives $0 + 346 \cdot 0 + 2 \cdot 0 = 0$ and so is homogeneous, but $a_n = a_{n-11} + n^2 - 5$ gives $0 + n^2 - 5 \neq 0$ and so is nonhomogeneous.

Definition 8.8.3. A recurrence relation with *constant coefficients* has coefficients of a_j terms that are not variable.

For example, $a_n = 342a_{n-2} + 978a_{n-62}$ has constant coefficients, whereas $a_n = (3n^2 - 2n)a_{n-1} + a_{n-7}$ does not.

Essentially, a linear homogeneous recurrence relation with constant coefficients has the form $a_n = c_1a_{n-1} + c_2a_{n-2} + \cdots + c_ka_{n-k}$. Believe it or not, there is an algorithm for finding a closed form for *any* linear homogeneous recurrence relation with constant coefficients. We will not explore it in full generality here because it's long (and at a certain point repetitive and boring), but we will deal with the simplest cases.

Now we will see how the content of this section is related to the title of the section. Recall that $a_n = 2a_{n-1}$ with $a_0 = 1$ has closed form $a_n = 2^n$. Similarly, the recurrence relation $a_n = qa_{n-1}$ generates the sequence $a_1, qa_1, q^2a_1, q^3a_1, \ldots,$ $q^na_1, \ldots,$ and has closed form $a_n = q^na_0$. This is a *geometric* sequence because consecutive terms are related by multiplication by a constant q (much as an arithmetic sequence has consecutive terms related by addition of a constant). The closed form is an exponential function of n. Thus, one might suspect (or at least hope) that a recurrence relation that is a sum of terms like qa_{n-1} (namely, a linear homogeneous recurrence relation with constant coefficients) would have a closed form that is a sum of exponential functions. And it turns out to be true!

First we will describe the algorithm that produces a closed form for a linear homogeneous recurrence relation with constant coefficients, then we will give an example of how to use the algorithm, and then we will explain why the algorithm works. First read the algorithm (you might not understand it just yet), then read Example 8.8.4 and compare it to the algorithm to see how to execute the algorithm.

How to find a closed form for a recurrence relation of the form $a_n = c_1a_{n-1} + c_2a_{n-2} + \cdots + c_ka_{n-k}$:

1. Rewrite $a_n = c_1a_{n-1} + c_2a_{n-2} + \cdots + c_ka_{n-k}$ as $x^n = c_1x^{n-1} + c_2x^{n-2} + \cdots + c_kx^{n-k}$.

2. Every term has at least an x^{n-k} in it, so divide through to get $x^{n-(n-k)} = x^k - c_1x^{k-1} + c_2x^{k-2} + \cdots + c_k$. This is called the *characteristic equation*.

3. Move all the terms to one side, as in $x^k - c_1x^{k-1} - c_2x^{k-2} - \cdots - c_k = 0$, to obtain a polynomial that we have some chance of factoring.

4. Find the roots $r_1, \ldots, r_k$ of this baby. This can be done by factoring into linear terms $(x - r_1) \cdot \cdots \cdot (x - r_k) = 0$ or by getting a computer or calculator to produce the roots.

5. Hope that $r_1, \ldots, r_k$ are all different because otherwise the rest of this algorithm won't apply, and hope that $r_1, \ldots, r_k$ are all real because we are not addressing complex numbers in a discrete mathematics textbook. (By the way, none of the r_j are 0 because if some $r_j = 0$, then we could factor x out of the characteristic equation, and that would have been canceled in step 2.)

6. Write out the closed-form equation $a_n = q_1r_1^n + q_2r_2^n + \cdots + q_kr_k^n$. All that's left is to figure out what the q_j are.

7. Go find the first k terms of the sequence $a_1, a_2, \ldots, a_k$ (you must have left them lying around *some*where…). Use these to generate the k equations

$$a_1 = q_1r_1^1 + q_2r_2^1 + \cdots + q_kr_k^1 = q_1r_1 + q_2r_2 + \cdots + q_kr_k,$$
$$a_2 = q_1r_1^2 + q_2r_2^2 + \cdots + q_kr_k^2,$$
$$\vdots \quad \vdots \quad \vdots$$
$$a_k = q_1r_1^k + q_2r_2^k + \cdots + q_kr_k^k,$$

and if you happen to have an a_0 lying around, you get a bonus simple equation

$$a_0 = q_1 r_1^0 + q_2 r_2^0 + \cdots + q_k r_k^0 = q_1 + q_2 + \cdots + q_k.$$

8. If you don't happen to have an a_0 lying around, figure out what value you can assign to a_0 that is consistent with your recurrence and initial values. That gives you the bonus simple equation.

9. Use lots of high-school-algebra symbolic manipulation to determine q_1, q_2, ..., q_k from the equations you generated in steps 7 and 8.

10. Plug these values back in to $a_n = q_1 r_1^n + q_2 r_2^n + \cdots + q_k r_k^n$ and rejoice in your closed-form formula.

Example 8.8.4. Consider the recurrence $a_n = 2a_{n-1} + 3a_{n-2}$ with initial terms $a_1 = 0, a_2 = 2$. (This should look familiar from Section 8.7. Do you have any conjectures as to what results we will find here?) Notice that it is linear and homogeneous and the coefficients are constant; it is in the form $a_n = c_1 a_{n-1} + c_2 a_{n-2}$, where $c_1 = 2$ and $c_2 = 3$. We will follow the algorithm.

1. We rewrite as $x^n = 2x^{n-1} + 3x^{n-2}$.

2. We divide through by x^{n-2} to get $x^2 = 2x + 3$.

3. We rewrite as $x^2 - 2x - 3 = 0$.

4. We factor as $(x-3)(x+1) = 0$ to see that the roots are $r_1 = 3, r_2 = -1$.

5. We do a tiny happy dance because $r_1 \neq r_2$ so we can continue.

6. Now we have $a_n = q_1 3^n + q_2 (-1)^n$ and need to determine q_1 and q_2.

7. Using the initial terms, $a_1 = 0 = q_1 3 - q_2$ and $a_2 = 2 = q_1 9 + q_2$.

8. We do not need any additional equations, but let us see how we might obtain one anyway. We have that $a_2 = 3a_1 + 2a_0$, or $2 = 2 \cdot 0 + 3a_0$, so $a_0 = \frac{2}{3}$. Then we have that $\frac{2}{3} = q_1 + q_2$.

9. From the first equation in step 7, we find that $q_2 = 3q_1$, which we can then use to see that $2 = 9q_1 + 3q_1 = 12q_1$, so that $q_1 = \frac{1}{6}$ and $q_2 = \frac{1}{2}$.

10. Finally, we arrive at $a_n = \frac{1}{6} 3^n + \frac{1}{2} (-1)^n$, which is the desired closed form.

Example 8.8.5 (the Binet formula). Consider now the recurrence $a_n = a_{n-1} + a_{n-2}$ with initial terms $a_1 = 1, a_2 = 1$. This is as linear and homogeneous as recurrences get, with coefficients so constant they're invisible. And if you haven't noticed yet, notice now: this is the Fibonacci sequence. Again we will follow the closed-form algorithm, though a bit less formally than before.

We start with $x^n = x^{n-1} + x^{n-2}$ and reduce to $x^2 = x + 1$. This becomes $x^2 - x - 1 = 0$, which does not have a nice factorization; however, the quadratic formula reveals that its roots are $\frac{1 \pm \sqrt{5}}{2}$. This tells us $a_n = q_1 (\frac{1+\sqrt{5}}{2})^n + q_2 (\frac{1-\sqrt{5}}{2})^n$.

We know that $a_1 = 1 = q_1(\frac{1+\sqrt{5}}{2}) + q_2(\frac{1-\sqrt{5}}{2})$ and $a_2 = 1 = q_1(\frac{1+\sqrt{5}}{2})^2 + q_2(\frac{1-\sqrt{5}}{2})^2$. A simpler equation would be nicer, so let's notice that $a_2 = a_1 + a_0$, i.e., $1 = 1 + a_0$ so $a_0 = 0$. This gives us $0 = q_1 + q_2$, or $q_2 = -q_1$. Substituting into the equation we got from a_1, we have $1 = q_1(\frac{1+\sqrt{5}}{2}) - q_1(\frac{1-\sqrt{5}}{2}) = q_1\sqrt{5}$. Thus, $q_1 = \frac{1}{\sqrt{5}}$ and $q_2 = \frac{-1}{\sqrt{5}}$.

Finally, we have $a_n = \frac{1}{\sqrt{5}}(\frac{1+\sqrt{5}}{2})^n - \frac{1}{\sqrt{5}}(\frac{1-\sqrt{5}}{2})^n$. This is known as the *Binet formula* for the Fibonacci numbers, after Jacques Binet (1786–1856), though he was by far not the first to have discovered the formula.

Now for an explanation of why the algorithm works. (It's kind of long but not too bad, and *really* interesting… you don't believe that, do you?) We will proceed by proving a lemma first.

> **Lemma 8.8.6.** *The formula $a_n = r_j^n$ is a closed form for the recurrence $a_n = c_1 a_{n-1} + c_2 a_{n-2} + \cdots + c_k a_{n-k} \iff r_j$ is a root of a_n's characteristic equation.*

This says that there is a different closed form for every root of the recurrence relation $a_n = c_1 a_{n-1} + c_2 a_{n-2} + \cdots + c_k a_{n-k}$. Don't freak out yet; notice that no mention was made of the initial conditions $(a_0, a_1, a_2, \dots)$. So Lemma 8.8.6 is not actually claiming there are multiple closed forms for any particular sequence.

Proof of Lemma 8.8.6: ($\Rightarrow$) Suppose that $a_n = r_j^n$ is a closed form for $a_n = c_1 a_{n-1} + c_2 a_{n-2} + \cdots + c_k a_{n-k}$. Then, $r_j^n = c_1 r_j^{n-1} + c_2 r_j^{n-2} + \cdots + c_k r_j^{n-k}$ by substitution. This implies that $r_j^k - c_1 r_j^{k-1} - c_2 r_j^{k-2} - \cdots - c_k = 0$. That is exactly the characteristic equation with r_j plugged in to x; because the left-hand-side evaluates to 0, r_j is a root of the characteristic equation.

($\Leftarrow$) Now suppose that r_j is a root of the characteristic equation of $a_n = c_1 a_{n-1} + c_2 a_{n-2} + \cdots + c_k a_{n-k}$. That means that when r_j is substituted for x in $x^k = c_1 x^{k-1} + c_2 x^{k-2} + \cdots + c_k$, the equation still holds. Thus, $r_j^k = c_1 r_j^{k-1} + c_2 r_j^{k-2} + \cdots + c_k$.

We can multiply through by r_j^{n-k} to obtain $r_j^n = c_1 r_j^{n-1} + c_2 r_j^{n-2} + \cdots + c_k r_j^{n-k}$, which exactly means that $a_n = r_j^n$ is a closed form for $a_n = c_1 a_{n-1} + c_2 a_{n-2} + \cdots + c_k a_{n-k}$. $\qquad\square$

Lemma 8.8.6 produces k closed forms, one for every root of the characteristic equation, or, equivalently, one for every term in the recurrence relation. We will now combine these closed forms into a masterpiece of equation-ness. If $r_1^n, r_2^n, \ldots, r_k^n$ are each closed forms for $a_n = c_1 a_{n-1} + c_2 a_{n-2} + \cdots + c_k a_{n-k}$, this means that

$$r_1^n = c_1 r_1^{n-1} + c_2 r_1^{n-2} + \cdots + c_k r_1^{n-k},$$

$$r_2^n = c_1 r_2^{n-1} + c_2 r_2^{n-2} + \cdots + c_k r_2^{n-k},$$

$$\vdots \qquad \vdots \qquad \vdots \qquad \vdots \qquad \vdots$$

$$r_k^n = c_1 r_k^{n-1} + c_2 r_k^{n-2} + \cdots + c_k r_k^{n-k}.$$

Let's multiply the first of those by q_1, the second by q_2 (and the jth by q_j), and the last by q_k, and add the whole shebang together. On the left-hand side, we get $q_1 r_1^n + q_2 r_2^n + \cdots + q_k r_k^n$. That's what the last step of our algorithm produces, so we hope it turns out that this *is* a closed-form formula for $a_n = c_1 a_{n-1} + c_2 a_{n-2} + \cdots + c_k a_{n-k}$. Now examine what happens with the right-hand side of the sum of the equations. We get $q_1 (c_1 r_1^{n-1} + c_2 r_1^{n-2} + \cdots + c_k r_1^{n-k}) + q_2 (c_1 r_2^{n-1} + c_2 r_2^{n-2} + \cdots + c_k r_2^{n-k}) + \cdots + q_k (c_1 r_k^{n-1} + c_2 r_k^{n-2} + \cdots + c_k r_k^{n-k})$, and by collecting terms, this is the same as $c_1 (q_1 r_1^{n-1} + q_2 r_2^{n-1} + \cdots + q_k r_k^{n-1}) + c_2 (q_1 r_1^{n-2} + q_2 r_2^{n-2} + \cdots + q_k r_k^{n-2}) + \cdots + c_k (q_1 r_1^{n-k} + q_2 r_2^{n-k} + \cdots + q_k r_k^{n-k})$. And yes, this is exactly what we want to see—inside the first set of parentheses is the $(n-1)$st version of our proposed a_n, inside the second set of parentheses is the $(n-2)$nd version of our proposed a_n, and more generally, inside each set of parentheses is the appropriately indexed version of the proposed a_n. This shows that our convoluted formula truly is a closed-form formula for the linear homogeneous recurrence with constant coefficients $a_n = c_1 a_{n-1} + c_2 a_{n-2} + \cdots + c_k a_{n-k}$.

In fact, it turns out that every possible closed form of a linear homogeneous recurrence with constant coefficients is of the form $a_n = c_1 r_1^{n-1} + c_2 r_2^{n-2} + \cdots + c_k r_k^{n-k}$, but that proof is a bit beyond this text. (If you know linear algebra, do try to figure it out!)

It probably seems super-strange to you that no matter how yucky the roots r_j are, integers always come out of the closed-form equation. They *have* to, of course—a_n is an integer sequence and we proved that this equation is a closed form for it—but it's still weird that, for example, all those $\sqrt{5}$s turn into Fibonacci numbers.

Check Yourself ────────────────────────

Do a representative sampling of these problems.

1. Which parts of the definition of a linear homogeneous recurrence relation with constant coefficients do each of these recurrences violate?

   (a) $a_n = na_{n-3} + 6$.

   (b) $a_n = a_{n-1}a_{n-43} + 23n$.

   (c) $a_n = 5n^2 a_{n-3} - (-1)^n a_{n-6}$.

2. Find the characteristic equation for ...

   (a) ... $a_n = 4a_{n-1}$.

   (b) ... $a_n = a_{n-1} - 3a_{n-2}$.

   (c) ... $a_n = 2a_{n-2}$.

3. Determine a_0 for each of these recurrences.

   (a) $a_1 = 3, a_n = 4a_{n-1}$.

   (b) $a_1 = a_2 = 1, a_n = a_{n-1} - 3a_{n-2}$.

   (c) $a_1 = 1, a_2 = 2, a_n = 2a_{n-2}$.

4. Determine the characteristic equation for $a_n = 2a_{n-1}$. What are its roots? Using this information and the initial condition $a_1 = 42$, determine a closed-form formula.

5. For each of the following recurrences, decide whether one could use (i) kth differences, (ii) the characteristic equation, or (iii) neither in order to find a closed form for the recurrence.

   (a) $a_n = 2a_{n-1} + 2$.

   (b) $a_n = a_{n-1} + 2a_{n-2} + 3a_{n-3} + 4a_{n-4}$.

   (c) $a_n = 3a_{n-3} + 3n$.

   (d) $a_n = a_{n-1} - 2n^2$.

8.9 Try This! Find Closed Forms for These Recurrence Relations!

For each of the recurrence relations given here, find a closed-form formula. You may wish to first identify the type of recurrence relation, then generate a few terms to see if intuition will produce a closed form quickly, and then proceed with your chosen closed-form-producing technique.

1. $a_1 = 2, a_n = a_{n-1} + n + 6.$

2. $a_0 = 3, a_1 = 10, a_n = 10a_{n-1} + 25a_{n-2}.$

3. $a_0 = 62, a_n = -a_{n-1}.$

4. $a_0 = -2, a_n = a_{n-1} + n - 7.$

5. $a_0 = 1, a_1 = 2, a_n = 4a_{n-2}.$

6. $a_0 = 0, a_n = a_{n-1} + 2n^2 - n.$

7. $a_0 = -3, a_1 = 1, a_n = 12a_{n-1} - 35a_{n-2}.$

8. $a_1 = 1, a_2 = 1, a_n = a_{n-1}a_{n-2} - 1.$

9. $a_0 = 1, a_1 = 2, a_2 = 3, a_n = 2a_{n-1} + 5a_{n-2} - 6a_{n-3}.$

8.10 Where to Go from Here

Recurrence relations are a class of examples of the more general process of recursion, in which one defines an object or procedure in terms of smaller or previous cases of the object or procedure. As a process, recursion is used throughout computer science and arises particularly in function definitions and algorithm execution. For example, binary search, which we will discuss in Chapter 10, is recursive in nature. Computer science courses in algorithm design and analysis will include further study of recursion, as will combinatorics courses in mathematics.

To learn more about producing closed forms for recurrences, see [5, Section 2.3]. The authors address linear homogeneous recurrence relations with constant coefficients for which the characteristic equation has multiple roots, some linear inhomogeneous recurrence relations, and more advanced techniques for simplifying complicated recurrence relations.

If you are particularly interested in Fibonacci numbers, you may want to learn more about number theory—in which case, see Chapter 16 and the resources given there. The journal *The Fibonacci Quarterly* publishes the latest research and exposition related to Fibonacci numbers. Back issues are freely downloadable via http://www.fq.math.ca/.

For integer sequences in general, no source beats the OEIS. (Many examples and exercises in this chapter derive from this source!) Integer sequences are a topic of active research, and in fact, the *Journal of Integer Sequences* is devoted to the area. Articles are available free online at http://www.cs.uwaterloo.ca/journals/ JIS/. Some chemists and physicists are also interested in integer sequences that count the number of ways of folding particular proteins, the number of benzenoid chains in a hydrocarbon structure, and the number of states in a Potts model.

Credit where credit is due: Section 8.8 and many problems in Section 8.14 were inspired by [1]. Section 8.6 was derived from notes written by Tom Hull. New problems at the start of Section 8.7 were suggested by a very helpful and gracious anonymous reviewer. Bonus Check-Yourself Problem 1 and Problem 30 of Section 8.14 were donated by Tom Hull. Obviously, many aspects of this chapter were inspired by Neil J.A. Sloane's OEIS! The unusual phrase on page 249 was the start of the first sentence of a story written by the author's father, about the origin of fractions: *Once upon a time, a long, long time ago now, so long ago that animals could talk and teachers were human beings, it was that long ago that there lived on the moon a fairy snow queen.*

8.11   Chapter 8 Definitions

Fibonacci numbers: Collectively these form an integer sequence defined by the recurrence $F_n = F_{n-1} + F_{n-2}$ with initial values 1, 1.

integer sequence: Integers listed in some order.

closed-form formula (or closed form): A rule for producing the nth term of a sequence given only the number n.

recurrence relation (or recurrence): A statement of the form $a_n = $ (some stuff, some of which involves $a_{\text{smaller than } n}$); a rule for generating more terms of a sequence by knowing only some of the previous terms.

recursion: The process of using a recurrence.

explicit formula: A closed-form formula, in contrast to the implicit expression of a recurrence.

recursing: The process of iterating a recurrence.

arithmetic sequence: A sequence with constant differences.

linear recurrence relation: **A** recurrence relation where none of the a_{n-j} terms are raised to powers other than 1 and none are multiplied by each other.

homogeneous recurrence relation: **A** recurrence relation that evaluates to 0 when 0 is plugged in to the a_j on the right-hand side.

constant coefficients: **A** recurrence relation with coefficients of a_j terms that are not variable.

linear homogeneous recurrence relation with constant coefficients: **A** recurrence with the form $a_n = c_1 a_{n-1} + c_2 a_{n-2} + \cdots + c_k a_{n-k}$.

geometric sequence: **A** sequence where consecutive terms are related by multiplication by a constant.

characteristic equation: The equation $x^k = c_1 x^{k-1} + c_2 x^{k-2} + \cdots + c_k$ associated to a linear homogeneous recurrence relation with constant coefficients $a_n = c_1 a_{n-1} + c_2 a_{n-2} + \cdots + c_k a_{n-k}$.

8.12   Bonus: Recurring Stories

Story 1.   People walk up stairs in many different ways. The author's mother usually climbs them one at a time; her father always climbs them two at a time. As a result, the author sometimes takes stairs one at a time and sometimes takes stairs two at a time and often combines these two approaches. How many ways are there for her to walk up a flight of n stairs?

First, of course, we must experiment. Check out $n = 1$; there is only one way to climb one stair (notice that it's one at a time). If there are two stairs, the author could climb them as 1-1 or as 2. That's two ways. If there are three stairs, she could climb them as 1-1-1 or as 1-2 or as 2-1. That's three ways. Lest you think $n = 1, 2, 3$ is enough experimentation, go to OEIS and discover that there are at least 14,709 different sequences that begin in this fashion. So at least let's look at $n = 4$. The author could take those stairs as 1-1-1-1 or 2-2 or 2-1-1 or 1-2-1 or 1-1-2. That's five ways. But sadly, the OEIS tells us that there are at least 2,262 sequences that begin with $1, 2, 3, 5$.

How can we determine which sequence we're dealing with? The answer is to try to create a recurrence, as this will specify the rest of the sequence. In other words, we want to break down the different ways the author could have gotten to the nth step into ways of getting to previous steps. We'll call the number of ways of climbing n stairs s_n. Let's see... if the author most recently climbed a single step, then there were $n - 1$ previous stairs and so s_{n-1} ways to climb them. If her last movement didn't go up a single step, she must have gone up two steps at once (there are no other ways to climb), so there were $n - 2$ previous stairs and s_{n-2} ways to climb them. By the sum principle, we add these two numbers of ways and discover that $s_n = s_{n-1} + s_{n-2}$.

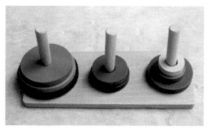

Figure 8.2. A Tower of Hanoi toy, before play (left) and during play (right).

Hopefully that recurrence looks familiar to you. Remember that if two sequences have the same recurrence relation and the same initial values, then they
are the same sequence. The only problem here is that the Fibonacci sequence starts
with $1, 1, 2, 3, 5 \ldots$, not with $1, 2, 3, 5 \ldots$. We can fix this by noting that $s_1 = F_2$
and $s_2 = F_3$, so in general $s_n = F_{n+1}$.

Let's unpack that process a little bit. In addition to creating a recurrence, we
provided a combinatorial explanation. Giving a combinatorial proof that the recurrence works assures us that it is the correct recurrence. This, in turn, allows us
to identify our sequence with a known sequence or to find a correct closed-form
formula.

Story 2. In many libraries and classrooms, one can find a toy consisting of a
board with three pegs and a pile of discs on one of the pegs. A small example is
shown in Figure 8.2. In playing with this toy, the goal is to move all discs from
one peg to a different peg, while only moving one disc at a time (from one peg
to another) and never placing a larger disc atop a smaller disc. This toy is known
as the Tower of Hanoi (though no one seems to know why Edouard Lucas (1842–
1891), its inventor, called it that).

Of course, once one has moved all discs from one peg to another according
to the rules, the question arises, How can this be done in the smallest number of
moves? And, for that matter, what *is* the smallest number of moves?

If you have not played with this toy, you should stop reading now, close the
book, and go play with the Tower of Hanoi for a while (perhaps at https://www.
mathsisfun.com/games/towerofhanoi.html). (However, if you actually followed
the directions in the previous sentence, you would not have made it to the end of
the sentence, or to this parenthetical remark.)

Now then, let us think about strategy. We think you will agree that what the
smallest number of moves *is* depends on the number of discs involved. Therefore,

let us denote the smallest number of moves it takes to move n discs from one peg to another by m_n.

Because we can only move one disc at a time, we certainly have to move the first $n-1$ discs off of the largest (nth, bottom) disc before we can move the largest disc onto another peg. The peg to which we move the largest disc must be empty, because we cannot move a larger disc onto a smaller disc, so all of the first $n-1$ discs must be moved to the same peg. That takes m_{n-1} moves. And for that matter, once we have moved the largest disc onto another peg, we have to put the first $n-1$ discs back onto it. That also takes m_{n-1} moves. So, in total (not forgetting the movement of the largest disc onto an empty peg), $m_n = m_{n-1} + 1 + m_{n-1} = 2m_{n-1} + 1$. Again, we constructed a recurrence using combinatorial proof.

Hmm... if we have but a single disc, it takes exactly one move to place that disc on a different peg. (We can't even be inefficient and use more than one move—the moment we make a single move, the disc is irrevocably on a different peg.) Therefore, $m_1 = 1$ and we can use the recursion we developed to write down a sequence: $1, 3, 7, 15, 31, \dots$. Use your intuition to find a closed form for this sequence. (The OEIS lists 53 sequences with this beginning, so you'll get no satisfaction there—unless you choose a closed form and prove by induction that it is correct.)

Story problems:

1. You are in a strange multifloor shopping mall. What's strange about it is the placement of the escalators. For every floor (except the first and second) there are two escalators that go down a single floor and three escalators that go down two floors at once, without letting shoppers get off at the intermediate floor. (The first floor has no down escalators, and the second floor has the expected two escalators that go down a single floor.) This makes navigation somewhat confusing. Still, for an n-floor shopping mall, how many ways are there to get from the top floor to the first floor so you can leave?

2. The puzzle Trench consists of a board and a number of tiles. There is a long groove cut into the board, $1''$ tall and quite long with lengths of $1''$ marked on it. The tiles are of two shapes, $1'' \times 1''$ squares and $1'' \times 2''$ right triangles. The goal of Trench is to find all possible ways of filling the groove with the tiles; a few examples are shown in Figure 8.3. Let us denote the number of ways of filling the first n inches of Trench as T_n.

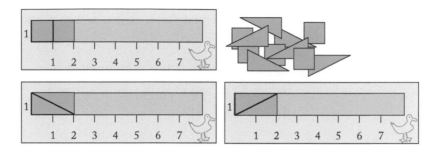

Figure 8.3. A mock-up of the puzzle Trench.

(a) What are the first few Trench numbers $T_1, T_2, T_3, T_4, \ldots$?

(b) Construct a recurrence for T_n and explain why it is correct.

(c) Find a closed form for your recurrence.

3. Now suppose that you want to solve the Restricted Trench puzzle. The setup is the same as the Trench puzzle of the previous problem, but only one direction of slant is allowed for the triangle's diagonal. Let us denote the number of ways of filling the first n inches of Restricted Trench as R_n.

(a) What are the first few Restricted Trench numbers $R_1, R_2, R_3, R_4, \ldots$?

(b) Construct a recurrence for R_n and explain why it is correct.

(c) Find a closed form for your recurrence.

4. When building a wall, all the bricks in a row should be directly next to each other and every brick should be set on top of exactly two other bricks. A few stacks of bricks are shown in Figure 8.4.

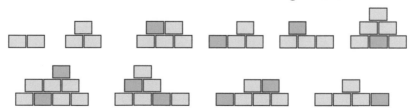

Figure 8.4. Some stacks of bricks.

We shall denote the number of different stacks of bricks with n bricks along the bottom by b_n.

(a) What are the first few brick-stack numbers b_1, b_2, b_3, b_4?

(b) Construct a recurrence for b_n and explain why it is correct.

8.13   Bonus Check-Yourself Problems

Solutions to these problems appear starting on page 611. Those solutions that model a formal write-up (such as one might hand in for homework) are to Problems 1 and 10.

1. Dandelions reproduce very quickly, as anyone who maintains a lawn knows. In fact, did you know that on any given day, if you went to your lawn and counted the dandelions, then the next day twice as many *new* dandelions will have emerged from the ground? Luckily, dandelions die after two days, so that helps keep the numbers down. Still, if on day 0 you had 1 dandelion, then on day 1 you would have 3 dandelions, on day 2 you'd have 8 dandelions, and then on day 3 you'd have 22 dandelions.

    (a) Write a recurrence equation for $d_n =$ the number of dandelions on day n.

    (b) Find a closed-form formula for d_n.

2. Generate the first 30 terms of the sequence $a_n = a_{n-1} + a_{n-2} - a_{n-3}, a_0 = 0, a_1 = 1, a_2 = 1$.

3. Suppose that $a_n = (-4)^n$, $a_n = 1$, and $a_n = 2^n$ are all closed forms for the same recurrence. Find a recurrence that fits this criterion and verify that it really does work for all three closed forms.

4. Consider the sequence $1, 3, 4, 7, 11, 18, 29, \ldots$.

(a) Find a recurrence that L_n satisfies.

(b) Prove that $L_n = F_{n-1} + F_{n+1}$.

5. Find a closed-form formula for the sequence $a_0 = -1, a_n = a_{n-1} + 3n + 1$.

6. Consider the recurrence relation $a_n = 3a_{n-1} - a_{n-2}$ with $a_0 = 0, a_1 = 1$. Generate some terms, make a conjecture as to what sequence this is, try to find the closed form, and try to explain what is going on here.

7. Consider the sequence $5, -3, 5, -3, 5, -3, 5, -3, 5, \ldots$. Find a recurrence for this sequence, and find two more (different) sequences that satisfy that recurrence.

8. Find a closed form for the sequence defined by the recurrence $a_n = -a_{n-1} a_{n-2} + 2, a_0 = 1, a_1 = 1$. How do things change if $a_0 = 0, a_1 = 0$?

9. Here is a characteristic equation: $x^5 + 4x^3 - 3x^2 - 1 = 0$. What is the associated recurrence?

10. Find a closed-form formula for the sequence $a_0 = 1, a_n = a_{n-1} + n^2 - 2n$.

8.14 Recurring Problems

1. Write $F_1 + F_3 + F_5 + \cdots + F_{2n-1}$ in summation notation. Then show that $F_1 + F_3 + F_5 + \cdots + F_{2n-1} = F_{2n}$.

2. Write $F_2 + F_4 + F_6 + \cdots + F_{2n}$ in summation notation. Then show that $F_2 + F_4 + F_6 + \cdots + F_{2n} = F_{2n+1} - 1$.

3. Prove that F_{3n} is even.

4. Write $F_1 - F_2 + F_3 - F_4 + \cdots + F_{2n-1} - F_{2n}$ in summation notation. Then show that $F_1 - F_2 + F_3 - F_4 + \cdots + F_{2n-1} - F_{2n} = 1 - F_{2n-1}$.

5. The first two terms of the Fibonacci sequence are $1, 1$. Find two other integers that, together with the recurrence $a_n = a_{n-1} + a_{n-2}$, do *not* generate Fibonacci numbers. Then, find two other integers that, together with the recurrence $a_n = a_{n-1} + a_{n-2}$, *do* generate Fibonacci numbers. What kinds of integers do this?

6. Consider the sequence $0, 1, 5, 12, 22, 35, 51, 70, 92, 117, 145, 176, \ldots$. Find both a recurrence and a closed form for this sequence.

7. Find a closed form for $a_0 = -1, a_1 = 1$, $a_n = 2a_{n-1} - a_{n-2}$.

8. Consider binary numbers with n digits (e.g., $1010, 00101, 011$). How many binary numbers of length n do not contain the substring 000? Denote this number by z_n; find a relationship between z_n, z_{n-1}, and (we're not going to tell you) in order to form an appropriate recurrence relation. *(Do not, oh, please, do not try to find a closed form for this recurrence.)*

9. Find a closed form for $a_1 = 2, a_n = a_{n-1} + 2n$.

10. What proof technique(s) was/were used in the proof of Lemma 8.8.6?

11. Find a recurrence that defines the sequence $1, 1, 1, 3, 5, 9, 17, 31, \ldots$. Now find a different sequence that satisfies this recurrence.

12. Look through the sequences you have generated since working on Section 8.4. Are there any that do not appear in the OEIS?

13. What proof method has been most commonly used in this chapter?

14. Find a closed form for $a_0 = -1, a_1 = -2, a_n = 4a_{n-1} - 3a_{n-2}$.

15. Find a recurrence that defines the sequence $2, 3, 5, 9, 17, 33, 65, \ldots$. Now find a different sequence that satisfies this recurrence.

16. **Mini-project: More Fibonacci identities than you can shake a stick at:** Prove each of the following identities.

   (a) $F_{2n} = F_n(F_{n+1} + F_{n-1})$.

   (b) $F_n \mid F_{2n}$.

   (c) $F_{2n} = F_n(F_n + 2F_{n-1})$.

   (d) $F_{2n} = (F_{n+1})^2 - (F_{n-1})^2$.

   (e) $F_{2n} = F_n(2F_{n+1} - F_n)$.

   (f) $(F_n)^2 + (F_{n+1})^2 = F_{2n+1}$.

   (g) $F_{(k+1)n} = F_{n-1}F_{kn} + F_{kn+1}F_n$.

   (h) $F_n = F_k F_{n-k+1} + F_{k-1}F_{n-k}$.

   (i) $F_n = F_k F_{n-k-1} + F_{k+1}F_{n-k}$.

   (j) $F_{3n} = F_{n+1}^3 + F_n^3 - F_{n-1}^3$.

   (k) $F_{n+1}^2 = 4F_n F_{n-1} + F_{n-2}^2$.

17. Find a closed form for the recurrence $a_n = 5a_{n-1} - 4a_{n-2}$, with $a_0 = 0$, $a_1 = 1$.

18. Suppose that you are presented with the equation $x^4 - 2x^2 + 3x - 3 = 0$. To what recurrence does this correspond?

19. Find a closed form for $a_1 = 1, a_n = a_{n-1} + 3$.

20. Suppose that $a_n = (-3)^n$ and $a_n = 4^n$ are both closed forms for the same recurrence. Find a recurrence that fits this criterion.

21. Consider the recurrence $a_n = a_{n-3} + 1$, with $a_0 = 1$, $a_1 = 0$, and $a_2 = 1$. Generate the first 30 terms of this sequence. Laugh delightedly. **Challenge:** find a closed-form formula for this sequence.

22. Suppose that you are given the roots to a characteristic equation, and they are -1, 2, and -3.

    (a) What is the characteristic equation?

    (b) To what recurrence does that correspond?

    (c) Find initial conditions so that $a_n = (-1)^n$ is the closed form for the recurrence. Use the recurrence to generate a few more terms to be sure your initial conditions work.

    (d) Find initial conditions so that $a_n = (-1)^n + 2^n$ is the closed form for the recurrence. Use the recurrence to generate three additional terms to be sure your initial conditions work.

    (e) Find initial conditions so that $a_n = (-1)^n + 2^n + (-3)^n$ is the closed form for the recurrence. Use the recurrence to generate three additional terms to be sure your initial conditions work.

23. Find a closed form and a recurrence relation for the sequence $0, 2, 8, 24, 64, 160, 384, 896, 2048, 4608, 10240, \ldots$.

24. Find a closed form for the sequence $0, 2, 10, 28, 60, 110, 182, 280, 408, 570, 770, \ldots$.

25. Consider the recurrence $a_n = 2a_{n-1} + a_{n-2} - 2a_{n-3}$.

    (a) The characteristic equation has three roots, r_1, r_2, r_3. What are they?

    (b) Find initial conditions so that $a_n = (r_1)^n$ is the closed form for the recurrence. Use the recurrence to generate a few more terms to be sure your initial conditions work.

    (c) Find initial conditions so that $a_n = (r_1)^n + (r_2)^n$ is the closed form for the recurrence. Use the recurrence to generate three additional terms to be sure your initial conditions work.

    (d) Find initial conditions so that $a_n = (r_1)^n + (r_2)^n + (r_3)^n$ is the closed form for the recurrence. Use the recurrence to generate three additional terms to be sure your initial conditions work.

26. Find a closed-form formula for the sequence given by $a_0 = 1, a_n = a_{n-1} + n^2 - 3n + 1$.

27. Find a closed form for the recurrence $a_0 = 1, a_n = a_{n-1} - 9n + 5$.

28. Find a closed form for the recurrence $a_n = 2a_{n-1} + a_{n-2} - 2a_{n-3}, a_0 = 1, a_1 = 0, a_2 = -1$.

29. Which of these are linear homogeneous recurrence relations with constant coefficients? For those that are not, which part(s) of the definition is/are violated?

    (a) $a_n = 2a_{n-2}a_{n-3} + 3na_{n-4} + 2$.

    (b) $a_n = 5a_{n-1} - 4a_{n-2} + 3a_{n-3} - 2a_{n-4} + a_{n-76}$

    (c) $a_n = 62a_{n-1} + 2^{n-1}$.

    (d) $a_n = a_{n-2}^3 + 15na_{n-5}$.

30. Find a closed form for the recurrence $a_n = a_{n-1} + 6a_{n-2}$, $a_1 = 2$, $a_2 = 8$.

31. Find a closed form involving $(?)^n$ for the recurrence $a_n = 2a_{n-1} + 2$, $a_1 = 4$.

32. Find a closed form for the sequence $111, 109, 105, 99, 91, 81, 69, 55, \ldots$.

33. Find a closed form for the recurrence $a_n = a_{n-1} - 4$, $a_0 = 126$.

34. Find a closed form for the sequence $12, 43, 74, 105, 136, 167, 198, \ldots$.

35. Find a closed form for the recurrence $a_n = a_{n-1} + 6n - 1$, $a_0 = -23$.

8.15   Instructor Notes

It has been a few weeks since students studied induction intensively, so they will probably need some review as they start their study of recurrences. Such a review can be built in to a warm-up interactive lecture that introduces the chapter. First, assign students to read Sections 8.1–8.3 as preparation for class. Then give an example of a simple recurrence ($a_n = 5a_{n-1}$, $a_0 = 2$ works well), have students generate the first few terms, propose a closed form, and prove that the closed form is correct. While doing the proof with them, recall the basic structure of base case, inductive hypothesis, and induction step and emphasize that the first terms of the sequence are base cases, while the recurrence itself is the key to completing the inductive step. You might also review a proof of a Fibonacci identity the students have seen (or give one that is not in the text). Then have the students work in groups on the problems in Section 8.4. It is useful to ask students to bring laptops (or other internet-capable devices) to class that day in order to use the OEIS. The problem in which students discover Fibonacci numbers hiding in Pascal's triangle is challenging, not because the discovery or proof is hard but because it is difficult for students to state their conjectures using mathematical notation.

For the second class, students could be assigned to read Sections 8.5 and 8.6. Start by asking for questions over the reading, and then arrange the students into groups to work on Section 8.7. The first problem will let students discover the connection between linear homogenous recurrences and geometric sequences, and thus prepare them for Section 8.8. Problem 2 is included as a transition from experimentation to recognition of recurrence types; a closed form can be found for that recurrence more quickly by using intuition than by using kth differences. The last problem is designed to have students begin to recognize recurrence formats for which closed forms can be found in a straightforward way. It is likely that students will have forgotten some of the details of solving simultaneous linear equations, so do not be surprised if this technique is where they get stuck.

After this, assign students to read Section 8.8. Some will simply take it on faith that the presented algorithm works, whereas others will struggle with the explanation. Decide how deeply you would like the students to understand this topic and perhaps start class by reviewing the material at that level. Start the students on Section 8.9 by discussing how to "type" recurrences and then letting them loose to work in groups. These problems take longer than you might expect because students may get bogged down in the arithmetic or symbolic details.

There is a question as to why we teach methods for finding closed forms for recurrences in a lower-level discrete mathematics course, given that WolframAlpha can do this quickly and for the low price of an internet connection. There are several cognitive aspects of the material that make it useful for these students. For example, the material reinforces the concept of *algorithm* in a concrete way. It links the familiar technique of solving simultaneous equations from high school with higher mathematics. More generally, acquiring facility with symbolic language (here, indexing and index substitution) is valuable. The linkage between a given symbolic notation and what it means is being reinforced when we do technical computations of some complexity. This, in turn, allows students to more easily recognize the situations in which the given symbolic notation applies.

You may choose to de-emphasize some of these methods; every class has different needs. Because this chapter hammers the connection between recurrence and induction, it would be a good idea in most classes to remind students that when two sequences have the same recurrence relation and initial terms, they are in fact the same sequence. This is mostly applicable when creating a recurrence to solve a problem but can also arise when a student finds a sequence and then discovers it is familiar.

Chapter 9

Cutting Up Food: Counting and Geometry

9.1 Introduction and Summary

There are lots and lots of applications of combinatorics to geometry (robot arms, minimal-length paths between sets of objects, origami, ...) but most of them require additional mathematical background and thus are too advanced for this course. So we will just investigate one example—but it's enough, as you will see! Interestingly, this family of geometric problems involves both recurrences and binomial coefficients, so it provides a review of some material from the previous two chapters, while also setting that material in a larger context.

We will begin with pizza (yum) and will want to know the largest number of pieces of pizza we can get using k cuts. We will then consider food of other dimensions (yams, spaghetti, hyperbeets, etc.).

9.2 Try This! Slice Pizza (and a Yam)

These problems are interrelated, so you will probably want to change which problem you are working on somewhat frequently. Insight from a later problem can help you finish an earlier problem.

Imagine a (round) pizza. We are going to slice it with a laser cutter, for example, as in Figure 9.1. It is expensive to use a laser cutter (but we need the accuracy), so we want to use the least number of cuts and obtain the largest number of pieces.

1. What is the largest number of pieces of pizza you can obtain using exactly three cuts? Try drawing a few pizzas of your own to experiment.

2. Cut up some more (drawn) pizzas. What is the largest number of pieces of pizza you can obtain using exactly four cuts?

Figure 9.1. Using two cuts garners at most four pieces.

3. Experiment to discover what the largest number of pieces of pizza is that can be obtained using exactly five cuts. You may find that you need to draw fairly large pizzas so your cuts don't blur into each other.

4. Fill in as much of this table as you can:

Number of cuts on a pizza	0	1	2	3	4	5	6	7	8
Maximum number of pieces possible			4						

It would be a good idea to collaborate with other groups of people to share information and compare answers.

5. Start conjecturing if you haven't already. What are the next terms in this sequence? Do you see a recurrence relation? How about a closed-form formula? Share your conjectures with others.

6. What properties do your sets of cuts need to have in order to obtain the maximum number of pieces?

7. Try to prove any conjectures available to you. Notice that because the only information you started with was a pizza and a cutting instrument, you will have to use that information in any proof you craft. (You may be able to show that some of your conjectures are logically equivalent to others, but to complete a combinatorial proof of a recurrence relation or closed-form formula, you must link back to the situation of the pizza.)

Now consider a large yam. Your mad scientist laboratory has a laser-cutter chamber in which you can mount the yam; the lasers are able to rotate around it so that cuts can be made at any angle. A cut yam is shown in Figure 9.2.

8. How many chunks of yam can you obtain when using exactly two cuts?

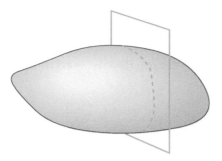

Figure 9.2. A singly cut yam.

9. What is the largest number of yam chunks obtainable using exactly three cuts? It is much more difficult to draw cut yams than to draw sliced pizza, but attempt it anyway.

10. What is the largest number of pieces of yam you can obtain using exactly four cuts? If you cannot get an exact number, then try to find upper and lower bounds on the number.

    (If you have access to a computer and would like a visual aid for your experiments or for communicating your ideas, see http://demonstrations.wolfram. com/CuttingSpaceIntoRegionsWithFourPlanes/—you can move planes such as those shown in Figure 9.3.)

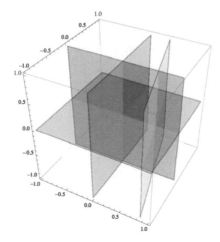

Figure 9.3. Four sample planes cutting space into regions.

9.3   Pizza Numbers

Hey! You! Don't read this unless you have worked through the problems in Section 9.2. I mean it!

By now, you have generated a sequence of numbers, $1, 2, 4, 7, 11, 16, \ldots$, that indicate the maximum number of pieces of pizza obtainable using n cuts. Let us denote the nth term of this sequence as p_n. You have probably conjectured a recurrence for p_n; for example, you might have noticed that the first differences begin as $1, 2, 3, 4, 5, \ldots$ and so suspect that $p_n = p_{n-1} + n$. (Be sure to check that your indexing is correct: the polynomial term must be n, not $n-1$ or $n+1$.) Using techniques from Chapter 8, you can find a closed form for this recurrence of $\frac{n^2}{2} + \frac{n}{2} + 1$.

That's lovely, but flawed. How do we know that the pattern of the first six numbers continues? How do we know that the conjectured recurrence is correct? Even if it is correct, does there exist a more interesting or insight-giving closed form? To answer these questions, we must analyze the situation and construct a combinatorial proof. Such a proof will assure us that the conjectured recurrence is correct and that the pattern observed in the sequence continues.

Let us begin to analyze the situation with ideal placement of the n cuts. How should they be placed so that the resulting number of pieces is maximized? You probably noticed that in the case of three cuts, if all go through a single point, only six pieces are obtained. If the three cuts intersect only pairwise, then seven pieces are obtained, and this is the maximum number. More generally, when three cuts pass through a single point, fewer pieces are obtained than when the cuts intersect pairwise. Additionally, placing a cut parallel to another cut yields fewer pieces than placing the cut so that it can intersect the previous cut. Lines that are placed so that no two are parallel and no three intersect at a single point are said to be *in general position*. Therefore, in order to maximize the number of pieces of pizza, we want the cuts to be placed in general position.

While we are getting all abstract and line-y, let's go all the way. In mathematical language, what you were doing in Section 9.2 was dividing a plane into regions; the fact that the pizza was allowed to be arbitrarily large means that it might have been as infinitely expansive as a plane. The cuts on the pizza correspond to lines that separate regions on the plane. In this view, our pizza drawings become those in Figure 9.4. We can now look at p_n, the maximal number of regions into which n

 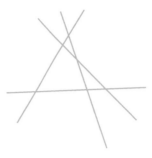

Figure 9.4. Without a crust, a collection of pieces of pizza (left) might as well be regions in the plane (right).

lines can divide a plane, from a recursive standpoint or from a closed-form standpoint. We will consider the closed-form standpoint in Section 9.5.

If we look at the situation recursively, then we want to consider what the relationship is between the maximum number of regions obtainable from n lines and from $n-1$ lines. The difference, of course, is the placement of that last line. Let us assume that we have $n-1$ lines placed such that the maximum number of regions is attained. How shall we place the last line?

Every time a line passes through a region, it cuts the region in two and produces an additional/new region. So we want to maximize the number of regions the last line passes through. Each region is bounded by lines, so we want to maximize the number of other lines that the last line intersects. The most that number can be is $n-1$ because that's how many lines there are. So the question becomes, "Can we achieve this number?" Always? Sometimes?

From geometry we know that two lines that are not parallel must intersect. Thus, as long as we place an nth line so that it is not parallel to any of the already placed $n-1$ lines, it will intersect each of them. The intersection may be quite far away, but it will exist. Moreover, we can nudge the angle of the line and slide the line in various directions to avoid having three lines intersect at a single point. (There are only finitely many intersections to avoid and an infinite number of different angles at which the nth line can be inclined.)

We may conclude that when placing an nth line, we can make it intersect $n-1$ existing lines. This creates n new regions (one "above" each line intersected and one "below" the lowest line), and therefore $p_n = p_{n-1} + n$. We know that this recurrence, together with the initial values of $p_0 = 1, p_1 = 2$, produces the closed form $p_n = \frac{n^2}{2} + \frac{n}{2} + 1$.

Make sure that you can do these problems, as they will prepare you for the more challenging ideas to come.

1. Prove by induction that $p_n = p_{n-1} + n$, $p_0 = 1$, $p_1 = 2$ has closed form $\frac{n^2}{2} + \frac{n}{2} + 1$.

2. How many regions of the plane result from four lines passing through a point? What if n lines pass through the point?

3. How many regions of the plane result from three lines passing through a point and a fourth line intersecting the others pairwise?

4. Draw two parallel lines and then two additional lines that intersect one of the parallel lines at a point. Make two nudges to place these four lines in general position.

9.4   Try This! Spaghetti, Yams, and More

The problems in this section are also interrelated, so you will again want to switch the problem(s) on which you are working from time to time. Insight from a later problem can help you finish an earlier problem.

It may seem a bit ridiculous, but sometimes you want to cut spaghetti with your laser cutter, as in Figure 9.5.

1. Consider a long strand of (one-dimensional) spaghetti. What is the largest number of pieces of spaghetti you can obtain using exactly four cuts? How about five cuts?

2. Fill in as much of this table as you can:

Number of cuts on a spaghetti strand	0	1	2	3	4	5	6	7	8
Maximum number of pieces possible	1								

Figure 9.5. Spaghetti is easy to cut and to eat.

3. What are the next terms in this sequence? Develop both a recurrence and a closed form for this sequence, perhaps labeling it as s_n. Prove that your recurrence and closed form are correct. Have you used induction yet?

4. Now return to questions 8–10 of Section 9.2. Fill in as much of this table as you can:

Number of cuts on a yam	0	1	2	3	4	5	6	7	8
Maximum number of pieces possible		2							

5. What properties do your sets of cuts need to have in order to obtain the maximum number of chunks of yam?

6. Revisit Problem 5 of Section 9.2 and try to use the understanding you gained in Problem 3 above to make further progress. Does this help at all with Problem 4?

7. Start conjecturing about yams if you haven't already. Do you see a possible recurrence relation for y_n? How about a closed-form formula? Try to prove any yammy conjectures you have. Again, because you began with only a yam and a laser cutter, this information must be used in your proofs.

8. Combine all the information you have into one table:

n	0	1	2	3	4	5	6	7	8
s_n	1								
p_n			4						
y_n		2							
$?_n$	1								

Does viewing this information together cause any new patterns to appear? What might go in that bottom row of the table? Have you any new conjectures? Do you see any connection to binomial coefficients?

9. Suppose you have a yam with $n-1$ cuts. Carefully slice the yam in two and look at the freshly cut faces. What do you see? Does this explain any earlier observations you made?

A laser cutter will cut through just about anything. It might even be able to cut a k-dimensional vegetable such as a hyperbeet. Consider this situation if you dare....

10. Suppose you want the maximum number of hyperbeet pieces to result from making n cuts. What properties should your sets of cuts have in order to obtain the maximum number of hyperbeet pieces?

11. Consider a hyperbeet with $n-1$ cuts. Carefully slice the hyperbeet in two and look at the freshly cut faces. What do you see?

12. Write a recurrence for the k-dimensional hyperbeet numbers $h_{n,k}$. Does it remind you of anything?

13. Conjecture a closed form for the k-dimensional hyperbeet numbers $h_{n,k}$.

9.5   Yam, Spaghetti, and Pizza Numbers

Hey! You! Don't even THINK about reading this section unless you have read Section 9.3 AND gone back and reconsidered the questions of Section 9.4!!! I mean it!

We're going to settle these yam questions for once and for all. First, let us deal with the structure of cuts through yams. In order to achieve the maximum number of chunks of yam, our cuts must again be in general position—but what does this mean for yam cuts? No two cuts may be parallel, no three cuts may meet in a line, and no four cuts may meet at a point.

To make the situation explicit, when we speak of cutting yams, we more precisely mean that we are arranging planes in three-dimensional space and counting the number of regions that result. Despite your best efforts, you may have had trouble drawing or even clearly visualizing the maximum number of regions obtainable with four cuts; hopefully Figure 9.6 will help. First notice that there are three planes at right angles to each other. These planes carve out eight regions. A fourth plane cuts through many of these eight regions. How many? Examine the perimeter of this fourth plane as shown in Figure 9.6: it passes through six regions. In addition, it passes through the "front" region, as can be seen by the pale triangular area there. So, in total the fourth plane passes through seven of the eight regions. This means there are $8 + 7 = 15$ regions possible. (In the *Mathematica* Demonstration "Cutting Space into Regions with Four Planes," available at http://demonstrations.wolfram.com/CuttingSpaceIntoRegionsWithFourPlanes/, you can create this sort of image and move the fourth cut around to see what happens at different angles.)

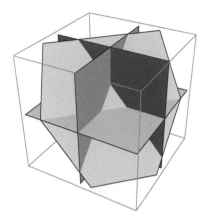

Figure 9.6. How many regions are demarcated by these four planes?

More generally, suppose we have $n - 1$ planes placed in general position in three-dimensional space. In order to create a recurrence for the sequence y_n, we will need to count the number of new regions created when an nth plane is added to the picture (in general position, of course, to achieve the maximum number of new regions). To that end, suppose such an nth plane has been added. For each region through which the nth plane passes, a new region is created. So, we need to count the number of regions through which the nth plane passes. Separate the entire space along the nth plane and look at one side of the cut (as though the nth plane were separating a potato into two hunks and you were examining one wet cut side—see Figure 9.7). What you see is a network of lines, separating the plane into regions. Those lines are cross sections of the $n - 1$ other planes that carve up the three-dimensional space, and the regions are cross sections of the solid regions through which the nth plane passes. Thus, we merely need to count the number

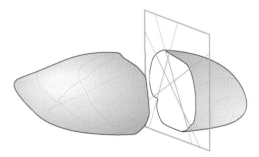

Figure 9.7. A freshly cut yam, with planar regions revealed.

Figure 9.8. A glowing line sweeps across some intersections.

of regions into which these $n-1$ lines divide the plane. But, *aha!*, these lines are in general position because the planes they represent are also in general position! So the number of regions is the corresponding pizza number p_{n-1}. In other words, we have a combinatorial proof that $y_n = y_{n-1} + p_{n-1}$.

Now, we have a polynomial closed form for p_n (and therefore for p_{n-1}), so we can find a closed form for y_n. We will allow you to do this computation in Problem 2 of Section 9.10. What's neat is that we can count regions directly and thereby obtain a closed form. To see how this goes, we will first complete such a direct count for the simpler case of p_n and then apply the same reasoning to y_n. Again, we use combinatorial proof.

Imagine that there are n lines dividing the plane into a maximum number of regions. Then, imagine a glowing line outside of the area that contains all the intersections (outside the pizza, as it were). The glowing line intersects $n+1$ regions because it is cut in n places. Now, sweep the glowing line across all the intersections; each time the glowing line sweeps across an intersection, it enters a new region. (See Figure 9.8, in which the glowing line is represented by a grey line.) If we count the number of intersections, we will then know how many regions there are. Ah, but the lines are in general position, so no two are parallel. That means that every pair of lines crosses, so there are $\binom{n}{2}$ intersections. (And no three lines intersect at a point, so this does not overcount.) Thus, the total number of regions is $p_n = \binom{n}{2} + n + 1$.

Now, on to yams. This time, we will move a softly glowing plane through all the intersections made by n planes in general position. When the glowing plane is outside all the intersections (outside the yam, so to speak), the glowing plane cuts through $\binom{n}{2} + n + 1$ three-dimensional regions because that is how many two-dimensional regions are inscribed by the n cross sections of planes (lines) on the glowing plane itself, and each such two-dimensional region is a flat face of an unbounded three-dimensional region. As the glowing plane passes an intersection, it moves into a new region, and so there are as many additional regions as intersections. And how many intersections are there? Because the planes are in general position, every three of them intersect at a point—so, there are $\binom{n}{3}$ of them. In total, we have $\binom{n}{3} + \binom{n}{2} + n + 1$ regions.

9.5.1   Let's Go for It! Hyperbeet Numbers

Now we will suppose we have a k-dimensional hyperbeet. We wonder how many hyperchunks are obtainable if we cut the hyperbeet with n hypercuts, and we will denote this maximal number of hyperchunks by $h_{n,k}$. Notice that we have two subscripts, the first for the number of hypercuts and the second for the number of dimensions of the hyperbeet. To make sense of the situation for hyperbeets, we will first return to cutting spaghetti. Indeed, this is the simplest situation of all, but it will yield insight that will help us generalize more effectively.

Examine the sequences we have generated thus far.

$$s_n = n + 1 : \qquad\qquad\qquad 1, 2, 3, 4, 5, \ldots.$$

$$p_n = \binom{n}{2} + n + 1 : \qquad\qquad 1, 2, 4, 7, 11, 16, \ldots.$$

$$y_n = \binom{n}{3} + \binom{n}{2} + n + 1 : \qquad 1, 2, 4, 8, 15, \ldots.$$

There are several patterns that can be observed here. One is that each of the sequences is the first differences of the sequence below it. This suggests that $p_n = p_{n-1} + s_{n-1}$ and $y_n = y_{n-1} + p_{n-1}$. We know both of these statements are true by our earlier combinatorial proofs. If this pattern continues, we will expect that $h_{n,k} = h_{n-1,k} + h_{n-1,k-1}$.

Another pattern is that the closed-form formulas seem to involve some binomial coefficients. Well, at least the formulas for p_n and y_n do. What about s_n? Note that $\binom{n}{1} = n$ and that $\binom{n}{0} = 1$. This means we can rewrite our closed forms as

$$s_n = \binom{n}{1} + \binom{n}{0},$$

$$p_n = \binom{n}{2} + \binom{n}{1} + \binom{n}{0},$$

$$y_n = \binom{n}{3} + \binom{n}{2} + \binom{n}{1} + \binom{n}{0}.$$

This suggests that $h_{n,k} = \binom{n}{k} + \cdots + \binom{n}{3} + \binom{n}{2} + \binom{n}{1} + \binom{n}{0}$.

We will shortly justify each of these patterns by using combinatorial proof. First, however, we must consider the placement of our hypercuts. Once again, in order to achieve the largest possible number of hyperchunks, we will need to place the hypercuts in general position. To see what this should mean, we will reason by analogy.

For pizza, no two cuts were parallel and no three intersected at a point.

For a yam, no two cuts were parallel, no three cuts intersected in a line, and no four intersected at a point.

Therefore, for a hyperbeet, no two cuts should be parallel, no three cuts should intersect in a $(k-1)$-dimensional cut, ..., no k cuts should intersect in a line, and no $k+1$ cuts should intersect at a point. This may seem an unwieldy definition (and it is), but it simply means that all intersections are suitably separate from each other. More brain-twisting-ly, what exactly *is* a hypercut? A cut in a plane is a line; a cut in three-dimensional space is a plane; and a cut in k-dimensional space is a $(k-1)$-dimensional flat space. It is quite difficult to imagine.

Now, let us justify the idea that $h_{n,k} = h_{n-1,k} + h_{n-1,k-1}$. Suppose we have $n-1$ hypercuts placed in general position in k-dimensional space. We must count the number of new regions created when an nth hypercut is added. We will suppose that an nth hypercut has been added in general position, so that we obtain the largest possible number of additional regions. As before, we need to count the number of regions through which the nth hypercut passes. If we separate the entire space along the nth hypercut—just as we did with our yam—and look at one side of the cut, we see cross sections of hyperregions and of hypercuts. A cross section of a hyperregion is $(k-1)$-dimensional (because the dimension has been reduced by one), and a cross section of a hypercut is similarly $(k-2)$-dimensional. Now, how many $(k-1)$-dimensional hyperregion boundaries appear on this hypercut? We know that there are n cross sections of hypercuts, and those cross sections are in general position. The taking of cross sections reduces everything by one dimension, so we have $h_{n-1,k-1}$ hyperregions appearing. Technically, this claim is true by a hidden inductive hypothesis. (You will be asked to formalize this in Problem 5 of Section 9.10.) Each of these hyperregions corresponds to a new region created by the nth hypercut cleaving an old region in two. So, we have the $h_{n-1,k}$ regions created by the first $n-1$ hypercuts, and to this we add the $h_{n-1,k-1}$ regions created by the nth hypercut. Thus, $h_{n,k} = h_{n-1,k} + h_{n-1,k-1}$.

We can also count directly to see that $h_{n,k} = \binom{n}{k} + \cdots + \binom{n}{3} + \binom{n}{2} + \binom{n}{1} + \binom{n}{0}$. Much as we did with pizza and with yams, we will move a softly glowing hypercut through all the intersections made by n hypercuts in general position. When the glowing hypercut is outside all the intersections (i.e., outside the hyperbeet), it cuts through $\binom{n}{k-1} + \cdots + \binom{n}{3} + \binom{n}{2} + \binom{n}{1} + \binom{n}{0}$ k-dimensional regions because that is how many $(k-1)$-dimensional regions are demarcated by the n cross sections of hypercuts on the glowing hypercut itself, and each such $(k-1)$-dimensional region bounds an unbounded k-dimensional region. Now we move the glowing hypercut through the hyperbeet. As the glowing hypercut passes an intersection, it moves into a new region, and so there are as many additional regions as intersections. Because the planes are in general position, every k of them intersect at a point— so, there are $\binom{n}{k}$ intersections. In total, we have $\binom{n}{k} + \cdots + \binom{n}{3} + \binom{n}{2} + \binom{n}{1} + \binom{n}{0}$ regions.

Check Yourself

Your brain may hurt, but please try these problems; they are gentle in comparison to the challenging material just presented.

1. Show that $\binom{n}{2} + n + 1 = \frac{n^2}{2} + \frac{n}{2} + 1 = \binom{n+1}{2} + 1$.

2. For what sequence is y_n the sequence of first differences? List the first five terms of that sequence.

3. What are the first seven terms of $h_{n,6}$?

9.6   Where to Go from Here

If you want to know more about geometric applications of combinatorics, seek out the field of computational geometry. Rarely are courses offered in this area at the undergraduate level. There is exactly one undergraduate-level text, and it's fairly new: *Discrete and Computational Geometry* by Satyan L. Devadoss and Joseph O'Rourke. For an extensive directory of applications of discrete and computational geometry, see David Eppstein's "Geometry in Action" page at http://www.ics.uci. edu/~eppstein/geom.html.

For more on the specific problem discussed in this chapter (counting regions defined by hyperplane arrangements), there are three articles you may wish to peruse.

- Seth Zimmerman, "Slicing Space," *The College Mathematics Journal*, Vol. 32, No. 2 (Mar. 2001), pp. 126–128.

  This is a treatment similar to that given in this chapter.

- Chungwu Ho and Seth Zimmerman, "On the Number of Regions in an *m*-dimensional Space Cut by *n* Hyperplanes," *The Australian Mathematical Society Gazette*, Vol. 33, No. 4 (Sept. 2006), pp. 250–264.

How many of the regions in our arrangement are bounded? What happens if the hyperplanes happen to all contain the origin? These questions are addressed in the paper. The authors' language is (unsurprisingly) more advanced than that used in this textbook, but most of the arguments are not significantly more sophisticated. In other words, if you're truly curious, you should have a look and you're likely to gain something.

🐦 Oleg A. Ivanov, "On the Number of Regions into Which n Straight Lines Divide the Plane," *The American Mathematical Monthly*, Vol. 117, No. 10 (Dec. 2010), pp. 881–888.

Suppose our n hyperplanes are not in general position. What numbers of regions are possible? This is the main topic of the paper. Be warned: while this paper is interesting, it does some of its calculations in projective space. Don't expect to understand the whole paper.

Credit where credit is due: The historical origin of this treatment of enumerating the number of regions determined by hyperplanes is George Polya's *Let Us Teach Guessing* video. Grover in Problem 13 of Section 9.10 lives on *Sesame Street*. Here is the full reference for the paper mentioned in Problem 20 and related problems in Section 9.10: Burkhard Polster, "YEA WHY TRY HER RAW WET HAT: A Tour of the Smallest Projective Space," *Mathematical Intelligencer*, Vol. 21, No. 2 (1999), pp. 38–43. It is available electronically at www.qedcat.com/articles/yea.pdf.

9.7   Chapter 9 Definitions

general position: In two dimensions, a placement of lines so that no two are parallel and no three intersect at a single point. In three dimensions, no two cuts may be parallel, no three cuts may meet in a line, and no four cuts may meet at a point. In k dimensions, no two cuts should be parallel, no three cuts should intersect in a $(k-1)$-dimensional cut, …, no k cuts should intersect in a line, and no $k+1$ cuts should intersect at a point.

hyperbeet: A k-dimensional root vegetable.

hypercut: A cut in k-dimensional space, so a $(k-1)$-dimensional flat space.

9.8   Bonus: Geometric Gems

Gem 1.   Believe it or not, if you toss 30 dots into a 1×1 square, at least two of the dots are no more than 0.29 units apart. *Whaaa???*, you say.

Figure 9.9. From left to right, $n = 4, 5, 6, 7$ points marked on a circle, with each pair joined by a line segment.

Consider this: If you divide that 1×1 square into a 5×5 grid of smaller squares, then by the pigeonhole principle, at least two dots must be in the same smaller square because $30 > 25$. The furthest two such dots can be from each other is the length of the diagonal of that small square, which is $\frac{\sqrt{2}}{5} < 0.285$.

Extension:

- 🐦 How few dots could you toss into a 1×1 square and still know that at least two of the dots are no more than 0.29 units apart?

- 🐦 Consider an $n \times n$ grid overlaid on a 1×1 square and fill in the blanks: ____ dots tossed into a 1×1 square include some pair of dots that are less than ____ units apart.

- 🐦 Prove that the statement (with blanks filled in) is true.

Gem 2. Draw a circle and arbitrarily mark n points on the circle. Connect each pair of points by a line segment. If no three line segments intersect at a single point, then how many intersections are there?

First look at Figure 9.9 to see what we're talking about. Essentially, we have drawn K_n with all the vertices on a circle. This reduces our original question to, "How many crossings are there interior to a K_n?" Interestingly, the answer is $\binom{n}{4}$ … but why? Examine the leftmost diagram in Figure 9.9; it shows that K_4 has a single crossing. One can begin with any crossing in K_n and see that it is part of a K_4 within the K_n. Because no two crossings overlap, this means that there is a one-to-one correspondence between K_4s and crossings. And every four points determine a K_4, so there are $\binom{n}{4}$ K_4s and thus $\binom{n}{4}$ crossings. Yeah!

Gem 3. Have you played the card game *SET*? It's totally fun. You can play a solitaire version of *SET* online (and get instructions) at http://smart-games.org/en/

set/start or https://www.lsrhs.net/faculty/seth/Puzzles/set/set.html, and there's
lots of information about the physical card game at http://www.setgame.com/set/.
Consider a version of this game we'll call mini-*SET*. Each card in the deck has
some symbols on it (one, two, or three ovals, diamonds, or squiggles). So every card has two attributes: which symbol it has and how many symbols it has. A
dealer lays out some cards, and then the players try to find three cards that have (the
same shape but different numbers of symbols) xor (different shapes but the same
number of symbols) xor (different shapes and different numbers of symbols). On
finding such a collection, a player calls out "mini-*SET*!" How many cards could
be dealt without a mini-*SET* turning up?

Whoa-hoah! The solution involves material from Chapters 1 and 5 *and* some
geometry! Get ready!

There are nine different cards. Each can be denoted by an ordered pair that
indicates the symbol type and the symbol number, i.e., the deck can be written as
$\{(o,1),(o,2),(o,3),(s,1),(s,2),(s,3),(d,1),(d,2),(d,3)\}$. Now, we are about to
find it convenient to use $\mathbb{Z}_3$ instead of $\{o,s,d\}$ and $\{1,2,3\}$, so we will instead use
a couple of one-to-one correspondences and write the deck as $\{(0,0),(0,1),(0,2),$
$(1,0),(1,1),(1,2),(2,0),(2,1),(2,2)\}$.

A mini-*SET* is characterized by all values of an attribute being the same, i.e.,
having $\{(k,?),(k,??),(k,???)\}$, and/or having all values of an attribute being different, i.e., having $\{(0,?),(1,??),(2,???)\}$, because there are only three values for
each attribute. In the all-the-same case, if we add the attribute values together,
we get $k+k+k = 3k \equiv 0 \pmod 3$. In the all-different case, adding produces
$0+1+2 = 3 \equiv 0 \pmod 3$. We therefore seek three cards where the sum of the
first coordinates and the sum of the second coordinates are both equivalent to 0
(mod 3). Examine the grid of cards shown in Figure 9.10; mini-*SET*s of cards
form lines on the grid. There are three horizontal lines, three vertical lines, three
left-diagonal lines, and three right-diagonal lines. Most of the diagonal lines don't
look like lines; they wrap around the grid and so appear to be chopped up. But
they're there.

By trial and error, you can probably find four cards that contain no mini-*SET*
(or if not, check out $\{(0,0),(0,1),(1,1),(1,2)\}$). Now, let us show that any five
cards we deal must contain a mini-*SET*. Suppose we have dealt five cards that do
not contain a mini-*SET*. (See? We're doing a proof by contradiction.) At most two
of the cards are on any line in the grid. Consider the three horizontal lines. One
of them contains two cards; a second one contains two more cards; and the third
line contains the fifth card (because it can't be on either of the other two horizontal
lines). Now, look at the card that's on a horizontal line all by itself. It is on three

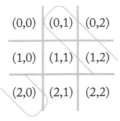

(0,0)	(0,1)	(0,2)
(1,0)	(1,1)	(1,2)
(2,0)	(2,1)	(2,2)

Figure 9.10. A not-very-liney-looking line indicated on the grid of mini-*SET* cards.

(0,0)	(0,1)	(0,2)
(1,0)	(1,1)	(1,2)
(2,0)	(2,1)	(2,2)

Figure 9.11. The four lines containing a card cover all nine mini-*SET* cards.

other lines (a vertical line, a left-diagonal line, and a right-diagonal line). Between those four lines, all nine cards are covered. (Check out Figure 9.11 to see it.)

Just to repeat it, we are looking at the card C that's on a horizontal line all by itself. The other four cards are on the other three lines that contain C, because they are not on the horizontal line that contains C. By the pigeonhole principle, there must be two of the four cards on one of the three lines that contain C. But (*wagging a finger*) no, no, no—that means there are three cards on that line (two of the four plus the one we started with) and that's a mini-*SET*! Contradiction.

Mega-challenge. Extend this analysis to the game not-as-mini-*SET* (with symbol type, symbol number, and symbol color as the attributes). This is quite involved and will require persistence! For assistance, see the paper "The Card Game Set," by Ben Davis and Diane Maclagan, in *The Mathematical Intelligencer*, Vol. 25, No. 3 (Fall 2003), pp. 33–40, a preprint of which is available online at http://homepages.warwick.ac.uk/staff/D.Maclagan/papers/set.pdf.

9.9   Bonus Check-Yourself Problems

Solutions to these problems appear starting on page 614. Those solutions that model a formal write-up (such as one might hand in for homework) are to Problems 3 and 5.

1. Let f_n be the maximum number of regions of four-dimensional space that are cut up by n three-dimensional cuts. What are f_0, f_1, f_2, f_3, f_4? And why?

2. If you cut a configuration with f_4 pieces with an additional cut, how many new pieces can you get?

3. Determine and explain a recurrence relation for f_n.

4. Determine and explain a closed form for f_n.

5. Use induction to prove that your closed form from Problem 4 is the correct closed form for your recurrence from Problem 3.

9.10   Problems That Combine Combinatorial Topics

1. How many regions of a plane are outside of a pizza?

2. The yam number recurrence is $y_n = y_{n-1} + \frac{(n-1)^2}{2} + \frac{n-1}{2} + 1$, with initial conditions $y_0 = 1, y_1 = 2$. Use the technique of kth differences to find a closed form for y_n.

3. Give an explanation in your own words for why $y_k = y_{k-1} + p_{k-1}$.

4. Under what conditions does making n cuts produce 2^n pieces?

5. Prove that $h_{n,k} = h_{n-1,k} + h_{n-1,k-1}$ using induction on k.

6. Prove that $h_{n,k} = \binom{n}{k} + \cdots + \binom{n}{3} + \binom{n}{2} + \binom{n}{1} + \binom{n}{0}$ using induction.

7. Consider the recurrence $h_{n,k} = h_{n-1,k} + h_{n-1,k-1}$. Suppose that $h_{1,0} = h_{1,1} = h_{n,0} = 1$ and that when $k > n$, $h_{n,k} = 0$. What familiar situation are we in?

8. Find a closed form for $a_0 = 0, a_n = a_{n-1} + \binom{n}{1}$.

9. Find a closed form for $a_0 = 0, a_n = a_{n-1} + \binom{n}{2}$.

10. Find a closed form for $a_0 = 0, a_n = a_{n-1} + \binom{n}{3}$.

11. Find a closed form for $a_0 = 0, a_n = a_{n-1} + \binom{n}{4}$.

12. Find a closed form for $a_0 = 0, a_n = a_{n-1} + \binom{n}{j}$ for j fixed.

13. Imagine a number line with the integers marked. Your old pal Grover starts at 0, and once per second takes a step to the left or a step to the right. (Being Grover, he probably sings a song about left and right as he does this.)

(a) Is it possible for Grover to take three steps and end up back at 0?

(b) Suppose that Grover ends his meandering at k. What can you say about the number of steps n that he took?

(c) How many ways are there for Grover to take four steps and end up back at 0?

(d) How many ways are there for Grover to take n steps and end up back at 0?

(e) How many ways are there for Grover to take n steps and end up at k?

14. The first few pentagonal numbers are shown in Figure 9.12. Create a sequence, find a recurrence, and find a closed form.

Figure 9.12. The first four pentagonal numbers.

15. The first few hexagonal numbers are shown in Figure 9.13. Create a sequence, find a recurrence, and find a closed form.

Figure 9.13. The first four hexagonal numbers.

16. Considering the previous two exercises, …

    (a) … draw the first four square numbers.

    (b) … draw the first four triangular numbers.

    (c) … find a recurrence and a closed form for the square numbers.

    (d) … find a recurrence and a closed form for the triangular numbers.

17. **Challenge:** Consider the *k-figurate* numbers. The first *k*-figurate number can be represented by a dot; the second *k*-figurate number can be represented by *k* dots that represent vertices of a *k*-gon. Find a recurrence and a closed form for the *k*-figurate numbers.

18. A *triangulation* of a convex polygon is a partition of that polygon into triangles that does not introduce any new vertices.

    (a) Consider a triangle. How many triangles are in a triangulation of a triangle?

    (b) There are different ways to triangulate a square. How many triangles does each have?

    (c) How many triangles does any triangulation of an *n*-sided convex polygon have? Prove that your response is correct.

19. We will triangulate (as in Problem 18) a convex polygon that has a convex polygonal hole cut out of its interior.

    (a) Draw a large triangle and within this draw (and shade) a smaller triangle. Triangulate the resulting polygonal flat doughnut. How many triangles do you obtain?

    (b) Draw a large pentagon and within this draw (and shade) a smaller triangle. Triangulate the resulting polygonal flat doughnut. How many triangles do you obtain?

    (c) Draw a large hexagon and within this draw (and shade) a smaller square. Triangulate the resulting polygonal flat doughnut. How many triangles do you obtain?

    (d) How many triangles does any triangulation of a polygonal flat doughnut have? Prove that your response is correct.

20. We are going to construct a *geometric structure* from a set of *lines* using a finite number of points in the plane. Suppose that we have four points a, b, c, d and that we consider a set of points to be a line if it contains exactly two points.

    (a) How many possible lines are there?

    (b) How many lines are parallel to the line ab?

    (c) How many lines intersect the line ab?

    (d) Suppose we want a geometric structure with no parallel lines (because they remind us of bottomless pits—see Burkhard Polster's paper "YEA WHY TRY HER RAW WET HAT"). We might start with the line ab and all lines that intersect it; however, we would need to be sure that no two of these lines are parallel. How many lines can our geometric structure have if no two are parallel? Which lines can we choose for our structure?

    (e) Now suppose further that we want exactly two lines to intersect at a point. Which lines can we choose for our structure?

(f) Try to draw the points and lines of this geometric structure.

21. As in Problem 20, we are going to create a geometric structure from a finite number of points in the plane. Suppose that we have six points a, b, c, d, e, f and that we consider a set of points to be a line if it contains exactly three points. We will select some lines, no two of which are parallel (and therefore, every pair of which intersect).

(a) How many possible lines are there?

(b) How many lines are parallel to the line abc?

(c) How many lines intersect the line abc? A pair of lines should only intersect at one point (otherwise they are in some sense curved).

(d) We need to make sure that none of the lines that intersect abc intersect each other at more than one point. How many lines intersect abc and intersect each other at no more than one point?

(e) Suppose we want a geometric structure with no parallel lines (because they remind us of bottomless pits). Can we select lines as in part (d) so that there are no pairs of parallel lines?

(f) Now suppose further that we want exactly two lines to intersect at a point. Which lines can we choose for our structure?

(g) Try to draw the points and lines of this geometric structure.

22. Let us generalize Problem 20. Suppose we wish to create a geometric structure

from k points in the plane and that we consider a set of points to be a line if it contains exactly r points.

(a) How many possible lines are there?

(b) How many lines are parallel to a given line?

(c) How many lines intersect a given line at exactly one point?

23. As in Problem 20, we are going to create a geometric structure from a finite number of points in the plane. Suppose that we have seven points a, b, c, d, e, f, g and that we consider a set of points to be a line if it contains exactly three points. We will select some lines, no two of which are parallel (and therefore, every pair of which intersect).

(a) How many possible lines are there?

(b) How many lines are parallel to the line abc?

(c) How many lines intersect the line abc? A pair of lines should only intersect at one point (otherwise they are in some sense curved).

(d) We need to make sure that none of the lines that intersect abc intersect each other at more than one point. How many lines intersect abc and intersect each other at no more than one point?

(e) Suppose we want a geometric structure with no parallel lines (because they remind us of bottomless pits). Can we select lines as in part (d) so that there are no pairs of parallel lines?

(f) Now suppose further that we want exactly three lines to intersect at a point. Which lines can we choose for our structure?

(g) Try to draw the points and lines of this geometric structure. You will need to draw at least one line in an unusual way in order to succeed. (This geometric structure is called the *Fano plane* and has many interesting properties—for example, the roles of the points and the lines can be switched.)

24. As in Problem 20, we wish to create a geometric structure from a finite number of points in the plane. Suppose that we have eight points a,b,c,d,e,f,g,h and that we consider a set of points to be a line if it contains exactly three points. We will select some lines, no two of which are parallel (and therefore, every pair of which intersect).

(a) How many possible lines are there?

(b) How many lines are parallel to the line *abc*?

(c) How many lines intersect the line *abc*? A pair of lines should only intersect at one point (otherwise they are in some sense curved).

(d) We need to make sure that none of the lines that intersect *abc* intersect each other at more than one point. How many lines intersect *abc* and intersect each other at no more than one point?

(e) Suppose we want a geometric structure with no parallel lines (because they remind us of bottomless pits). Can we select lines as in part (d) so that there are no pairs of parallel lines? Explain.

25. **Challenge**: As in Problem 20, we wish to create a geometric structure from a finite number of points in the plane. Suppose that we have eight points a,b,c,d,e,f,g,h and that we consider a set of points to be a line if it contains exactly four points. We will select some lines, no two of which are parallel (and therefore, every pair of which intersect).

(a) How many possible lines are there?

(b) How many lines are parallel to the line *abcd*?

(c) How many lines intersect the line *abcd*? A pair of lines should only intersect at one point (otherwise they are in some sense curved).

(d) We need to make sure that none of the lines that intersect *abcd* intersect each other at more than one point. How many lines intersect *abcd* and intersect each other at no more than one point?

(e) That's pretty lousy. Redo part (d) with nine points a,b,c,d,e,f,g,h,i and see if that makes the situation better.

(f) Do you expect the use of ten points a,b,c,d,e,f,g,h,i,j to resolve this issue? Why or why not?

9.11 Instructor Notes

There are only two days of class work provided in this chapter in order to leave room for an exam day at the end of the combinatorics section of the course. Certainly if there is enough time in your particular course, let students explore more!

When Polya addresses the enumeration of regions determined by hyperplanes in *Let Us Teach Guessing*, he does so using interactive lecture. The treatment given in this text uses his basic structure but breaks Polya's content into a series of guided exercises that students can complete collaboratively.

Very little introduction is needed before students can begin to experiment. Draw a circle on the board, mention that it is a pizza, and explain that you want to get the largest number of pieces possible for a given number of cuts. Then draw a couple of arbitrary cuts on the pizza. Be sure the cuts do not pass through the center of the pizza to quickly dispel the assumption that cuts must meet in the center. Mention that it's okay to have tiny pieces of pizza and that the pizza can be arbitrarily large (and do not be surprised if you need to repeat this information later). Ask whether students have any questions about the setup before turning them loose to experiment.

For these problems, students tend to start working in pairs to generate data and later cluster into groups to discuss conjectures. Students will be, on the whole, more productive if instead of completing problems in a strictly sequential manner, they attempt lots of problems before finishing many of them. Therefore, as you manage the group work, do encourage students to consider new questions if they seem to be bogged down at all. It is helpful, after the students have been working for a few minutes, to draw a table on the board such as is presented in Problem 4 of Section 9.2. Fill this in as students/groups report results to you. Add also a heading "Conjectures:" on a separate part of the board, and create a list of student conjectures as you hear them made while circulating among groups. Seeing the results and conjectures of other groups will spur students to verify, extend, and investigate further. This class may feel a bit chaotic, but that is to be expected. Moreover, it is not important that students reach any conclusions by the end of the first class period; it will all come together in the second class period (with your guidance, if necessary).

One caution is that students are likely to jump to conclusions: They will believe that they have found a pattern, but one incorrect number has led them down a blind alley. Or, they will conjecture a correct recurrence and find a closed form for it, but they will not have justified that this recurrence corresponds to the maximal number of pieces of pizza obtainable with n cuts. Be sure to regularly ask groups to justify their conclusions and to explain why their results answer the original question.

In general, students are able to draw and conceptualize pizza diagrams quickly. However, they have trouble drawing three-dimensional diagrams that illustrate ideas they can see clearly in their heads. Be aware that this will be a source of frustration for them. Some students find it challenging to visualize cuts in three dimensions as well.

Additionally, when students get stuck on this topic, they get *very* stuck. Expect that you will do a lot of individualized triage to keep groups on track. Instead of the usual prompting and questioning that you might do for particular problems in a Try This! section, here you will likely need to prompt and question across problems and at a higher cognitive level.

It is reasonable to expect that a class will have collectively completed a pizza-number table and started on a yam-number table by the end of the first class. They find it difficult to count the number of yam pieces obtainable with four cuts and will likely be stuck on this at the end of the class. Let them be stuck and ponder. It is advantageous to let the idea of thinking in different dimensions sink in, so as you wrap up the work on the first day, make sure to summarize the data generated by the students, the questions addressed by the students, and any conjectures proposed by groups of students. Review the questions for three dimensions and suggest that students think about these same questions in more dimensions as preparation for the next class. If they have made enough progress, assign them to read Section 9.3 as well.

You might begin the second class meeting by introducing spaghetti numbers and seeing how quickly they generate a table and conjectures before letting them loose to tackle Section 9.4. Be sure to leave at least 15 minutes at the end of class to wrap things up: review the conjectures and reasoning the class has produced, and relate this to the more general ideas of binomial coefficients and recurrences from previous chapters.

Theme II Supplement: Problems on the Theme of Combinatorics

These problems could be used for studying for (or writing!) in-class or take-home exams, or just for more enrichment. (The problems cover Chapters 6–9.) They are not given in any particular order. Well, they have been intentionally mixed up so that they are *not* in chapter order, so that the solver cannot use the ordering of the problems as a clue in solving them.

1. If three cats like tuna, five cats like salmon, two cats like both tuna and salmon, and one cat does not like fish at all, how many cats have been consulted?

2. Consider the word PEACH. We will make lists from the letters in PEACH, with repetition allowed.

    (a) How many three-letter lists are there that don't begin with C or don't end in P?

    (b) How many four-letter lists are there in which the sequence of letters P, E, A appears in that order?

    (c) How many three-letter lists are there in which the letters C, H appear consecutively?

3. There is a colony of gnats who all wear hats. Each gnat wears a green, white, or red hat (they are Italian gnats). Moreover, a single gnat wears a green or red hat, and a partnered gnat wears a white hat—and partnered gnats always hold hands when they fly. Gnats often line up while flying.

    (a) How many different hat-color strings are possible for two gnats with hats flying in a line?

    (b) How many different hat-color strings are possible for three gnats with hats flying in a line?

    (c) Consider the sequence g_n of hat-color strings for n gnats with hats flying in a line. Write a recurrence and find a closed form for g_n.

4. Prove that the sum of the entries in the nth row of Pascal's triangle is 2^n using induction.

5. Prove that the number of 1s in the binary representation of $n \in \mathbb{N}$ is counted by the function $b(n)$, defined as $b(2n) = b(n)$, $b(2n+1) = b(n) + 1$, $b(1) = 1$.

6. For each of the following closed forms, write out the first several terms of the sequence (at least five) and use this to create a recursion for the sequence.

(a) $a_n = 5n - 2$.

(b) $a_n = 3^n + 1$.

(c) $a_n = 7n - 6$.

7. Figure TII.1 shows a 2×4 grid that has been tiled with four 1×2 rectangles.

(a) How many ways are there to tile a 2×1 grid with 1×2 rectangles? (The rectangles may be rotated so as to be 2×1 rectangles.)

(b) How many ways are there to tile a 2×2 grid with 1×2 rectangles?

(c) How many ways are there to tile a 2×3 grid with 1×2 rectangles?

(d) How many ways are there to tile a 2×4 grid with 1×2 rectangles?

(e) How many ways are there to tile a $2 \times n$ grid with 1×2 rectangles? Make and prove a conjecture.

Figure TII.1. A 2×4 grid (left) tiled with four 1×2 rectangles (right).

8. For each of the following recurrence relations, write out the first several terms of the sequence (at least five) and use this to find a closed form for the sequence.

(a) $a_1 = 5; a_n = 5a_{n-1}$.

(b) $a_1 = 2; a_n = 5a_{n-1}$.

(c) $a_1 = 6; a_n = 3 + a_{n-1}$.

9. The organization Red Delicious—Crappy Apples (RDCA) distributes at farmer's markets a variety of brochures on the virtues of various varieties of apples and, of course, on the vices of the hated Red Delicious apple. Passersby are invited to take as many different brochures as they want, but no more than one of each kind. If there are n different brochures available, how many selections of brochures might a passerby take? (Note: RDCA was a real organization in the mid-1990s but did not do significant outreach. It had at least three members.)

10. Find a closed form for the recurrence $a_n = a_{n-1} + 4a_{n-2} - 4a_{n-3}$ and initial values $a_0 = 4, a_1 = 2, a_2 = 10$.

11. Consider $a_n = 2^n - 1$ and form the new sequence $s_n = a_1 + a_2 + \cdots + a_n$. Fill in the blanks in this table:

n	1	2	3	4	5	6
a_n	1	3				
s_n	1	4				

What is the relationship between the bottom two rows? Use this relationship to write a recurrence relation for s_n.

12. How many ways are there to place a row of nine coins (all from the same country) on the edge of a table?

13. Find a closed form for $a_1 = 1, a_n = a_{n-1} + n + 1$.

14. Solve the recurrence relation $a_1 = 1$, $a_n = na_{n-1}$.

15. How many ways are there to distribute 5 past-their-prime vegetables (a squash, a crown of broccoli, a tomato, a clump of Swiss chard, and a large radish) to 14 different chickens?

16. How many nonnegative solutions are there to the equation $w_1 + w_2 + w_3 + w_4 = 6$?

17. How many edges does a k-regular graph with n vertices have? Explain.

18. At the Delicious Pie Bakery, there are 18 banana-containing pies, 23 chocolate-containing pies, 12 peanut-butter-containing pies, 15 chocolate-banana-containing pies, 10 peanut-butter-chocolate-containing pies, and 3 chocolate-peanut-butter-banana pies. There are 27 pies total. How many peanut-butter-banana-containing pies are there?

19. An upper-level math class has 12 students. 3 of them live off campus and 9 live on campus. How many ways can a project group of three students be chosen so that it has two on-campus members and one off-campus member? Explain *briefly*.

20. How many ways are there to distribute 28 doses of dye to 12 skeins of (currently ugly but soon to be lovely) yarn?

21. What percentage of (theoretical, seven-digit) phone numbers have all digits distinct?

22. At the ice-cream store from Chapter 1, there are five flavors of ice cream left (peppermint, hoarhound, chocolate malt, gingerbread, and squirrel) and you want to order three quarts. In Chapter 1, you counted the number of ways to do so by brute force; in Chapter 6, you divided this into cases and used choice numbers. Solve this problem once again, but this time using the techniques of Chapter 7.

23. How many seven-digit telephone numbers...

    (a) ... begin with 538–?

    (b) ... begin with 538– and have four different numbers in the remaining digits?

    (c) ... begin with 538– and contain a 9, a 6, and a 2?

24. A dance company has ten members. How many different ways can a choreographer choose six dancers for a dance piece?

25. According to Dave Perkins, a chain reaction in the iPhone game *Drop 7* can lead to scores of $7, 39, 109, 224, 391, 617, \ldots$. Assuming this integer sequence continues in the fashion it started, what should the next number be?

26. Twelve hundred students at Ördek University were surveyed about the Summer Olympics: 620 wanted to watch pole vault, 730 wanted to watch gymnastics, and 460 were interested in the decathlon. Additionally, 170 wanted to watch the opening ceremonies and nothing else. It also turns out that 270 of the students favor both pole vault and decathlon, 330 students are fans of both pole vault and gymnastics, and 240 intend to view both gymnastics and decathlon. How many students like gymnastics or like decathlon? How many students want to watch all three sports?

27. A pastry shop opens across the street from Haddad Library, and of course, it is very popular. The very first day, 80 people buy apple turnovers, 80 buy blueberry muffins, and 80 buy chocolate croissants; 30 buy both apple turnovers and blueberry muffins, 30 buy both apple turnovers and chocolate croissants, and 20 buy both blueberry muffins and chocolate croissants; and 20 people buy apple turnovers, blueberry muffins, *and* chocolate croissants. How many people bought only apple turnovers? Give jus-

tifications for your answer, once using Venn diagrams and again without using Venn diagrams.

28. Figure TII.2 shows a polyhedron with 30 vertices. (Note that at each vertex, two triangles and two hexagons meet.) How many edges does it have? Explain, perhaps by overcounting carefully.

Figure TII.2.   Reproduction of an oil painting of a venerable icosidodecahedron.

29. Solve the recurrence relation $a_1 = 0$, $a_2 = 4, a_n = 5a_{n-1} - 6a_{n-2}$.

30. Show that $\displaystyle\sum_{j=0}^{m} (-1)^j 10^{m-j} \binom{m}{j} = 9^m$.

31. How many anagrams (including nonsensical anagrams) are there of the word CREEPIER? Explain *briefly*.

32. Examine the identity $\binom{n}{k} = \frac{n}{k}\binom{n-1}{k-1}$.

   (a) Verify the identity's veracity using the factorial formula for binomial coefficients.

   (b) Now find a combinatorial proof that the identity holds by counting the same thing in two different ways.

33. In a sample of 100 students, 43 like avocados, 71 like radishes, and 36 like olives in their salads. Each student likes at least one vegetable. If 26 students like both avocados and radishes, 16 students like avocados and olives, while 22 like radishes and olives, how many students like all the ingredients in an avocado, radish, and olive salad?

34. A robot can only walk forwards. Each step it takes is of length 1 foot or 2 feet. So, for example, the robot may walk 6 feet by taking four steps as lengths $2, 2, 1, 1$ or as lengths $1, 2, 1, 2$. The sequence r_n is defined as the number of ways the robot can walk n feet. What are r_1, r_2, r_3, r_4? Find a recurrence relation for r_n. Give a combinatorial explanation for why this is the correct recurrence relation.

35. Use the choice notation identity $\binom{n}{k} = \binom{n-1}{k-1} + \binom{n-1}{k}$ to evaluate $\binom{n}{0} - \binom{n}{1} + \binom{n}{2} - \cdots \pm \binom{n}{n}$.

36. A computer science department has eight faculty members, three of whom have the last name Jeong. How many ways can two representatives to the Faculty Council be chosen so that one representative has the last name Jeong and the other does not?

37. We will use the letters a, b, c, d, e, f to form words of length 4. No word can have two of the same letter.

   (a) How many words can be made?

   (b) How many words end in ef?

   (c) How many words contain ef as consecutive letters?

   (d) How many words contain both e and f, but not necessarily consecutively?

   (e) How many words contain an e or an f (or both)?

38. A basket contains ten numbered balls; seven are purple and three are teal. You pull four balls out of the basket.

    (a) How many different sets of four balls could you pull out of the basket?

    (b) How many sets of four balls are all purple?

    (c) How many sets of four balls contain at least one teal ball?

39. How many ways are there to deal a standard deck of 52 cards to seven players?

40. How many different seven-digit phone numbers begin with 231- and contain exactly one 9?

41. In how many ways can you grab 5 pens from a pile of 20 pens?

42. Write an algorithm for generating Pascal's triangle.

43. Solve the recurrence relation $a_1 = 12$, $a_n = a_{n-1} - 7$.

44. You are captured by an ogre who ties you to a chair at a table and prepares a diabolical puzzle for you. Squatting across from you, the ogre shows you 15 identical chocolate chips and an eye-dropper full of sparkling liquid—*poisonous* sparkling liquid, says the ogre.

    (a) Under the table, the ogre drips the liquid onto four of the chips. In how many ways can the ogre arrange the 11 unpoisoned and 4 poisoned chips in a line on the table? It is impossible to tell a poisonous chip from a nonpoisonous one.

    (b) The ogre forces you to choose and consume two of the chips. Of the total number of choices, how many will contain at least one poisoned chip?

45. Prove the binomial theorem by induction.

46. A small skunk stumbles upon a plover's nest that contains six eggs. The skunk is only able to choose three of the six eggs to take with hir; how many choices total does ze have? (Fortunately for the eggs, the plover returns while the skunk is calculating and chases the skunk away.)

47. Give an expression for the number of banana splits you could order from 31 Flavors, assuming there are three different scoops in a banana split and that there are actually 31 flavors to choose from. (Here, there are no toppings and we do not care in what order the scoops land in the dish.) Compute this number directly. Compute this number recursively. Explain how you did each computation.

48. Farmer Jimenez hides 12 decorated duck eggs, all colored differently, and then lets hir three weasels out of a cage to hunt for them. In how many ways can the weasels find the eggs? Bear in mind that it is possible for a weasel to find no eggs, or all 12, and so on.

49. Is the following statement true or false: $\binom{n-1}{2} = \sum_{i=1}^{n} i$? If the statement is true, prove it. If it is false, revise it and then prove it.

50. The nutritious snack shelf contains a box of Nature Valley Trail Mix bars (two flavors) and several bags of Crispi Rolls (four flavors). How many ways are there to pack two nutritious snacks for a walk?

Credit where credit is due: Problem 3 was inspired by a set of stories the author's father told when going for his evening walk near her house. Problem 33 is in honor of a student (Jackie Kajos) who did *not* like lettuce. Problem 47 refers to the old subname of the Baskin-Robbins ice cream chain. Problem 48 was derived from one of Dave Perkins's exam problems; Problems 44 and 46 were donated by Dave Perkins.

Part III
Theme: Graph Theory

Chapter 10 🐦🐦🐦🐦🐦🐦🐦🐦🐦

Trees

10.1   Introduction and Summary

You may remember from Chapter 3 that a tree is a connected graph with no cycles. Trees are one of the most popular classes of graphs, because they have so many applications in computer science and operations research and because they are so useful in pure mathematics proofs. We will see a pure mathematics use for trees in Chapter 11, and in this chapter, we will concentrate on aspects of trees that lead to applications. First we will investigate spanning trees—given a graph, can you find a subgraph that is a tree and covers all the vertices of (spans) the graph? If so, how? What if the graph has values assigned to the edges, and we want the lowest (or highest) total value in our spanning tree? Most of our approaches will involve a particular type of algorithm (called greedy, but parsimonious in nature), so we will briefly discuss its use across graph theory.

Then we will look at how to use trees to search efficiently through data. For this purpose, a tree can correspond either to decisions made when searching or to the organization of the data itself; binary trees (they have exactly two edges emanating "downwards" from each vertex) are frequently used for searching and storage of data. We will introduce a type of forest that solves many matching problems; finally, we will discuss backtracking as a general way of efficiently finding solutions to problems with constraints. The Bonus section introduces the branch-and-bound algorithm for finding best solutions to systems of linear equations in lots of binary variables (and depends on Bonus Section 7.9).

10.2   Basic Facts about Trees

It has been many chapters since we studied graphs, but hopefully you will recall (from Section 10.1) that a tree is a connected graph with no cycles. You may also recall from Example 4.2.4 that if a tree has n vertices, then it has $n-1$ edges. Let us show that a partial converse is true.

Theorem 10.2.1. *If a connected graph G has n vertices and n − 1 edges, then G is a tree.*

Proof: Suppose G is connected, has n vertices, and has $n − 1$ edges. We will proceed by contradiction; suppose that G is not a tree. Because G is connected, G must have at least one cycle. If we remove one edge e_1 from this cycle, $G_1 = G \setminus e_1$ is still connected. If G_1 has a cycle, then remove an edge e_2 to obtain $G_2 = G_1 \setminus e_2$; we see that G_2 is still connected. Continue in this way to obtain a graph G_k with no cycles; we know that G_k is connected, so G_k is a tree. Additionally, G_k is G with k edges removed, and so it has $n − 1 − k$ edges. However, G_k has n vertices, so because it is a tree, it has $n − 1$ edges. This is a contradiction unless $k = 0$, in which case G has no cycles to begin with… and thus, G is a tree. □

Well, there's another partial converse…

Theorem 10.2.2. *If a graph G with no cycles has n vertices and n − 1 edges, then G is a tree.*

Proof: Suppose G is *acyclic* (i.e., it has no cycles), has n vertices, and has $n − 1$ edges. We will proceed by contradiction; suppose that G is not a tree. Because G is acyclic, G must not be connected. Consider any connected component of G. It is a tree because it is connected and acyclic, so it has k vertices and $k − 1$ edges. Now consider two connected components of G; each is a tree, and they respectively have k_1, k_2 vertices and $k_1 − 1, k_2 − 1$ edges. Together, they have $k_1 + k_2$ vertices and $k_1 − 1 + k_2 − 1 = (k_1 + k_2) − 2$ edges. Considering all r components of G, we have $k_1 + k_2 + \cdots + k_r = n$ vertices and $(k_1 + k_2 + \cdots + k_r) − r = n − 1$ edges. Therefore, $r = 1$ (so G is connected) and G is a tree. □

Corollary 10.2.3. *If a graph G has n vertices and* fewer *than n − 1 edges, then G is not connected.*

Proof: Suppose G has n vertices and has $s < n − 1$ edges. We will proceed by contradiction; suppose that G is connected. If G has no cycles, then by definition it is a tree and has $n − 1$ edges, which is a contradiction. Therefore, suppose that G has at least one cycle. Using the technique in the proof of Theorem 10.2.1, we may produce the acyclic graph G_k that has n vertices and $s − k$ edges. Because G_k is connected as well as acyclic, it is a tree and therefore $s − k = n − 1$. However, $s − k \leq s < n − 1$, which is a contradiction. □

> Summary. Together, Theorems 10.2.1 and 10.2.2 and Corollary 10.2.3 tell us that trees are the connected graphs with the least number of edges—which, if you thought about it, perhaps you had the sense should be the case, but now we are *sure*.

Theorem 10.2.4. *Every tree T with at least two vertices has at least one leaf. (Recall that a leaf is a vertex of degree 1.)*

Proof: Suppose, for the sake of contradiction, that T is a tree with no leaves. Then every vertex of T must have degree 0 or degree greater than 1. Because T is connected, no vertex can have degree 0. Thus, every vertex of T has degree greater than 1. Consider some vertex v of T. Walk along T, using a new edge at each step and labeling the vertices $v_1, v_2, \ldots$. If you walk to a vertex that already has a label, then T has a cycle and this is a contradiction. If you never walk to an already labeled vertex, then the path must end... in a leaf. □

Theorem 10.2.5. *Every tree T with at least two vertices has at least two leaves.*

Proof: Suppose, for the sake of contradiction, that T is a tree with only one leaf ℓ. (We know it has at least one leaf from Theorem 10.2.4.) Begin at ℓ and walk along T, using a new edge at each step and labeling the vertices $v_1, v_2, \ldots$. If you walk to a vertex that already has a label, then T has a cycle and this is a contradiction. If you never walk to an already labeled vertex, then the path must end ... in another leaf. □

We'll need these facts later. Hopefully, they have warmed up your brain to work with graphs again.

Check Yourself

Enjoy these quick problems.

1. Draw all trees on five vertices.

2. What must be true about the degree sequence of a tree?

3. Suppose a graph G has 364 vertices and 365 edges. Can it be a tree?

4. Suppose a graph G has 28 vertices and 27 edges. Must it be a tree?

5. Suppose a connected graph G with 432,894,789 vertices has no cycles. How many edges might it have?

6. Draw a tree that has exactly two leaves.

7. Draw a tree that has exactly three leaves.

10.3   Try This! Spanning Trees

Every graph contains at least one tree as a subgraph (for example, a single vertex or a single edge with both its incident vertices), but usually it has lots and lots of trees as subgraphs. In the following exercises, we will think about the largest possible trees we could find as subgraphs, namely, those trees that contain all the vertices of a given graph. Such a tree is called a *spanning tree*. Figure 10.1 shows three graphs along with two example spanning trees for each one.

1. Give an example of a subgraph of one of the graphs in Figure 10.1 that is not spanning.

2. Give an example of a subgraph of one of the graphs in Figure 10.1 that is a spanning subgraph but not a tree.

3. Show that every connected graph has at least one spanning tree by giving an algorithm for finding one.

4. Did your algorithm begin with just the vertices, or did it begin with the whole graph? Find a second algorithm that begins differently than your first.

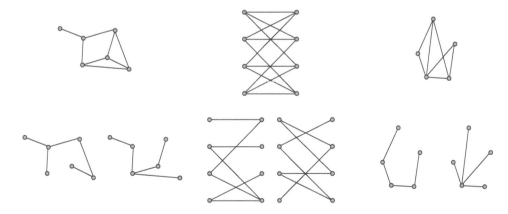

Figure 10.1. The upper graphs contain the lower spanning trees.

5. Prove that your algorithms work. That is, show that the output is a tree and that the tree includes all the vertices of the original graph.

6. Show that every graph, connected or not, has a spanning forest.

Let's move to reality for a little while (though it will still be abstracted reality). Why would we want to find a spanning tree? Suppose that we need to build an oil pipeline: it transports oil from several wells to a processing center, and it must be built parallel to existing roads so that in the event of a leak it can be reached and repaired quickly. It will be cheapest if the total length of pipeline is as short as possible. So, there is a graph in which vertices are oil wells and a processing center and in which edges are roads. In this case, distances along roads matter, as well as which roads connect which oil wells. More generally, we may have situations in which costs of transport or lengths of cable or amounts of energy used are important, in addition to adjacency. Therefore, whatever physical network we have that is represented by a graph will have labels on its edges to denote the costs or distances or energies. These are referred to as *weights*, and such a labeled graph is called a *weighted* graph. (Technically it is edge-weighted; one could, after all, weight vertices.) Figure 10.2 shows a weighted graph and three of its spanning trees.

7. Compute the total weight of each of the spanning trees shown in Figure 10.2. Which has the smallest weight? Is that the minimum possible weight? If not, construct a minimum-weight spanning tree.

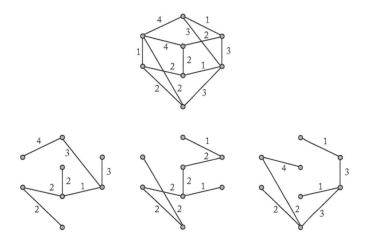

Figure 10.2. Three spanning trees of a weighted graph.

8. Develop an algorithm for finding a minimum-weight spanning tree in a connected graph.

9. Did your algorithm begin with just the vertices, or did it begin with the whole graph? Find a second algorithm that begins differently than your first.

10. Try to prove that your algorithms work, in the sense that (a) they produce trees and (b) they produce a total weight that is smallest. (This is pretty challenging, so do not be surprised if you can only come up with part of a proof.)

10.4   Spanning Tree Algorithms

Hey! You! Don't read this unless you have worked through the problems in Section 10.3. I mean it!

Spanning trees are used all the time in computer science. For example, doing a search on a graph (see Bonus Section 5.8) requires a spanning tree. So, it is important to have algorithms for finding spanning trees.

Chances are that in Section 10.3 you came up with the following two algorithms (or at least similar algorithms) for finding a spanning tree.

Finding a spanning tree: Start big

1. Begin with a connected graph G.

2. Consider a duplicate of G and name it H.

3. Pick an edge, any edge, of H and call it e.

4. If $H \setminus e$ is connected, remove e and rename $H \setminus e$ as H; otherwise, mark e as necessary (and don't consider it again).

5. Pick any edge of H not marked as necessary and call it e. If there are no unmarked edges left, rename H as T and be done.

6. Go to step 4.

Your algorithm was probably less formal than this one. But this algorithm still leaves a bit to be desired—how do you pick an edge of a graph? As a human, you can close your eyes, put your finger down on the graph, and see which edge your finger landed on after you open your eyes. A computer, however, cannot do this.

On the other hand (on the other bit? circuit? processor?), a graph must be stored as some sort of computer-readable structure in order for a computer to perform any operations on it. Two common (but not super-efficient) ways of storing a graph are as an adjacency matrix and as a pair of lists (vertices, edges). (See Section 3.7.2 for a bit more on this topic.) In either case, the data are ordered and so the computer can choose the first edge in the stored structure. Moreover, testing $H \setminus e$ for connectedness requires a separate algorithm of its own. So, the algorithm presented here is neither very formal nor complete.

In this chapter (and similarly in Chapter 12), there will be several algorithms presented, but we will not specify how we would get a computer to perform steps of the algorithm. Such implementation is in the purview of computer science; for example, some aspects of implementation are dependent on the programming language and data types used. However, for each algorithm we will point out some aspects of implementation a coder should consider.

Finally, how do we know the start-big spanning tree algorithm works? We begin with a connected graph and at each stage retain connectedness; the result has no cycles, as otherwise there would be an edge not marked as necessary. Thus, the algorithm produces a tree. Additionally, the tree must be spanning because we began with all vertices and took no action that would remove one from the tree (because at each stage the resulting graph is connected). Performing this algorithm on each component of a nonconnected graph will produce a spanning forest.

Now, the previous algorithm started big. We could also start small.

Finding a spanning tree: Start small

1. Begin with a graph G.

2. Grab a copy of the null graph (no vertices or edges) and name it H.

3. Pick an edge, any edge, of G and call it e.

4. If $H \cup e$ is a tree, add e and its vertices to H and rename $H \cup e$ as H; otherwise, mark e as superfluous.

5. Pick an edge of G incident to H and not marked as superfluous and call it e. If there are no unmarked edges left, rename H as T and be done.

6. Go to step 4.

How do we know this algorithm works? We need to be sure that the resulting graph T is connected, has no cycles, and includes all the vertices of G. Because at each stage we only consider edges incident to H, each intermediate graph is connected; and, because at each stage we only add edges that retain tree-ness, no cycles are created. Thus, T is a tree. Suppose, however, that there is a vertex v of G not included in T. It must be incident only to edges e_i marked as superfluous (as otherwise it would have been added to T at some stage), and therefore $T \cup e_i$ is not a tree for any i. This means that $T \cup e_i$ must contain a cycle (as e_i is incident to T, we know $T \cup e_i$ is connected), and therefore $v \in T$.

Now let us consider finding spanning trees in graphs with weighted edges.

Example 10.4.1 (of a spanning and weighted spanning tree). In the center of Figure 10.3 is a weighted graph. At left, we have the graph with a spanning tree highlighted, but this is not a minimum-weight spanning tree; at right, the unique minimum-weight spanning tree is highlighted.

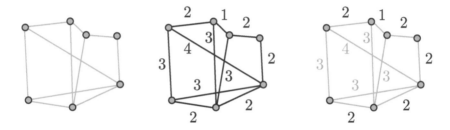

Figure 10.3. A spanning tree highlighted, a weighted graph, and a minimum-weight spanning tree highlighted.

Example 10.4.2. Minimum-weight spanning trees are eminently practical. Suppose you want to create a high-speed computer cluster from existing machines so that the processing power of the machines can be utilized around the clock. In order to do so, you will need to network the computers together using physical cables for maximum speed. At the University of Universe City (UUC), the computer science and mathematics departments have agreed to pool their computing resources for a cluster. The two departments are in adjacent buildings, one of which has two floors. Only some connections will be physically possible (because of various obstacles between computers, such as lead-filled walls and the like). The network is shown in Figure 10.4. Which segments should you pick in order to use the least amount of expensive cable? A minimum-weight spanning tree will answer your question: it assures that every computer can communicate with every other, avoids superfluous cable, and is cheapest. Note that every spanning tree must include the cable that runs between the two buildings.

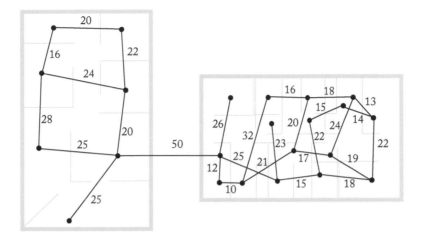

Figure 10.4. An edge-weighted graph representing possible routes for high-speed cabling.

We will now examine the two most famous algorithms for finding minimum-weight spanning trees. You probably came up with one of these algorithms but not the other.

Finding a minimum-weight spanning tree: Kruskal's algorithm

1. Begin with a connected edge-weighted graph G. Order the edges in increasing order of weight as $e_1, \ldots, e_n$.

2. Let e_1 along with its vertices be called H and set $j = 2$.

3. If $H \cup e_j$ has no cycles, then rename $H \cup e_j$ as H; otherwise, do nothing.

4. If $|E(H)| = |V(G)| - 1$, output H as the desired tree; otherwise, do nothing.

5. Increment j by 1 and go to step 3.

Kruskal's algorithm is straightforward to perform by hand, but more detail would be needed in order to implement it. For example, the edges of G must be ordered by weight, and this is a separate procedure. (One inefficient possibility is bubble sort; see Bonus Section 6.12.) Additionally, we must create a procedure for determining whether a graph is acyclic. One way to make such a determination for $H \cup e_j$ is to check whether e_j's endpoints are already in H or not. If they are, then perform a further check to see whether they're in the same connected component

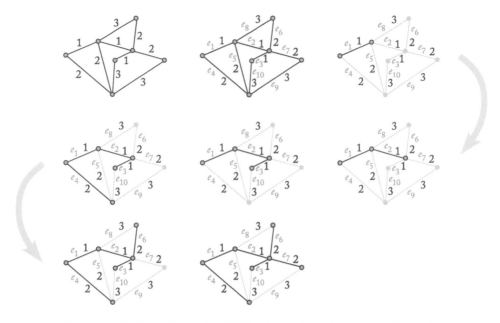

Figure 10.5. A small graph with Kruskal's algorithm executed upon it.

of H. Of course, for this check to work, we need to know what the connected components of H are; how do we determine them? Notice that there is a large gap between understanding Kruskal's algorithm and implementing it.

Example 10.4.3. Figure 10.5 shows Kruskal's algorithm executed on a small graph.

If you would like to examine additional examples, see https://www-m9.ma. tum.de/graph-algorithms/mst-kruskal/index_en.html, which shows Kruskal's algorithm executed step by step on some customizable graphs, and http://students. ceid.upatras.gr/~papagel/project/kruskal.htm, which generates a variety of graphs with step-by-step executions of Kruskal's algorithm.

Proof that Kruskal's algorithm works: We will use the notation w_e to denote the weight on edge e and the notation $w(T)$ to denote the total weight of a tree.

We need to show that the resulting H is a tree, meaning that (a) it is connected and (b) it has no cycles, and we also need to show that the result has the smallest possible weight. Criterion (b) is taken care of for us by step 3 of the algorithm; we always maintain acyclicity. The algorithm only finishes when $|E(H)| = |V(G)| - 1$, at which point Theorem 10.2.2 guarantees that H is a tree and thus connected.

It was clear that our unweighted spanning tree algorithms would terminate, as we finished after examining every edge of the original G. However, Kruskal's algorithm terminates exactly when $|E(H)| = |V(G)| - 1$; is it possible we could consider every edge and still have $|E(H)| < |V(G)| - 1$?

Let's suppose Kruskal's algorithm does not terminate. Then, either H is not connected or it does not span G. In either case, there must be two vertices v_1 and v_2 of G between which there exists a path in G but not in H. Consider the edges of this path in G. Now, what happened to them in Kruskal's algorithm? When each was encountered, its addition to H would have caused a cycle. In other words, both vertices of the edge already belonged to H. As this is true for every edge on a path between v_1 and v_2, that means that v_1 and v_2 are in H, which is a tree. Therefore, there *is* a path in H between v_1 and v_2, which is a contradiction.

Finally, we need to show that Kruskal's algorithm does produce a minimum-weight spanning tree! (Doubt arises because perhaps we have several edges of the same weight, and if we addressed those edges in a different order, we might be led to choose a different set of edges later, perhaps of lower total weight.) We will do a proof by contradiction. Thus, we will assume that there is a spanning tree with lower total weight than that of H.

Because G has a finite number of spanning trees, we can sort them by total weight. First, we will consider those with minimum total weight, and then among those we will choose one with the largest number of edges in common with H. Call our selected spanning tree T. By assumption, $w(T) < w(H)$.

Consider the edges of T and H. We will label the edges of H in the order Kruskal's algorithm picked them from $e_1, \ldots, e_n$ and call them h_1, h_2, etc. We are not given an ordering on the edges of T because we do not know how T was produced, so we will label the edges of T in increasing order by weight, and within a sequence of edges of the same weight, we will list those from H first (and in the same order as in H). Using this labeling, $t_1 = h_1, t_2 = h_2$, and so forth, until T and H first differ, at which point $t_k \neq h_k$.

Consider $T \cup h_k$, and notice that it has $V(G)$ edges; it's still connected, so it must have a cycle Y. Any edge of Y can be removed without disconnecting $T \cup h_k$. And, at least one edge y of Y is not any of $h_1, \ldots, h_{k-1}$ because otherwise $h_1, \ldots, h_{k-1}, h_k$ would have formed a cycle in H. This means that $w(h_k) \leq w(y)$. So, consider $(T \cup h_k) \setminus y$. It is a spanning tree because it contains all vertices of G and is a tree by Theorem 10.2.2. Plus, $w((T \cup h_k) \setminus y) = w(T) + w(h_k) - w(y) \leq w(T)$. If $w((T \cup h_k) \setminus y) < w(T)$, then this contradicts the minimum-weight-ness of T. If instead $w((T \cup h_k) \setminus y) = w(T)$, then we have created a minimum-weight spanning tree with more edges in common with H than T had, and this contradicts the way we chose T. Thus, we have in any case a contradiction. $\square$

Our next algorithm dates back to 1930 (though Prim wrote about it in the late 1950s), which is pretty recent (compared to, say, the development of calculus).

Finding a minimum-weight spanning tree: Prim's algorithm

1. Let G be a connected edge-weighted graph. Find the edges of least weight in G and pick one of them. Name it *ankka*.

2. Let *ankka* along with its vertices be called P.

3. Look at all the edges of G that have exactly one vertex in P. Among these, pick one of least weight and name it e. If $P \cup e$ has no cycles, then add e and its other vertex to P, and rename $P \cup e$ as P. Otherwise, mark e as *bad* so you don't look at it again.

4. If $V(P) = V(G)$, then we're done and output P. Otherwise, go to step 3.

This version of Prim's algorithm is very human-oriented. Usually when Prim's algorithm is implemented, it begins with an arbitrary choice of vertex rather than an arbitrary choice of edge in order to avoid unnecessary pre-processing of the graph. Still, the method for finding the set of edges with exactly one vertex in P depends heavily on how the graph is stored in a computer.

Example 10.4.4. Figure 10.6 shows Prim's algorithm executed on a small graph.

If you would like additional worked examples, see https://www-m9.ma.tum. de/graph-algorithms/mst-prim/index_en.html, which shows Prim's algorithm executed step by step on some customizable graphs, and http://students.ceid.upatras. gr/~papagel/project/prim.htm, which generates graphs on which Prim's algorithm is worked step by step.

Proof that Prim's algorithm works: First, we will show that Prim's algorithm terminates. Suppose it does not. Then $|V(P)| < |V(G)|$, and yet there is no edge of G with exactly one vertex in P. That means that every edge of G not in P either has both vertices in P or has neither vertex in P. If there is an edge with neither vertex in P, then there exists no path connecting either vertex to any vertex in P; otherwise, there would be an edge on that path with exactly one vertex in P. This is a contradiction to G being connected.

Prim's algorithm definitely produces a tree because at every stage P is connected and no cycle is ever formed: let's show it gives a minimum-weight spanning tree.

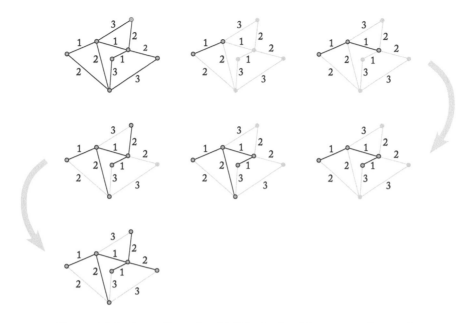

Figure 10.6. A small graph with Prim's algorithm executed upon it.

We know that G has a minimum-weight spanning tree—after all, there are only a finite number of spanning trees, so we can select those with minimum total weight. So let's look at all the minimum-weight spanning trees and choose one that has the maximum number of edges in common with P. We will call this tree T.

We can label the edges of P by the order in which we added them. (Think of this as having the weights written in black and the order-added labels written in teal.) Consider $P \setminus T$ and check out the edge that was picked first for P. (It has the lowest teal number.) Call that edge e. Now consider $T \cup e$. It has a cycle C (because $T \cup e$ has $V(G)$ edges) and e is on that cycle.

Now look at the growing P just before we added e and call it P_e—basically, this means we are going to consider Prim's algorithm up to the point when P diverged from T. Every vertex of G is either in $V(P_e)$ or in $V(G \setminus P_e)$. One vertex of e is in $V(P_e)$ and the other vertex of e is in $V(G \setminus P_e)$. If you start in P_e and go across e into $V(G \setminus P_e)$, and then follow the cycle C around, you have to get back into $V(P_e)$ at some point in order to get to the start of the cycle again. So that means there is some other edge in C that has one vertex in $V(P_e)$ and the other vertex in $V(G \setminus P_e)$. Let's call that other edge e_{hwyaden}.

Now back in the moment of Prim's algorithm, just before adding e, we had a pile of edges we were considering adding, and e was one of the available edges with lowest weight. And because e_{hwyaden} goes between P_e and the rest of G, it must be that e_{hwyaden} is in that pile of edges under consideration. Therefore, the weight of e_{hwyaden} is larger than or equal to the weight of e.

Check out T again. The graph $T \cup e$ has a cycle and e_{hwyaden} is on that cycle, so $(T \cup e) \setminus e_{\text{hwyaden}}$ is a spanning tree and $w((T \cup e) \setminus e_{\text{hwyaden}}) \leq w(T)$. If $w((T \cup e) \setminus e_{\text{hwyaden}}) < w(T)$, that's a contradiction to T having minimum weight, so $(T \cup e) \setminus e_{\text{hwyaden}}$ must have the same weight as T.

But wait a sec! The tree $(T \cup e) \setminus e_{\text{hwyaden}}$ has one more edge in common with P than T does, which is also a contradiction—we chose T to have the maximum number of edges in common with P!

Heck, yeah—that means that P was a minimum-weight spanning tree to begin with.

So Prim's algorithm works. $\qquad\square$

Now we will address an algorithm you may or may not have envisioned.

Finding a minimum-weight spanning tree: Start big

1. Begin with a connected edge-weighted graph G. Order the edges in decreasing order of weight as $e_1, \ldots, e_n$.

2. Consider a duplicate of G and name it J. Set $k = 1$.

3. If e_k is contained in a cycle of J, remove e_k and rename $J \setminus e_k$ as J; otherwise, do nothing.

4. If $|E(J)| = |V(G)| - 1$, output J as the desired tree; otherwise, do nothing.

5. Increment k by 1 and go to step 3.

In practice, algorithms such as Kruskal's and Prim's that build trees are preferred to algorithms that reduce from a graph to a spanning tree. To see why, examine and contrast the verification steps: in Kruskal's algorithm, one checks to see whether adding an edge will create a cycle; in Prim's algorithm, one finds all edges incident to the growing tree; in a reduction algorithm, one checks to see whether a particular edge is in a cycle. The graphs being checked in the early stages of a reduction algorithm are larger than those checked in the later stages of a building algorithm and so take longer to verify.

We close by noting that most graphs have lots and lots of spanning trees; as a result, it would not be practical to look at all possible spanning trees and pick one with minimum weight.

10.4.1 Greedy Algorithms

Kruskal's and Prim's algorithms are both examples of the same approach to solving a problem. Let's see their common structure:

1. Select a place to start. (Both algorithms begin with an edge of lowest weight.)

2. Look nearby. (Kruskal: look at the next edge in the list. Prim: look at all edges incident to the current tree.)

3. See which nearby things may help to solve the problem. (Kruskal: check that adding the next edge wouldn't create a cycle. Prim: only look at edges with exactly one endpoint in the current tree.)

4. Choose the best option from those that are nearby. (Kruskal: the pre-ordering of the edges takes care of this. Prim: find an edge of lowest weight.)

5. Check to see whether the problem has been solved. (Kruskal: compare number of edges to $|V(G)| - 1$. Prim: compare number of vertices to $|V(G)|$.)

6. If problem is solved, be done; if not, go to step 2.

This structure describes a general class of algorithms called *greedy algorithms*. As [5] put it, such algorithms are called greedy because they grab the best thing available at every stage. (For problems whose goal is to minimize something, this name doesn't make intuitive sense; it seems that in these cases the algorithms are *parsimonious* instead because they are grabbing the least stuff possible.)

Most of the time, a greedy/parsimonious algorithm works pretty well. For example, we will see in Chapter 12 that one can be used to find the shortest distance between two vertices and in Chapter 13 that one does a decent job of coloring graphs. But sometimes they're just terrible. Here's an example.

Example 10.4.5. In the graph of Figure 10.7, your assignment is to find two edges that cover all four vertices and have minimum weight.

Figure 10.7. A small but lopsided graph.

Hopefully you have picked the two middle (crossing) edges for a total weight of 6. But if you used a greedy/parsimonious algorithm? Hah! You'd start with the weight-2 edge and then be forced to pick the weight-735 edge, for a very non-minimum total weight of 737.

Check Yourself ───

Do about half of these problems.

1. An unweighted graph could be considered an edge-weighted graph with all edges of the same weight (perhaps 1). What happens if you run Prim's algorithm on it?

2. What happens if you run one of the spanning tree algorithms for unweighted graphs on an edge-weighted graph?

3. If an edge-weighted graph has several edges of the same weight, there will be more than one way to order the edges while still having them in increasing order of weight. What difference do these orderings make to Kruskal's algorithm?

4. Prim's algorithm does not specify an edge of G with which to start. What would happen if you ran Prim's algorithm twice, but starting with different edges?

5. In Example 10.4.3, would the same spanning tree have resulted if the labels were switched on edges e_4 and e_6?

6. In Example 10.4.4, at what stage could one have made a choice of edge that would have resulted in a different spanning graph?

10.5   Binary Trees

Let's begin with a definition.

Definition 10.5.1. A *binary tree* is a tree that

🐦 has a distinguished vertex called the *root*, usually drawn at the top of the tree (in opposition to the way natural trees are),

🐦 has at most two edges growing "downwards" from each vertex, and

🐦 has a designation of "left" or "right" for each edge, with edges usually drawn as descending leftwards or rightwards.

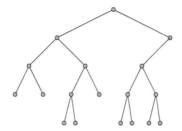

Figure 10.8. An incomplete binary tree.

A sample binary tree is shown in Figure 10.8.

The *k*th *rank* of a binary tree is the collection of vertices that are distance k from the root. A *complete binary tree* has exactly two edges growing "downwards" from each vertex, except for those leaves in the bottom rank. (In graph theory, the adjective "complete" means that wherever an edge *can* be, it *is*.)

Example 10.5.2. Let us link complete binary trees with one of our first examples of counting proofs with the product principle, namely, that a finite set E with n elements has 2^n subsets. There, we noted that for each element e_i of E there were two possibilities: either e_i is in a given subset or it's not. In our proof, this corresponded to whether e_i filled one of n blanks or not. In terms of binary trees, this corresponds to following one of two different edges of a binary tree. In Figure 10.9, the nodes and edges are labeled to show the two subset possibilities for each e_i.

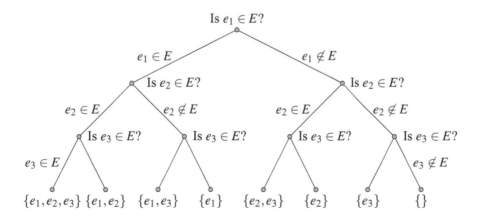

Figure 10.9. This complete binary tree has $2^3 = 8$ nodes in the bottom rank.

Example 10.5.3 (of an unexpected use for binary trees). You—yes, you! We're talking to you. You, personally, have an ancestor who has an ancestor who is the ancestor of both their genetic parents—within the last 1,000 years. That is, some ancestor A of yours, with parents P_1, P_2, has some earlier ancestor B from whom both P_1 and P_2 descended. All within the last millennium. No kidding! We're about to prove it (by contradiction), in case you don't believe this wild claim.

Suppose not. How on earth can we express this mathematically? Well, check this out: Your ancestry is basically a complete binary tree. You had two genetic parents (vertices adjacent to your vertex), who each had two genetic parents (your four genetic grandparents), who each had two genetic parents (your eight genetic great-grandparents), and so forth and so on. But if there are two vertices in this tree that represent the same person B, then that person has two different paths of descendants that meet first in some ancestor A of yours, and A's parents are both descendants of B. So if there isn't such a person B, your ancestry is a complete binary tree with all vertices distinct.

According to the US Centers for Disease Control, the mean age at which a child-bearing woman has her first child is 26.6 (as of 2016, http://www.cdc.gov/nchs/fastats/births.htm). Many mothers have multiple children, so we might estimate that the current average length of time represented by an edge between two ranks of the binary tree is 32 years; however, even a few hundred years ago, the mean age at first birth would have been much earlier, and so we may overestimate the average edge time-length as 28 years. There are at least 35 ranks in your ancestry tree over the last 1,000 years, for a total of at least $2 + \cdots + 2^{35} = 68,719,476,734$ ancestors.

As of this writing, the world had just under 7.5 billion people in it. The upper-bound estimate for the cumulative population of the world for the last 1,000 years (using data from a US Census page that sadly no longer exists) is 45,814,000,000. So there have definitely been fewer people alive over the last 1,000 years than the number of nodes in your ancestral tree for the past 1,000 years. Therefore, at least two of those nodes represent the same person, and so one of your ancestors has an earlier mumble mumble (reread the start of the problem for a precise statement)!

Binary trees can be used to store data efficiently and effectively, with one datum per node. A binary tree can also be used to encode an effective search algorithm, and that is what we will discuss next.

Imagine some set of data that is ordered, such as a dictionary (ordered alphabetically) or an employment database (ordered by ID number). We would like to search for some item x (in a dictionary, *xylophone*; in an employment database, 242424).

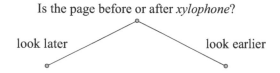

Figure 10.10. A human searches a dictionary like this.

Let's look at the dictionary example. As humans, we know that x is near the end of the alphabet, so we open the dictionary near the end. If the page we see lists words that come before *xylophone*, we know to look later in the dictionary. If the page we see lists words that come after *xylophone*, we know to look earlier in the dictionary. That's like the start of a binary tree, as shown in Figure 10.10.

We basically repeat this same process (so it is a recursive algorithm!) by opening a new page in the dictionary, and depending on whether the page shows words that come after or before *xylophone*, we flip to an earlier or later page.

Now, what should we have a computer do? (We should *not* have it check every piece of data, in order.) The computer doesn't "know" that *xylophone* should appear "near the end" of the dictionary. (Nor can it flip to an arbitrary page.) It has to follow the same procedure for every word that it is given. So let's divide the alphabet in half, and then in half again…. Figure 10.11 shows the first two steps in an algorithm a computer can follow in looking for a word in an electronic dictionary.

After determining that it needs to look in the *x*s, the computer will look at the second letter of the word and follow a further path in the binary decision tree until it finds the word in question at a leaf.

But what if we search for a nonsense string, like *xjeipo*? The computer will come to a decision point, such as, "Is the second letter (*a* or *e*) or is it *i*?" and find that neither choice is appropriate. (There are no dictionary words that begin with

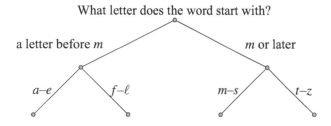

Figure 10.11. The first two ranks of decisions in a binary decision tree for a dictionary.

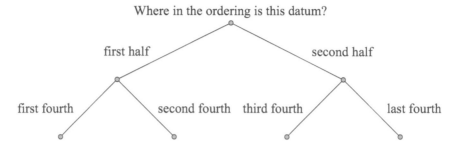

Figure 10.12. The start of a decision tree.

xf, xg, xh, or xj.) It will have to generate an error, which will probably return to the user a message such as, "Word is not in the dictionary."

In general, Figure 10.12 indicates how a *binary decision tree* works.

The diagram shows only the first two decisions; usually there are lots and lots of decisions to be made. This turns out to be an efficient way to search ordered data sets. In practice, this type of decision process is implemented by storing the data directly on the nodes of the tree.

Definition 10.5.4. A *binary search tree* has a datum associated with each node such that in the ordering of the data set, the datum occurs earlier than any of the data downwards and to the right and occurs later than any of the data downwards and to the left.

This sounds a bit strange, but Figure 10.13 shows a sample binary search tree for the micro-dictionary $aaa, ab, baa, baba$. Here is how one might "read" Figure 10.13 to search for the word *baba*: Look at the root. It contains *ab*. We want *baba*, which is later in the data set, so let's go to the right. The node we get to contains *baa*, which is earlier in the data set than *baba*, so we should go to the right again. Oh, hey, this next node contains *baba*! Aces, we're done.

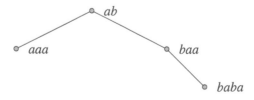

Figure 10.13. A micro-dictionary search tree.

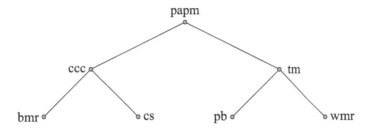

Figure 10.14. An alphabetical search tree for seven cat toys.

It is advantageous to have as few ranks as possible in a search tree, as the number of ranks gives the largest number of search steps it will take to find the desired piece of data.

Example 10.5.5 (of a binary search tree vs. a binary decision tree). A binary search tree stores data on all of the vertices, whereas a binary decision tree stores queries on the non-leaf vertices. We will show a binary search tree and a binary decision tree for the same data set, a collection of seven cat toys {*teal mouse, blue milk ring, catnip candy cane, paper bag, white milk ring, pink-and-purple mouse, catnip snake*}, abbreviated {tm, bmr, ccc, pb, wmr, papm, cs} for convenience.

First, Figure 10.14 shows a binary search tree that has ordered the data alphabetically. Next, Figure 10.15 shows a binary decision tree for identifying one of the cat toys.

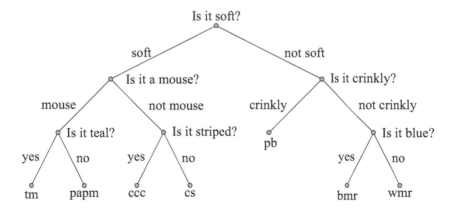

Figure 10.15. This binary decision tree helps us identify cat toys.

Check Yourself ───

Doing all of these problems will ensure that you understand the definitions introduced in this section.

1. Suppose that a binary decision tree for set membership is labeled consistently (i.e., "left" indicates an element is in the set and "right" indicates an element is not in the set). What subset will be assigned to the leftmost leaf? ... the rightmost leaf?

2. Consider a language with only two letters (*a* and *b*), and a binary decision tree that encodes dictionary ordering for short words (no more than five letters long) in this language. What is the practical meaning indicated by the tree being incomplete?

3. In Example 10.5.3, what principle allows us to conclude that two nodes must represent the same person?

4. Placing *baa* at the root, draw a binary search tree for the micro-dictionary *aaa*, *ab*, *baa*, *baba*.

───

10.6   Try This! Binary Trees and Matchings

1. Consider the maze shown in Figure 10.16.

   (a) Create a binary decision tree that describes all possible ways to proceed through the maze, with left-leaning edges corresponding to left turns and right-leaning edges corresponding to right turns.

   (b) One way to find a path through a maze from entrance to exit is to walk in and always keep your right hand on the wall. Trace such a path on your tree.

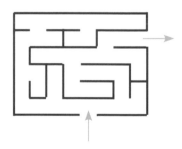

Figure 10.16. Meditate upon this maze.

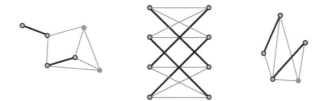

Figure 10.17. Three graphs (do they look familiar?) with matchings highlighted.

2. Create a binary decision tree for a robot to use so it can determine which of the current US coins (penny, nickel, dime, quarter, half-dollar, or dollar) you have just offered it. The robot can use its sensors but cannot directly recognize the coins.

A *matching* in a graph G is a subgraph of G in which every vertex has degree 1. In other words, it's a pile of edges (with vertices included)—examples are shown in Figure 10.17. A *perfect matching* in a graph G is a matching that includes all of $V(G)$.

3. Which matchings in Figure 10.17 are perfect matchings?

4. Find a perfect matching for each graph in Figure 10.17, or explain why no perfect matching exists.

5. Examine Figure 10.18, which shows three trees. If possible, find a perfect matching in each tree; if not, explain why no perfect matching exists.

6. List at least two criteria that (when present) prevent a tree from having a perfect matching.

7. Does every bipartite graph have a perfect matching?

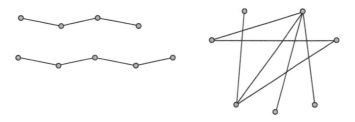

Figure 10.18. Do any of these trees have perfect matchings?

10.7   Matchings

At first you might wonder what matchings are doing in a chapter about trees. But in fact, a matching is a type of forest, so matchings fit right in! Matchings are incredibly useful. (Yes, yes, that is said about many topics in this text. One might as well say "discrete mathematics is useful"—and it is.)

Example 10.7.1. A Fish of the World class has eight students who are doing research papers. In the library reserve area, there is a shelf of 12 ichthyology books. Each student has a list of the reserved books that are pertinent to hir topic. Can all eight students work in the library at the same time?

The answer to this question can be found by searching for a matching in a graph: Assign each student to a vertex, and assign each book to a vertex. Draw an edge from a student to each book on hir list of pertinent sources. If there is a matching that includes all eight students, then they can all work in the library at the same time.

Example 10.7.2. At the University of Kačica (in Slovakia), advising for the Spring semester is done in one week during the Fall semester. Every student submits a list of time slots in preference order. How can we schedule all students for advising appointments so that the most preferences are honored?

We will again search for a matching, but this time our edges have weights: Assign each student to a vertex and each appointment slot to a vertex. Draw an edge from each student to every appointment slot listed, and weight it with hir preference (1 for first choice, 2 for second choice, etc.). Now, we seek a matching *of minimum weight* that includes all student vertices.

Example 10.7.3. In Mx. Tanaka's ballet class, students must pair up for assisted stretching. Is it possible to have every dancer paired with a friend?

In this case, we create a graph where the vertices are dancers and edges represent friendship. If the graph has a perfect matching, then every dancer can stretch with a friend.

We will solve problems patterned after Examples 10.7.1–10.7.3 in Section 10.13.

Matchings are also used in other parts of mathematics. They appear in graph colorings (see Chapter 13), for example. There is an immense amount that can be said about matchings, and far too little of it is at the level of this text. So, for now you must be satisfied with simply knowing what a matching is and how to recognize and locate one by hand.

Try these three quickies.

1. What is the relationship between the number of edges in a matching and the number of vertices in that matching?

2. Is it possible for a graph with an odd number of vertices to have a matching?

3. Is it possible for a graph with an odd number of vertices to have a perfect matching?

10.8 Backtracking

Consider an 8×8 grid. We would like to place eight coins on squares of the grid so that no two are on the same row, same column, or same diagonal. (Here, we mean any $45°$ diagonal.) A partial example is given in Figure 10.19. We could try all possible ways of putting eight coins on the grid and hope that we eventually run into one that works. This is terrible, though, because it might require as many as $\binom{64}{8} = 4,426,165,368$ attempts.

Okay, we can do better. We know that no two coins can be in the same column, so we can make a list of eight numbers, each indicating the position the coin has in the corresponding column. Example: $(3,2,5,6,2,7,1,8)$ does not work because there are two coins in the second row (in the second and fifth columns)... and there are two coins on the same diagonal. The number of length-8 lists with eight numbers possible in each slot is $8^8 = 16,777,216$, which is more than 100 times better than our first attempt, but we can do better still, and pretty easily! To avoid having two coins in the same row, let's require that the eight numbers in the list be different. That means that each list of eight numbers is just a permutation of the integers from 1 to 8, and there are $8! = 40,320$ of those (a 1,000-fold improvement).

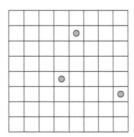

Figure 10.19. An 8×8 grid with coins placed in the second row/fifth column, fifth row/fourth column, and sixth row/eighth column.

Now we would just have to check each of those to see whether it has two coins on the same diagonal or not.

What all of these methods have in common is that we first generate a placement of coins on the grid and then check to see whether it satisfies our criteria (no two coins on the same row, column, diagonal). *Backtracking* is more efficient: As we place coins on the grid, let's check to see that we satisfy the criteria after placing each coin. Then, if we have (for example) two coins placed that satisfy the criteria and placing a third in a particular square violates a criterion, we just have to pick up that third coin and try a different square. (Of course, if it turns out that no placement of the third coin works, then we have to pick up the second coin as well and try a different square... and so forth.) More generally, backtracking is an effective approach whenever a problem has potential solutions that can be expressed as finite sequences. The start of a sequence is checked for validity; if it is valid, then the partial sequence is extended, and if it is invalid, the partial sequence is discarded. In this way, one avoids checking many potential solutions that have invalid starts.

A general algorithm for backtracking:

1. Start with an empty list.
2. Append the first possible solution element.
3. Check to see whether this partial solution is valid.
4. If so, check to see whether the solution is complete; if so, terminate; if not, continue.
5. If so, append the earliest possible sequence element and go to step 3; if not, remove the last sequence element and replace it with the next possible sequence element.

How does backtracking relate to trees? (They are the theme of the chapter, after all.) Construct a tree for the coins-on-a-grid problem as follows: the root node is labeled with the empty list (); in the rank below are eight nodes, each labeled with one of the lists $(1), (2), (3), (4), (5), (6), (7), (8)$; and in the rank below that are 64 nodes, labeled with length-2 lists $(1,1), (1,2), (1,3), \ldots, (8,6), (8,7), (8,8)$; and so forth on down until the eighth rank, which has 8^8 length-8 lists. (This is *not* a binary tree!) Each node represents a partial solution to the problem, and in this case we are labeling with the positions of coins in columns. Figure 10.20 shows a snip of the tree; notice that the lists $(2,1), (2,2), (2,3)$ are absent because we know they are impossible configurations. In creating our tree, we make "impossible" nodes only if we can't tell they represent impossible configurations. Once we discover

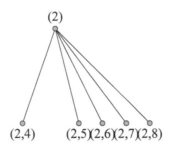

Figure 10.20. A snip from a tree of partial solutions.

an impossibility, we erase that node and the subtree below it. This is an example of a *backtracking tree*.

This is what makes backtracking nice; we create a tree that reflects a hierarchical organization of conceivable solutions to a problem and then cut off branches of the tree as we discover they lead to invalid (or, depending on the type of problem, suboptimal) solutions.

The simplest (and most usual way) to approach such a tree is to go down/left repeatedly until we find a solution or a conflict. If we find a solution, we're done. If we find a conflict, then we go back up one rank and then down/right. (If there are lots of choices of "right," we choose the leftmost of those edges.) This procedure is known as a *depth-first search*, and you can learn more about it in Bonus Section 5.8. Notice that this is the same strategy one uses in the always-keep-your-right-hand-on-the-wall approach to finding the exit of a maze. In the case of the maze, backtracking is no more efficient than simply searching the whole tree because there is no opportunity to discover a potential solution is invalid before hitting a dead end.

It turns out that for the eight-coins-on-an-8×8-grid problem, a backtracking tree is a lot more efficient than any of our original methods—there are 2,057 nodes in it. (That number was calculated by computer.)

Check Yourself

There are only two problems here; try them both.

1. We claimed that the column-position list $(3, 2, 5, 6, 2, 7, 1, 8)$ had two coins on the same diagonal. Which two and why?

2. Why is this approach called backtracking?

10.9   Where to Go from Here

Trees are not usually studied in their own right; however, they form an excellent class of examples for any graph-theoretic investigation. They are an indispensable tool in computer science because so many data sets and solution spaces can be modeled using trees. If you want to read more about the theory and implementation of spanning tree algorithms, greedy algorithms, binary trees, and backtracking, check out [5].

The introduction to matchings given here leads to Hall's matching theorem (which sometimes masquerades under the unfortunate and less descriptive name "Hall's marriage theorem"), which gives conditions under which a perfect matching exists for a bipartite graph. Bipartite graphs are a natural model for most real-life situations in which one wants to create a one-to-one correspondence between two sets. Many websites and most advanced discrete mathematics and graph theory texts contain a statement, a proof, and applications of Hall's matching theorem. Should you want massive amounts of information on matchings, to the tune of having swallowed an elephant, *Matching Theory* by László Lovász and Michael Plummer will suit your purposes admirably. Matchings are an active area of research in pure mathematics. For example, recent papers have given improved estimates of a lower bound for the number of perfect matchings of a 3-regular graph (in terms of the number of vertices of the graph).

Credit where credit is due: Section 10.3 was inspired by [3]. The proof that Kruskal's algorithm works was adapted from Gábor Hetyei's proof. The proof that Prim's algorithm works was adapted from [7]. Example 10.4.5 was adapted from a discussion on pages 216–218 of [4]. Sections 10.8 and 10.11 were adapted and expanded from Section 6.6.1 of [5]. A hat tip to Tom Hull for the ideas for Bonus Check-Yourself Problem 4 and Problem 1 in Section 10.6!

10.10   Chapter 10 Definitions

acyclic: An acyclic graph has no cycles.

tree: A connected graph without cycles.

spanning tree: A tree that contains all the vertices of a given graph; it is the largest possible tree that is also a subgraph.

weights: Labels on the edges and/or vertices of a graph that often denote costs or distances or energies.

weighted graph: A graph labeled with weights.

greedy algorithm: An algorithm that selects the best option available at every stage.

parsimonious algorithm: An algorithm that has the goal of minimizing something and so selects the smallest or least option available at every stage.

root: A distinguished vertex.

binary tree: A tree with a root, at most two edges growing "downwards" from each vertex, and a designation of "left" or "right" for each edge.

rank: The collection of vertices that are the same distance k from the root.

complete binary tree: A binary tree with exactly two edges growing "downwards" from each vertex, except for those leaves in the bottom rank.

binary decision tree: A binary tree with a two-option question associated with each node, answers to this question associated to the edges incident to that node, and data associated with the bottom-rank leaves.

binary search tree: A binary tree with a datum associated with each node such that in the ordering of the data set, the datum occurs earlier than any of the data downwards and to the right and occurs later than any of the data downwards and to the left.

matching: A subgraph of a graph in which every vertex has degree 1. In other words, it's a pile of edges (with vertices included).

perfect matching: A matching that includes all vertices of a graph.

backtracking: An approach to solving problems whose solutions can be expressed as finite sequences.

backtracking tree: A rooted tree of partial potential solutions to a problem, with rank k hosting length-k partial potential solutions.

depth-first search: A method of searching a tree that starts by going down and left until either no down/left edges remain or the searched-for item is found. When no down/left edges remain, the search goes back up until it can go down/right.

10.11 Bonus: The Branch-and-Bound Technique in Integer Programming

If you have not yet read Bonus Section 7.9, please do so before approaching this section.

The branch-and-bound technique is a type of backtracking applied to the extremely interesting problems of integer programming. (There is no bias here.) Recall that an integer programming problem consists of a bunch of linear equations where all of the variables x_i have to be integers plus some function that needs to be maximized or minimized. As is often done, we will assume that all variables are binary. Recall also that while integer programming problems are hard to solve, linear programming problems are easy to solve quickly by computer, and that for any integer programming problem, we have an associated linear programming problem obtained by removing the requirement that the variables have to be integers.

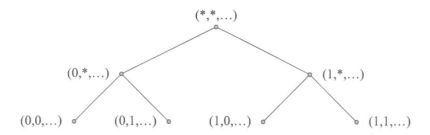

Figure 10.21. The start of a branch-and-bound tree.

Here's what we do to make a branch-and-bound tree for integer programming problems. For the purposes of this explanation, let us examine only the first two binary variables (x_1 and x_2) of our hundreds of variables. We will also assume that we want to maximize a function denoted $M((x_1, x_2, \dots))$. Figure 10.21 shows a start to a branch-and-bound tree. Each node of this tree is marked with a possible value for $(x_1, x_2, \dots)$. (Notice that because we have binary variables, the tree itself is also binary.) Where a variable is assigned the value $*$, that means we consider it to be unknown. So here is how we do branch-and-bound for integer programming (described for a whole pile o' variables):

- Start at the root of the tree. None of the variables has a specific value, so solve the linear programming problem associated to the original integer programming problem. If there is no solution $(x_1, x_2, \dots)$, we're done for and there's no solution to the integer programming problem either. (Consider: if there are no solutions at all, then there certainly aren't any solutions that are integers.) If there is a solution $(x_1, x_2, \dots)$, go on. (Also, if $x_1, x_2, \dots$ happen to be integers, we are super-lucky and done with the problem.)

- Set $x_1 = 0$, which means going one rank down and to the left. The variables $x_2, \dots$ can be anything (not necessarily integral), so plug $x_1 = 0$ into all of our equations and solve for $x_2, \dots$ using linear programming.

  - If there is no solution $(0, x_2, \dots)$, cut off the node labeled $(0, *, \dots)$ (and everything under it) and go up-right/down to $(1, *, \dots)$. (See Generic step below for what to do next.)

  - If there is a solution $(0, x_2, \dots)$, go down a rank to $(0, 0, \dots)$. (See Generic step below for what to do next.)

- Generic step: Solve the linear programming problem corresponding to the integer programming problem at this node (where some variables are fixed).

If there is no solution, cut off the node and everything under it and backtrack. If there is a solution, go down a rank. Repeat this step until all variables are fixed.

🐦 When at a node where all variables are fixed, plug in the variables and see if this is a solution to the equations.

- ⊙ If it's not a solution, backtrack.
- ⊙ If it is a solution, compute $M((x_1, x_2, \dots))$ and make note of the value (and the solution it goes with). Compare this value to that for any other integer-valued solutions that have already been found. Keep the best one and toss the rest out.
- ⊙ Backtrack and do the Generic step.

🐦 Keep going until you run out of nodes.

Want to see how this works on an example? Go to http://www.diku.dk/hjemmesider/ansatte/pisinger/KNAPDEMO/ and choose option 10. This opens two windows: one holds the tree and the other shows blocks that can be used to fill a bag. You can have the software step through the algorithm, or you can explore the tree via the "Show entire tree" button.

Yes, the branch-and-bound algorithm is long and convoluted. And it has to go through much of an entire binary tree. But guess what? It's the best algorithm out there for solving integer programming problems! Well, okay, it's not the *best*, but everything else that's better is just a refinement of it in some way. Seriously.

10.12   Bonus Check-Yourself Problems

Solutions to these problems appear starting on page 615. Those solutions that model a formal write-up (such as one might hand in for homework) are to Problems 4, 5, and 8.

1. Find two different spanning trees of the graph shown at left in Figure 10.22.

2. Find two different minimum-weight spanning trees of the graph shown at right in Figure 10.22. Are there more?

3. Find, if possible, a perfect matching in each of the graphs shown in Figure 10.22.

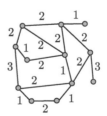

Figure 10.22.   A graph and an edge-weighted graph.

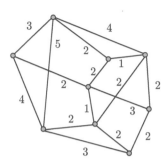
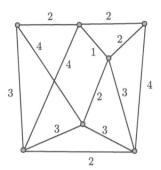

Figure 10.23. Two edge-weighted graphs.

4. Prove that for $n \geq 3$, every n-vertex tree has at most $n - 1$ leaves.

5. Create a binary search tree for the mini-dictionary {*block, black, brack, bract, brace, trace, race, ace, mace, maze, maize, baize*}.

6. Find a minimum-weight spanning tree of the graph shown at left in Figure 10.23 using Kruskal's algorithm.

7. Create an efficient binary decision tree for identifying members of the set {*coat, mittens, hat, scarf, duck, boots*}.

8. Prove that in any tree with at least two vertices, any two vertices are connected by a unique minimum-length path.

9. Use backtracking to find all the ways to add numbers from {1, 2, 3, 4, 5} to get 8.

10. Find a minimum-weight spanning tree of the graph shown at right in Figure 10.23 using Prim's algorithm.

10.13   Tree Problems

1. Draw all trees on seven vertices.

2. Show that the average degree of a tree is less than 2.

   Explain how this result provides a proof that every tree has at least one leaf.

3. List at least eight spanning trees and their corresponding total weights for the graph in Figure 10.24. How many minimum-weight spanning trees does this graph have?

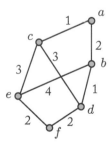

Figure 10.24. A weighted graph.

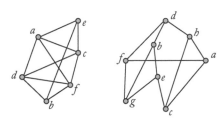

Figure 10.25. Two graphs.

4. Find a spanning tree of the graph at left in Figure 10.25.

5. Find a spanning tree of the graph at right in Figure 10.25.

6. Find a minimum-weight spanning tree of the right-hand graph in Figure 10.26.

7. Find a minimum-weight spanning tree of the left-hand graph in Figure 10.26 using Kruskal's algorithm.

8. Find a minimum-weight spanning tree of the left-hand graph in Figure 10.26 using Prim's algorithm.

9. Show that every tree is bipartite. (Create a proof different from any you created earlier in this course.)

10. Figure 10.27 gives a friendship graph for Example 10.7.3.

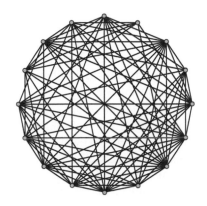

Figure 10.27. A friendship graph for a dance class.

Is it possible for every dancer to stretch with a friend?

11. In the start-big minimum-weight spanning tree algorithm, we removed edges from cycles. We could have instead chosen to remove edges that do not disconnect the tree. Do these two algorithms ever differ?

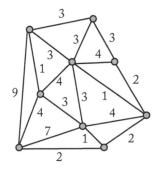

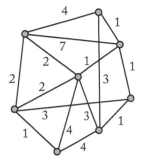

Figure 10.26. Two weighted graphs.

12. Is the start-big minimum-weight spanning tree algorithm more like the opposite of Kruskal or more like the opposite of Prim?

13. An unweighted graph could be considered an edge-weighted graph with all edges of the same weight (perhaps 1). Rewrite Kruskal's algorithm to work with unweighted graphs.

14. Find a minimum-weight spanning tree of the graph given in Example 10.4.2. First use any method you like, then do it using Kruskal's algorithm, and then again using Prim's algorithm. How do your minimum-weight spanning trees differ from each other?

15. Create a reasonable definition for a *trinary* tree.

16. Create a binary decision tree that reflects the way a coin-sorting machine deals with standard US coins (penny, nickel, dime, quarter, half-dollar, and dollar).

17. Create a binary search tree for these library titles. (All but one are real.)

    A Duck is a Duck
    Enslaved by Ducks
    Mini Ducks Songbook
    Fowl-Weather Friends
    Big Dig Ducks
    Regarding Ducks and Universes
    Domesticated Ducks
    Pocketful of Poultry: Chickens,
    *   Ducks, Geese, Turkeys*

18. After Example 10.7.1, we examine the case of a class on The Reality of Ducks. Below are listed the students and the books on reserve each seeks. (All titles are of real books.) Can all students work simultaneously?

🦆 Amit: *Waterfowl Earth Cover Selection Analysis within the National Petroleum Reserve, Alaska*; *Dabblers & Divers: A Duck Hunter's Book.*

🦆 Beth: *The New Duck Handbook: Ornamental and Domestic Ducks: Everything about Housing, Care, Feeding, Diseases, and Breeding, with a Special Chapter on Commercial Uses of Ducks*; *Determinants of Breeding Distributions of Ducks*; *Red Fox Predation on Breeding Ducks in Midcontinent North America.*

🦆 Carla: *Species, Age and Sex Identification of Ducks Using Wing Plumage*; *Ducks of North America and the Northern Hemisphere.*

🦆 Dmitri: *The Wood Duck and the Mandarin: The Northern Wood Ducks*; *Ducks of North America and the Northern Hemisphere*; *Red Fox Predation on Breeding Ducks in Midcontinent North America.*

🦆 Eglantine: *Why Ducks Do That: 40 Distinctive Duck Behaviors Explained & Photographed*; *Autumn Passages: A Ducks Unlimited Treasury of Waterfowling Classics*; *Ruddy Ducks & Other Stifftails: Their Behavior and Biology.*

🦆 Fatima: *The New Duck Handbook: Ornamental and Domestic Ducks: Everything about Housing, Care, Feeding, Diseases, and Breeding, with a Special Chapter on Commercial Uses of Ducks*;

Red Fox Predation on Breeding Ducks in Midcontinent North America.

🦆 Gertrude: *Cavity Nesting Ducks*; *Waterfowl Earth Cover Selection Analysis within the National Petroleum Reserve, Alaska*; *Ruddy Ducks & Other Stifftails: Their Behavior and Biology.*

🦆 Hiroto: *Red Fox Predation on Breeding Ducks in Midcontinent North America*; *Dabblers & Divers: A Duck Hunter's Book.*

19. Now suppose that the information given in the previous problem is listed in preference order, i.e., the book the student desires most is listed first. Use these preferences to create a weighted graph, and try to find a minimum-weight matching.

20. In Figure 10.28 we have a puzzle with partial solution $(1, *, *, *, 2, *, *, *, *)$. The rules are that we must place a 1, 2, and 3 in every row and every column. Follow a backtracking procedure to solve the puzzle. How many possible solutions are there?

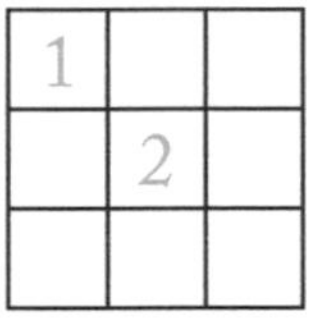

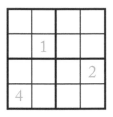

Figure 10.28. A partially solved puzzle.

21. Consider the mini-sudoku puzzle of Figure 10.29, in which each row, column, and quadrant needs to contain the numbers 1, 2, 3, 4. List the branches of

a tree that are cut off by the numbers already placed in the mini-sudoku.

Figure 10.29. A partially solved mini-sudoku.

22. Here are the clues for the crossword puzzle in Figure 10.30.

   1 Down: the best color (hm…, blue, teal, rose, grey, ecru, jade?)

   2 Down: has lots of water (sea, tub, bay?)

   1 Across: also has lots of water (ocean, sewer, river, storm, cloud?)

   2 Across: and also has lots of water (tank, well, lake?)

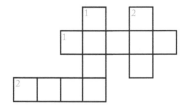

Figure 10.30. An empty crossword puzzle.

Use backtracking to find all possible solutions to the crossword, given the potential answers revealed above!

23. Show that a graph is connected if and only if it has a spanning tree.

24. How many different binary search trees can be made with three pieces of data? What about with four pieces of data?

25. In Section 10.8, the tree of partial solutions could have had as many as $1 + 8 + 64 + \cdots + 8^8 = 19{,}173{,}961$ nodes.

    (a) We knew that node $(2,2)$ corresponded to a nonsolution. How many nodes did we remove from the tree by removing $(2,2)$?

    (b) We also know that $(2,1),(2,3)$, and all other pairs of the form $(k, k-1)$, $(k,k),(k,k+1)$ correspond to nonsolutions. How many such nodes are there?

    (c) How many nodes, total, does removal of these length-2 nodes cause to be removed?

    (d) So, just removing the length-2 nodes that we know are no good, how many nodes are left in the tree of partial solutions?

26. Find three different spanning trees of the graph shown in Figure 10.31.

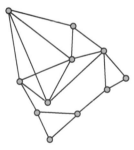

Figure 10.31. A run-of-the-mill graph.

27. Find two different minimum-weight spanning trees of the graph shown in Figure 10.32.

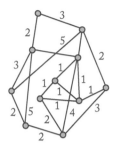

Figure 10.32. An ordinary working graph.

28. Find a minimum-weight spanning tree of the graph shown in Figure 10.33 by using Kruskal's algorithm.

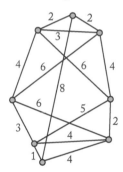

Figure 10.33. A graph that likes chocolate.

29. Find a minimum-weight spanning tree of the graph shown in Figure 10.33 by using Prim's algorithm.

30. Find a spanning tree that is not a minimum-weight spanning tree of the graph shown in Figure 10.33.

31. Create a binary search tree for the set of words *whisker, lollipop, pixie, nudibranch, peapod.*

32. Create a binary decision tree to distinguish between the objects *whisker, lollipop, pixie, nudibranch, peapod.*

33. Find two different matchings of maximum size in the graph shown in Figure 10.31.

34. Find two different perfect matchings in the graph shown in Figure 10.33.

35. You have nine tiles of height one, three of which are one unit long, three of which are two units long, and three of which are three units long. Use backtracking to find all the ways to tile a 6-unit-long rectangle of height one.

10.14   Instructor Notes

With this chapter, the book (and course) move into the graph theory theme. Remember that your students haven't studied graph theory for more than a month, and so they will probably not remember much about graphs at first. As preparation for the first class spent on this chapter, ask them to read Sections 10.1 and 10.2 and do the associated Check Yourself problems.

Because it's always a good time to review induction, an excellent warmup for this chapter is to re-present an inductive proof that a tree with n vertices has $n - 1$ edges. Then proceed to Section 10.3. Students working in groups are likely to finish all but the last problem in Section 10.3 during a class period, especially if you can nudge them to avoid getting bogged down in producing proof that their algorithm(s) accomplish the intended goals. If you have a particularly computer-science-y class, it's worth asking students how their algorithms would proceed on graphs stored in particular ways (lists, adjacency matrices, etc.). This, of course, presumes they have some experience with data storage.

Here are two ways you might want to use the second and third class periods (and of course, other excellent ways exist!), depending on what you wish to emphasize in your class. If you feel that it is important that students understand the proofs that Kruskal's and Prim's algorithms produce minimum-weight spanning trees, then it would make sense to lecture on these proofs on the second class day. Another approach is to leave these proofs as optional and emphasize backtracking and optimization, lecturing on these on the third class day.

Assign the students to read Sections 10.4 (perhaps making optional the proofs that Kruskal's and Prim's algorithms work) and 10.5 and do the Check Yourself problems as preparation for the second class day. On that day, begin by asking for questions on the reading. If you are going to lecture on Kruskal's and Prim's algorithms, do so. You will probably have a sense, at this point in the course and after the first day's work, whether your students need practice in executing Kruskal's and Prim's algorithms or not. If they do, then draw a few edge-weighted graphs on the board (just invent them on the spot and make sure they are not already trees) and ask students to break into groups and run both algorithms on them.

Next, set your students the task of working on the problems in Section 10.6. (This may be on the second or third class day depending on your plan.) Now (and perhaps from now on), when dividing students into groups for in-class work, use a binary sort: for example, divide them by height; divide into the two groups taller and shorter than $5'\ 6''$, then divide at $5'\ 3''$, $5'\ 9''$, etc. This reinforces the concept, rearranges groups, and allows for some levity.

After this class, assign students to read Sections 10.7 and 10.8 and do the Check Yourself problems. Depending on your plan, this may be the end of your time on trees, you may have students continue working on Section 10.6, or you may ask students to read Bonus Section 10.11 and lecture over this material.

Chapter 11

Euler's Formula and Applications

11.1   Introduction and Summary

Euler's formula is one of the coolest topics in all of mathematics. Seriously! And the proof is fun, too—it can be told as a story… so that's what we'll do. In fact, we will mix things up a bit and introduce the proof of the formula before we (actually you) state it! There are lots and lots of other proofs of Euler's formula, and you are encouraged to read about them after we have generated our proof.

First, however, we will set the mathematical stage by exploring planar graphs a bit.

11.2   Try This! Planarity Explorations

If we can draw a graph in the plane (on a piece of paper, on the blackboard, etc.) without edges crossing, then the graph is *planar*. A graph can be planar but be drawn in a nonplanar way—just make one of the edges curly and wild and long so that it crosses every other edge of the graph. Recall that K_n is the complete graph on n vertices, so that all possible edges are present, and that $K_{n,m}$ is the complete bipartite graph with n vertices in one part and m vertices in the other part.

1. Try to draw K_4 twice, once with at least two edges crossing and once with no edges crossing. Can you do it?

2. Try to draw K_5 twice, once with at least two edges crossing and once with no edges crossing. Can you do it?

3. In the standard way of drawing $K_{m,n}$, one draws m dots in a row and, a bit below this row, draws another n dots in a row. Then one connects each of the m dots to each of the n dots. Make a standard drawing of $K_{2,4}$. Do you think there is a different drawing of $K_{2,4}$ with no edges crossing?

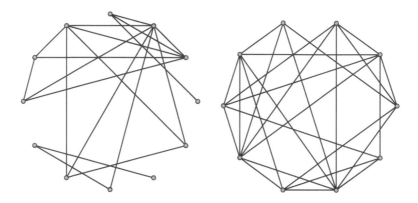

Figure 11.1. Planar or not?

4. Make a standard drawing of $K_{3,3}$. Do you think there is a different drawing of $K_{3,3}$ with no edges crossing?

5. Is the graph shown at left in Figure 11.1 planar? If so, give a planar drawing. (GeoGebra files for Figure 11.1 are available for your use at http://www.toroidalsnark.net/dmwdlinksfiles.html.)

6. Is the graph shown at right in Figure 11.1 planar? If so, give a planar drawing. (GeoGebra files for Figure 11.1 are available for your use at http://www.toroidalsnark.net/dmwdlinksfiles.html.)

A planar drawing of a planar graph has regions called *faces*. These are contiguous areas of the plane bounded by edges. The area around a planar graph is also considered a face and is sometimes called the exterior face or the outer face. (The reasons for this will be made clear in Bonus Section 11.11.)

Now we will explore *sizes* of faces, measured by the number of edges bounding them. There is exactly one slightly tricky aspect of counting edges in this fashion: if an edge is on one face twice (instead of separating two different faces as often happens), then it contributes to the size of the face twice as well. See Figure 11.2 for examples.

7. Find two different planar drawings of the left-hand graph of Figure 11.2, each of which has a face of size 6. How many faces, total, does each drawing have?

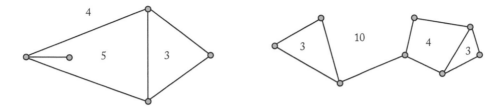

Figure 11.2. Two planar graphs with the sizes of their faces marked.

8. Find two different planar drawings of the left-hand graph of Figure 11.2, each of which has exterior face of size 3. How many faces, total, does each drawing have?

9. Find three new planar drawings of the right-hand graph of Figure 11.2. What are the face sizes for each drawing? How many faces, total, does each drawing have?

10. Add up the sizes of the faces in each graph drawing you have made. What seems to be true about the sum of the face sizes of a graph? Explain why this is so.

11.3   Planarity

Hey! You! Don't read this section unless you have worked through the problems in Section 11.2. I mean it!

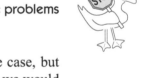

You have probably concluded that K_5 is nonplanar. Indeed, this is the case, but proving it directly is quite difficult. Given the tools we currently have, we would need to show that no matter how we draw K_5—no matter where the vertices are, no matter where we place the edges—we cannot draw all edges without forcing a crossing. In addition to the sheer number of cases we would have to deal with, there is the challenge of making a clear and rigorous argument about drawing. In Section 11.6 we will use a new tool to create a simple way to show that some nonplanar graphs are, indeed, nonplanar. (It will not work on all nonplanar graphs, however!) In contrast, it is relatively easy to demonstrate that a planar graph is planar; we just need to provide a planar drawing of the graph. GeoGebra can help you find a planar drawing of a graph, just as back in Chapter 3 it helped you determine whether two graphs were isomorphic.

You may have observed that every planar drawing of a graph has the same number of faces. We will understand why in Section 11.5. Now, however, we can prove that the sum of the sizes of the faces is independent of which planar drawing we examine.

> **Theorem 11.3.1.** *For any graph G drawn without edges crossing, the sum of the sizes of the faces is equal to $2|E(G)|$.*

Proof: We use a combinatorial proof. Notice that if an edge is between two faces, it counts once for each face, and if an edge is on one face, it counts twice for that face. Thus, each edge contributes twice to the sum of the sizes of the faces, and therefore, the sum of the face sizes is twice the number of edges.   □

Check Yourself

Do all one of these problems.

1. Go to http://planarity.net; enjoy.

11.4   A Lovely Story

Hey! You! Don't read this section unless you have special permission from your instructor, or unless you have no instructor. I mean it!

Once upon a time, a long, long time ago now, so long ago that animals could talk and teachers were human beings, that long ago, off the coast of Japan there was an island. On Iki Island there were fields, mainly for rice but also for some other crops. The residents built a network of stone walls to delineate the boundaries between the fields. This was not the only purpose of the stone walls: they also served as walkways across the island, so that residents could pass from one part of the island to another without trampling any crops. Additionally, seeing fields from slightly above gave a different and longer perspective on how to optimize the planting. Now, Iki Island was fairly close to the mainland (if one considers a much larger island to be a mainland) of Japan, and what with regular wars over land (humans are so violent), there was always a chance of invasion. Besides, many of the residents of Iki Island were pirates, and of course, some other Japanese citizens wanted revenge for the pirates' plundering. So every so often along the walls, there were platforms from which one could see quite far and watch for intruders.

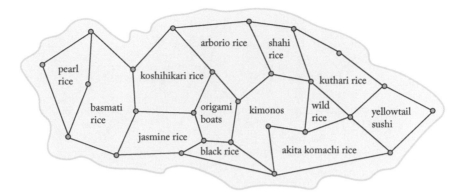

Figure 11.3. A map of the fields on Iki Island, with watch platforms marked as small circles.

Now, the pirates of Iki Island were very specialized pirates: they were Japanese rice pirates. They stole rice from all over the world so that they could grow and breed rice of their own. As a result of excellent piracy, on Iki Island they grew black rice, jasmine rice, arborio rice, akita komachi rice, koshihikari rice, basmati rice, pearl rice (or sticky rice), kuthari rice, shahi rice, and wild rice. (See Figure 11.3 for a map of the fields of rice.)

One day, as the pirates were peacefully tending their rice fields, a tsunami came and crashed through the stone wall separating the beach from the akita komachi rice field and flooded the field completely. It was a terrible day, for more tsunami arrived: soon the black rice, jasmine rice, and basmati rice fields were flooded. The pirates began to panic. They feared that they would all drown if they did not reach safety; what if all the walls were crushed beneath tsunami? Meanwhile, more tsunami came and broke through to the wild rice and koshihikari rice fields. These pirates were not supremely intelligent, as they might otherwise have noticed that not *all* the walls could fall; once both sides of a wall are flooded, there is not enough pressure from water for it to break.

The head pirate, Mauler of Rules, began shouting to the others. "Four Mule Laser, get to the boats!" For the pirates had developed a special waterproof rice paper from which to fold excellent boats. Just then, a tsunami freed the yellowtail sushi (which were, truth to tell, overfished anyway), and a wall protecting the pearl rice fell to another tsunami. "Lo, Surreal Fume!" exclaimed one pirate to another. "Lo, Flu Measurer," replied hir friend, "To the boats, before they are carried away in yet another flood!" Luckily the stone walls about the origami boat field held, though the arborio rice and kuthari rice fields were flooded. A Fuller Mouser and

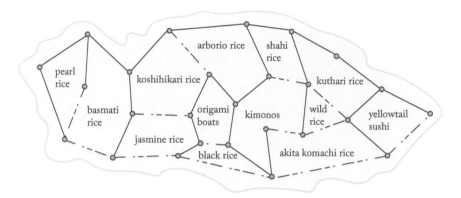

Figure 11.4. A map of the fields on Iki Island after the tsunami tragedy.

Usual for Mr. Lee raced to the boat area just in time to secure the last few boats, for a tsunami broke through and all were set afloat in a flood. Too late, Lemurs Are Foul realized what was happening. "Not the kimonos!" ze cried, for a tsunami had just crushed one of the walls protecting their prized kimono field. With a last blast of gusto, the tsunami broke through to the shahi rice field, flooding it and rendering the entire island suffused with water. At this, Allure of Serum remarked, "Your name is far too apt, Realm So Rueful."

The pirates sadly floated about on their boats and feared that more tsunami would drown them, but none came. (See Figure 11.4 for a map of the destruction.) Over the following days, the waters receded, leaving only ruined crops on wet fields. The pirates dragged a couple of origami boats to the sea, and Mauler of Rules took Lemurs are Foul and Usual for Mr. Lee off on an expedition to plunder rice to plant. Ze assigned the remaining pirates to rebuild the stone walls while ze led the expedition. Over time, the walls were rebuilt and the fields reseeded with many kinds of delicious rice and exquisite kimonos, and all was well.

Please don't think this is the only sea tragedy in history. There are many more, some recent. For example, back in the year 2000, global warming caused the Isle of Ventenese to flood, but luckily Al Gore enacted legislation that turned back the waters. A year later, a network of clog and tulip fields in the Netherlands flooded until enough people held their fingers in the dykes.

11.5 Or, Are Emus Full?: A Theorem and a Proof

Look again, or look for the first time, at Figure 11.3. Notice that the stone walls with platforms could be considered the edges and vertices of a graph. What properties does this graph have? It is connected and none of the edges cross. Thus, our graph is planar and the regions delineated by the edges are both faces of our graph and the fields in our story.

Now, what happened in our story in terms of graph theory? We began with a connected planar graph and removed edges until no cycles remained. (Every face became flooded.) Additionally, we did not allow the removal of an edge that would disconnect the graph—only edges contained in cycles could be removed. In other words, we removed edges from our planar graph until we were left with a spanning tree. Then we replaced the edges we'd removed in order to obtain the original graph. That summary is pretty short compared to our story... and it still doesn't tell us what we were proving.

Let us make a couple of combinatorial observations. The number of vertices of the planar graph is the same before and after edges are removed. The number of edges removed is the same as the number of faces in the original graph. These suggest that our mystery theorem has something to do with counting. Now, we shall reveal it:

Theorem 11.5.1 (Euler's formula). *For any connected planar graph G with faces $F(G)$, $|V(G)| - |E(G)| + |F(G)| = 2$.*

Proof: We begin with a connected planar graph G and identify a spanning tree T of G (for example, by using the procedure explained on page 315 in Chapter 10). Let us abbreviate $|V(G)|$ as v_G, $|E(G)|$ as e_G, and $|F(G)|$ as f_G; let us likewise abbreviate $|V(T)|$ as v_T, $|E(T)|$ as e_T, and $|F(T)|$ as f_T. Each time we remove an edge not in T, one cycle is broken and two faces merge. (This is true even when we break down "exterior" walls, as they separate the exterior face from interior faces.) When we break all cycles, a single face remains. Thus, we remove $f_G - 1$ edges from G to achieve T. We know that $e_T = v_T - 1$ from Example 4.2.4, and rewriting this we have $v_T - e_T = 1$. The spanning tree has exactly one face, so $f_T = 1$ and adding gives us $v_T - e_T + f_T = 2$. Now, no vertices were removed or even changed in the process of converting G to T, so $v_T = v_G$. We already noted that $e_G - (f_G - 1) = e_T$. Substituting into $v_T - e_T + f_T = 2$, we have $v_G - (e_G - (f_G - 1)) + f_T = 2$, which simplifies to $v_G - e_G + (f_G - 1) + 1 = 2$; this becomes the desired result, $v_G - e_G + f_G - 1 = 2$. $\square$

This is by far not the only proof of Euler's formula! See http://www.ics.uci.edu/~eppstein/junkyard/euler/ for at least 18 other proofs. (And play with one of these proofs at http://demonstrations.wolfram.com/ProvingEulersPolyhedral FormulaByDeletingEdges/.) Pretty darned neat, eh?

Verify your understanding with these quickies.

1. Verify Euler's formula for K_4. (Be sure to draw K_4 without edges crossing.)

2. Draw K_3. Count the number of vertices, edges, and faces. How many edges must you remove to obtain a spanning tree? Do so. Count the number of vertices, edges, and faces of the spanning tree. Verify Euler's formula for K_3 and for the spanning tree you obtained.

3. Verify Euler's formula for W_6, the wheel with five spokes.

4. Explain why every planar drawing of a graph has the same number of faces.

11.6   Applications of Euler's Formula

Euler's formula can be used to prove some surprising results. Here is one example.

Theorem 11.6.1. *If G is simple, planar, and connected and has at least three vertices, then $|E(G)| \leq 3|V(G)| - 6$.*

Why is this surprising? Because it relates only $|E(G)|$ and $|V(G)|$ and makes no mention of $|F(G)|$. One would expect that the faces of G would come into play somewhere in the theorem statement (instead, they arise in the proof). The consequence of requiring G to be planar is that faces must have some indirect role. Here is a very useful fact.

Theorem 11.6.2. *If G is simple, planar, and connected and has at least three vertices, then $3|F(G)| \leq 2|E(G)|$.*

Proof: Because G is simple, no cycle can have fewer than three edges. Therefore, each face of G has at least three edges. As an inequality, this statement is $3 \leq$ (number of edges bounding a face). If we make such an inequality for each face of G, the sum of the left-hand sides is $3|F(G)|$. When we sum the right-hand sides, we obtain $2|E(G)|$ because each edge borders two faces (or occurs twice on the boundary of one face). Thus, $3|F(G)| \leq 2|E(G)|$ as desired.   □

Here is a second useful fact.

Theorem 11.6.3. *If G is simple, planar, connected, and has no 3-cycles, then $4|F(G)| \leq 2|E(G)|$.*

Proof: Because no cycle can have fewer than four edges, each face of G has at least four edges. As an inequality, this statement is $4 \leq$ (number of edges bounding a face). If we make such an inequality for each face of G, the sum of the left-hand sides is $4|F(G)|$. When we sum the right-hand sides, we obtain $2|E(G)|$ because each edge borders two faces (or occurs twice on the boundary of one face). Thus, $4|F(G)| \leq 2|E(G)|$ as desired. ☐

In the exercises, you will be asked to generalize Theorems 11.6.2 and 11.6.3. But now, we can proceed with the proof of Theorem 11.6.1.

Proof of Theorem 11.6.1: Because G is planar, we know by Theorem 11.5.1 that $2 = |V(G)| - |E(G)| + |F(G)|$. From Theorem 11.6.2 it follows that $|F(G)| \leq \frac{2}{3}|E(G)|$. Thus, $2 \leq |V(G)| - |E(G)| + \frac{2}{3}|E(G)| = |V(G)| - \frac{1}{3}|E(G)|$, which rewrites to $6 \leq 3|V(G)| - |E(G)|$, and this is equivalent to the desired statement. ☐

Using Theorem 11.6.3, we can generalize Theorem 11.6.1:

> **Theorem 11.6.4.** *If G is simple, planar, connected, has no 3-cycles, and has at least three vertices, then $|E(G)| \leq 2|V(G)| - 4$.*

Proof: Because G is planar, we know that $2 = |V(G)| - |E(G)| + |F(G)|$. From Theorem 11.6.3 it follows that $|F(G)| \leq \frac{1}{2}|E(G)|$. Thus, $2 \leq |V(G)| - |E(G)| + \frac{1}{2}|E(G)| = |V(G)| - \frac{1}{2}|E(G)|$, which rewrites to $4 \leq 2|V(G)| - |E(G)|$, and this is equivalent to the desired statement. ☐

In Section 11.2 you convinced yourself that it is impossible to draw K_5 without edges crossing; in other words, K_5 is nonplanar. But how can we prove it?

> **Theorem 11.6.5.** *Both K_5 and $K_{3,3}$ are nonplanar.*

Proof: Check *this* out. We will proceed by contradiction. Suppose that K_5 and $K_{3,3}$ are planar. They are both simple. The smallest cycle in K_5 has length 3. By Theorem 11.6.1, $|E(G)| \leq 3|V(G)| - 6$. We know that K_5 has five vertices and $\binom{5}{2} = 10$ edges. Therefore, $10 \leq 15 - 6 = 9$, which is a contradiction. And, we know that $K_{3,3}$ has six vertices and $3 \cdot 3 = 9$ edges, so $9 \leq 18 - 6 = 12$, which is... *not* a contradiction.

However, the smallest cycle in $K_{3,3}$ has length 4. By Theorem 11.6.4, $|E(G)| \leq 2|V(G)| - 4$. We know that $K_{3,3}$ has six vertices and $3 \cdot 3 = 9$ edges. Therefore, $9 \leq 12 - 4 = 8$, which *is* a contradiction. Much better.

Figure 11.5. The superposition of the two left-hand graphs is shown at right to make explicit that K_5 has thickness 2.

And there's more:

Theorem 11.6.6. *If G is simple, planar, and connected, then G has at least one vertex of degree no more than 5.*

Proof: Suppose not. That is, suppose every vertex of G has degree at least 6. Then for each vertex, we may write $6 \le$ (the degree of the vertex). If we add up all such statements, we get $6|V(G)|$ on the left-hand side, and we get the sum of all the degrees in the graph on the right-hand side. In Chapter 3 you proved the handshaking lemma (Lemma 3.5.1 on page 77), which tells us that the sum of all the degrees in a graph is equal to $2|E(G)|$. So we have $6|V(G)| \le 2|E(G)|$ or $3|V(G)| \le |E(G)|$. Now, if G has at least three vertices, then by Theorem 11.6.1, $|E(G)| \le 3|V(G)| - 6$. Putting these facts together, we have that $3|V(G)| \le |E(G)| \le 3|V(G)| - 6$, which cannot be true. Contradiction!

However, if G has fewer than three vertices, it can only be a lone vertex or K_2. The theorem holds for these two examples, so we're done. □

It's time for a practical application!

Imagine a circuit board. It has these metal dots that go all the way through the board, and on each side there are metal curves that connect some of the dots. Other dots are connected by components (such as resistors or transistors). The metal curves do not touch—that would cause a short circuit. So each side of the circuit board shows a planar graph with the dots playing the role of vertices and the metal curves playing the role of edges. However, the graph formed by the metal dots and the metal curves from *both* sides of the circuit board is probably not planar. It is two planar graphs glued together at their vertices, so we say that this graph has *thickness* 2. Figure 11.5 shows that K_5 has thickness 2. Notice that one way to visualize thickness of a graph G is as a drawing of G with edges in different colors, where no two edges of the same color cross.

We have already been interested in which graphs are planar and which are not; now we are interested in which graphs can be printed on circuit boards and which cannot. We say that a planar graph has thickness 1 and a circuit-boardar graph has thickness 2. In general, the thickness of a graph G is denoted $t(G)$, and it is the smallest number of planar graphs that can be glued together at their vertices to form G.

Euler's formula gives a couple of lower bounds for the thickness of a graph (and thus quickly tells us whether we should bother trying to draw that graph on a circuit board):

Proof of Theorem 11.6.7: By Theorem 11.6.1, if G is planar, then $|E(G)| \leq 3|V(G)| - 6$, or $\frac{|E(G)|}{3|V(G)|-6} \leq 1$. If we have two graphs G_1 and G_2, then $\frac{|E(G_1)|}{3|V(G_1)|-6} + \frac{|E(G_2)|}{3|V(G_2)|-6} \leq 2$. If additionally we have that $|V(G_1)| = |V(G_2)| = v$ and we name $|E(G_1)| + |E(G_2)| = e$, then this becomes $\frac{e}{3v-6} \leq 2$. This arithmetic generalizes to any finite number of graphs $G_1, \ldots, G_k$. We may thus think of the ratio $\frac{|E(G)|}{3|V(G)|-6}$ as measuring the smallest number of planar subgraphs whose superposition could comprise G. Of course, a number of subgraphs must be an integer, so we must round up and the smallest number must be $\lceil \frac{|E(G)|}{3|V(G)|-6} \rceil$. This is the smallest possible thickness, so $\lceil \frac{|E(G)|}{3|V(G)|-6} \rceil \leq t(G)$. $\qquad\square$

Theorem 11.6.7. *If G is a simple graph with at least three vertices, then $t(G) \geq \lceil \frac{|E(G)|}{3|V(G)|-6} \rceil$.*

Check Yourself

Make sure you understand all the details of the theorems in this section by doing these problems.

1. Why is the constraint $|V(G)| \geq 3$ necessary in Theorem 11.6.1?

2. Draw a nonsimple graph that violates Theorem 11.6.6.

3. Theorem 11.6.3 requires that G have no 3-cycles. This requirement could be replaced with the constraint that G be drawn with no faces of size 3. Why is this a weaker constraint?

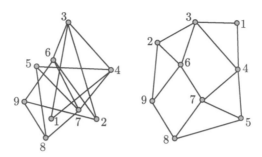

Figure 11.6. Nonplanar and planar drawings of the same graph.

11.7 More Planarity-ish Examples

Example 11.7.1 (of a planar graph). Figure 11.6 shows a planar graph; at left is a nonplanar drawing of the graph and at right is a planar drawing of the graph.

Example 11.7.2 (of a nonplanar graph). Figure 11.7 shows a nonplanar graph with a $K_{3,3}$ subgraph highlighted in teal. One part of $K_{3,3}$ has dark teal vertices and the other has pale teal vertices. (Because the graph contains $K_{3,3}$, we know it is nonplanar by Theorem 11.6.5.)

Example 11.7.3 (of a nonplanarity proof using Euler's formula). You are shown a graph that looks suspiciously like a scribbly duck, and asked whether or not it is planar. A quick count shows that it has 23 vertices and 65 edges. *Fine,* you think, *it could be planar. After all,* $23 - 65 + 44 = 2$, *so it just needs* 44 *faces.* But somehow you feel unsettled by trying to draw all those faces. *Maybe there's a consequence of Euler's formula that will help? Ah, yes, Theorem 11.6.1*—and $65 \not\leq 3 \cdot 23 - 6$, so we now know the scribbly duck graph cannot be planar.

Figure 11.7. A nonplanar graph with highlighted $K_{3,3}$ subgraph.

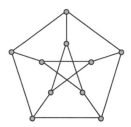

Figure 11.8. The best graph.

11.8 Try This! Applications of Euler's Formula

Enjoy working on these problems with peers.

1. Prove that for a planar graph with k components, $|V(G)| - |E(G)| + |F(G)| = 1 + k$.

2. Prove that the Petersen graph (shown in Figure 11.8) is not planar.

3. Show that if every face of a planar graph has four edges, then $|E(G)| = 2|V(G)| - 4$.

4. For which n is K_n planar and for which n is K_n nonplanar? Make and prove a conjecture.

5. For which m, n is $K_{m,n}$ planar and for which m, n is $K_{m,n}$ nonplanar? Make and prove a conjecture.

6. Check out Figure 11.9 for an interesting way to represent the torus (surface of a doughnut):

   (a) Convince yourself that the left and right diagrams really are two representations of the same thing.

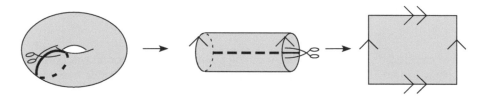

Figure 11.9. Cutting a doughnut skin.

(b) Draw a few graphs on copies of the torus. (Make sure to cross each of the rectangle edges so that you're genuinely *using* the torus.)

(c) Try out Euler's formula on these graphs. Does it still hold? If not, does some other formula hold?

(d) **Special challenge**: Prove your conjecture.

11.9   Where to Go from Here

To learn more about planar graphs, first consult the easy-to-read [24]. You can continue learning about planar graphs in a graph theory course and on to the level of research—within the few months prior to this writing, papers were published on finding minimum-weight partly spanning trees of planar graphs, degree sequences of parts of bipartite planar graphs, decomposing planar graphs into forests, coloring planar graphs (see Chapter 13 for what this means), planar graphs without Hamilton cycles (see Chapter 12 for what this means), and drawing planar graphs with few slopes. There is lots of research on (and even conferences devoted to) graph drawing in general; both computer scientists and mathematicians participate in this work.

Euler's formula generalizes to other surfaces (such as multiholed doughnut skins). Find out more in [7]. Euler's formula also generalizes to polyhedra and higher-dimensional objects (called polytopes)! Understanding the statement and proof of this generalization requires a course in linear algebra and then taking a course in or reading a text on convex geometry (such as *An Introduction to Convex Polytopes* by Brøndsted).

There really is an Iki Island. And they really do grow rice there (see http://www.iki-island.net/). And historically, there really were pirates inhabiting the island! (See http://www.iki.co.jp/cat14/?page_id=46 and engage Google Translate.)

Credit where credit is due: The global-warming proof of Euler's theorem was presented by Matthew Riddle in July 2000, and the Netherlander proof was presented by Abraham Flaxman in July 2001. Section 11.6 was adapted from [24]. The first sentence of Section 11.4 heavily overlaps the first sentence of a story written by the author's father, about the origin of fractions: see page 267 and also page 249.

11.10 Chapter 11 Definitions

planar: **A** graph that can be drawn in the plane (on a piece of paper, on the blackboard, etc.) without edges crossing.

face: **A** contiguous area of the plane bounded by edges, in a planar drawing of a planar graph.

size (of a face): The number of edges bounding the face. (Sometimes one edge appears twice on the boundary of a face, in which case it is counted twice.)

thickness (of a graph): For a graph G with n vertices, the smallest number of n-vertex planar graphs that can be stacked up (aligning the vertices) to form G.

11.11 Bonus: Topological Graph Theory

If prior to this point you were aware of any basic facts about the author of this text, you surely suspected you wouldn't be able to read the whole thing without getting a bit of proselytizing as to the coolness of her favorite portion of mathematics.

In ordinary graph theory, there are two types of objects (vertices and edges), and only combinatorial structure (adjacency) matters. How the graph is drawn is irrelevant to its structure. The way in which topological graph theory differs is that we encode additional information by drawing graphs without edges crossing. You are already familiar with planar graphs, and we will generalize this idea in a moment. First, we should set planar graphs into a slightly different context: they would more correctly be called *spherical* graphs (though no one does call them that) because topologically speaking, the plane might as well be a sphere! Figure 11.10 shows that the parts of a plane we perceive as going off to infinity correspond to the "back" of a sphere. This allows us to see that what we call the exterior face is just a face like any other, as long as we rotate the sphere.

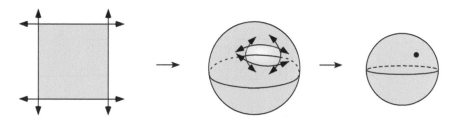

Figure 11.10. A plane can be bent around and its infinite boundary contracted to a single point on a sphere.

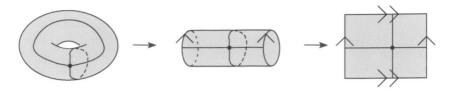

Figure 11.11. A graph with one vertex and two edges (both are loops) is embedded on the torus.

Is there a way of drawing *non*planar graphs so that the edges do not cross? Yes, as long as surfaces other than the sphere are considered; we call such drawings *embeddings*. In Figure 11.9 we saw a second surface, the torus, and how to represent it in a flat way. Figure 11.11 shows a small graph drawn on the torus. The arrows indicate a sort of gluing, so if an edge seems to go off the boundary, it just comes back on across the way. We want to have flat representations of surfaces so that we can more easily embed graphs on them—it gets confusing to have to draw a lot of dotted lines.

Faces of a graph can only be defined relative to an embedding, so what the faces are depends on how the graph is drawn and on what surface it is drawn. The faces of a graph embedding are the regions of the surface carved out by the graph, so for the graph G embedded on the surface S, the collection of faces is $S \setminus G$. (Notice now that Theorem 11.3.1 does not mention on which surface the graph G is embedded, nor does its proof invoke planarity.)

We will only consider *cellular* embeddings, in which every face is mooshable into a disk. (In other words, faces cannot have punctures or doughnut holes.) This is the most common type of embedding studied in topological graph theory. The embedding depicted in Figure 11.11 is, in fact, cellular and has exactly one face, as shown in Figure 11.12.

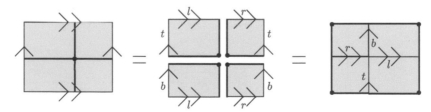

Figure 11.12. The face of the Figure 11.11 embedding is a cell, which we can see by assembling its four parts into a whole. The top halves of the arrowed edges are marked t, and likewise the bottom halves are marked b, the left halves are marked l, and the right halves are marked r.

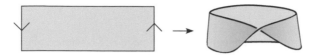

Figure 11.13. A Möbius band has a single twist.

Figure 11.13 gives another example of a surface, perhaps familiar to you, along with its standard flat representation.

> Short activity:
>
> 1. Attempt to embed K_5 on the torus. Is it possible?
>
> 2. Attempt to embed K_5 on the Möbius band. Is it possible?
>
> 3. Attempt to embed $K_{3,3}$ on the torus. Is it possible?
>
> 4. Attempt to embed $K_{3,3}$ on the Möbius band. Is it possible?
>
> 5. Attempt to embed the Petersen graph on the torus. Is it possible?
>
> 6. Attempt to embed the Petersen graph on the Möbius band. Is it possible?

If you enjoyed this activity, you may also like playing with http://demonstrations. wolfram.com/ToroidalWrapping/.

There is a classification of topological surfaces in which we measure how many holes (like a torus) and how many twists (like a Möbius band) the surface has. A summary of this classification may be found at http://www.math.ohio-state. edu/~fiedorow/math655/classification.html, and details may be found in the paper http://new.math.uiuc.edu/zipproof/zipproof.pdf.

So, one of the big questions in topological graph theory is how to determine on which surfaces a graph can be cellularly embedded. We do have a characterization for which graphs are planar and which aren't; this is Kuratowski's theorem, which requires terminology beyond this text to state correctly and mathematics *far* beyond this text to prove. In essence, it states that every nonplanar graph has a copy of K_5 or $K_{3,3}$ in it in some way. There is a similar list of forbidden smallest graphs for the torus, though at last count (January 2016) it was 17,473 graphs long. For further study of topological graph theory, see *Topological Graph Theory* by Gross and Tucker.

11.12   Bonus Check-Yourself Problems

Solutions to these problems appear starting on page 618. Those solutions that model a formal write-up (such as one might hand in for homework) are to Problems 2, 7, and 9.

1. Compute the thickness of K_6.

2. Check out Figure 11.14 to see an image of an annulus (or washer).

Figure 11.14. **An annulus. Or a washer. Who knows?**

(a) Draw a few graphs on annuli (that's the plural of *annulus*). The rule here is that you have to cover the annulus edges with graph edges (and vertices) so that you don't have partial faces.

(b) Try out Euler's formula on these graphs. Does it still hold? If not, does some other formula hold?

(c) Prove your conjecture.

3. Is the complement of any star graph planar? Are *all* complements of star graphs planar? Justify your responses.

4. Can there exist a planar graph with degree sequence $(1,2,2,2,3,5,5,6)$?

5. Could the graph at left in Figure 11.15 be planar?

6. The graph at right in Figure 11.15 is definitely planar. How many faces does a planar drawing of this graph have?

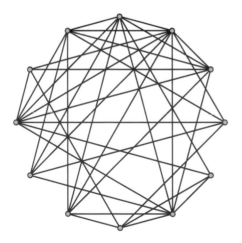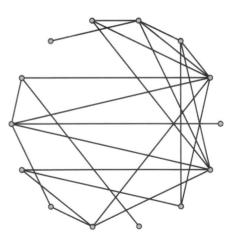

Figure 11.15. Two random graphs courtesy of *Mathematica*.

7. Prove that a connected planar graph has exactly one face if and only if it is a tree.

8. How many vertices must a 4-regular planar graph with 12 faces have?

9. Can a planar graph with nine vertices and all faces of size 4 be k-regular for any k?

10. Compute the thickness of the nonplanar Grötzsch graph, shown in Figure 11.16.

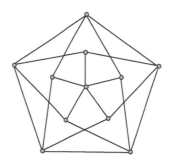

Figure 11.16. I am named after Herbert (Camillo) Grötzsch.

11.13 Problems about Planar Graphs

1. Without looking at David Eppstein's page of many proofs of Euler's formula, make your own proof by using induction on the number of edges.

2. In Section 11.6, it was proved that $|E(G)| \leq 3|V(G)| - 6$. Under what conditions is $|E(G)| = 3|V(G)| - 6$?

3. Prove that there is no graph with six vertices, ten edges, and all vertices of the same degree.

4. The graph G_L at left in Figure 11.17 is planar. How many faces does a planar drawing of G_L have?

5. The graph G_R at right in Figure 11.17 is planar. How many faces does a planar drawing of G_R have?

6. Suppose a connected planar graph P has every vertex of degree at least 3, and every face of size at least 3. Can P have fewer than six edges?

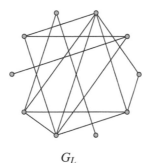

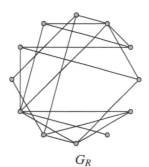

G_L G_R

Figure 11.17. Two planar graphs.

7. Suppose a connected planar graph P has every vertex of degree at least 3 and every face of size at least 3. Can P have exactly seven edges?

8. Prove that the graph shown in Figure 11.18 is nonplanar.

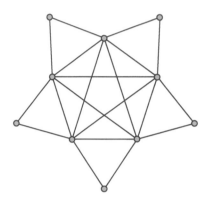

Figure 11.18. See how nonplanar I am?

9. The *girth* of a graph is the length of its smallest cycle. Use this to generalize the statements and proofs of Theorems 11.6.2 and 11.6.3.

10. Use the results of the previous problem to generalize the statements and proofs of Theorems 11.6.1 and 11.6.4.

11. Compute the thickness of $K_{3,3}$.

12. Compute the thickness of the Petersen graph.

13. Give an example of a 4-regular planar graph and an example of a 4-regular nonplanar graph.

14. Prove that the graph shown in Figure 11.19 is nonplanar.

15. Compute the thickness of $K_{4,4}$.

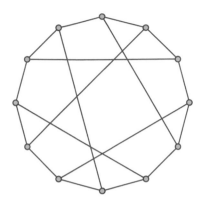

Figure 11.19. Sort of symmetric and nonplanar all the way.

16. Generalize Theorem 11.6.6 slightly: prove that every simple, planar, connected graph G has at least three vertices of degree no more than 5.

17. Might the graph in Figure 11.20 be planar?

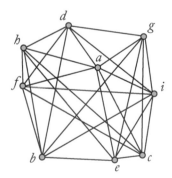

Figure 11.20. Am I planar?

18. The goal of this problem is to use Euler's formula to list all possible regular polyhedra. (In this case, *regular* means that every face has the same number of edges and every vertex has the same degree.)

(a) Convert the idea of a regular polyhedron into a graph in some way.

(b) Find an upper bound on the possible degrees of the vertices.

(c) Find a lower bound on the possible degrees of the vertices.

(d) Find a relationship between the number of vertices and the number of edges.

(e) Find a relationship between the number of faces and the number of edges.

(f) Now use Euler's formula and solve some equations.

(g) Find graphs, and polyhedra, that correspond to your solutions.

19. Write a story proof of Euler's formula involving ducks.

20. Prove that the graph shown in Figure 11.21 is nonplanar.

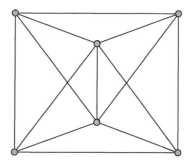

Figure 11.21. I am totally nonplanar.

21. Prove that if G has at least 12 vertices and is simple, then G and $\overline{G}$ cannot both be planar. (Recall that $\overline{G}$ is the complement of G. See page 84 for the definition.)

22. Prove that if G has 11 vertices and is simple, then G and $\overline{G}$ cannot both be planar.

23. Can you find a graph with seven vertices such that G and $\overline{G}$ are both planar?

24. Is the complement of a 6-cycle planar?

25. Is the complement of an 8-cycle planar?

26. Which wheel graphs, if any, have planar complements?

27. Find two examples of nonplanar graphs where every vertex has either degree 3 or degree 4 (and yes, both must be present).

28. The Wagner graph, shown in Figure 11.22, is nonplanar. What is its thickness?

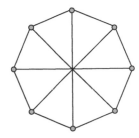

Figure 11.22. People sometimes call me a Möbius ladder.

29. Consider the graphs in Figures 10.2, 10.3, and 10.4. Which are planar and which are nonplanar? Justify.

30. Find two examples of planar graphs where all vertices except one have the same degree.

31. Challenge: Find four infinite families of planar graphs where each family member has all but one vertex of the same degree.

32. Is the Royle graph shown in Figure 11.23 planar or not? Justify your conclusion.

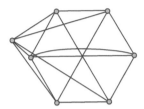

Figure 11.23. **Royally Royle.**

33. How many faces does a planar drawing of a simple connected graph with 8 vertices and 17 edges have?

34. How many faces does a planar drawing of a simple connected graph with 12 vertices and 32 edges have?

35. Suppose G is planar, every vertex has degree at least 3, and every face is either 5-sided or 6-sided. What is the least number of 5-sided faces possible?

11.14   Instructor Notes

This chapter is written with a highly discovery-based approach to planarity and Euler's formula. For the first class meeting, have students jump right in—ask them to read Section 11.1 and then work in groups on the planarity exercises in Section 11.2. They will likely complete all the problems with enough time remaining to present their results to their classmates. Assign students to read Section 11.3 and do the Check Yourself problem as preparation for the next class. (Note that this assigns them to play Planarity at http://planarity.net!)

For the second class day, I suggest an approach that is unusual and perhaps a bit daring or even unwise: you present a disguised proof (or story) of Euler's formula without stating the theorem and then elicit a more mathematical statement and proof from the students. (Tips are given below.)

An alternate approach is to draw a planar graph on the board and ask students to draw their own planar graphs individually or in pairs. Have them draw a bunch of randomly placed vertices and then add edges without allowing any to cross. Next, ask them to count the vertices and edges and faces, and ask them to compute $v - e + f$. Get them to report so that you build up a table, and so that it appears surprising when one row of the table turns out to be all 2s. Then state the theorem.

Every class deserves its own customized version of a story proof. For this reason, you may wish to forbid your students from ever reading Section 11.4; or, you may wish to assign your students to read it well after you have treated the customized version of the proof in class. Sometimes there are themes or running jokes in a class that are obvious candidates for inclusion in the story structure. At other times, one may ask for some elements to use and run with them. (Having pirates, dragons, and a sea battle all at once was somewhat challenging for me because the battle was to take place away from the island. The request for sea pirates *and* a space attack was particularly difficult for me to honor. In comparison, the Evil Nomadic Horde was an easy villain to incorporate in a proof.) These elements form the frame of the story. Begin by drawing a planar graph on the board, on top of an island or what have you, and start placing labels in faces. Some labels will suggest themselves from the story frame or pop up in your mind, but you might (I usually do) ask the students to tell me what is (kept in this room of the castle, grown in this field, protected in this prison). Whatever disaster or attack occurs can also be orchestrated by the students. On request, they will tell you which walls are crushed and in what order, and an eraser can be used to remove most of each of these walls. Sometimes a student will suggest removing a wall that would disconnect the graph, but I claim this is not possible because of (water pressure, invader superstition) and mention it casually rather than making a big deal of it. The students can be quite helpful in knowing in what order to restore the walls.

At the end of this exercise, I ask the students what theorem we have proved and what the proof is. This has varying degrees of success, and depending on how much time remains in class, one may take different approaches. Here are a few possibilities.

1. Ask the students to mention what they think are salient mathematical facts. They will often notice the reduction to a spanning tree. It is harder for them to notice planarity of the graph because it's so much a part of the story, but doing the Section 11.2 planarity exercises ahead of time may subtly call planarity to their communal attention. It probably takes about half an hour to elicit the bulk of the proof and theorem statement from the students.

2. Tell the students to count the number of vertices, edges, and faces of the graph.

3. Some student volunteers that the proof is of Euler's formula (and names it straightforwardly).

After the statement of Euler's formula has come out, be sure to review the story in context of proving Euler's formula. That is, repeat the basics of the story and translate each into a statement that is a step in the proof.

Such a story/proof review is an excellent way to begin the third class. As preparation for that class, ask students to read Sections 11.5 and 11.6 (and perhaps Section 11.4) and do the Check Yourself problems. Ask for questions, and then start students working in groups on the problems of Section 11.8. They will likely take an entire class period.

Chapter 12

Graph Traversals

12.1 Introduction and Summary

A traversal is a way of visiting every desired place. (To traverse means to go across or through.) For example, a traversal of a house would be a path that includes every room. We are concerned with graph traversals… because we are studying graph theory. Sometimes we want to traverse every edge of a graph exactly once, but we don't mind visiting some vertices multiple times. This is called an Euler traversal. Sometimes we want to traverse every vertex of a graph exactly once (in which case we cannot traverse any edge more than once). This is called a Hamilton traversal. For each sort of traversal, we call it a *circuit* if we need to end up back where we started, and a *trail* or path otherwise. We are going to develop conditions on a graph that tell us when it has an Euler circuit or trail. There are almost no useful conditions on a graph that tell us when it has a Hamilton circuit or path! This is all related to a very practical problem called the Traveling Salesperson Problem (TSP for short)—the problem is for the traveling salesperson to visit all the cities in an area to sell stuff and to find the shortest route to save on fuel. Le sigh; there is no good general solution. Strangely enough, there is a nice way to find the shortest route between any two particular points, and so we will learn about that.

12.2 Try This! Euler Traversals

Just to keep everything clear, let us define our terms.

Definition 12.2.1. An *Euler traversal* visits every edge of a graph exactly once and may visit some vertices of the graph more than once. (Most Euler traversals are not paths because of repeated vertex visitation, but they are *trails*.) An *Euler circuit* is an Euler traversal whose starting and ending vertices are the same.

Figure 12.1 gives an example of an Euler traversal that is not an Euler circuit and an example of an Euler circuit. The edges are numbered in the order a particular

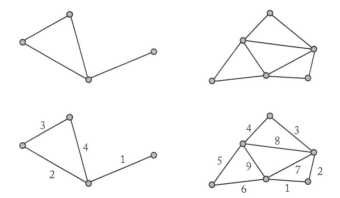

Figure 12.1. At left, a graph with an Euler traversal; at right, a graph with an Euler circuit.

traversal visits them. Notice that the order really does matter! If we tried to visit the edges of the left-hand graph in Figure 12.1 by starting with 4–2–3, we could never reach edge 1; similarly, in the right-hand graph a trail beginning with 1–2–7–6–5–9 will end without reaching edges 3, 4, or 8.

Now it is time to experiment.

1. Draw six different graphs, each with a different number of vertices. Check to see whether each graph has an Euler circuit or not. Do any of your graphs have an Euler traversal but not an Euler circuit?

2. Draw a graph that has an Euler circuit.

3. Draw a graph that does not have an Euler circuit but does have an Euler traversal.

4. Draw a graph that has neither an Euler circuit nor an Euler traversal.

5. Conjecture a necessary condition for a graph to have an Euler circuit (i.e., if a graph has an Euler circuit, this condition *must* hold).

6. Make a conjecture about what property (or properties) a graph needs to have (or not have) for it to be guaranteed to have an Euler circuit.

7. Make similar conjectures about Euler traversals.

8. Try to prove your conjectures.

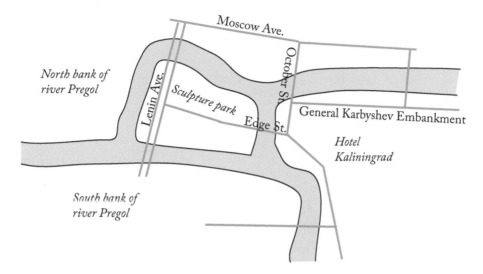

Figure 12.2. Map of Kaliningrad near the city center.

12.3 Euler Paths and Circuits

Hey! You! Don't read this unless you have worked through the problems in Section 12.2. I mean it!

The river Pregol flows through Kaliningrad, Russia. Back in the day, it was the city of Königsberg, Prussia. Legend has it that there were two islands in the Pregol and seven different bridges in Königsberg that joined these islands to the banks of the river (and the banks of the river to each other). The Bridges of Königsberg problem asks whether one can walk over each of the seven bridges and return to one's starting point without crossing any bridge twice. As is so often the case, legend got things a bit wrong because one of those islands is actually an island on which one might walk (it contains a sculpture park), but the other is about 4 km long and is the first in a series of four long islands about 30 km in total length that split the Pregol into north and south branches. Additionally, all seven of the bridges in question were blown up in World War II (some were rebuilt). Near the little island, things currently look as in Figure 12.2. Two of those river-spanning lines might not be bridges, but there are certainly structures there that go across the Pregol.

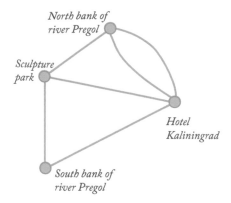

Figure 12.3. Graph of Kaliningrad bridges near the city center.

We might as well model this situation as a graph, shown in Figure 12.3. Notice that when there is more than one route between two locations, we draw more than one edge between the location-representing vertices.

Perhaps your experimentation in Section 12.2 has led you to conjecture that whether an Euler circuit or traversal exists is related to the degrees of the vertices of the graph. If so, you are correct. You would also be correct if you suspected that Euler circuits and traversals are named after someone. In this case, it's the famous mathematician Leonhard Euler. He lived in the 1700s and was the first to officially solve the Bridges of Königsberg problem.

Theorem 12.3.1. *(1) A connected graph G has an Euler circuit $\Longleftrightarrow$ every vertex of G has even degree.*

(2) A connected graph G has an Euler traversal but not an Euler circuit $\Longleftrightarrow$ G has exactly two vertices of odd degree.

Part of the proof of Theorem 12.3.1 is straightforward, and part is a bit technical. So, before diving in, we will give an example of how the proof works to produce an Euler circuit in an all-vertices-of-even-degree graph.

Example 12.3.2. We will build an Euler circuit on a graph that has all vertices of even degree. The process is shown in Figure 12.4. We start at some vertex (shown in grey) and walk arbitrarily around the graph until we run out of edges to walk along, and we number our edges as we go. In this case, we show the walk in grey and label our edges 1–9. Then we look around for a vertex we have already visited that has some unused edges. Here, we pick the vertex between edges 1 and 2. (Again, this vertex is shown in grey.) We walk arbitrarily along unused edges until we run out of edges again. This time, we walked along three new grey edges

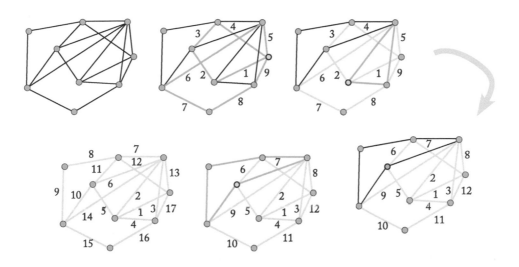

Figure 12.4. Building an Euler circuit in a graph with all even vertices.

(and our old edges are shown in light grey). Now, we incorporate this new walk into our enumeration by using our old numbering until we get to the start vertex of the new walk, number along the new walk, and then increase the labels on the rest of the old walk to match. In Figure 12.4, this is shown in the fourth diagram, and the numbers 2-3-4 are inserted after edge 1, while edges 2–9 become relabeled as 5–12. There are some edges we have not yet visited, so we find a second vertex we have already visited that has some unused edges (here, between edges 5 and 6), walk along the remaining unused edges, and then incorporate the newest walk into the old numbering.

Proof of Theorem 12.3.1: ($\Rightarrow_1$) We give a direct proof. Suppose G has an Euler circuit. Let us call the circuit C and call its start/endpoint v. Look at some generic vertex, which we will call w. Is the degree of w even or odd? As we travel along C, each time we encounter w, we go along an edge when we approach w and an edge when we leave w. Thus, each visit to w contributes 2 to its degree. (Even if we travel along a loop, we count 2 towards the degree of w.) Therefore, w has even degree in C, and because the edges of G and C are the same, w has even degree in G. We chose w arbitrarily, so *every* vertex of G has even degree.

($\Leftarrow_1$) Our proof combines constructing an algorithm with contradiction to show the algorithm works. Suppose that every vertex of G has even degree. Pick any vertex of G (let's call the one we picked v) and start meandering around the graph, marking edges as we visit them (so that we do not revisit an edge). Keep meandering until we run into a vertex that has no unvisited edges. Either we have just formed a circuit (though possibly not an Euler circuit) and landed at v, or we have not formed a circuit and instead landed at some other vertex z. Suppose we have landed at z. Then, each time we passed through z we used an even number of edges, and arriving at z (but not leaving z) we used a single edge. Therefore, z must have odd degree, which is a contradiction. So instead of landing at z, we must have instead landed at v. Landing at v means we have formed a circuit. If this is an Euler circuit, we are done. If not... this means that there are edges of G not in our circuit. ⋆ Choose a vertex a on our circuit that has unused edges—if there is not one on our circuit, then the graph must be disconnected—and repeat the procedure given at the start of this paragraph to construct a circuit of unused edges that begins and ends at a. This new circuit can be melded with our old circuit as follows. Start at v, go along Old Circuit to a, go along the entirety of New Circuit, and continue along Old Circuit from a back to v. Again, the melded circuit might be an Euler circuit (in which case we're done) or it might not be an Euler circuit (in which case we have more to do). If it's not an Euler circuit, go to ⋆ and repeat until no edges of G are unused. Because G is finite, we will eventually run out of edges. (That is, our algorithm terminates.)

$(\Rightarrow_2)$ We give a direct proof. Suppose G has an Euler traversal but not an Euler circuit. Thus, G has a trail that includes all edges of G. By the reasoning in $(\Rightarrow_1)$, we know that every vertex of G, except perhaps those beginning and ending the traversal, must have even degree. Now consider the start of the traversal, at vertex v. As we leave v, we mark one edge. Each time we return to v, we leave it again (as otherwise it would not be the start of a traversal that is not a circuit), and this uses two more edges. In total, v must have odd degree. The same reasoning applies to the ending vertex of the traversal, and it is the other vertex of odd degree, for a total of exactly two vertices of odd degree.

$(\Leftarrow_2)$ We construct a traversal as follows. G has exactly two vertices of odd degree; call them v_1 and v_2. Add a fake edge to G joining v_1 and v_2, so that G_{fake} has all vertices of even degree. Therefore, by the previous theorem G_{fake} has an Euler circuit. By its circuitous nature, the Euler circuit may be considered to start at any vertex with any incident edge (just go once 'round the circuit from there), so we may as well assume that it begins with v_1 and then travels along the fake edge. If we delete the fake edge from the circuit, we are left with a trail that visits all edges of G, begins with v_2, and ends with v_1. This is an Euler traversal. □

An actual, real-life use for Euler traversals in biology is presented in Bonus Section 12.9. If you would like to practice finding Euler circuits, go to http://digitalfirst.bfwpub.com/math_applet/euler_curcuit.html where there are four sample graphs for your circuit-finding pleasure.

Make sure you understand Theorem 12.3.1 by doing these problems.

1. Does the graph in Figure 12.3 have an Euler traversal? ... an Euler circuit?

2. Does K_5 have an Euler traversal? ... an Euler circuit?

3. Does K_6 have an Euler traversal? ... an Euler circuit?

4. Which cycle graphs have Euler traversals? ... Euler circuits?

5. **Challenge:** Draw three graphs with the same number of vertices, that differ from each other by at most two edges, and where one graph has an Euler circuit, one has only an Euler traversal, and one has no Euler traversal.

12.4 Dijkstra's Algorithm, with sides of Hamilton Circuits and the Traveling Salesperson Problem

A natural question that ought to arise in your head after studying Euler traversals is, "Does anyone study graph traversals that visit each vertex of a graph but perhaps not all of its edges?" Answer: Yes.

Definition 12.4.1. A *Hamilton traversal* visits every vertex of a graph exactly once. A *Hamilton circuit* is a Hamilton traversal whose starting and ending vertices are adjacent. (Every Hamilton circuit is a cycle.)

The mathematician after whom Hamilton traversals are named is William Hamilton, who lived in the 1800s. Because he was around a full century after Euler, we always study Hamilton traversals after studying Euler traversals. (Just kidding.) There are some theorems of the form *if G has (some property), then G has a Hamilton circuit*, but most are boring or impractical—the *(some property)* part is generally as difficult to verify as it is to just look at *G* and figure out whether it has a Hamilton circuit by hand. So we will present only one of these theorems here; it was proved by Dirac in 1952.

Theorem 12.4.2. *Let the simple graph G have $n \geq 3$ vertices. If the degree of every vertex is more than $\frac{n}{2}$, then G has a Hamilton circuit.*

We defer the proof to Bonus Section 12.11.

Notice that Theorem 12.4.2 requires that a graph have lots and lots of edges compared to the number of vertices; however, plenty of graphs with not so many edges still do have Hamilton circuits. The situation is somewhat depressing.

To make our outlook more bleak, let's consider graphs with weights on their edges. Now we not only want to know whether a graph *has* a Hamilton circuit, but we want to find the *shortest* Hamilton circuit. This is called the *Traveling Salesperson Problem*, or *TSP* for short. (The problem has nothing to do with teaspoons.) And to make our situation downright painful, guess what? It turns out that TSP has been proven to be as computationally difficult (from the standpoint of algorithm efficiency) as any known problem. (That's a friendly way of saying that it's NP-complete without explaining what NP-complete means, as that concept is beyond the scope of this book.) Worst of all, TSP has lots and lots of practical applications (public transportation routes, plans for moving heavy equipment among farms that share it, directing a factory machine to apply a bunch of rivets, …), so the world does need to deal with it. And thus, there is a website (http://www.math.uwaterloo.ca/tsp/) that is an authoritative source on practical advances related to TSP. Go there. It's cool.

Let's get over the horror of unsolved and computationally complex problems by instead addressing a real-world problem that not only has been solved but has a solution we can handle ourselves. We will spend the rest of this section cogitating on how to find the shortest path between two vertices on an edge-weighted graph. The algorithm we will use was found by Edsger Dijkstra, who not only has the letters *i-j-k* in a row in his name but is also Dutch. (His name is pronounced *d eye kstra* in English.) Powerfully, this algorithm finds the shortest distance from a given vertex to *every* other vertex in a graph. Here we go:

How to find the shortest path between two vertices using Dijkstra's algorithm:

1. Get ready by locating at least two colored pens, finding a graph that has weighted edges, and labeling the vertices with letters in one color.

2. Circle the start vertex s.

3. Look at the weights of the edges incident to s. For the smallest weight, go along one edge corresponding to that weight and tag the adjacent vertex with (w_e, s)—in a different color than used for labeling the vertices—where w_e is the weight of the edge connecting the vertex to s. This means that the new vertex is tagged with the shortest distance from s and the previous vertex in the shortest path (in this case, it's just s).

4. Consider all the tagged vertices as a collective. For each tagged vertex v, compute the shortest distance from s to each of its untagged neighbors. That is, add the weight of each incident edge to the number in v's tag to produce a list of distances for the vertex, and select the smallest for each vertex.

5. Determine the smallest distance d among all the lists for all the tagged vertices.

6. Each occurrence of d corresponds to a tagged vertex connected to an untagged vertex. Tag each of these untagged vertices with (d, v), where v is the label on the corresponding tagged vertex. (If an untagged vertex could have two tags (d, v_1) and (d, v_2), it does not matter which tag is selected.)

7. If there are untagged vertices, go to step 4.

The result is that every vertex in the graph is tagged with the length of the shortest path from s as well as the previous vertex in that shortest path. In order to determine the shortest path from s to any particular vertex, simply follow the tags backwards from that vertex to s.

Example 12.4.3. We will find the shortest distance from a circled start vertex s to a boxed end vertex e using Dijkstra's algorithm. Each sub-diagram of Figure 12.5 gives a list of possible tags for untagged neighbors of tagged vertices, and current tags are shown in teal.

Lest you think Dijkstra's algorithm does not solve a real-world problem, go to Google Maps and find the shortest route from Boston to Philadelphia. Then revise your conception.

Example 12.4.4. We will find the shortest distance from a start vertex s to every other vertex of a graph using Dijkstra's algorithm. Each step of the algorithm is shown in Figure 12.6, which uses the same markings as in Example 12.4.3.

For other examples of Dijkstra's algorithm in action, check out https://www-m9. ma.tum.de/graph-algorithms/spp-dijkstra/index_en.html, and http://students. ceid.upatras.gr/~papagel/project/kef5_7_1.htm. (These use directed graphs to account for one-way traffic.)

A proof that Dijkstra's algorithm works does not provide any more insight than an informal explanation, so we will not provide a proof here. Instead, consider that

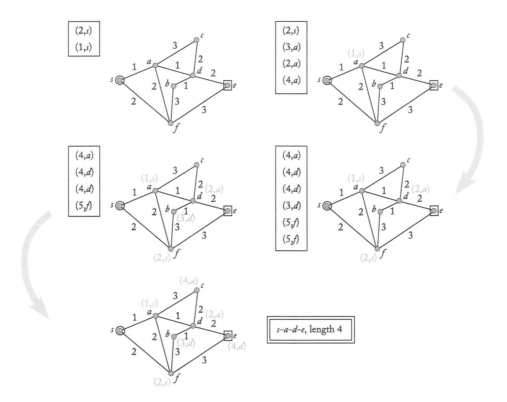

Figure 12.5. Dikjstra's algorithm finds the shortest path from s to e.

at each iteration, Dijkstra's algorithm examines all possible paths from s to nearby unmarked vertices. Among these possible paths, the algorithm marks the set of shortest paths. Marking the entire set of shortest paths eliminates the possibility that the order in which vertices are marked will make a difference in the total length of the paths. And, checking all possible paths means that the algorithm is exhaustive and so it cannot have missed a shorter way to reach a given vertex. (If this paragraph did not make sense to you, review Example 12.4.4 and then reread the paragraph.)

While it is tempting to try it, we cannot use Dijkstra's algorithm to solve TSP. The temptation is to pick two vertices and find the shortest route between them, then delete that route from the graph and find the shortest route remaining, and then join the two routes into a Hamilton circuit. However, the two shortest routes might not cover all the vertices. Worse yet, deletion of the shortest route might disconnect the graph so that there is no route available to complete the circuit.

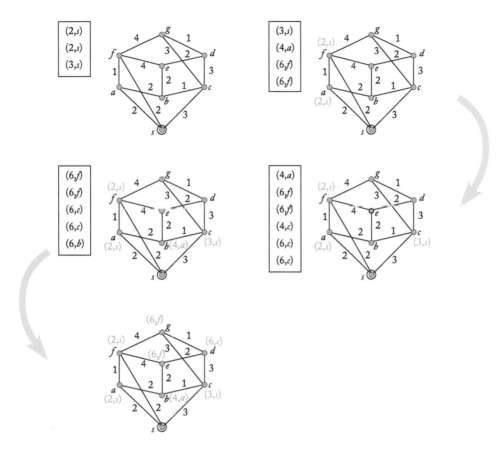

Figure 12.6. Dijkstra's algorithm finds the shortest path from s to every other vertex.

Check Yourself

These exercises will help you understand how to execute Dijkstra's algorithm, so please do them.

1. In the third subdiagram of Figure 12.5, why are there two labels $(4, d)$?

2. In the first list of distances computed by Dijkstra's algorithm, how many tags are present?

3. In the third subdiagram of Figure 12.5, there are two labels $(5, f)$. Neither is placed on the graph in the following step... so why is only one of them left in the list in the fourth subdiagram?

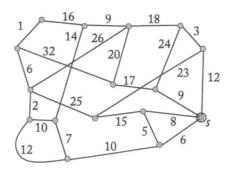

Figure 12.7. I desire Dijkstra.

12.5   Try This!—Do This!—Try This!

It's time for you to try using Dijkstra's algorithm yourself.

1. Use Dijkstra's algorithm to find the shortest path from s to every other vertex in the graph shown in Figure 12.7.

2. Perhaps you recall from Example 1.3.2 that the Restaurant Quatre-Étoile offers prix fixe meals and that one of the available dishes is Foie Gras Falafel with Fig Fondue. The local geese have heard about this abomination (it is made from marinated goose liver!!), and in solidarity a coalition of ducks has joined them to protest the restaurant! However, they are not very fast walkers (they can't fly while holding signs). Can you help them by finding the shortest route from the pond to the restaurant? A map of the area is shown in Figure 12.8.

   Now you should come together as a large group. Select four people to go to the front of the room (bonus if there are four whose first names begin with A, B, C, and D) and two to write on the board.

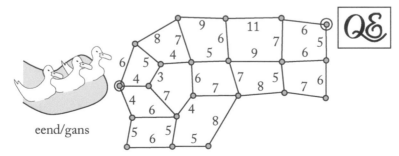

Figure 12.8. A map of the area near the duck/goose pond and the Restaurant Quatre-Étoile.

A, B, C, and D should stand in a row at the front, left to right. One of the board writers should record this ordering (ABCD). Now, two of the A, B, C, D people who are next to each other should switch, and a board writer should record the new ordering. Your communal goal is to see if you can reach every possible ordering, with no repeats, by just switching two next-to-each-other people… again and again. (Such a switch is called an *adjacent transposition*.) Go for it—and record the orders as you go!

After ten minutes or so of this fun, whether or not you have achieved your goal, break into groups to work on the following problems.

3. How many possible orderings of A, B, C, D are there?

4. What is the connection of this situation to graph theory? That is, there is a graph here, so…

   (a) … what are the vertices of this graph? (How many will there be?)

   (b) … which vertices are adjacent to form edges of this graph?

   (c) … what are the degrees of the vertices?

   (d) … what structure do you seek in this graph?

5. Construct the graph you have just described.

6. Using this graph, try to achieve your list-of-possible-orderings goal.

12.6 Two More Examples

Example 12.6.1 (of building an Euler circuit). Figure 12.9 shows the process of building an Euler circuit from a graph. (A) First, we pick a start vertex and travel

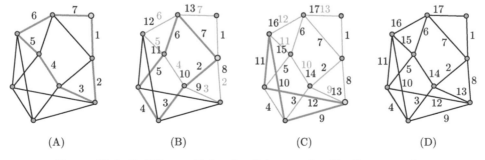

Figure 12.9. Building an Euler circuit in a graph with all even vertices.

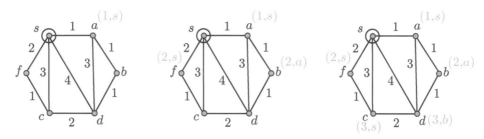

Figure 12.10. Executing Dijkstra's algorithm on a small graph.

around the graph until we end where we started, labeling as we go. (B) Seeing no unused edges at our start vertex, we pick the next available vertex with unused edges and travel from there; we then incorporate this new path in our numbering. (C) We repeat the process used at (B) and exhaust the edges of the graph. (D) Finally, we present the original graph with complete Euler circuit.

Example 12.6.2 (of executing Dijkstra's algorithm). Figure 12.10 shows the process of finding the distance from one vertex to all others of a small graph. The tags at each step are shown in teal.

12.7 Where to Go from Here

Bonus Section 12.9 gives a practical application of Euler traversals to biology. Unsurprisingly, Euler traversals are used throughout computer science, and variants on Euler traversals are studied in discrete mathematics research. While Euler traversals are useful, the study of Euler traversals themselves is essentially complete (and you have completed it), so they are more of a tool than a direction of research.

On the other hand, Hamilton traversals and TSP are active areas of research. Research articles on the existence of Hamilton traversals for particular classes of graphs are frequently published in discrete mathematics journals. For lots more on TSP, see http://www.math.uwaterloo.ca/tsp/. Some decent approximations to TSP solutions are shown at http://demonstrations.wolfram.com/AlgorithmsFor FindingHamiltonCircuitsInCompleteGraphs/. The website https://www-m9.ma. tum.de/games/tsp-game/index_en.html will let you try to make your own solutions for TSP examples, and describes several algorithms for generating solutions.

Dijkstra's algorithm is fairly efficient, so it is often used in practice for applications such as IP routing... but it is not fast enough to be used in Google Maps. (See http://googleblog.blogspot.com/2007/11/road-to-better-path-finding.html for verification.)

The graph you constructed in Section 12.5 is the vertices and edges of the *permutahedron*. It is one of a family of permutatopes; they are studied in abstract algebra classes and in convex geometry classes. And they are cool.

Credit where credit is due: The idea for the class activity involving the permutahedron was contributed by Karl Schaffer. Section 12.9 was adapted from [9], and Section 12.10 was informed by [2]. The Jim addressed on page 386 is James McCoy of the Starship Enterprise. Bonus Check-Yourself Problem 7 was suggested by Tom Hull. Problems 26 and 27 were donated by Karl Schaffer. Problem 31 was inspired by an EPS-ungroup bug combined with the insight of Ollie Levy into duck social networking; Problem 35 was inspired by Sam Oshins's character who was inspired in turn by this book (see page 148).

12.8   Chapter 12 Definitions

Euler traversal: A trail that visits every edge of a graph exactly once and may visit some vertices of the graph more than once. It is named after Leonhard Euler (1707–1783), who is regarded by many as one of the greatest mathematicians of all time.

Euler trail: An Euler traversal that is not a circuit.

Euler circuit: An Euler traversal whose starting and ending vertices are the same.

Hamilton traversal: A trail that visits every vertex of a graph exactly once. It is named after William Rowan Hamilton (1805–1865), who may be best known for carving an equation into a bridge.

Hamilton path: A Hamilton traversal that is not a circuit.

Hamilton circuit: A Hamilton traversal whose starting and ending vertices are adjacent.

Hamilton cycle: A Hamilton circuit.

Traveling Salesperson Problem: The problem of finding the *shortest* Hamilton circuit in a graph.

TSP: Abbreviation for Traveling Salesperson Problem.

adjacent transposition: A permutation that switches two elements that are next to each other in an arrangement.

12.9   Bonus: Digraphs, Euler Traversals, and RNA Chains

Proteins are encoded in living cells with ribonucleic acid (RNA) chains. There are four nucleobases used in this encoding, namely, adenine (A), cytosine (C), guanine (G), and uracil (U). Thus, an RNA chain may be denoted by a string of letters from the alphabet A, C, G, U. There are some rules for the use of this alphabet; the nucleobases come in triples called *codons*, and there is one codon (AUG) that always starts a chain and three (UAA, UAG, UGA) that can end a chain. (All chains given in this section were constructed from a list of actual codons.) One hypothetical RNA chain would be AUGCAGCCUAUGGGAAAAUAG. Almost all RNA chains are unpronounceable.

In order to sequence an RNA chain, biologists hit it with one or more enzymes to break it into shorter bits. (They can't just read the letters off of the chain. Please don't ask why. *I'm a mathematician, not a biologist, Jim!*) One of these enzymes breaks a chain after every G, and another one breaks a chain after every C and after every U. Our hypothetical RNA chain would be broken into AUG, CAG, CCUAUG, G, G, AAAAUAG by the G-enzyme and into AU, GC, AGC, C, U, AU, GGGAAAAU, AG by the CU-enzyme. But of course, in reality these chain bits would not be in a nice order; they would be all jumbled together. The mathematical task is to deduce the RNA chain sequence from the enzyme-broken bits. Naively, there are $6! = 720$ different chains that could correspond to our G-broken example and $8! = 40,320$ different chains that could correspond to our CU-broken example. We can narrow this down quite a bit.

First, notice that we can determine the end of the chain: In our example, one of the CU-bits ends with G. That has to be the end bit of the chain because the CU-enzyme didn't break it. More generally, there will be at most one CU-bit ending in A or G and at most one G-bit ending in A, C, or U. At worst, there will be one of each type of bit ending in A, but because both bits have to end the chain, one will be a sub-bit of the other and all is well.

Now for the graph theory. We'll take each G-bit and break it up with the CU-enzyme; so, for example, CCUAUG becomes C, C, U, AU, G. Similarly, we'll take each CU-bit and break it up with the G-enzyme; so, for example, AGC becomes AG, C. We will make a directed edge corresponding to each G-bit and each CU-bit by only paying attention to the first and last resulting bits of the double-broken sequences, so the examples from the previous two sentences become $C \xrightarrow{\text{CCUAUG}} G$ and $AG \xrightarrow{\text{AGC}} C$. Then, we glue this all together into a directed graph. For the hypothetical RNA chain we've been using here, we obtain the digraph shown in Figure 12.11. (Notice that we do not include the fragments that are only single-broken and not double-broken.) Then we try to find an Euler traversal of the digraph that ends with the known end-bit. Each Euler traversal can be "read" by listing the edge labels in the order we traverse them, eliminating the first/last letters in common. If we are lucky, there is only one Euler traversal. For the digraph of Figure 12.11, there are multiple Euler traversals—we must start our reconstruction with AUG, but could then continue as AUGGGAAAAUAGCCUAUGCAG, as AUGCAGC-CUAUGGGAAAAUAG, or as AUGCCUAUGGGAAAAUAGCAG.

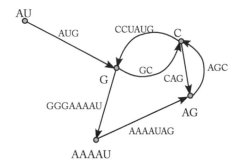

Figure 12.11. The digraph corresponding to our hypothetical RNA chain.

Short activity:

1. Chop up these chains with the G-enzyme.

    (a) AUGGCUCACUAUGGCAUACUCUAA.

    (b) AUGUUGUUCCAAAGUUGA.

    (c) AUGCGACCCGUUACCAAGUAG.

2. Start over and chop the three chains from the previous problem, but use the CU-enzyme.

3. Identify the end-bit for each of the following chopped chains.

    (a) (G-enzyme) CACG AUG AG A UACCG UAUUG AACACCG.
        (CU-enzyme) C C AU U AC AC GAAC GA GU GAGU GC AC AU.

    (b) (G-enzyme) UUCACAAAAG AUUAA AUG.
        (CU-enzyme) U U AA C AC AU GU AAAAGAU.

    (c) (G-enzyme) UG CAG CCAAG AUG UUAG UAUG.
        (CU-enzyme) C U AU AG AGU AU GC GC GU AAGU.

    (d) (G-enzyme) G G UG AUG AUG UAG UAUUUCUAACG.
        (CU-enzyme) U U AG AU AU CU GU GU GAU GGGU AAC.

4. Attempt to reconstruct the four chains for which fragments are given in the previous problem. Do any have a unique reconstruction?

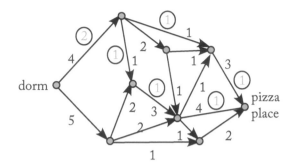

Figure 12.12. Driving to dinner with ducks.

12.10   Bonus 2: Network Flows

Two major branches of combinatorial optimization are linear/integer programming, which we introduce in Bonus Section 7.9 (and expand on in Bonus Section 10.11), and network flows, which we introduce right here. A network flow is a type of labeling of an edge-weighted directed graph. One real-life application of network flows is using microclimate information for a mountain to figure out where best to relocate endangered plant species so they are likeliest to survive expected climate change. (Yes. Really. Aaron Archer and Steven Phillips, who at the time worked for AT&T, did that research with some ecologists. It is as cool as it sounds. The paper is "Optimizing Dispersal Corridors for the Cape Proteaceae Using Network Flow," by Steven J. Phillips, Paul Williams, Guy Midgley, and Aaron Archer, in *Ecological Applications*, Vol. 18, No. 5 (2008), pp. 1200–1211. The article is available online at http://www.klamathconservation. org/docs/phillipsetal2008.pdf.)

We will begin with an example.

Example 12.10.1. It is a little-known fact that ducks love pizza. (Note: not actually true.) On Take-Your-Duck-to-Dinner night, several students who live in the same dorm decide to take their ducks to the pizza place. There are too many students and ducks for a single car, so they take two cars. Figure 12.12 shows local streets and alleyways represented as edges on a graph (the vertices are intersections), with the number of pizza-place-wards lanes given as edge weights. The circled numbers represent the number of cars traveling simultaneously on the streets they label. As you can see from the figure, the two cars take different routes to the pizza place. We think of the traffic as flowing from the dorm to the pizza place.

We can now define a network flow.

Definition 12.10.2. A *network flow* is an edge-weighted directed graph such that

▪ there is a vertex with no edges pointing towards it, known as a *source*, and

▪ there is a vertex with no edges pointing out of it, known as a *sink*,

along with a labeling of the edges such that

▪ for every vertex other than the source and sink, the sum of the labels on the edges pointing in equals the sum of the labels on the edges pointing out, and

▪ the label on an edge does not exceed the capacity (weight) of the edge.

In Example 12.10.1, the source is the dormitory and the sink is the pizza place. The circled numbers label the flow, and the unlabeled edges may be understood to be labeled as having 0 flow. We usually think of the labeling in a network flow as the flow itself; it indicates the amount of traffic "flowing" across each edge of the graph. This also means we also usually think of the edge-weighted directed graph as the network... even though "network" can also just mean "graph."

Definition 12.10.3. We call the sum of the flow labels pointing to a vertex the *flow in* and the sum of the flow labels pointing out of a vertex the *flow out*.

A common network flow problem is to determine the maximum amount of traffic that can be sent across the network at any one time.

Example 12.10.4. Two ridiculous networks are shown in Figure 12.13. In the left-hand network, it appears we could send a lot of traffic out of the sink (more than

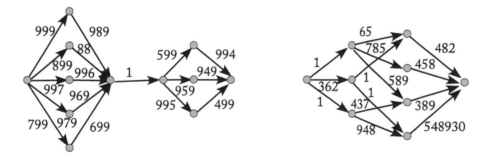

Figure 12.13. Ridiculous networks.

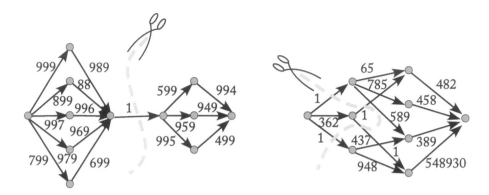

Figure 12.14. Cutting ridiculous networks.

3,600 cars or cabs or cats or cans or caps!), but all traffic must flow through the capacity-1 edge in the middle of the graph, so the maximum flow across this network at any one time is 1. The right-hand network seems better at first, but we can only send four things across at once! We can only send one thing across each of the two capacity-1 edges emanating from the source. The capacity-362 edge is all but useless because only two things can make it out of its end vertex at any one time. And that's that.

We can find some upper bounds on the maximum flow across a network: certainly we cannot send more stuff than the flow out of the source, nor can we send more stuff than the flow in to the sink. Figure 12.14 points out that if we can separate the source from the sink by cutting through some edges, then the total capacity across the cut edges is an upper bound for the maximum flow. After all, every bit of stuff sent from source to sink would have to go across *one* of those edges.

Notice that in Example 12.10.4, we found the maximum flows of the ridiculous networks, and these maxima happen to match the small-capacity cuts we exhibit in Figure 12.14. This is no coincidence, but instead the famous and useful max-flow/min-cut theorem! We will not prove it here because while the proof is not difficult, it is a bit long and a bit technical (but also more than a bit interesting). An elementary exposition can be found in [2], and *Integer and Combinatorial Optimization* by George Nemhauser and Laurence Wolsey has a treasure trove of clearly written information on network flows at the advanced level.

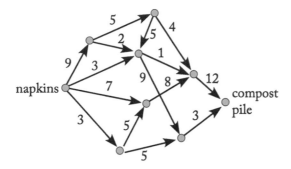

Figure 12.15. Lunch carts across downtown Račja.

Some problems:

1. Find a network flow in Example 12.10.1 in which both cars travel along the same route.

2. From the flow out of the source and the flow into the sink in Example 12.10.1, we know that the maximum flow is at most 9. Is there a flow that achieves this upper bound?

3. Consider the network shown in Figure 12.15. It reflects the paths lunch carts can take through the downtown Račja (Slovenia) area during the lunch rush, from a napkin supplier to a compost pile.

   (a) If you send nine lunch carts along the capacity-9 edge from the source, can they simultaneously reach the sink?

   (b) Find a flow that sends a dozen lunch carts from source to sink, all at the same time.

   (c) Starting with the flow you just devised, can you identify any paths with unused capacity along which additional lunch carts can be sent? By how much can you increase the total flow in this fashion?

4. You'd like to download the six most recent episodes of the television show *Exile on Eggs Isle*, so the hosting server must perform a check to see whether there is enough network bandwidth to send all the bits of

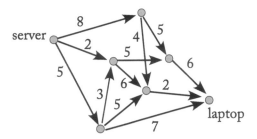

Figure 12.16. Is there enough bandwidth?

file at the same time. A sample network is shown in Figure 12.16, with vertices representing intermediate servers and edge capacities having units of Gb (there are 8 Gb in 1 GB).

(a) Using the source or sink, find an upper bound on the throughput from the server to your laptop.

(b) Find a smaller upper bound on the throughput by finding a set of edges that separate the source from the sink. You may need two pairs of scissors.

(c) Find a flow that achieves your new upper bound.

(d) If the file is 1,750 MB, will the server start the download or will it return a Bandwidth Exceeded error?

12.11   Bonus 3: Two Hamiltonian Theorems

Earlier, in Section 12.4, we stated Theorem 12.4.2: *Let the simple graph G have* $n \geq 3$ *vertices. If the degree of every vertex is more than* $\frac{n}{2}$, *then G has a Hamilton circuit.* We promised a proof of this theorem; here it is.

Proof: Let us proceed with a proof by contradiction. Suppose that G has no Hamilton circuit, but each vertex of G has $deg(v) > \frac{n}{2}$. If we added as many edges as possible to G, we would have K_n, and K_n certainly has a Hamilton circuit. So consider a worst-case scenario; $G \subseteq G'$, where G' has the same vertices as G but has more edges—so many edges, in fact, that having just one more would create a Hamilton circuit. That is, for some vertices v_a, v_b, G'-with-$\{v_a, v_b\}$ has a Hamilton circuit v_1-v_2-...-v_a-v_b-...-v_n-v_1. Therefore, G' has a Hamilton path $P = v_b$-...-v_n-v_1-v_2-...-v_a. By definition, this includes all the vertices. And because $n \geq 3$, v_a

has degree more than $\frac{3}{2}$ and therefore degree at least 2. So v_a is adjacent to some other vertex of G', and perhaps lots of other vertices of G'; we shall pick one and call it v_i.

Now for a strange-sounding claim. No matter which v_i we pick, v_{i+1} cannot be adjacent to v_b. If v_{i+1} *is* adjacent to v_b, we can make the Hamilton circuit v_b-(follow P)-v_i-v_a-(follow P in reverse)-v_{i+1}-v_b. But, this contradicts the statement that G' has no Hamilton circuit.

How is this useful? It means that, in the ordering of the Hamilton path, every vertex adjacent to v_a is followed by one that is *not* adjacent to v_b. Also, there are at least $\frac{n}{2}$ vertices adjacent to v_a appearing in the Hamilton path (because every vertex in G' is on the path and v_a has degree at least $\frac{n}{2}$). This means that there are at least $\frac{n}{2}$ vertices *not* adjacent to v_b appearing in the Hamilton path—and also v_b is not adjacent to v_b, so at most there are $n - \frac{n}{2} - 1$ vertices adjacent to v_b. Wait! That's a contradiction to the assumption that every vertex of G' has degree at least $\frac{n}{2}$! Therefore G' must actually have a Hamilton circuit after all.

That means we could take one of the non-G edges out of G' and have a new graph G'' that is in the worst-case scenario. If we run through the proof again, we get another contradiction. Repeating this process means we eventually show that G itself has a Hamilton circuit. □

This theorem and its proof exemplify one of the many excellent ways in which mathematics advances: Dirac proved Theorem 12.4.2 in 1952. In 1960, Ore was able to generalize by reading Dirac's proof carefully. Let's try to see what he saw.

1. Where exactly in the proof is the condition *the degree of every vertex is more than $\frac{n}{2}$* used?

2. What condition on v_a, v_b is needed for the numerical contradiction in the proof to work out?

3. Combine parts of your answers to the previous two questions to make a substitute for the condition *the degree of every vertex is more than $\frac{n}{2}$*, as follows. Your condition should include the phrase *for every two* _____ *vertices* and something about their degrees.

4. Use this criterion to show that v_a has degree at least 2.

5. Your answer to Problem 3 probably included a "$> n$" somewhere. Can you make it work with a "$\geq n$" instead? (If so, you have derived Ore's theorem and proof!)

12.12   Bonus Check-Yourself Problems

Solutions to these problems appear starting on page 620. Those solutions that model a
formal write-up (such as one might hand in for homework) are to Problems 3 and 7.

1. An *n-prism* graph is constructed by
putting one (slightly smaller) *n*-cycle C_n
inside another, and adding edges to join
the vertices of one C_n to the other radially. (We do need $n \geq 3$.) See Figure
12.17 for an example. Do any *n*-prism
graphs have Euler circuits? What about
Hamilton circuits?

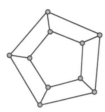

Figure 12.17.   I am a proud 5-prism
graph.

2. List all possible orderings of ABC (how
many are there?). Associate each of
these orderings to a vertex of a graph.
Add an edge when two orderings differ
only by an adjacent transposition.

(a) What is the degree sequence of this
graph?

(b) Does it have an Euler circuit or trail?

(c) Does it have a Hamilton circuit or
trail?

(d) Is it planar?

(e) What are the answers to the previous
questions if we also consider the first
and last letters to be adjacent?

3. Look at the graphs in Figure 10.23 on
page 340. Does either have a Hamilton
circuit? … Hamilton traversal? … Euler circuit? … Euler traversal?

4. Again examine Figure 10.23 on page
340. For each graph, compute the shortest distance from the lower-right vertex
to all other vertices. (Tip: Dijkstra is a
good choice here.)

5. Do any of the graphs in Figure 12.18
have Hamilton circuits? What about
Hamilton traversals?

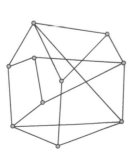
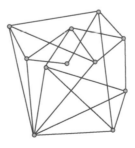

Figure 12.18. Three graphs. Yup.

6. Do any of the graphs in Figure 12.18 have Euler circuits? What about Euler traversals?

7. For which values of m, n does $K_{m,n}$ have a Hamilton circuit?

8. The towns Gesund and Reichtum are near each other in a tourism district. In each town, all but two of the intersections are four-way stops. In Gesund, there is a five-way stop and a "T" intersection (a three-way stop), and in Reichtum there are two five-way stops. Currently, there is no direct road between Gesund and Reichtum. The tourism bureau wants to build a road so that they can create and advertise a Tour of the Towns, which will take tourists down every road of Gesund and of Reichtum without repetition. What advice can you give the tourism bureau?

9. In the metropolis of Altana, the Traffic Council has decreed that cars in the flying lanes must pay twice the tolls of ground-based cars (because of the additional fuel needed for flying police). What is the cheapest way to get from point a to point b? A map showing skyways in grey and ground-roads in black is shown in Figure 12.19—those dots are toll stations where you pay for the segment you've just traveled.

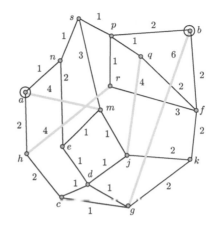

Figure 12.19. A map of Altana toll plazas.

10. Can you take one walk and cover every road in the map of Snakeland given in Figure 12.20 exactly once?

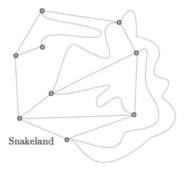

Figure 12.20. A map of Snakeland.

12.13 Problems with Traversing

1. Can you trace over the entirety of Figure 12.21 without lifting your writing instrument from the paper (or tracing any line twice)?

Figure 12.21. A classic envelope exercise.

Figure 12.22. Map of the original locations of Königsberg bridges.

2. Given a graph with an Euler traversal but not circuit, how can you produce a graph with an Euler circuit?

3. Consider the complete graph K_n and any two vertices a and b in it.

   (a) Why are there paths of every length from 1 to $n - 1$ joining a and b?

   (b) Find a graph other than K_5 that has the property that there exist two vertices that have paths of lengths $1, 2, 3, 4$ joining them.

   (c) While you're at it, find one for every $n \geq 4$.

4. Figure 12.22 shows the original placement of the Bridges of Königsberg.

   Could someone walk over each of the seven bridges and return to the starting point without crossing any bridge twice? Why or why not? (Notice that legend *has* to be wrong—no one is going to walk all that distance.)

5. Do any trees have Euler traversals?

6. Tragedy has occurred at the Lovely Estate (shown in Figure 12.23), and "… James Bomb, the internationally known detective, former notary public, current

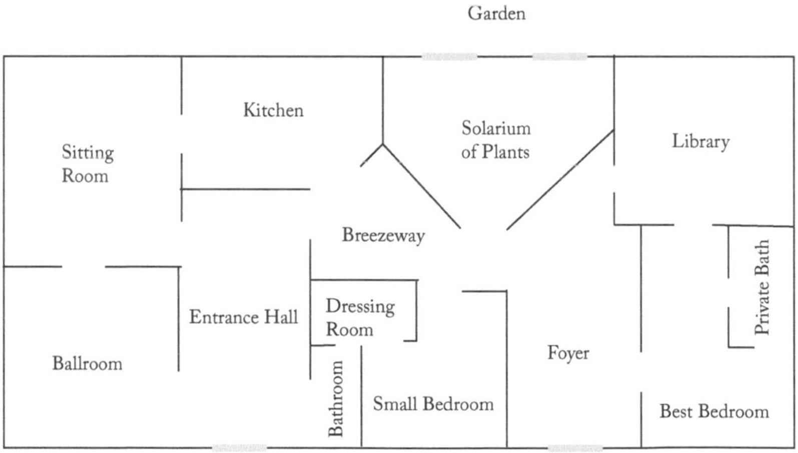

Garden

Kitchen

Sitting Room

Solarium of Plants

Library

Breezeway

Entrance Hall

Dressing Room

Foyer

Private Bath

Ballroom

Bathroom

Small Bedroom

Best Bedroom

Figure 12.23. A floor plan of the Lovely Estate.

assistant manager of Chicken Delight, and part-time graph theorist, has been called in to investigate" [6]. A person dressed in red (or was hir dress soaked with blood?) was found dead on the floor of the small bedroom with a bloody knife and a lead pipe nearby. James Bomb has secured the perimeter and brought all persons on the grounds into the sitting room for questioning. There are two individuals found leaving the house whom the elder Lovelys do not recognize, so Bomb questions them first.

The first person says, "I am a door inspector—it's what I do. We have to make sure that the hinges are tight so the doors won't fall on people, the doorknobs are secure so they can't come off in anyone's hand, and the locks work. I've just been through every door in the house exactly once (and had to repair two locks and tighten three doorknob screws) and now everything is in working order."

The second person says, "I am from the cleaning service. I visited each room exactly once to clean it—after all, I would not want to step on any clean floors, nor would I want to get additional dirt on my shoes in going from room to room."

James Bomb clears hir throat. What is ze about to announce? Is it possible that either the first or the second person is telling the truth?

7. Are there any wheel graphs that have Euler traversals?

8. Look again at the graph shown in Figure 11.20 on page 366. Does it have an Euler traversal? Does it have an Euler circuit? How about a Hamilton circuit or traversal?

9. In Section 12.5, you created a graph that had vertices labeled with orderings of ABCD, edges labeled with adjacent-letter switches, and a Hamilton circuit. Make a planar drawing of this graph or show that there is no planar drawing. You may find GeoGebra useful for experimentation.

10. IGS, an international group of scientists, proposes that a network of canals on Mars be dredged in preparation for irrigation and terraforming. Figure 12.24 shows a map of the network in

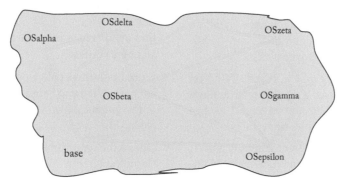

Figure 12.24. Map of proposed dredging of Mars canals.

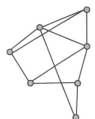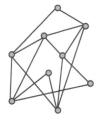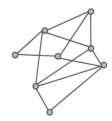

Figure 12.25. We three graphs of Chapter 12 are.

question. You have been asked to consult with IGS on some practical matters.

(a) It is very expensive to move dredging equipment over land. Is it possible to plan the dredging so the equipment can be delivered from Earth to the base and be near or at the base at the end of dredging?

(b) Mars is windy and so there will need to be maintenance of the dredged canals until terraforming is well underway. Part of the IGS proposal suggests that there be observation stations built at every canal intersection; when terraforming is complete, these can be converted into water-flow-control stations. Is there a way to visit all of the observation stations in one trip?

11. Look at the graphs in Figure 10.25 on page 341. Does either have a Hamilton circuit? ... Hamilton traversal? ... Euler circuit? ... Euler traversal?

12. Look at the graphs in Figure TIII.2 on page 436. Does either have a Hamilton circuit? ... Hamilton traversal? ... Euler circuit? ... Euler traversal?

13. Consider two different graphs G and H, each of which has an Euler circuit. Construct the graph F by adding an edge

connecting some vertex g of G to some vertex h of H. Is there anything notable about F?

14. Back at the University of Universe City (see Example 10.4.2), you have been asked to network together the computer just inside the door of the Computer Science building (on the right in Figure 10.4 on page 317) with the computer in the upper-right-hand room. What is the smallest amount of cable you can use?

15. Do any of the graphs in Figure 12.25 have Hamilton circuits? What about Hamilton traversals?

16. Do any of the graphs in Figure 12.25 have Euler circuits? What about Euler traversals?

17. Run Dijkstra's algorithm on the graph shown in Figure 10.24 on page 341 to find the distance from the top vertex to all other vertices.

18. Find the shortest path from the bottom-left vertex to the top-right vertex in each of the graphs shown in Figure 10.26 on page 341.

19. Does either graph in Figure 12.26 have a Hamilton circuit? What about a Hamilton traversal?

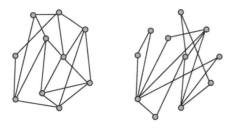

Figure 12.26. Jazz hands!

20. Does either graph in Figure 12.26 have an Euler circuit? What about an Euler traversal?

21. Find the shortest distance from the bottom vertex to the top vertex of the graph shown in Figure TIII.3 on page 436.

22. Suppose G has exactly four vertices of odd degree. What Euler-ish and traversal-ish property does G have? Prove your conjecture by constructing a graph G' with an Euler traversal or circuit and reasoning about the relationship between G' and G.

23. You know how there are fire hydrants on most blocks? They need to be connected to a water-pumping station in order to get water, and the pipes leading to them have to be laid along streets so that they can be serviced easily (and so they stay on public land). Laying pipe is expensive so the city would like to lay the least length of pipe possible. Figure 12.27 shows a city neighborhood with fire hydrants and a local water-pumping station. Each block is 5 units wide and 4 units tall.

(a) Make a graph corresponding to this situation.

(b) Put appropriate weights on the edges of your graph.

(c) Find the minimum length of pipe needed to supply the fire hydrants with water.

24. Hamilton (yes, the same Hamilton after whom Hamilton circuits were named) marketed a puzzle he called *The Icosian Game*, back in the 1850s. Figure 12.28 shows the playing board, which had holes where vertices are shown. There was also a bin with 20 labeled pegs. The challenge given was, Can you place each peg into a hole so that following

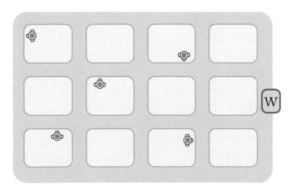

Figure 12.27. A grid of streets with fire hydrants and a water-pumping station.

the pegs in order makes a trip around the entire board, back to where you started? (Well, *can* you?)

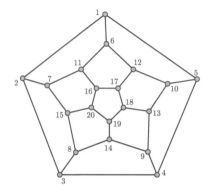

Figure 12.28. The playing board for *The Icosian Game.*

25. **Challenge:** Another game was marketed with the board for *The Icosian Game* as in Figure 12.28 (but laid out over the head of a mushroom-like object; see http://puzzlemuseum.com/month/picm02/200207icosian.htm). It was called *The Traveler's Dodecahedron: A Voyage Round the World.* In this case, two players start at some peg of the graph and walk to make a path of length four pegs. The goal is to complete a Hamilton circuit starting with this walk. Can it always be done?

26. Draw a continuous closed loop on a sheet of paper, such that every intersection point has an even number of arcs leaving or entering the point. Prove that the regions created may be colored using two colors such that no two regions that share an arc are the same color.

27. Now draw *n* closed loops, possibly overlapping, as in Problem 26. Can the resulting configuration of regions still

be 2-colored such that no two regions that share an arc are the same color? Prove this or find a counterexample.

28. Find an Euler circuit in the graph shown in Figure 12.29.

Figure 12.29. YAG: Yet Another Graph.

29. A *Möbius ladder* graph is constructed by joining the top of an *n*-rung ladder to the bottom after performing a half twist. (Of course, $n \geq 2$.) See Figure 12.30 for an example. Do any Möbius ladder graphs have Euler circuits? What about Hamilton circuits?

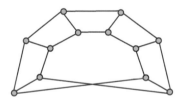

Figure 12.30. I am a twisted 6-ladder.

30. Is there a Hamilton circuit in either graph shown in Figure 10.22 on page 339? What about a Hamilton traversal?

31. Ghost Duck (see Figure 12.31) has connected with Ether Duck on Mergansr and wants to pick hir up on the way to their upcoming date. What is Ghost Duck's shortest path through the underworld?

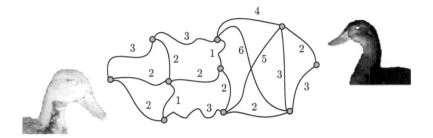

Figure 12.31. A map of the underworld near the residences of Ghost Duck and Ether Duck.

32. Does any graph in Figure 3.23 on page 87 have a Hamilton circuit?

33. Look at the graphs in Figure 3.21 on page 87. Does either have a Hamilton circuit? … Hamilton traversal? … Euler circuit? … Euler traversal?

34. Look at the graphs in Figure 3.22 on page 87. Does either have a Hamilton circuit? … Hamilton traversal? … Euler circuit? … Euler traversal?

35. The ever-intrepid Pvaanzba Ohaf wants to dine at the infamous Flayed Finger restaurant. (We will not discuss what Mx. Ohaf plans to eat at this establish-ment.) Figure 12.32 shows a map of the area near Pvaanzba's hotel. What is the shortest distance Pvaanzaba Ohaf can walk to reach the Flayed Finger?

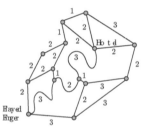

Figure 12.32. A graph with teal vertices.

12.14   Instructor Notes

As preparation for the first class on traversals, have students read Section 12.1. Start them out at the beginning of class by grouping them (a binary sort on mother's first name is a fine idea for deciding who works together) and having them start on Section 12.2. These problems will take most of the class period; students will quickly make and prove correct conjectures about when graphs do *not* have Euler traversals, and they will quickly make conjectures about when graphs *do* have Euler traversals, but they are unlikely to come up with proofs.

For the second class day, have students read Sections 12.3 and 12.4 and attempt the Check Yourself problems. Begin class by running Dijkstra's algorithm on a small example graph. Then ask for questions over the reading, and start students working on Section 12.5

in groups. With at least one-half hour left in the class, guide students through the Do This! part of Section 12.5. That is, have four students come to the front of the room and try to walk through all 24 orderings by performing adjacent transpositions only. When this no longer seems like a highly productive group activity, have them break into groups again to work on the remainder of the problems, and assign them to try to finish making the permutahedron graph as homework for the third class meeting.

Depending on the composition and progress of your particular class, you may wish to use the third class day in different ways. If they are having trouble with Dijkstra's algorithm, then do another example or two and create some graphs on which they can practice using Dijkstra's algorithm. If they are on top of this material, either use the opportunity for further work on previous material or ask students to prepare by reading a Bonus section (from this or another chapter) and spend class time with students working in groups on the problems therein. You might also have them experiment more to find the Hamilton circuit in the permutahedron.

Some students or classes may find GeoGebra useful in doing experiments or in typesetting their homework because it has the capability to drag vertices around, highlight paths, label vertices, and add textual notes.

Chapter 13

Graph Coloring

13.1 Introduction and Summary

Using crayons or markers, we can color the vertices of a graph or the edges of a graph (or both, but that's not done very often). Most of the time, we only care about proper colorings, which for vertices means that no two vertices connected by an edge can be the same color. (And for edges, it means that no two edges that touch a vertex can be the same color. Some of the author's research is on edge colorings.) Colorings are not just fun but also useful—certain types of information-scheduling problems (e.g., wireless communication) can be solved using graph colorings, as can register allocation issues (in computer science). (Truthfully, graph colorings are only practical when there are few types of constraints on scheduling or allocation problems and when there are no preferences that should be respected. In reality, linear and/or integer programming is used for highly constrained applications. See Bonus Section 7.9 for information on linear and integer programming!)

13.2 Try This! Coloring Vertices and Edges

A *coloring* of the vertices (or edges) of a graph G is technically a function $c : V(G) \to C$ (or $c : E(G) \to C$), where C is a set of colors. But we treat it less formally here, and simply think of a coloring as an assignment of colors to a graph's vertices (or edges).

13.2.1 Vertex Coloring

Definition 13.2.1. A graph G is *properly vertex colored* if each vertex of G is assigned a color such that no two adjacent vertices have the same color.

Figure 13.1 shows three graphs that all look vaguely like K_5 but that are not K_5. One of them may be familiar to you.

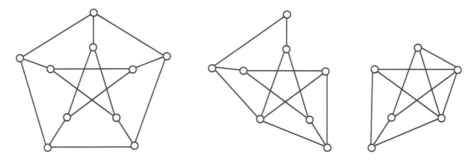

Figure 13.1. Three of K_5's first cousins.

1. Find the smallest number of colors needed to properly vertex-color each of the above graphs. Note that the vertices are drawn so that you can fill them in with colors, should you happen to have colored pens/pencils with you (and should you be willing to write in a book, unless of course you are working from a photocopied page, in which case there should be no problem... do you think it is possible to have a parenthetical remark that is longer than the parent sentence?).

2. Prove that you used the very smallest number of colors to properly color the vertices of these graphs. For each graph, argue that it is not possible to use fewer colors than you *did* use and still have a proper coloring. (What proof technique have you used?)

3. Can you come up with a truly useless lower bound for the number of colors needed to properly vertex-color a graph? Find a slightly better lower bound.

13.2.2 Edge Coloring

Definition 13.2.2. **A graph** G is *properly edge colored* if each edge of G is assigned a color such that no two edges incident to the same vertex have the same color.

Figure 13.2 shows three graphs that all look vaguely like K_5 but that are not K_5. You may recognize one of them.

4. Find the smallest number of colors needed to properly edge color each of the graphs in Figure 13.2. (Proofs are not needed at this stage.)

5. Can you come up with a super-silly lower bound for the number of colors needed to edge-color a graph? How about a less-silly lower bound?

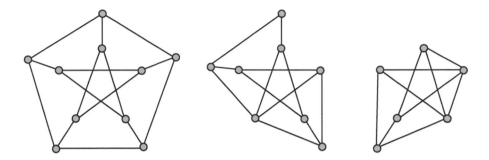

Figure 13.2. Three graphs with their vertices dyed to look younger.

13.2.3 More on Vertex Coloring

Figure 13.3 shows two highly vertex-colorable graphs.

6. Properly vertex-color these graphs.

7. For each graph, argue that it is not possible to use fewer colors than you *did* use and still have a proper coloring.

13.2.4 More on Edge Coloring

8. Go back to Figure 13.2 and *prove* that you did use the smallest number of colors to color the edges of these graphs.

9. Can you come up with a ridiculous upper bound for the number of colors needed to properly edge color a graph? Can you come up with a somewhat better upper bound?

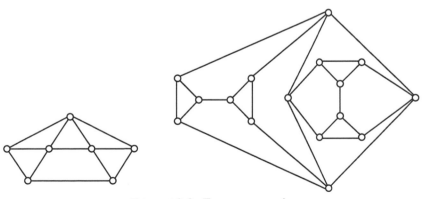

Figure 13.3. Two more graphs.

13.3   Introduction to Coloring

Hey! You! Don't read this unless you have worked through the problems in Section 13.2. I mean it!

Graph coloring problems can be practical; let's do a quick example.

Example 13.3.1. Almost any town has several radio stations. In order to avoid listening problems, the radio stations broadcast on different frequencies. But there are only so many frequencies allocated for public use by the Federal Communications Commission (FCC), and in a large enough area, some frequencies will need to be reused. How might we know which stations can use the same frequency?

Assign each station to a vertex, and join vertices when they are within broadcast range of each other. Certainly, two stations that are not in broadcast range of each other can use the same frequency without interference, but two stations in broadcast range of each other should not use the same frequency. So we color the vertices properly—not allowing adjacent vertices to get the same color—and this tells us which stations can use the same frequency; it's the stations of the same color.

Figure 13.4 shows a fake graph of real radio stations; any two stations whose broadcast areas overlap in our fake world are joined by an edge. At left in Figure 13.4, we start by coloring the excellent KUNI teal. Then we notice that none of KRNI, KUNY, WNEK, or WMUA can have the same color. If we tried to color all of them grey, we would have a problem because KRNI and KUNY are adjacent, and KUNY and WNEK are adjacent. So we will color KRNI and WNEK dark grey and color KUNY and WMUA light grey in the center diagram of Figure 13.4. This leaves WFCR and WRIU; both are adjacent to WNEK and so neither can be colored dark grey; moreover, WRIU is adjacent to WMUA so it cannot be light grey. But we can color WRIU teal and color WFCR light grey, and we do so at right in Figure 13.4. We have used three colors, and this is the smallest number possible because there are three mutually adjacent vertices in the graph.

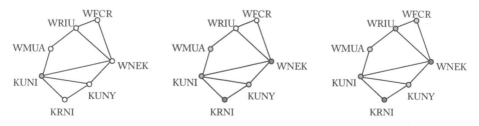

Figure 13.4. Radio stations that are real, but in reality do not have this relationship… at all.

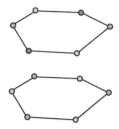

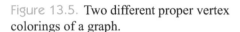

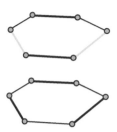

Figure 13.5. Two different proper vertex colorings of a graph.

Figure 13.6. Two different proper edge colorings of a graph.

Definition 13.3.2. The *chromatic number* of a graph G is the smallest number of colors needed to properly color the vertices of G; we denote it by $\chi(G)$. A graph is *k-colorable* (or *k-vertex-colorable*) if it can be properly colored using k vertex colors, and it is $\chi(G)$-*chromatic*.

Notice that for any $k \geq \chi(G)$, the graph G is k-colorable. At the top in Figure 13.5, we see a 3-coloring of a 2-chromatic graph (and at the bottom, we exhibit the 2-chromaticity of the graph), so that the graph is also 3-colorable. If you would like to practice coloring the vertices of graphs properly, try http://digitalfirst. bfwpub.com/math_applet/graph_coloring.html or http://bcs.whfreeman.com/ webpub/Ektron/fapp9e/MathApplets/GraphColoring.html.

Definition 13.3.3. The *chromatic index* of a graph G is the smallest number of colors needed to properly color the edges of G; we denote it by $\chi'(G)$. A graph is *k-edge-colorable* if it can be properly colored using k edge colors, and it is $\chi'(G)$-*chromatic*.

Notice that for any $k \geq \chi'(G)$, the graph G is k-edge-colorable. At the top in Figure 13.6, we see a 3-edge coloring of a 2-edge chromatic graph (and at the bottom, we exhibit the 2-edge chromaticity of the graph), so that the graph is also 3-edge-colorable.

In order to determine the chromatic number (or index) of a graph, two things are required: First, you need to exhibit a coloring that uses only the proposed chromatic number (or index) of colors. And second, you must show that using a smaller number of colors leads to two adjacent vertices (or incident edges) being assigned the same color. In other words, you need to do a proof by contradiction. We'll show this process on a fact you likely noticed while working on the problems in Section 13.2.

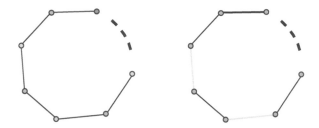

Figure 13.7. Proper vertex and edge colorings of an odd cycle.

Example 13.3.4. We will prove that the chromatic number of an odd cycle is 3, as is the chromatic index of an odd cycle. First, we exhibit 3-vertex and 3-edge colorings of a generic odd cycle in Figure 13.7. Then, we argue that two colors do not suffice. Suppose we may only use the colors teal and grey. Start with any vertex (or edge) and color it teal. Color an adjacent vertex (or incident edge) grey. Now, this grey vertex (or edge) is adjacent (or incident) to exactly one uncolored vertex (or edge), so that uncolored vertex (or edge) has to be colored teal. In this fashion, the vertices (or edges) must alternate colors around the cycle. However, because there are an odd number of vertices (or edges), the last vertex (or edge) to be colored must be the same color as the first vertex (or edge) to be colored. These two vertices are adjacent (or edges are incident), so they may not have the same color. This is a contradiction.

Example 13.3.5 (of chromatic number and chromatic index computation). We will compute the chromatic number and chromatic index of the 5-prism graph G pictured in Figure 12.17 on page 394. Because G contains an odd cycle, $\chi(G) \geq 3$, and we exhibit a 3-vertex coloring of G in Figure 13.8, so $\chi(G) \leq 3$. Therefore $\chi(G) = 3$. Because G has a vertex of degree 3, $\chi'(G) \geq 3$, and we exhibit a 3-edge coloring of G in Figure 13.8, so $\chi'(G) \leq 3$. Therefore $\chi'(G) = 3$.

   Actually, we can compute the chromatic number and chromatic index of the n-prism graph Pr_n. In the case that n is even, $\chi(Pr_n) = 2$; we can alternate colors on the outer cycle, and use the same kind of coloring on the inner cycle but with

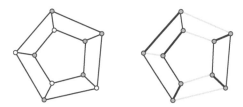

Figure 13.8. A proper vertex coloring and a proper edge coloring of the 5-prism graph.

the colors reversed. In the case that n is odd, $\chi(Pr_n) = 3$; we can use the same scheme as in Figure 13.8. When n is even, $\chi'(Pr_n) = 3$; we can alternate colors on the outer cycle, and use the same coloring on the inner cycle, and use the third color on all of the struts. In the case that n is odd, $\chi'(Pr_n) = 3$; we can use the same scheme as in Figure 13.8.

So far, we have (basically by default) been considering coloring simple graphs. Notice that multiple edges do not have any effect on vertex coloring as they do not change adjacency but that the presence of a loop renders a graph not vertex-colorable as there is a vertex that is adjacent to itself! Similarly, a loop renders a graph not edge-colorable as that edge is incident to itself. Multiple edges increase the number of edge colors needed at a vertex.

13.3.1 Coloring Bounds

In this section we will discuss lower and upper bounds on the chromatic number and chromatic index of a graph. For example, you probably noticed that any graph needs at least one color for either vertex or edge coloring (at least, if the graph has any vertices or edges) and can't possibly need more colors than vertices (or edges). However, these bounds are maximally useless; we can do better.

From Example 13.3.4, we know that if G contains an odd cycle, then $\chi(G) \geq 3$ and $\chi'(G) \geq 3$. But we can do better still and improve our lower bounds: notice that a 3-cycle has three mutually adjacent vertices and, therefore, requires three vertex colors.

Lower bounds. A graph containing n mutually adjacent vertices (that is, a copy of K_n) will need at least n vertex colors. So if G contains K_n, then $\chi(G) \geq n$. As you noticed in Section 13.2.3, the converse is not true—even if a graph does *not* contain a copy of K_n, it may still need n or more vertex colors. The corresponding lower bound for edge coloring is that if the maximum degree of a vertex in G is $\Delta(G)$, then $\chi'(G) \geq \Delta(G)$ because there are $\Delta(G)$ edges that are incident at some vertex.

Now we will give some reasonable upper bounds for $\chi(G)$ and $\chi'(G)$. It turns out that the most straightforward way to discover these bounds is by seeing what works in the proofs!

Theorem 13.3.6. *Let G be a simple graph with largest degree $\Delta(G)$. Then* $\chi(G) \leq \Delta(G) + 1$.

Figure 13.9. Deleting a vertex from a graph, coloring the remainder, and replacing the vertex.

Proof: We will use induction (it's been awhile, eh?) on the number of vertices. As a base case, note that if $|V(G)| = 1$, only one color is needed and $1 \leq 0 + 1$. To be sure, if $|V(G)| = 2$, at most two colors are needed and $2 \leq 1 + 1$. Our inductive hypothesis is that any simple graph G with $k \leq n$ vertices is $(\Delta(G) + 1)$-colorable. So, suppose G has $n + 1$ vertices. Pick any vertex v and delete it (recall that the incident edges go right with it). See Figure 13.9 for a demonstration of this process. The remaining graph $G \setminus v$ has only n vertices, so the inductive hypothesis applies and the vertices of $G \setminus v$ can be colored using at most $\Delta(G \setminus v) + 1$ colors. Now replace v (and its incident edges) so that G has all vertices colored except v.

Question: How are $\Delta(G \setminus v)$ and $\Delta(G)$ related? Answer: Any vertex other than v or its neighbors has the same degree in G as it does in $G \setminus v$. Any neighbor of v has degree one more in G than in $G \setminus v$ because of the edge connecting it to v. Finally, v may have degree much higher than $\Delta(G \setminus v)$. This tells us that $\Delta(G) \geq \Delta(G \setminus v)$.

But let's see what happens when we try to color v. There are at most $\Delta(G \setminus v) + 1$ colors used by the neighbors of v. Conveniently, v has at most $\Delta(G)$ neighbors. Therefore, at worst $G \setminus v$ was colored using $\Delta(G)$ colors. Only one more is needed to color v, so G can be colored using at most $\Delta(G) + 1$ colors; therefore, $\chi(G) \leq \Delta(G) + 1$. □

It turns out that the constraint that G is simple is unnecessary (we could allow multiple edges), but it sure does simplify the proof. Also notice that there are many graphs for which Theorem 13.3.6 way overestimates: consider the star graph S with n spokes, which has $\Delta(S) = n$ but $\chi(S) = 2$.

Theorem 13.3.7. *Let G be a simple graph with largest degree $\Delta(G)$. Then $\chi'(G) \leq 2\Delta(G) - 1$.*

Proof: We will use induction again, but this time on the number of edges. As a base case, note that if $|E(G)| = 1$, only one color is needed and $1 \leq 2 \cdot 1 - 1$. Similarly, if $|E(G)| = 2$, only two colors are needed and $2 \leq 2 \cdot 2 - 1$. Our in-

Figure 13.10. Deleting an edge from a graph, coloring the remainder of the edges, and replacing the edge.

ductive hypothesis is that any simple graph G with $k \leq n$ edges is $(2\Delta(G) - 1)$-edge-colorable. So, suppose G has $n + 1$ edges. Pick any edge e and delete it. See Figure 13.10 for a demonstration of this process. The remaining graph $G \setminus e$ has only n edges, so the inductive hypothesis applies and the edges of $G \setminus e$ can be colored using at most $2\Delta(G \setminus e) - 1$ colors. Now, examine the vertices near (the missing) e. At worst, both of them have degree $\Delta(G \setminus e)$, and at worst, every edge is a different color, so at most $2\Delta(G \setminus e)$ colors appear on the edges incident to the vertices that would touch e (if e were there). That means that if we reinsert e, we have used at most $2\Delta(G \setminus e) + 1$ colors. It remains to translate this to G. Every vertex of $G \setminus e$, except for the two incident to e, has the same degree as it does in G. Those two vertices incident to e have degree one higher in G than in $G \setminus e$. So, if (as in the worst-case scenario) they had degree $\Delta(G \setminus e)$ in $G \setminus e$, they will have degree $\Delta(G)$ in G and $\Delta(G) = \Delta(G \setminus e) + 1$. Therefore, $\Delta(G \setminus e) = \Delta(G) - 1$, and the number of colors we used is at most $2\Delta(G \setminus e) + 1 = 2(\Delta(G) - 1) + 1 = 2\Delta(G) - 1$, as desired. $\square$

13.3.2 Applications of Vertex Coloring

A classic application of graph coloring is the storage of chemicals.

Example 13.3.8. The Egg Yolk Wasn't Here building (see http://www.seakingdom. net/blog/wp-content/uploads/2011/01/eggyolkgraffito.jpg and http://www. flickr.com/photos/lunapark/3781387208/) is the warehouse for a large chemical supply company (in our dreams) called EYWH. While it might at first make sense to arrange the chemicals in alphabetical order by name, that is not safe—many chemicals are highly reactive, and so should not be stored near each other in case of spillage, bottle breakage, etc. Instead, EYWH supplies us with a list of chemicals they want to store, with chemical interactions noted. We convert this into a graph: each chemical is assigned a vertex, and every chemical interaction becomes an edge in the graph. Then, we find a proper vertex coloring with the smallest possible number of colors (let's say k). Each color designates a collection of chemicals that may be stored in proximity to each other. We then give a report to EYWH: it consists of k lists of chemicals, along with the information that the chemicals in each list must be stored in a region of the warehouse (perhaps a floor or a wing of a floor) that is protected from each other region in the warehouse.

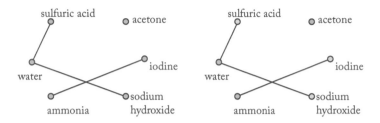

Figure 13.11. Chemical interactions determine chemical storage.

That involves too much data to show how it works in practice. However...

Example 13.3.9. EYWH receives deliveries of chemicals every day. These are un-loaded from trucks or tankers to a receiving area, and from there they are forklifted to the appropriate storage place in the warehouse. To prevent explosions resulting from spillage or accidents, there are several small holding bays in the receiving area. Every day, the receiving manager receives a list of deliveries scheduled for the day and to which bay each delivery must be unloaded. The logistics manager for EYWH (who is of course trained in graph colorings!) creates the list for July 12, 2010 as follows. There are supposed to be deliveries of ammonia, sodium hy-droxide, iodine, sulphuric acid, and acetone at various points during the day. The logistics manager knows that ammonia and iodine react to create the explosive nitrogen triiodide and that sodium hydroxide and sulphuric acid both react with water to produce intense heat. This suggests the graph and vertex-colored graph shown in Figure 13.11. From this, we know that we only need two holding bays, but one of them must be very dry.

Notice that we can use vertex coloring to solve a problem when we are able to model conflicts (with chemicals, explosive interactions) as edges in a graph. A less classical, but more pervasively practical, application of vertex colorings is to traffic-light cycles.

Example 13.3.10. Imagine, if you will, a road intersection with a traffic light. Ev-ery so often the traffic light changes, and after a while, the pattern of greens and reds and yellows on the traffic light repeats. This is a traffic-light cycle. Now imagine that for each lane of traffic, there is a light pattern that shows green for that lane and red for all other lanes. Eventually, the light turns yellow for that lane and red for all other lanes and then switches to have a different lane shown green (with all the rest shown red). The traffic-light cycle has as many patterns as lanes. It takes forever to get through this intersection and drivers are very annoyed! City Hall is inundated with calls. There are letters to the editor of the local newspaper about the terrible traffic situation.

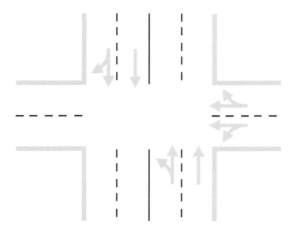

Figure 13.12. One of the many intersections of Broad St. in Philadelphia with a one-way street (perhaps Arch?).

Now imagine the same intersection, but this time the traffic light has patterns that are green for all lanes but one. Traffic proceeds quickly, but with many collisions. An ambulance has to stand by in the area to take away the wounded. Lawyers sue the city for negligence. There are letters to the editor of the local newspaper about the terrible traffic situation.

Let's avoid these situations—by finding a way to allow as many lanes to have green lights as possible at the same time, while also keeping drivers from colliding with each other. If you guessed that we would use vertex coloring to address the problem, you are correct; and, you *should* have guessed this because this example *is* in the section on applications of vertex coloring. We must first create a graph to color. A sample street intersection is shown in Figure 13.12.

In this situation, what are the conflicts that we can model with graph edges? They are potential car collisions. For example, we do not want a car traveling east to have a green light at the same time as a car traveling north. Therefore, we will let the lanes be vertices, and we will let potential lane-occupant collisions be edges. As shown in Figure 13.13, we can number the lanes and use these numbers as vertex labels and see which paths of travel intersect to find edges in the corresponding graph.

Sometimes an intersection is sufficiently busy that it's very difficult to turn left. In such a situation, we have a protected left turn, indicated by a green arrow on the traffic light. When such an option exists, we consider that traffic path as a separate vertex. You can explore this possibility in Problem 7 of Section 13.10.

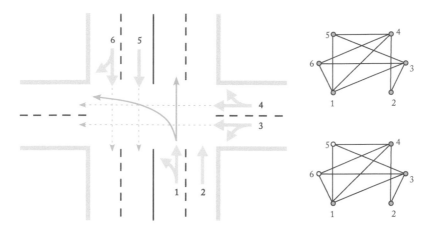

Figure 13.13. Cars traveling in lane 1 intersect paths of travel from lanes 3, 4, 5, and 6 (left). The graph corresponding to this intersection along with a proper vertex coloring (right).

Finally, we consider an application that is the subject of much research.

Example 13.3.11. Traditionally, wireless networks function by having wireless access points that distribute information to nearby machines (robots, laptops, phones). However, there are also wireless networks in which each machine distributes information to nearby machines. Sometimes the information's destination is a neighbor machine, and sometimes the information needs to "hop" through several machines to reach its destination. Thus, these wireless networks are called *multi-hop* networks. They are also called *ad hoc* networks (because the machines may move around), *sensor* networks (because the machines might be sensors that primarily report information about their local environments), and *mesh* networks (because the machines may be spaced out so as to provide coverage of an area).

Instead of having all communication regulated by an access point, the machines in an ad hoc network have to communally regulate their communication. One protocol for regulating communication is called Time Division Multiple Access (TDMA). The way this works is that time is divided into many parts so that there is a schedule of which machines can send information at what times. Machines that are far enough away from each other can send information at the same time (thereby having multiple access to a time slot).

To have efficient communication, one wants to give as many machines access to each time slot as possible and to thereby give each machine access to as many time slots (in a given hour, let's say) as possible. We will create a graph where each machine is assigned to a vertex, and any two machines that are within communication range of each other will be connected by an edge. (In practice, a second graph may be used, where each machine has a vertex but an edge connects two machines that have distance 1 or 2 (or k) on the original graph. This is because, for example, two machines that have a common communication neighbor might both transmit messages that arrive simultaneously at the neighbor. Then the neighbor would get completely confused. So this type of simultaneous transmission is usually prohibited.)

A proper vertex coloring of this graph (with the least number of colors possible used) will tell us which machines can send information at the same time. The number of colors corresponds to the number of different time slots needed. Of course, in reality such communication schedules need to be made quickly and may need to be remade on the fly as machines move around. (This is the case with networks of battlefield sensors.) For this reason, it is of practical interest to find efficient vertex coloring algorithms, so research in this area is active.

Check Yourself

These problems are easy, fun, and quick. Do them!

1. What is $\chi(K_n)$?

2. Find a graph with $\chi(G) = 1$.

3. Find the chromatic number and index of a path of length 5.

4. Find the chromatic number and index of a cycle of length 423.

5. **Challenge:** Create a graph G for which $\chi(G) > \chi'(G)$.

13.4   Try This! Let's Think about Coloring

Here we have four multipart problems. They are completely independent of each other, so start with your favorite topic!

1. Determine $\chi'(K_3)$, $\chi'(K_4)$, $\chi'(K_5)$, $\chi'(K_6)$, and $\chi'(K_7)$.

   (a) Conjecture the value of $\chi'(K_n)$.

   (b) Come up with ways to edge-color K_n that verify your conjecture.

   (c) Can you prove your conjecture?

2. Recall that a bipartite graph has the property that the vertices can be grouped into two parts so that neither part has internal edges. A complete bipartite graph $K_{m,n}$ has all possible edges between a part with n vertices and a part with m vertices.

   (a) Determine $\chi(K_{2,2})$, $\chi(K_{2,3})$, $\chi(K_{3,3})$, $\chi(K_{3,4})$, and $\chi(K_{3,5})$.

   (b) What is $\chi(K_{m,n})$? Prove it.

   (c) For B any bipartite graph, what is $\chi(B)$? Prove it.

   (d) Determine $\chi'(K_{2,2})$, $\chi'(K_{2,3})$, $\chi'(K_{3,3})$, $\chi'(K_{3,4})$, and $\chi'(K_{3,5})$.

   (e) Conjecture the value of $\chi'(K_{m,n})$.

   (f) Can you prove your conjecture?

3. We will take a quick sojourn into algorithms for coloring.

   (a) Design a greedy (parsimonious) algorithm for coloring the vertices of a graph.

   (b) Your algorithm addresses the vertices in some order. Try it out on the graph in Figure 13.14. In fact, try your algorithm with each of the vertex orderings given in Figure 13.14. Does your algorithm give the optimal coloring in each case?

   (c) Design a greedy (parsimonious) algorithm for coloring the edges of a graph.

   (d) Your algorithm addresses the edges in some order. Try it out on the graph in Figure 13.15. In fact, try your algorithm with each of the edge orderings given in Figure 13.15. Does your algorithm give the optimal coloring in each case?

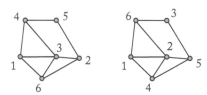

Figure 13.14. Two different orderings of the vertices of a graph.

Figure 13.15. Two different orderings of the edges of a graph.

4. Recall that a matching in G is a subgraph with all vertices of degree 1 and that a perfect matching is a subgraph with all vertices of degree 1 that includes all vertices of G.

   (a) Find as many different perfect matchings as you can in a 6-cycle graph.

   (b) Find as many different perfect matchings as you can in K_4.

   (c) Create a 4-regular planar graph and properly edge-color it.

   (d) How does the concept of perfect matchings relate to edge coloring of k-regular graphs?

13.5 Coloring and Things (Graphs and Concepts) That Have Come Before

Hey! You! Don't read this unless you have worked through the problems in Section 13.4. I mean it!

13.5.1 Let's Color the Edges of Complete Graphs

Theorem 13.5.1. *For n even, $\chi'(K_n) = n - 1$, and for n odd, $\chi'(K_n) = n$.*

Proof: First notice that because every vertex of K_n has degree $n - 1$, $\chi'(K_n) \geq n - 1$. Therefore, if we exhibit an edge coloring of K_n that uses $n - 1$ colors, we will have shown that $\chi'(K_n) = n - 1$. Consider n even. Draw K_n as follows: place one vertex in the center and the remaining (odd number of) vertices in a ring around the center vertex. Then draw edges as shown at right in Figure 13.16. (Notice that there are $\frac{n}{2}$ of them.) Give these edges the same color. Now rotate this edge configuration by $\frac{1}{n-1}$, and give this new set of edges a second color. Continue in

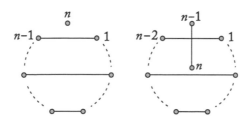

Figure 13.16. Matchings that lead to edge colorings of K_n for n odd (left) and n even (right).

this fashion until you have drawn $\frac{n}{2} \cdot \frac{1}{n-1}$ edges in $n-1$ colors, and notice that you have completed all $\frac{n(n-1)}{2}$ edges of K_n (exactly once each)! We can make a similar coloring for K_n when n is odd. See the left diagram of Figure 13.16. In this case, each color has $\frac{n-1}{2}$ edges and there are n different rotations (n ways to have an un-edge-colored vertex), so a total of $\frac{n(n-1)}{2}$ edges are drawn in n colors. How do we know this is the optimal coloring, though? Maybe there's a more clever proper coloring that only uses $n-1$ colors. Well, suppose such a coloring exists. Then, because each color can only be at each vertex once, there are no more than $\frac{n}{2}$ edges of a given color. But that's not an integer, because n is odd—so there are no more than $\frac{n-1}{2}$ edges of each color, and *that* means there are only $(n-1)\frac{n-1}{2}$ edges total. And that's a contradiction.   □

13.5.2 Let's Color Bipartite Graphs

Theorem 13.5.2. *A graph G is bipartite if and only it is 2-vertex-colorable.*

Proof: ($\Rightarrow$) Suppose G is bipartite. Name the two parts T and P. Color the vertices in T teal and the vertices in P purple. This is a proper coloring because no vertex in the teal (or purple) part is adjacent to any other vertex in the teal (or purple) part.

($\Leftarrow$) Suppose G is 2-vertex-colorable. Without loss of generality, let the two colors be teal and purple. Collect the teal vertices and consider them to be one part, and let the remaining (purple) vertices be a second part. With this structure, G is bipartite; no vertex in the teal (or purple) part is adjacent to any other vertex in the teal (or purple) part.   □

From Theorem 13.5.2, it follows that all trees are bipartite: Given a tree, color some vertex v teal, and all its neighbors purple, and all of the distance-2 neighbors of v teal, and so forth. Because every tree is acyclic, there will never be a coloring conflict. Similarly, we now know that even cycles are also bipartite. This leads to a new characterization of bipartite graphs.

Theorem 13.5.3. *A graph G is bipartite $\Longleftrightarrow$ G has no odd cycles.*

Proof: ($\Rightarrow$) Suppose G is bipartite and has an odd cycle. The odd cycle requires three colors by Example 13.3.4, and this contradicts Theorem 13.5.2.

($\Leftarrow$) Suppose G has no odd cycles. By Theorem 13.5.2, it is sufficient to provide a proper 2-vertex coloring of G. We propose a coloring and then show that it is proper. Choose any vertex v of G. Color it teal. For each other vertex of G, color it purple if the shortest path to v has even length and teal if the shortest path to v

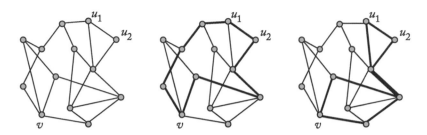

Figure 13.17. A non-bipartite graph (left) with shortest paths from v to u_1, u_2 that form a cycle (center) and with shortest paths from v to u_1, u_2 that share an edge (right).

has odd length. Now we will proceed by contradiction; assume that the coloring is not proper, so that there exist two vertices u_1 and u_2 that are adjacent and have the same color. That means that either the shortest path from u_1 to v has even length and so does the shortest path from u_2 to v, or the shortest path from u_1 to v has odd length and so does the shortest path from u_2 to v. Walk from v to u_1 to u_2 to v using the shortest paths possible at each stage. This uses an odd number of edges because it's either even $+ 1 +$ even $=$ even $+ 1 =$ odd or odd $+ 1 +$ odd $=$ even $+ 1 =$ odd. Now, this is probably not a cycle because the shortest paths from v to u_1 and u_2 may have some edges in common (see Figure 13.17). However, if we delete any edges common to those two paths, we are left with some cycles. Each common edge was counted twice in our walk-edge-count, so we still have an odd number of edges between those cycles. That means at least one of the cycles must be odd; this is a contradiction. □

In Section 13.4, you likely conjectured that $\chi'(K_{m,n})$ is m or n (whichever is larger). We will now prove a more general result:

Theorem 13.5.4. *For bipartite G, $\chi'(G) = \Delta(G)$.*

Proof: We shall induct on the number of edges of G. As a base case, consider G such that $|E(G)| = 1$, in which case $\Delta(G) = 1$ and exactly one color is needed. To be sure, let's examine a second base case of $|E(G)| = 2$, in which case either there are two vertices with a double edge or there are three vertices with two edges that form a path. Either way, $\Delta(G) = 2$ and exactly two colors are needed. The inductive hypothesis is that any bipartite graph G with $k \leq n$ edges can be edge-colored using exactly $\Delta(G)$ colors. So, we consider a bipartite graph with $n + 1$ edges, and we remove one (let's say e, with endpoints v_1 and v_2). This gives us a bipartite graph $G \setminus e$ with n edges, so by the inductive hypothesis we can edge-color $G \setminus e$ with $\Delta(G \setminus e)$ colors. Now, $\Delta(G) \geq \Delta(G \setminus e)$ and, in particular, either

$\Delta(G) = \Delta(G \setminus e)$ (if there is a vertex of degree $\Delta(G)$ that is not incident to e) or $\Delta(G) = \Delta(G \setminus e) + 1$ (if every vertex of degree $\Delta(G)$ is incident to e).

In the case where $\Delta(G) = \Delta(G \setminus e)$, we know that in $G \setminus e$ neither v_1 nor v_2 has degree $\Delta(G \setminus e)$, as each has lower degree than it did in G. Therefore, there is some one of the $\Delta(G \setminus e)$ colors not used at v_1, and also a color not used at v_2. If these are the same color, for example, teal, then color e teal and be done. If (for example) v_1 has no teal edge but *does* have a purple edge and v_2 has no purple edge but *does* have a teal edge, then we must be a bit crafty. Travel along the graph, starting at v_1, and alternate between purple and teal edges. This path cannot reach v_2 because (a) if it did, it would have to have odd length because v_1 and v_2 are in different parts, and (b) that would mean it started and ended with purple edges, but purple is not used at v_2. So, switch the colors of the edges along this path so that v_1 has no purple edge but *does* have a teal edge. (This does not ruin the coloring because any vertex on the path had exactly one edge that was teal and one that was purple for the coloring to be proper.) Now, neither v_1 nor v_2 has an incident purple edge, so color e purple... and we're done.

On the other hand, if $\Delta(G) = \Delta(G \setminus e) + 1$, then $G \setminus e$ has been edge-colored with $\Delta(G) - 1$ colors. This means we have an extra color to play with—so we will color e a new color and we're done.   □

13.5.3   Add a Condition, Get a Different Bound

We know that $\chi(G) \le \Delta(G) + 1$. But what if $\Delta(G)$ is huge (like 10,000)? Consider, for example, the graph in Figure 13.18. It has $\Delta(G) = 12$, but its vertices can be colored with only two colors. If G happens to be planar (which this G happens to be), we get a better bound.

> **Theorem 13.5.5.** *Every simple planar graph can be vertex-colored with at most six colors.*

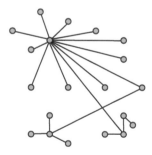

Figure 13.18. A graph G with $\chi(G) = 2$ but $\Delta(G) = 12$.

Proof: We proceed (yet again) by induction. We will induct on the number of vertices of the graph. Our base cases are all planar graphs with six or fewer vertices, as any graph with six or fewer vertices can be vertex-colored with at most six colors. Our inductive hypothesis is that any planar graph with $k \leq n$ vertices can be vertex-colored with no more than six colors. Consider any planar graph G with $n + 1$ vertices. Now, to proceed with the inductive step, we need to use the restriction that G is planar. (Hey, that's why it's in the statement of the theorem.) We will go all the way back to Theorem 11.6.6, which says that G must have a vertex of degree at most 5. Pick one such vertex, name it *Anatinus*, and yank it out. $G \setminus Anatinus$ has only n vertices, so the inductive hypothesis holds, and $G \setminus Anatinus$ can be vertex-colored with no more than six colors. Now return *Anatinus* to its home, and because *Anatinus* has degree at most 5, its neighbors are colored with no more than five colors. This means that if we need an additional color for *Anatinus*, we can use a sixth color (either one that is already used to color $G \setminus Anatinus$ or a new one) and G has been vertex-colored with at most six colors. Wahoo! □

13.5.4   Greedy Matchings

Okay, that section title is misleading. (But it's goofy!) Really, we will talk briefly about greedy algorithms and coloring and then talk briefly about matchings and edge coloring.

As we promised in Section 10.4.1, we will address the utility of greedy algorithms in coloring. Here is one greedy algorithm for coloring vertices (or edges) of a graph.

1. Order the vertices (or edges) of the graph as $v_1, \ldots, v_n$ (or as $e_1, \ldots, e_n$).

2. Make a list of colors, namely color 1, color 2, ..., color n.

3. Color the first vertex (or edge) with color 1.

4. Consider the next vertex (or edge) in the list, and give it the lowest-numbered color that is not already in use on one of the vertex's neighbors (or one of the edges incident to this edge).

5. If all of the vertices (or edges) are colored, be done. If not, go to step 4.

As you surely noticed in Section 13.4, a greedy algorithm for coloring does not always give an optimal coloring! In fact, it can sometimes give an awful coloring.

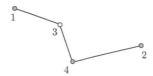

Figure 13.19. A pathetic little path graph.

Example 13.5.6 (of a greedy algorithm gone bad). When the vertices of a path of length 4 are ordered as in Figure 13.19, a greedy algorithm proceeds as follows: Vertex 1 is assigned color *teal*. Vertex 2 is not adjacent to any colored vertices, so it can also be assigned color *teal*. Vertex 3 is adjacent to a vertex with color *teal*, so we must use color *white*. Vertex 4 is adjacent to vertices colored *teal* and *white*, so we are forced to use color *grey*. That is, we have used three colors on a path of length 4 when only two are needed for an optimal coloring.

However, most of the time a greedy algorithm produces a pretty decent coloring, so sometimes greedy algorithms are used in practice.

> **Matchings and edge colorings.** If you look at the edges of just one color, it is a matching—no edge can touch another edge of the same color, so each vertex can have degree at most 1 in that color! So an edge coloring is a union of matchings. If every one of the matchings is a perfect matching, then the coloring is proper (and the graph is regular). For this reason, some proofs about edge coloring can be done using perfect matchings.

Check Yourself ————————————————————————————

Verify your understanding with these quickies.

1. What is $\chi'(K_{578349})$?

2. What is the shortest length a cycle can be in a bipartite graph?

3. What is $\chi'(K_{42,87})$?

4. What proof method(s) is/are used in the proof of Theorem 13.5.2?

5. Let G be planar. When is the upper bound on $\chi(G)$ better from planarity than from $\Delta(G) + 1$?

6. Use Theorem 13.5.4 to determine $\chi'(K_{m,n})$.

7. **Challenge:** Create a graph (other than the one in Example 13.5.6) on which a greedy algorithm produces a truly awful vertex coloring.

13.6   Where to Go from Here

The simplest way to learn more about graph coloring is to take a graph theory course; consult your nearest math department member to see when or whether such a course will be offered. Almost every graph theory textbook will have some material on coloring; you may enjoy reading [24] and [25] in particular. For more on coloring in graph theory *and* across the rest of mathematics, see *The Mathematical Coloring Book: Mathematics of Coloring and the Colorful Life of Its Creators* by Alexander Soifer.

In this chapter, you developed several upper and lower bounds on the chromatic number and chromatic index of a graph. There are better upper bounds! Look for Brooks's theorem and Vizing's theorem in particular; these results may surprise you. Their proofs are not difficult, but are rather technical.

Graph coloring is an active area of research in pure mathematics, and algorithms for finding graph colorings are an active area of research in computer science. Here are a few paper titles published in discrete mathematics journals in the middle of 2018: "List star chromatic index of sparse graphs"; "Maximizing the number of x-colorings of 4-chromatic graphs"; "Berge-Fulkerson coloring for $C_{(8)}$-linked graphs"; "Thoroughly dispersed colorings"; "Planar graphs have two-coloring number at most 8"; "Chromatic index determined by fractional chromatic index."

Credit where credit is due: Example 13.3.11 was adapted from "Coloring Unstructured Wireless Multi-Hop Networks," by Johannes Schneider Roger Wattenhofer, 28th ACM Symposium on Principles of Distributed Computing (PODC), Calgary, Canada, August 2009. Jillian Bakke provided the information on chemical interactions. The traffic-light problems were inspired by [7]. Bonus Check-Yourself Problem 8 and Problem 31 of Section 13.10 are about the real MathILy Week of Chaos; see http://www.mathily.org/facts. html. Problem 34 of Section 13.10 was inspired by the TV series *Get Smart* and *The Adventures of Rocky and Bullwinkle and Friends*, and Sam Oshins's character who was inspired in turn by this book (see page 148).

13.7   Chapter 13 Definitions

coloring: An assignment of colors to a graph G's vertices (or edges); technically, it is a function $c : V(G) \to C$ (or $c : E(G) \to C$), where C is a set of colors.

proper edge coloring: Each edge of G is assigned a color such that no two edges incident to the same vertex have the same color.

proper vertex coloring: Each vertex of G is assigned a color such that no two adjacent vertices have the same color.

chromatic number: The smallest number of colors needed to properly color the vertices of G; we denote it by $\chi(G)$.

k-colorable: A graph that can be properly colored using k vertex colors; also called *k-vertex-colorable*.

k-chromatic: A graph that can be properly colored using *no fewer than k* vertex colors; G is always $\chi(G)$-chromatic.

chromatic index: The smallest number of colors needed to properly color the edges of G; we denote it by $\chi'(G)$.

k-edge-colorable: A graph that can be properly colored using k edge colors.

k-edge-chromatic: A graph that can be properly colored using *no fewer than k* edge colors; G is always $\chi'(G)$-chromatic.

13.8   Bonus: The Four-Color Theorem

Surprise! We are not going to give a correct proof of the four-color theorem. We will state it, though.

Theorem 13.8.1. *Every simple planar graph G has $\chi(G) \leq 4$.*

This is not the original statement of the four-color theorem, but it is logically equivalent to the original statement. The four-color theorem has a fascinating history. It was first posed in 1852 but not proved until 1976. The first correct proof was done exhaustively by computer; this proof was simplified to a more reasonable number of cases in 1995. However, even the simpler proof is still not checkable by humans. For this reason, the proof is somewhat controversial, and work continues to try to find a better proof (one that is checkable by hand or that convincingly explains *why* the four-color theorem is true). There are two excellent books you might want to read in order to learn more of the history of the four-color theorem and about the people and mathematics involved. These are *Four Colors Suffice* by Robin Wilson and *Graphs, Colourings, and the Four-Color Theorem* by Robert Wilson (no, they are not related). *Four Colors Suffice* is written for the general public, and *Graphs, Colourings, and the Four-Color Theorem* is a well-written textbook full of enticing problems. Robin Thomas, one of the mathematicians who produced the simplest known proof of the four-color theorem, gives a brief history of the four-color theorem and a somewhat technical explanation of the 1995 proof at http://people.math.gatech.edu/~thomas/FC/fourcolor.html.

So as not to leave you completely bereft of additional mathematical knowledge, we will prove the five-color theorem, and we will give a famous incorrect proof of the four-color theorem.

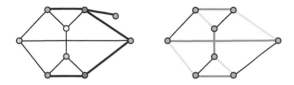

Figure 13.20. At left, a teal-charcoal Kempe chain is highlighted, and at right, a teal-charcoal edge Kempe chain is highlighted.

Theorem 13.8.2. *Every simple planar graph G has $\chi(G) \leq 5$.*

We will prove this theorem *twice* because both proofs are so interesting. For the first proof, we need an additional bit of terminology.

Definition 13.8.3. A *Kempe chain* is a maximal sequence of vertices that alternate between two given colors. An *edge Kempe chain* is a maximal sequence of edges that alternate between two given colors. See Figure 13.20 for examples; notice that a Kempe chain may be a tree.

Proof of Theorem 13.8.2: We will induct on the number of vertices, much as in the proof of Theorem 13.5.5. The set of simple planar graphs with five or fewer vertices forms our base cases, as no such graph could need more than five colors for its vertices. Our inductive hypothesis is that any simple planar graph with $k \leq n$ vertices may be colored with five or fewer colors. Now, consider a simple planar graph G with $n+1$ vertices. We know from Theorem 11.6.6 that G must have a vertex of degree at most 5. Find such a vertex, call it v, and yank it out of the graph so that we are left with $G \setminus v$. Because $G \setminus v$ has only n vertices, the inductive hypothesis applies and it may be colored with at most five colors. Now we will reinsert v and examine the neighbors of v. If the neighbors collectively use no more than four colors, all is well and we may use the fifth color for v. However, if the neighbors use all five colors among them... we must be quite artful. The situation is shown in Figure 13.21.

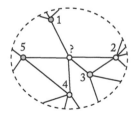

Figure 13.21. The colorful neighborhood of v.

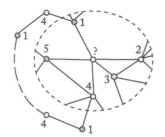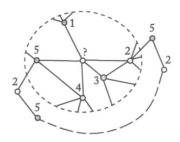

Figure 13.22. Two of the many Kempe chains that must join neighbors of v.

Our first move will be to examine the Kempe chains emanating from each of v's neighbors. If, for example, the neighbor with color 1 is part of a color-1–color-2 Kempe chain that ends before reaching the neighbor with color 2, then we can just switch the colors on the Kempe chain and use color 1 for v. (If the neighbor with color 1 is not adjacent to a vertex of color 2, then the Kempe chain has length 1 and we can just change the color of the color-1 neighbor.) Therefore, if any of v's neighbors is part of a color-c–color-d Kempe chain that does not reach any of v's other neighbors, we can switch the colors on the Kempe chain and thereby reduce the number of colors used by v's neighbors to four. The remaining color can be used for v.

So, let us consider the remaining case, where every Kempe chain emanating from each of v's neighbors reaches one of v's other neighbors. In particular, this means that there is a color-1–color-4 Kempe chain and a color-2–color-5 Kempe chain, as shown in Figure 13.22. Here's the artful bit: these two Kempe chains have to cross. There's just no way for the color-2–color-5 chain to get from the color-2 neighbor to the color-5 neighbor without going around the color-1 neighbor or the color-4 neighbor. That would be all well and good if G weren't a planar graph, but G *is* a planar graph. So the two Kempe chains can't have edges that cross. So, they must cross at a vertex. But what color is that vertex? It's part of the color-1–color-4 Kempe chain, so it must be color 1 or color 4. And it's part of the color-2–color-5 Kempe chain, so it must be color 2 or color 5. But that's a contradiction, because a vertex can't have two different colors. Therefore, this situation cannot arise! And that means this case can never happen. So G can be colored with only five colors. □

A different proof of Theorem 13.8.2: We will do a sneaky proof by induction, using the same base cases and inductive hypothesis as in the previous proof of Theorem 13.8.2. By Theorem 11.6.6, we know that G must have a vertex x of degree at most 5. The neighbors of x cannot be mutually adjacent or we'd have a K_5 subgraph—and then G would be nonplanar by Theorem 11.6.5. Thus, we may consider two edges e_1 and e_2 that are incident to x and whose non-x ends are not adjacent to each other. If we contract these edges, their non-x ends merge into x. The resulting graph is still planar and has no loops (but might have multiple edges, which matters not for vertex coloring) and thus is 5-vertex-colorable. Let's say that x is teal in this coloring. Un-contract e_1 and e_2 (that is, expand out those edges) so we have G again. Color the other (non-x) ends of e_1 and e_2 teal. We will need to recolor x as it is now adjacent to two teal vertices. However, x's other three-or-fewer neighbors are not teal, so there's at least one color left over for x—use it and we're done. Slick, eh? □

In the early 1900s, a "proof" of the four-color theorem was published, and the flaw in the argument was not found for more than a decade after its publication. The author was Alfred Bray Kempe (1849–1922), after whom Kempe chains were named. (He also did important work in geometric linkages.) Here is one version of his "proof."

Essence of a flawed proof of Theorem 13.8.1: We proceed by induction, using as base cases all simple planar graphs with four or fewer vertices. Our inductive hypothesis is that any simple planar graph with $k \leq n$ vertices may be colored with four or fewer colors. Now, consider a simple planar graph G with $n + 1$ vertices. We know from Theorem 11.6.6 that G must have a vertex of degree at most 5. Find such a vertex, call it v, and remove it. The remaining graph $G \setminus v$ may be colored using only four colors, by the inductive hypothesis. Now restore v. At worst, its five neighbors have four different colors.

If any Kempe chains emanating from any of v's neighbors could be switched so that only three colors appear among v's vertices, then we can color v the remaining fourth color. So let us assume that this is impossible, and every Kempe chain emanating from a neighbor of v includes some other neighbor of v.

Now, the situation is essentially that shown at left in Figure 13.23. We may assume that the two neighbors of the same color are not next to each other in order

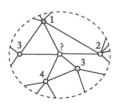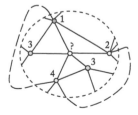

Figure 13.23. A configuration of neighbors of v.

around v, as we could do some Kempe switch to move them apart. Therefore, we have two neighbors of v that are the same color; traveling in one direction around v, they have one neighbor between them, and traveling in the other direction around v, they have two neighbors between them.

In our first proof of Theorem 13.8.2, the crucial point was that there had to be two Kempe chains that used four colors between them and intersected. However, that need not be the case here! Consider the left diagram of Figure 13.23; a color-1–color-3 Kempe chain does not have to intersect a color-2–color-4 Kempe chain. (Note that switching one of two separate color-1–color-3 Kempe chains will either produce two vertices of color 1 or simply permute the locations of the existing colors around v.) Similarly, a color-1–color-4 Kempe chain need not intersect a color-2–color-3 Kempe chain.

On the other hand, check *this* out—we have a color-1–color-4 Kempe chain and a color-2–color-4 Kempe chain as shown at right in Figure 13.23. Together, these block out the color-3–color-2 Kempe chain emanating from the left color-3 vertex and the color-3–color-1 Kempe chain emanating from the right color-3 vertex. So we switch those two Kempe chains and thereby get rid of the two color-3 neighbors of v, and color v with color 3.   ☐

Brain-breaking question. What is the flaw in Kempe's false proof of the four-color theorem?!

> Brain-enhancing activity: Let's avoid breaking our brains. Look at the graph given in Figure 13.24. All of the following questions refer to this figure.
>
> 1. Which vertex plays the role of v in Kempe's false proof?
>
> 2. Make a one-to-one correspondence between the colors of v's neighbors in Figure 13.24 and the numbering of colors used in our version of Kempe's false proof.
>
> 3. Identify the different Kempe chains emanating from neighbors of v. Which chains play the roles of the color-1–color-4 Kempe chain and the color-2–color-4 Kempe chain in Kempe's false proof?
>
> 4. Which chains play the roles of the color-3–color-2 Kempe chain and the color-3–color-1 Kempe chain in Kempe's false proof?

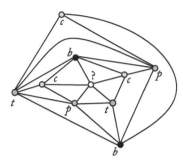

Figure 13.24. A wacky mostly-4-colored graph. (The colors are black, purple, teal, and chartreuse.)

5. Switch one of the chains you identified in the previous problem. Now switch the other chain you identified in the previous problem. What happens?

6. Kempe wanted both chains switched at once. Try this; what happens?

7. Try to explain the flaw in Kempe's false proof of the four-color theorem.

If you would enjoy 4-coloring some graphs and maps, proceed to http://www.nikoli.com/en/take_a_break/four_color_problem/ (warning: uses Flash!) and to http://demonstrations.wolfram.com/FourColoringPlanarGraphs/.

13.9 Bonus Check-Yourself Problems

Solutions to these problems appear starting on page 624. Those solutions that model a formal write-up (such as one might hand in for homework) are to Problems 2, 4, and 5.

1. Find the chromatic number and chromatic index of the graph shown in Figure 10.3 on page 316.

2. Prove that if $\chi(G) \geq 3$, then G must contain an odd cycle.

3. Find the chromatic number and chromatic index of each graph shown in Figure 10.22 on page 339.

4. Find the chromatic number and chromatic index of the graph shown in Figure 11.16 on page 365.

5. Let G be a planar graph with smallest cycle length (girth) 6. Let $v_G = |V(G)|$, $e_G = |E(G)|$, and $f_G = |F(G)|$.

   (a) Develop an inequality that relates f_G to e_G.

   (b) Use this to show that
   $$2e_G \leq 3v_G - 6.$$

(c) Show that G must have a vertex of degree less than 3.

(d) Prove that $\chi(G) \leq 3$. (Hint: use induction.)

6. Without doing any actual coloring, give quick lower and upper bounds for the chromatic number and chromatic index of the graph shown in Figure 13.25.

7. Find the chromatic number and chromatic index of the Snakeland map graph shown in Figure 12.20 on page 395.

8. During the Week of Chaos at MathILy 2016, there were five timeslots for classes and four classes offered in each timeslot. Six instructors taught three classes each, and the director taught two classes. Create a potential class schedule.

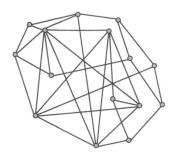

Figure 13.25. A semi-mysterious graph.

9. Find the chromatic number of each graph shown in Figure 12.18 on page 394.

10. Find the chromatic index of each graph shown in Figure 12.18 on page 394.

13.10   Colorful Problems

1. Find the chromatic number of each graph shown in Figure 3.3 on page 73.

2. Find the chromatic index of each graph shown in Figure 3.3 on page 73.

3. If a graph G is isomorphic to another graph H, then is it true that $\chi(G) = \chi(H)$?

4. Find $\chi(K_{2,2,3})$; $K_{2,2,3}$ is shown in Figure 3.9 on page 78. What is $\chi(K_{r,s,t})$? How about $\chi(K_{r,s,t,u,v,w,x,y,z})$?

5. Without doing any actual coloring, give quick lower and upper bounds for the chromatic number of the graph shown in Figure 13.26.

6. Without doing any actual coloring, give quick lower and upper bounds for the chromatic index of the graph shown in Figure 13.26.

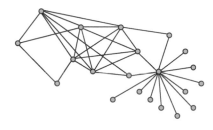

Figure 13.26. A kind of yucky graph.

7. In Example 13.3.10, we had a busy city street that intersected a one-way street. Suppose the traffic signal includes a protected left turn, and remodel the traffic pattern as a graph. How many vertex colors are needed for this graph? Is that more colors, fewer colors, or the same number of colors as was needed before? Now, look at your graph. Can

you change the coloring (but still keep it proper) so that traffic flow will improve?

8. Find the chromatic number of the graph shown in Figure 13.27.

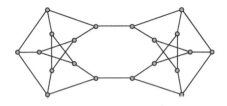

Figure 13.27. A double-Petersenish graph (okay, it's called a Blanuša snark).

9. Find the chromatic index of the graph shown in Figure 13.27.

10. For G as shown in Figure 13.27, use a greedy algorithm to color the edges.

11. Give an algorithm for properly edge-coloring $K_{m,n}$ with the least possible number of colors.

12. True or false: A graph G can be 2-vertex colored $\iff$ G has only even cycles. Explain.

13. True or false: A graph G can be 3-vertex colored $\iff$ G has all cycles of lengths that are multiples of 3. Explain.

14. If a graph G is isomorphic to another graph H, then is it possible that $\chi'(G) \neq \chi'(H)$?

15. Figure 13.28 shows an actual street intersection. Create a graph and properly vertex-color it to find the smallest number of patterns needed in the traffic-light cycle.

Does it give an optimal coloring? If not, then use color switching along paths or trees to improve the coloring.

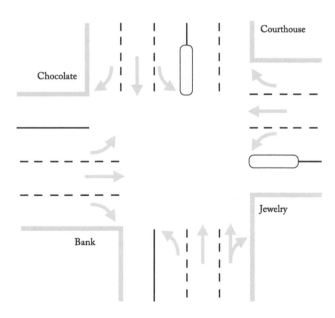

Figure 13.28. The intersection of Main Street with Pleasant/King Street in Northampton, MA.

16. Find the chromatic number and chromatic index for the two graphs shown in Figure 3.22 on page 87.

17. Without doing any coloring, give good lower and upper bounds for the chromatic number of the graph shown in Figure 13.29.

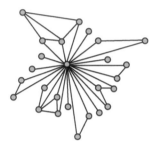

Figure 13.29. A nicely drawn but wacky graph.

18. Suppose G has a Hamilton circuit H. How many colors are required to vertex-color H?

19. Find $\chi(G)$ for G as shown in Figure 13.30.

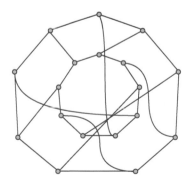

Figure 13.30. A round graph (it doesn't have a name-y name).

20. For G as shown in Figure 13.30, use a greedy algorithm to color the vertices.

Does it give an optimal coloring? If not, then use color switching along paths or trees to improve the coloring.

21. **Challenge:** Show that $\chi'(G) = 3$ for G as shown in Figure 13.30.

22. Let D be the graph shown in Figure 13.31. What are $\chi(D)$ and $\chi'(D)$?

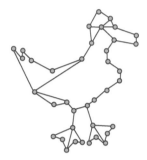

Figure 13.31. A duck graph.

23. Figure 13.32 shows a real street intersection. Create a graph and properly vertex-color it to find the smallest number of patterns needed in the traffic-light cycle. You will have at least one low-degree vertex. How is this useful in terms of traffic flow?

24. Show that if G is 3-regular and has a Hamilton circuit, then $\chi'(G) = 3$.

25. For any connected G, what is the relationship between $\chi(G)$ and $\chi'(G)$?

26. Every tree T has $\chi(T) = 2$. (This is because every tree is bipartite.) Construct a tree (that is not a path) and order the vertices of that tree such that when you color the vertices using a greedy/parsimonious algorithm, you need at least three colors.

27. Extend the idea in our example of a greedy algorithm gone bad (Example 13.5.6): given any $k \geq 3$, construct

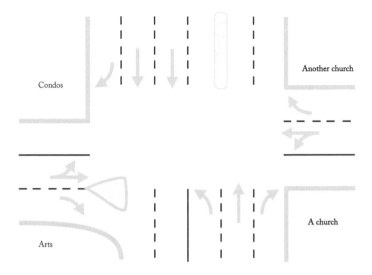

Figure 13.32. The intersection of Main Street with South/State Street in Northampton, MA.

a tree and order the vertices of that tree such that when you color the vertices using a greedy/parsimonious algorithm, you need at least k colors.

28. Compute the chromatic number and chromatic index for the Möbius ladder of size n, as defined on page 400 and shown in Figure 12.30 on page 400.

29. Find $\chi'(R)$ for the Royle graph shown in Figure 11.23 on page 368.

30. Find the chromatic index of the graph shown in Figure 12.29 on page 400.

31. Here is a realistic MathILy Week of Chaos scenario. Brian has planned classes on primes, complexity, and ordinality; Max's classes are on randomness, game theory, and complex analysis; sarah-marie will teach about knots and combinatorial optimization; Hannah is leading projective geometry, fractals, and long division; Tom's classes are combinatorics, strange geometries, and generating functions; Cynthia's one and only class is on tropical geometry. Max has asked Tom to observe his class. Brian, Max, and sarah-marie all want to take Cynthia's class. Is it possible to schedule the classes so that there are five timeslots, with three classes offered in each slot?

32. Find the chromatic number and the chromatic index of the graph shown in Figure 10.31 on page 344.

33. Find the chromatic index of the graph shown in Figure 10.33 on page 344.

34. Hard-working master spy Pvaanzba Ohaf is doing 'round-the-clock surveillance and needs to schedule hir six novice spies (004, Agent 99, Maxwell, Spy X, Boris, Natasha) so that there are two on each 4-hour shift, subject to the

agency rule that no two spies can work together if their names share any letters. Please produce a schedule for Mx. Ohaf.

35. Find the chromatic number and the chromatic index of the graph shown in Figure 11.22 on page 367.

13.11   Instructor Notes

There are only two (very) full days of class work provided in this chapter in order to leave room for a review day at the end of a course. However, it is easy to extend the chapter to three days, as detailed presently.

Ask students to read Section 13.1 as preparation for the first day of class. Students take to coloring like ducks to water, so they need do no further preparation for the first day. Simply take a poll in class: Who prefers vertices? Who prefers edges? Collect the vertex preferrers into working groups and the edge preferrers into different working groups, and distribute those students with no preference evenly. Then ask everyone to work on Section 13.2, with those who are vertex-focused beginning with Section 13.2.1 and those who are edge-focused beginning with Section 13.2.2. As they complete these problems (it should take 15–20 minutes for them to do so), have vertex workers present their results publicly for the benefit of the edge workers, and vice versa. With the remaining time in class, have the students return to working on Sections 13.2.3 and 13.2.4. They will take another 15 or so minutes to work through these, so you may need to stop them in order to have enough time to exchange ideas before the end of the class.

For follow-up, ask students to review the problems in Section 13.2 that they didn't work on in class. Ask them to read Section 13.3—mention that there is additional material there—and do the Check Yourself problems. Start the second class day by asking for questions, and then have students start working in groups again. You may want them to finish Sections 13.2.3 and 13.2.4 before continuing on to Section 13.4. The problems are written to be independent of each other, so if you wanted to save time and still have all problems addressed, you could have different groups work on different problems and present the results to each other. It's likely that this set of problems will take one to two class periods to work through. It is worth mentioning explicitly (at the end of the second class or, if you have a third, at the start of the third) that greedy algorithms are a decent way to color graphs but do not necessarily result in optimal colorings.

Finally, have students read Section 13.5 as reinforcement and extension.

Theme III Supplement: Problems on the Theme of Graph Theory

These problems could be used for studying for (or writing!) in-class or take-home exams, or just for more enrichment. (The problems cover Chapters 10–13.) They are not given in any particular order. Well, they have been intentionally mixed up so that they are *not* in chapter order, so that the solver cannot use the ordering of the problems as a clue in solving them.

1. Find $\chi(G)$ and $\chi'(G)$ for the two graphs shown in Figure 3.24 on page 87.

2. Compute $\chi(W_n)$ for W_n, the wheel with n total vertices.

3. Compute $\chi'(W_n)$ for W_n, the wheel with n total vertices.

4. Figure TIII.1 shows an actual street intersection. Create a graph and properly

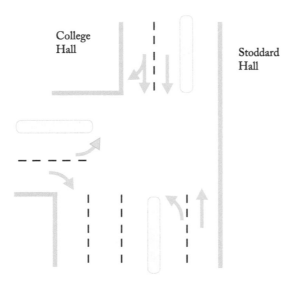

Figure TIII.1. The intersection of Main/Elm Street with West Street in Northampton, MA.

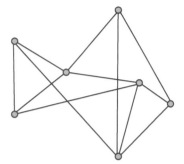

Figure TIII.2. Two graphs.

vertex-color it to find the smallest number of patterns needed in the traffic-light cycle. You will have at least one low-degree vertex. How is this useful in terms of traffic flow?

5. Find every simple graph G with five vertices and the property that if exactly one edge e is added, the resulting graph $G \cup e$ has an Eulerian circuit.

6. Find a spanning tree of the graph at left in Figure TIII.2.

7. Find a spanning tree of the graph at right in Figure TIII.2.

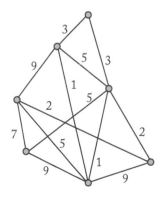

Figure TIII.3. A weighted graph.

8. Find a minimum-weight spanning tree of the graph shown in Figure TIII.3, once using Kruskal's algorithm and once using Prim's algorithm.

9. True (prove) or false (give counterexample)?

(a) A graph with more vertices than edges is connected.

(b) A graph with fewer edges than vertices is not connected.

(c) A graph with $|E(G)| < |V(G)| - 3$ has no cycles.

(d) Two trees with the same degree sequence are isomorphic.

10. Draw three different binary search trees for the micro-dictionary $ace, base, bat, cat, rat, tat$.

11. Find several perfect matchings of the Petersen graph.

12. Suppose G is connected and k-regular and has no Eulerian circuit. Prove that if $\overline{G}$ is connected, then $\overline{G}$ has an Eulerian circuit.

13. Might the simple graph G_L on the left in Figure TIII.4 be planar? (GeoGebra files for Figure TIII.4 are available for

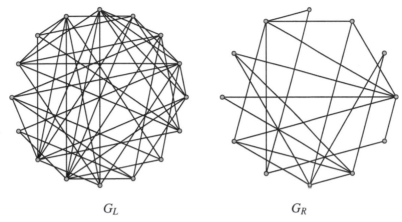

G_L G_R

Figure TIII.4. Could we be planar?

your use at http://www.toroidalsnark. net/dmwdlinksfiles.html.)

14. Might the simple graph G_R on the right in Figure TIII.4 be planar? (GeoGebra files for Figure TIII.4 are available for your use at http://www.toroidalsnark. net/dmwdlinksfiles.html.)

15. Give an example of a graph that has an Euler circuit but not a Hamilton circuit. Explain.

16. Draw a binary decision tree for sorting playing cards.

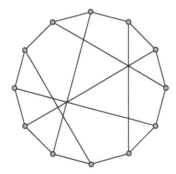

Figure TIII.5. Round like a wheel but still nonplanar.

17. Prove that the graph shown in Figure TIII.5 is nonplanar.

18. Give a greedy algorithm for making change at a cash register.

19. Give an example of a graph that has a Hamilton circuit but not an Euler circuit. Explain.

20. Figure TIII.6 shows a network of canals around artificial islands in a pond, created as a duck playground. White Duck (pictured in Figure TIII.6) swims aimlessly around the playground and then goes to see Grey Duck. White Duck says, "I swam through each of the 16 canal intersections at the playground exactly twice. In fact, here's the order in which I swam: a–o–n–h–k–g–b–m–i–e–p–f–c–l–d–j–c–g–n–h–j–k–d–l–p–i–e–f–b–o–a–m." "The canal intersections aren't usually labeled," replies Grey Duck, "so I don't exactly know the path you took. But you know what? I think you can't be quite right about where you swam." White Duck pauses to think, then responds, "Ah, yes, you're

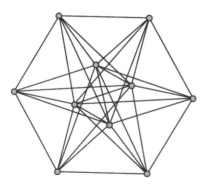

Figure TIII.6. **The duck playground canals.**

right. I switched two of the letters in my order." Can you figure out what White Duck's error was?

21. Prove that the graph shown in Figure TIII.7 is nonplanar.

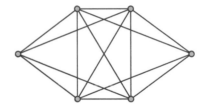

Figure TIII.8. Nonplanar-ness personified.

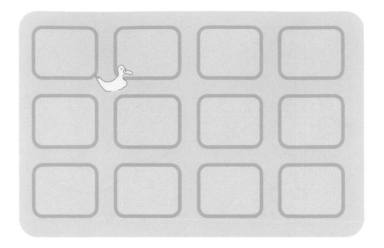

Figure TIII.7. I can be drawn on a doughnut with no crossings….

22. Prove that the graph shown in Figure TIII.8 is nonplanar.

23. Give an example of a nonplanar graph with chromatic number 3.

24. Let G be an n-vertex rooted tree, where each vertex has either 0 or k descendants. Given a fixed k, for which values of n is this possible?

25. Give an example of a graph with largest degree equal to twice the chromatic number.

Does there exist a graph with chromatic number equal to twice the largest degree? If so, give an example, and if not, explain why not.

26. A *bridge* is an edge of a graph G whose removal disconnects the graph. Prove that an edge e is a bridge if and only if it is contained in every spanning tree of G.

27. Does there exist a planar graph with five vertices, ten edges, and seven faces?

28. Let connected simple G have vertices of degrees $4,4,4,5,5,5,6,6,6,7,7,7$. Prove that G is nonplanar.

29. Reza and Rania play a guessing game. Reza picks a whole number in the 0–15 range. How many yes-or-no questions does Rania need to ask Reza in order to determine the number?

30. Rania thinks that the guessing game played with Reza in Problem 29 was too easy. Rania says, "Reza, this time you are allowed to lie to me—but just once—when answering my questions about your number."

    (a) Come up with a general strategy for Rania to use in determining Reza's number.

    (b) Is there a way that Rania can phrase questions so as to obtain, from a given question, the desired information—whether or not Reza is lying?

31. Does there exist a graph with largest degree equal to twice the chromatic index? If so, give an example, and if not, explain why not.

32. Prove or give a counterexample: every subgraph of a nonplanar graph is nonplanar.

33. Consider a sequence of numbers $(d_1, d_2, \ldots, d_{n-1}, d_n)$ such that the sum of the d_i is $2n - 2$, and for consecutive entries d_i, d_{i+1}, it is true that $d_i \le d_{i+1}$.

    (a) What is the largest value that d_1 can have? d_2?

    (b) What is the smallest value that d_n can have?

    (c) Let this sequence be the degree sequence of a graph G. Must G be a tree?

    (d) **Challenge:** Does there always exist a tree with this sequence as its degree sequence?

34. Prove, using induction, that every connected graph contains a spanning tree.

35. Give an example of a graph that does not have an Euler circuit, but does have an Eulerian trail/path (No explanation is needed; just make a drawing of the graph.)

36. Let G be a simple graph with degree sequence $3,4,4,4,5,6,6$. Prove that G is nonplanar.

37. Identify a spanning tree of the graph shown in Figure TIII.9.

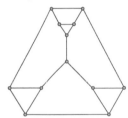

Figure TIII.9. I need a spanning tree.

38. Consider the star graph S_n. In Figure TIII.10, for reference, is S_5.

    (a) Find the chromatic number of S_n. Justify *briefly*.

    (b) Find the chromatic index of S_n. Justify *briefly*.

Figure TIII.10. Consider me.

39. Find a Hamilton *path* on the Petersen graph, shown in Figure TIII.11.

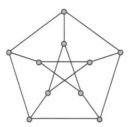

Figure TIII.11. My friend... it's been so long since I've seen you.

40. In general, which is larger: $\chi(tree)$ or $\chi'(tree)$? Explain.

41. Find a minimum-weight spanning tree of the graph shown in Figure TIII.12 (and mention which algorithm you used to produce it).

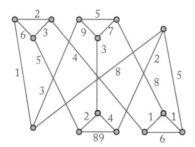

Figure TIII.12.   Help me find a minimum-weight spanning tree!

42. You hang a $3 \times 3 \times 3$ cube of pressed birdseed outside your window in order to attract avian wildlife. A grackle comes every day and eats a $1 \times 1 \times 1$ cubelet of birdseed. You wonder... can it eat the entire cube of birdseed by always eating a cubelet adjacent to the cubelet it ate the previous day?

43. Find $\chi(G)$ and $\chi'(G)$ for the graph shown at left in Figure 3.5 on page 76.

44. Consider the n-cube, where $n \geq 1$. (You may think of this as a graph or as a geometric object.) The vertices of the n-cube are labeled with length-n binary strings, and there is one vertex for every such string. Two vertices $u = (u_1, \ldots, u_n)$ and $v = (v_1, \ldots, v_n)$ are adjacent if and only if u and v differ in exactly one position.

    (a) What is the degree sequence for the n-cube? Explain.

    (b) Let e_n be the number of edges in the n-cube. What are e_1, e_2, e_3, e_4?

    (c) Find a recurrence relation for e_n and explain the presence of each term.

    (d) Find and prove a closed form for e_n. (You may wish to use overcounting rather than using the recurrence relation.)

    (e) For which n is the graph of the n-cube planar? Justify.

45. Find a Hamilton circuit on the graph of the soccer ball (shown in Figure TIII.13).

46. At the Chemical Storage Unit, every day the receiving manager uses a list of deliveries scheduled for the day with the holding bay to which each delivery is assigned. You, as a representative of the Graph Theorists Union, are employed to create this list. Here are the deliveries for the day:

> Methanol
> Potassium hydroxide
> Iodine
> Sodium
> Ammonia
> Nitric acid

You know that ammonia and iodine react to create the explosive nitrogen triio-

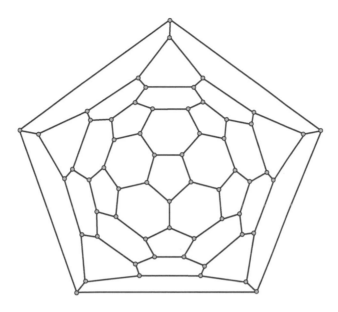

Figure TIII.13. I feel the need for a Hamilton circuit.

dide and that potassium hydroxide and nitric acid both react with water to produce intense heat. So how many holding bays are necessary?

47. Use backtracking to find all the ways to add numbers from $\{1,1,2,2,3\}$ to get 6.

48. Let G be simple, planar, connected, and 3-regular. If f_k is the number of k-sided faces, prove that $3f_3 + 2f_4 + f_5 - f_7 - 2f_8 - 3f_9 - \cdots = 12$.

49. Find an Euler circuit in the graph shown in Figure TIII.14.

50. The Duck Lab has six 3D printers, labeled anatra, canard, lacha, papra, rosë, and utka. Because vibrations from one 3D printer will affect the print of another 3D printer, most of the printers cannot be running at the same time. Be-

cause of careful placement within the lab, the following pairs of printers can be run at the same time: anatra and papra, anatra and rosë, anatra and utka, canard and rosë, and lacha and papra. The Duck Lab needs to make one print on each printer, and each takes an hour to complete the job. What is the shortest amount of time in which this can be done?

Figure TIII.14. A graph with teal vertices.

Credit where credit is due: Problem 20 was adapted from [11]. Problem 50 was contributed by Tom Hull.

Part IV
Other Material

Chapter 14

Probability and Expectation

14.1   Introduction and Summary

You may have some previous experience with probability. If so, please put it aside and approach this chapter with a fresh and open mind.

Because this is a discrete mathematics book, we are going to restrict ourselves to discrete probability. Effectively, we will only consider situations in which there are a finite number of events that may happen. This makes a lot of things easier. In fact, some things are so much easier that they appear to simply be counting problems! However, they aren't *just* counting problems. The concepts that lead to these enumerations have much deeper meaning. If you approach such problems as merely computations, you may fall prey to misconceptions. Focus instead on the big ideas and let the computations be the pleasurable end of reducing harder problems to easier problems.

We begin by explaining what probability is and how to calculate the most basic of probabilities. Then the idea of random variable is introduced, and we define the expected value of a random variable. Within discrete probability, expected value is a fundamental concept that underpins almost all further study of probability. The first Try This! lets you practice calculating basic probabilities and expected values. With this experience with random variables in hand, you can then learn about calculating more complicated probabilities—for values of random variables under various constraints. A second Try This! gives many fun conditional probability calculations; you'd better remember PIE to be able to complete them! We then return to expected value so that you can get a glimpse of the most common use of probability in discrete mathematics—the probabilistic method.

To avoid potential confusion, our strategy will be to use a precise and detailed approach. A benefit of this comparatively elaborate introduction is to use language that is consistent with more advanced treatments of probability, so that when you encounter them (as surely you will!) you will recognize the terminology and ideas.

14.2   What Is Probability, Exactly?

According to English dictionaries, a *probability* is the likelihood that some given event will occur. So in mathematics, the study of probability is the study of likelihoods of events. In order to focus on discrete probability (hey, this *is* discrete mathematics), we will only consider situations where there are a finite number of different events that might occur.

Example 14.2.1. When flipping a fair coin, it is equally likely that it will land heads up and that it will land tails up. The probability of it landing heads up is $\frac{1}{2}$, as is the probability of it landing tails up. When rolling a fair die, it is equally likely that it will land with any particular face pointing towards the viewer. If the die has six faces, then the probability of rolling a 2 is $\frac{1}{6}$ (as is the probability of rolling a 1, 3, 4, 5, or 6). If the die has eight faces, then the probability of rolling a 2 is $\frac{1}{8}$ (as is the probability of rolling a 1, 3, 4, 5, 6, 7, or 8).

Definition 14.2.2. A *state space* is the set of different *states* or configurations that a system might have. It is also called a *sample space*.

Example 14.2.3 (of state spaces). For an ordinary light switch, the state space is $S_L = \{on, off\}$. For a coin to be flipped, the state space is $S_C = \{heads, tails\}$. Rolling a fair six-sided die has state space $S_D = \{roll\ 1, roll\ 2,\ roll\ 3, roll\ 4, roll\ 5, roll\ 6\}$. A system with a light switch and a coin has state space $S_{LC} = \{on/heads, on/tails, off/heads, off/tails\}$. We may also think of this state space as the Cartesian product $S_L \times S_C = \{(on, heads), (on, tails), (off, heads), (off, tails)\}$.

Notice that a system cannot be in more than one state at the same time. This means that states are *exclusive*. This is different from the possible *events* that may occur. An event describes a collection of states; it is a subset of the state space.

Example 14.2.4. Using a fair die, we know that the probability of rolling an even number is $\frac{1}{2}$. Here, the event is *rolling an even number*. Another die-related event is *rolling a number greater than* 3 (for which the probability is also $\frac{1}{2}$ on a six-sided die). Notice that these two events are not exclusive: the state *roll 4* is a possibility for each of the two events.

Example 14.2.5 (of states vs. events). Suppose you toss two fair coins in the air. When they land and flop, there might be two heads, one head and one tail, or two tails. This might suggest that the state space is $\{head/head, head/tail, tail/tail\}$. But that's not true—the two coins are distinct, so the states are $\{head_1/head_2, head_1/tail_2, tail_1/head_2, tail_1/tail_2\}$. Our original list of coin-landing possibilities was a list of events rather than states.

Example 14.2.5 exhibits but one of the many ways in which we can fool ourselves when approaching probability problems. Let us define probability precisely so that we may avoid errors. The definition may seem technical, but it translates into our common-sense understanding of probability.

Definition 14.2.6. Let $P : \mathscr{P}(S) \to [0,1]$ be a function from all subsets of a state space to the unit interval $\{x \mid 0 \le x \le 1\}$. The function P measures probability if the following are true:

- $P(S) = 1$, that is, *some* event definitely occurs. In other words, the probability that the system is in *some* state is one.

- $P(\emptyset) = 0$, or there is no possibility that nothing happens. In other words, the probability that the system is not in any state is zero.

- For any element $s \in S$, $P(s) \ge 0$ and likewise for any subset (event) $E \subset S$, $P(E) \ge 0$. That is, it does not make sense to have a negative probability.

- Whenever two events E_1, E_2 are exclusive, meaning they cannot happen at the same time, $P(E_1 \text{ or } E_2) = P(E_1) + P(E_2)$. Note that any two states are exclusive, so their probabilities always have this property.

These conditions are known as the *probability axioms*.

A consequence of Definition 14.2.6 is that probabilities are unitless. Notice that when we measure the probability of an event, or subset of states, we are actually measuring the probability that at least one of those states occurs. You can also think of this as the probability that (for example) s_1 or s_2 or s_3 occurs, in the logic sense of *or* (see Chapter 2).

Example 14.2.7. Examining the light switch again, $P(\{on, off\}) = 1$ and $P(\emptyset) = 0$ because the switch is either on or off. It is possible that a switch is on or off, so $P(on), P(off) \ge 0$; because a switch cannot be both on and off, we know that $P(on, off) = P(on) + P(off)$ or, in other words, $P(on) + P(off) = 1$, which is what we would think should be true.

Now, how do we decide what $P(on)$ and $P(off)$ are? This depends on our specific light switch! If the light switch is in a common room with many windows, it would be reasonable to say $P(on) = .5$ and $P(off) = .5$ because the lights will be on whenever it's dark. If the light switch is in a little-used basement, then it would be reasonable to say $P(on) = .1$ and $P(off) = .9$ (or even $P(on) = .05$ and $P(off) = .95$, or...). It is only when we know events occur equally often that we can assign them the same probabilities.

Just to reinforce a point: it's easy to get confused when trying to reason intuitively about probability. Seeming paradoxes can arise. (Their study is very interesting but does not pertain to this course.) So if you find yourself in a probability situation that does not make sense, first ask yourself, What is your state space? What are the events? Have you assigned probabilities so that the axioms hold?

Check Yourself ──────────────────────────────

You should do all of these problems to make sure you are ready for the ensuing sections.

1. What is the probability of rolling a 12 using a fair 20-sided die?

2. List the elements of the state space for flipping three fair coins at once.

3. List the elements of the state space for rolling an eight-sided die.

4. Determine the probability for each state in Example 14.2.5. What is the probability of getting one head and one tail? To what subset of the state space does this event correspond?

5. **Challenge:** Invent your own situation and list the elements of the corresponding state space.

14.3 High Expectations

In order to consider events in a more general way, we will find the next definition essential.

Definition 14.3.1. A *random variable* is actually a function (yes, even though it is called a variable) $X : S \to N$ from a state space S to a finite set of real numbers N. (This makes sense because S is also finite.) Usually N has units related to the situation at hand.

Example 14.3.2. Consider the state space of possible rolls of an eight-sided die, S_8. The random variable $O : S_8 \to \{0,1\}$ indicates when the roll is an odd value; that is, $O(roll\ 1) = 1$, $O(roll\ 2) = 0$, $O(roll\ 3) = 1$, $O(roll\ 4) = 0$, $O(roll\ 5) = 1$, $O(roll\ 6) = 0$, $O(roll\ 7) = 1$, and $O(roll\ 8) = 0$.

Another random variable returns the value of a roll, so we define $V : S_8 \to \{1,2,3,4,5,6,7,8\}$ to give $V(roll\ k) = k$.

We can also define $T : S_8 \to \{0,1\}$, which indicates whether or not the roll is a 2. That is, $T(roll\ 1) = 0, T(roll\ 2) = 1, T(roll\ 3) = 0, T(roll\ 4) = 0, T(roll\ 5) = 0, T(roll\ 6) = 0, T(roll\ 7) = 0$, and $T(roll\ 8) = 0$.

When we see the expression $X(s)$ (where X is a random variable and $s \in S$), it represents a number. Conversely, when we see the expression $X = k$, this defines a subset of S—it's those $s \in S$ such that $X(s) = k$. In other words, $X = k$ defines an event on the state space.

Example 14.3.3. Using the notation from Example 14.3.2, $O = 1$ represents $\{roll\ 1, roll\ 3, roll\ 5, roll\ 7\} \subset S_8$. Similarly, $V = 1$ represents $\{roll\ 1\} \subset S_8$, $T = 1$ represents $\{roll\ 2\} \subset S_8$, and $T = 0$ represents $\{roll\ 1, roll\ 3, roll\ 4, roll\ 5, roll\ 6, roll\ 7, roll\ 8\} \subset S_8$.

You can think of a random variable as assigning weights to states (like assigning weights to vertices or edges in a graph). *However,* a random variable represents a spectrum of possibilities (the possible values in N), and in this sense it is variable.

You might think of a random variable X as being like a whirlwind: it's indeterminate and constantly spinning around, but with only a finite number of whirling things it can drop when it stops. When we have $X = value$, that's like having all states s with $X(s) = value$ glow green. In this visualization, $P(X = value)$ makes sense because it's the probability that when you reach into the whirlwind, you pull out some state that evaluates to *value*. And, $P(X)$ must be 1 because you can pull *something* out of the whirlwind.

Example 14.3.4. Consider four ducks that live together: one is white, one is white with grey spots, one is grey with white spots, and one is black with white spots. Together, all subsets of these four ducks form our state space C_d.

We may define a random variable $W : C_d \to \{0, 1, 2, 3, 4\}$ that counts the number of ducks in a given subset that have white on them. For a single duck, $W(d) = 1$ if the duck in question has some white, and $W(d) = 0$ if the duck has no white on it. For a subset A of the ducks, $W(A) = |A|$ because every one of our ducks has some white on it.

Similarly, define the random variable $WH : C_d \to \{0, 1\}$ as $WH(d) = 1$ if the duck in question is all white, and $WH(d) = 0$ if the duck is not all white. Then for a subset A of the ducks, $WH(A) = 1$ if the all-white duck is in A, and $WH(A) = 0$ if the all-white duck is not in A.

Definition 14.3.5. The *probability distribution* of a random variable X is a display of all possible values for X with the corresponding probabilities.

Example 14.3.6. Let us determine the probability distribution of the random variable W defined in Example 14.3.4. To do this, we need to know how to evaluate $P(W = k)$ for the possible values of k. Suppose it is equally likely that we see any subset of the ducks. There are $2^4 = 16$ subsets of the four ducks, and of those one has zero ducks, four contain one duck, six contain two ducks, four contain three ducks, and one contains all four ducks. Thus, $P(W = 0) = \frac{1}{16}$, $P(W = 1) = \frac{4}{16}$, $P(W = 2) = \frac{6}{16}$, $P(W = 3) = \frac{4}{16}$, and $P(W = 4) = \frac{1}{16}$. This list is one way of giving the probability distribution. We might also use the following table.

Number of ducks with white ($W = k$)	Associated probability ($P(W = k)$)
0	$\frac{1}{16}$
1	$\frac{4}{16}$
2	$\frac{6}{16}$
3	$\frac{4}{16}$
4	$\frac{1}{16}$

To obtain the probability distribution for the random variable WH, we may look at the duck subsets and discover that there are eight subsets that include the all-white duck and eight subsets without the all-white duck. Thus, $P(WH = 0) = \frac{1}{2} = P(WH = 1)$, which we could show as in Figure 14.1.

Definition 14.3.7. The *expected value* of a random variable X is denoted $\mathbb{E}[X]$ and is defined as $\mathbb{E}[X] = \sum_{k \in N} kP(X = k)$. Here, N is the target space of X, so the expected value formula is adding up the probabilities of all possible values of the random variable, each weighted by that value.

It is often useful to produce the probability distribution for X before computing $\mathbb{E}[X]$.

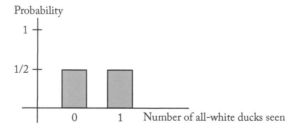

Figure 14.1. The uninteresting probability distribution for the random variable WH.

Example 14.3.8. Let us compute the expected value for each of the two random variables defined in Example 14.3.4. By definition, $\mathbb{E}[W] = 0P(W = 0) + 1P(W = 1) + 2P(W = 2) + 3P(W = 3) + 4P(W = 4)$. We use the probability distribution from Example 14.3.6 to obtain $\mathbb{E}[W] = 0 \cdot \frac{1}{16} + 1 \cdot \frac{4}{16} + 2 \cdot \frac{6}{16} + 3 \cdot \frac{4}{16} + 4 \cdot \frac{1}{16} = 2$ white ducks. The practical interpretation is that we expect to see two ducks when we go visiting.

Similarly, $\mathbb{E}[WH] = 0 \cdot \frac{1}{2} + 1 \cdot \frac{1}{2} = \frac{1}{2}$ white duck. That is, half the time we expect to see the white duck.

In Definition 14.3.7, we say, "the expected value formula is adding up the probabilities of all possible values of the random variable, each weighted by that value." This statement could be rephrased to say, "expected value is the sum of all the probabilities of all the states, each weighted by the value of the random variable at that state." This phrasing differs by focusing on the states rather than on the values of the random variable. In symbols, this says expected value is $\sum_{s \in S} X(s)P(s)$. Does this truly give the same results as Definition 14.3.7?

Example 14.3.9. Let's return to S_8. By Definition 14.3.7, we have $\mathbb{E}[O] = 0 \cdot P(O = 0) + 1 \cdot P(O = 1) = 0 \cdot \frac{1}{2} + 1 \cdot \frac{1}{2} = \frac{1}{2}$. What does $\sum_{s \in S_8} O(s)P(s)$ give us?

$$\sum_{s \in S_8} O(s)P(s) = O(roll\ 1)P(roll\ 1) + O(roll\ 2)P(roll\ 2) + O(roll\ 3)P(roll\ 3)$$

$$+ O(roll\ 4)P(roll\ 4) + O(roll\ 5)P(roll\ 5) + O(roll\ 6)P(roll\ 6)$$

$$+ O(roll\ 7)P(roll\ 7) + O(roll\ 8)P(roll\ 8)$$

$$= 1 \cdot \frac{1}{8} + 0 \cdot \frac{1}{8} + 1 \cdot \frac{1}{8} + 0 \cdot \frac{1}{8} + 1 \cdot \frac{1}{8} + 0 \cdot \frac{1}{8} + 1 \cdot \frac{1}{8} + 0 \cdot \frac{1}{8}$$

$$= 4 \cdot \frac{1}{8} = \frac{1}{2} = \mathbb{E}[O].$$

So far, so good.

We'll compute for the other two random variables we defined on S_8 as well:

$$\mathbb{E}[V] = 1 \cdot P(V = 1) + 2 \cdot P(V = 2) + 3 \cdot P(V = 3) + 4 \cdot P(V = 4)$$

$$+ 5 \cdot P(V = 5) + 6 \cdot P(V = 6) + 7 \cdot P(V = 7) + 8 \cdot P(V = 8)$$

$$= 1 \cdot \frac{1}{8} + 2 \cdot \frac{1}{8} + 3 \cdot \frac{1}{8} + 4 \cdot \frac{1}{8} + 5 \cdot \frac{1}{8} + 6 \cdot \frac{1}{8} + 7 \cdot \frac{1}{8} + 8 \cdot \frac{1}{8}$$

$$= (1 + 2 + 3 + 4 + 5 + 6 + 7 + 8)\frac{1}{8} = 36 \cdot \frac{1}{8} = \frac{9}{2},$$

and on the other hand

$$\sum_{s \in S_8} V(s)P(s) = V(roll\ 1)P(roll\ 1) + V(roll\ 2)P(roll\ 2) + V(roll\ 3)P(roll\ 3)$$

$$+ V(roll\ 4)P(roll\ 4) + V(roll\ 5)P(roll\ 5) + V(roll\ 6)P(roll\ 6)$$

$$+ V(roll\ 7)P(roll\ 7) + V(roll\ 8)P(roll\ 8)$$

$$= 1 \cdot \frac{1}{8} + 2 \cdot \frac{1}{8} + 3 \cdot \frac{1}{8} + 4 \cdot \frac{1}{8} + 5 \cdot \frac{1}{8} + 6 \cdot \frac{1}{8} + 7 \cdot \frac{1}{8} + 8 \cdot \frac{1}{8} = \frac{9}{2}$$

as before. Excellent.

Now, $\mathbb{E}[T] = 0 \cdot P(T = 0) + 1 \cdot P(T = 1) = 0 \cdot \frac{7}{8} + 1 \cdot \frac{1}{8} = \frac{1}{8}$ by Definition 14.3.7. Using the state-focused computation instead, we have $T(roll\ 1)P(roll\ 1) + T(roll\ 2)P(roll\ 2) + T(roll\ 3)P(roll\ 3) + T(roll\ 4)P(roll\ 4) + T(roll\ 5)P(roll\ 5) + T(roll\ 6)P(roll\ 6) + T(roll\ 7)P(roll\ 7) + T(roll\ 8)P(roll\ 8) = 0 \cdot \frac{1}{8} + 1 \cdot \frac{1}{8} + 0 \cdot \frac{1}{8} + 0 \cdot \frac{1}{8} + 0 \cdot \frac{1}{8} + 0 \cdot \frac{1}{8} + 0 \cdot \frac{1}{8} + 0 \cdot \frac{1}{8} = \frac{1}{8}$. Yup! The same.

It seems like the two ways of looking at expected value give the same results... at least for the examples we checked. How can we see that these two perspectives are secretly the same? We'll have to prove it.

> Lemma 14.3.10. $\displaystyle\sum_{k \in N} kP(X = k) = \sum_{s \in S} X(s)P(s).$

Proof: Notice first that $P(X = k)$ is $\sum_{s \text{ such that } X(s)=k} P(s)$. In words, the probability that X takes on the value k is the sum of the probabilities of the states at which X has the value k. This makes sense because any two states are exclusive (they cannot happen at the same time).

Now, we have that $\sum_{k \in N} kP(X = k) = \sum_{k \in N} k \sum_{s \text{ such that } X(s)=k} P(s)$. Here the inner sum does not depend on k, so we can push k inside to obtain the expression $\sum_{k \in N} \sum_{s \text{ such that } X(s)=k} kP(s)$. But wait! *Inside* the inner sum, each occurrence of k is linked to a particular s because we have restricted to s with $X(s) = k$. Therefore, $k = X(s)$ and we may rewrite as $\sum_{k \in N} \sum_{s \text{ such that } X(s)=k} X(s)P(s)$. Now the double sum can be seen as a single sum as follows. The inner sum adds over a batch of k-related states and then the outer sum adds the batches (over k), which is also just adding over all possible states s. We now arrive at $\sum_{s \in S} X(s)P(s)$ and our proof is complete. $\square$

Lemma 14.3.10 is useful because in some situations it's simpler to compute $P(s)$ for all states s, and in others (for example, C_d), it's easier to compute $P(X = k)$ for all values k (that is, the probability distribution of X).

We could view a state space and random variable as a box containing a whirl-wind of indeterminate states. When we open the box, we see what state the system is in (and can figure out what corresponding value the random variable has). Then, when we compute expected value, we are asking what result we expect when we open the box—*not* what *state* we expect, but what *value for the random variable* we expect. This is because different states can give the same value to a random variable. This also means we can expect a result that corresponds to no individual state.

Example 14.3.11. Suppose we have two excitable children Adrian and Maryam who have chairs a and m. Every few minutes some happy music starts, and the two children bop around until a bell rings, at which time they sit down again in the nearest chairs. Either Adrian sits in chair a and Maryam sits in chair m (state Aa/Mm) or Adrian sits in chair m and Maryam sits in chair a (state Am/Ma). How many children do we expect will be sitting in their own chairs at any given time?

First, let us identify a random variable: it's the number of children sitting in their own chairs at some time. Next, we identify the state space $S = \{Aa/Mm, Am/Ma\}$. We can now figure out what values the random variable should take on for each element of the state space. So, $X(Aa/Mm) = 2$ children sitting in their own chairs and $X(Am/Ma) = 0$ children sitting in their own chairs. More formally, we have that $X : S \to \{0, 2\}$, and the units on $\{0, 2\}$ are *children sitting in their own chairs*.

Because the children are excitable and boppy, it is equally likely that the first one to sit down will land in hir own chair or in the other chair. Thus, $P(Aa/Mm) = P(Am/Ma) = .5$. These states are exclusive and form the whole state space, so $P(Aa/Mm) + P(Am/Ma) = 1$.

Now we can compute the expected value $\mathbb{E}[X]$. By definition, $\mathbb{E}[X] = 0P(X=0) + 2P(X=2)$. When $X = 2$, we can only have the state Aa/Mm and so $P(X=2) = \frac{1}{2}$. Likewise, when $X = 0$ we must have the state Am/Ma and so $P(X=0) = \frac{1}{2}$. This gives $\mathbb{E}[X] = 2 \cdot \frac{1}{2} = 1$. We can also use Lemma 14.3.10 to compute $\mathbb{E}[X]$. We have only two states, so there will be only two terms: $X(Aa/Mm)P(Aa/Mm) + X(Am/Ma)P(Am/Ma) = 2 \cdot 0.5 + 0 \cdot 0.5$ children sitting in their own chairs $= 1$ child sitting in hir own chair. Notice that this expected value does not correspond to any possible state of the state space! Expected value can be weird.

Let's go back to the start of our question again. We asked, "How many children do we expect will be sitting in their own chairs at any given time?" The word *expect* told us we should do an expected value calculation. But we also could have asked,

"What's the average number of children who will be sitting in their own chairs at any given time?" For a discrete probability situation this, too, would have been answered by our expected value calculation. We will explore why this is generally true in Problem 6 of Section 14.12. For this case, we can calculate the average number of children sitting in their own chairs as follows: an average consists of the sum of the values of all possible states divided by the total number of states. Here, that gives us $\frac{2+0}{2} = 1$.

Check Yourself

Do at least one of the first three of these problems.

1. Compute the expected value for each of the random variables defined in Example 14.3.4, assuming that you always see all four ducks when you go visiting. Compute the probability distribution, use the definition of expected value, and then use Lemma 14.3.10.

2. Define a random variable B describing the number of black ducks you see when you visit the ducks from Example 14.3.4.

    (a) What is N?

    (b) Give $P(B = n)$ for each $n \in N$ (the probability distribution of B).

    (c) Compute $\mathbb{E}[B]$.

3. Define a random variable G, the number of grey ducks you see when you visit the ducks from Example 14.3.4, and compute its expected value.

4. **Challenge:** For the state space you created in the Check Yourself Challenge in Section 14.2, define at least one random variable.

14.4 You Are Probably Expected to Try This!

Explore the relationships between probability and expectation with your peers.

1. Suppose you have two fair dice, a 12-sided die and a four-sided die.

    (a) Describe the state space of rolling these two dice.

    (b) What is $P((roll\ 11, roll\ 3))$?

    (c) What is $P((roll\ 11, roll\ 3)$ or $(roll\ 12, roll\ 2))$?

    (d) What is $P(roll$ a total of $14)$?

  (e) What is $P(roll$ a total not equal to 14)?

  (f) Define a random variable F that describes whether or not the total rolled is 14.

  (g) Compute $\mathbb{E}[F]$.

2. Suppose you have two fair six-sided dice, one red and one blue.

  (a) Describe the state space of rolling these two dice.

  (b) Consider the event of rolling an even sum, and define a random variable E that describes whether or not this event happens.

  (c) Compute $\mathbb{E}[E]$.

  (d) Consider the number of odd numbers rolled, and define a random variable O that counts the number of odd numbers rolled.

  (e) Compute $\mathbb{E}[O]$. (Be sure to note the units on the result.)

  (f) What changes, if anything, if both of your dice are red?

3. In early December 2011, the Yarn Harlot felt "a mitten thing coming on" and, with eight colors of yarn—orange, red, yellow, lime green, forest green, robin's egg blue, indigo, and black—knitted a pile of mittens (true story). Suppose she knitted eight pairs of mittens, one left mitten and one right mitten of each base color, and placed them in the mitten box.

  (a) Describe the state space of pulling a mitten from the box.

  (b) What is $P(\text{left mitten})$?

  (c) What is $P(\text{cool color of mitten})$? (Greens, blues, and purples are cool colors. Reds, oranges, and yellows are warm colors.)

  (d) What is $P((\text{left mitten}) \text{ or } (\text{cool color of mitten}))$?

  (e) What is $P((\text{left mitten}) \text{ and } (\text{cool color of mitten}))$?

  (f) Design your own random variable R on this state space and compute $\mathbb{E}[R]$.

14.5   Conditional Probability and Independence

Consider a single fair die with six sides and state space $\{roll\ 1, roll\ 2, roll\ 3, roll\ 4, roll\ 5, roll\ 6\}$. There is a natural random variable X measuring the number of pips showing after rolling the die, so that $X(roll\ k) = k$ pips. However, there are some other random variables that can be defined on this state space. Consider Y indicating the parity of the number of pips showing after rolling the die, Z indicating whether the number of pips showing after rolling the die is less than 5, and W revealing whether the die rolls a 3 or not. For these three random variables, we have

$$Y(roll\ k) = \begin{cases} 1 & k \text{ is even,} \\ 0 & k \text{ is odd;} \end{cases} \quad Z(roll\ k) = \begin{cases} 1 & k \geq 5, \\ 0 & k < 5; \end{cases} \quad W(roll\ k) = \begin{cases} 1 & k = 3, \\ 0 & k \neq 3. \end{cases}$$

We can measure probabilities of random variables having particular values. So, for example, we could compute $P(X = 1)$, which is the same as $P(roll\ 1) = \frac{1}{6}$. We could also compute $P(Y = 1)$, which is the same as $P(roll\ 2 \text{ or } roll\ 4 \text{ or } roll\ 6)$ $= P(roll\ 2) + P(roll\ 4) + P(roll\ 6) = \frac{1}{2}$. Similarly, $P(Z = 1) = P(roll\ 5 \text{ or } roll\ 6)$ $= P(roll\ 5) + P(roll\ 6) = \frac{1}{3}$ and $P(W = 1) = P(roll\ 3) = \frac{1}{6}$.

Now, let us consider $P(Y = 1 \text{ and } Z = 1)$. The only state in which both $Y = 1$ and $Z = 1$ is when a 6 is rolled. Therefore, $P(Y = 1 \text{ and } Z = 1) = \frac{1}{6}$.

Compare the probability of rolling an even value given that the only values possible are 5 and 6 to the probability of rolling an even value when all six values 1, 2, 3, 4, 5, and 6 are possible; in each case, half of the values are even. We have $P(Y = 1 \text{ and } Z = 1) = \frac{1}{2} \cdot \frac{1}{3} = \frac{1}{6}$ and $P(Y = 1 \text{ and } s \in S) = \frac{1}{2} \cdot 1 = \frac{1}{2}$. Thinking of this information differently, we see that

$$\frac{P(Y = 1 \text{ and } Z = 1)}{P(Z = 1)} = \frac{\frac{1}{6}}{\frac{1}{3}} = \frac{1}{2} \quad \text{and} \quad \frac{P(Y = 1)}{P(s \in S)} = \frac{\frac{1}{2}}{1} = \frac{1}{2}.$$

This more closely reflects what we were trying to measure, which is the probability of something happening (in this case, $Y = 1$) when the state space is restricted (in the first case, to states where $Z = 1$ and in the second case, to all states (which is not a restriction at all)).

Definition 14.5.1. The *conditional probability* that event E_1 happens, given that E_2 definitely occurs, is denoted $P(E_1 | E_2)$ and is measured by $\frac{P(E_1 \text{ and } E_2)}{P(E_2)}$.

Example 14.5.2. Continuing with our die-roll random variables, let us compute the probability that $W = 1$ given that $Y = 1$. Intuitively, we expect that this probability should be zero, because 3 is not even. Indeed, there are no states where $W = 1$ and $Y = 1$, so the numerator of $\frac{P(W=1 \text{ and } Y=1)}{P(Y=1)}$ is zero.

Now let us compute $P(Y = 1 | W = 0)$. There are three of the six states for which $Y = 1$ and $W = 0$, so $P(Y = 1 \text{ and } W = 0) = \frac{1}{2}$. There are five states for which $W = 0$, so $P(W = 0) = \frac{5}{6}$. Thus, $P(Y = 1|W = 0) = \frac{3}{5}$. We could also directly notice that if $W = 0$, there are five states to consider, of which three have the property $Y = 1$.

We will also compute the probability that $W = 0$ given that $Z = 0$. There are four states for which $Z = 0$, of which three (rolling a 1, 2, or 4) have $W = 0$; thus, $P(W = 0|Z = 0) = \frac{3}{4}$. We could instead use the definition and note that there are three states when $W = 0$ and $Z = 0$, and six states total, so $P(W = 0 \text{ and } Z = 0) = \frac{1}{2}$. There are four states for which $Z = 0$, so $P(Z = 0) = \frac{2}{3}$. The quotient of these fractions is $\frac{3}{4}$.

Finally, we will compute $P(Z = 0|W = 0)$. We know that $P(W = 0 \text{ and } Z = 0) = \frac{1}{2}$. There are five states for which $W = 0$, so $P(W = 0) = \frac{5}{6}$. Thus, $P(Z = 0| W = 0) = \frac{3}{5}$.

Example 14.5.3. The dread disease Dread Disease (DD) occurs in 0.000004 of the population. (It's rare—four out of every million people get it.) Thankfully, we can test for DD using ToDD (the Test of Dread Disease). If someone has DD, then ToDD is positive 95% of the time. (This measure is called the *sensitivity* of a test.) If someone *does not* have DD, then ToDD is positive 3% of the time. (The measure of someone *not* having a disease testing *negative* is called the *specificity* of a test.) Given that someone has a positive ToDD, what is the chance this person has DD?

First, we translate to probability language: we seek $P(\text{DD}|\text{ToDD}+)$. To compute this, we need to know $P(\text{DD and ToDD}+)$ and $P(\text{ToDD}+)$. However, we are not given either of these quantities! How do we compute them?

To find $P(\text{DD and ToDD}+)$, we multiply the incidence of DD in the population (0.000004) by the percentage of DD-ridden people who test positive (95%), and get 0.0000038. This basically means that 38 of every 10 million people have DD and test positive for DD. Finding $P(\text{ToDD}+)$ is slightly more challenging. Notice that someone who tests positive either has DD or doesn't, and these are exclusive of each other, so $P(\text{ToDD}+) = P(\text{DD and ToDD}+) + P(\text{not-DD and ToDD}+)$. Thus we compute $P(\text{ToDD}+) = 0.0000038 + (1 - .000004)(.03) = 0.0300037$, so slightly more than 3% of the population tests positive for DD.

Putting this all together, $P(\text{DD}|\text{ToDD}+) = 0.0000038/0.0300037$, which is 0.000126651 or .012%, so we may conclude that the chance of someone with a positive ToDD has about a one-in-ten-thousand chance of having DD. (That's much higher than the four-in-a-million chance in the general population, however!) A fun fact—or really, not-so-fun fact—studies show that physicians consistently and wildly misestimate the answer to this type of question; see "Simple tools for understanding risks," *BMJ* (2003), available at https://www.ncbi.nlm.nih.gov/pmc/articles/PMC200816/.

Example 14.5.4. We have a group of friends, of whom $\frac{2}{3}$ own cats. A local veterinary office sends out vaccination reminder postcards; $\frac{3}{4}$ of the cat owners receive these postcards and $\frac{1}{5}$ of the friends who do not own cats receive these postcards.

What fraction of the friends receive postcards?

Let's see. Informally, three-quarters of the $\frac{2}{3}$ cat-owning friends receive postcards, which is $\frac{1}{2}$ of the friends. Also, $\frac{1}{5}$ of the $\frac{1}{3}$ non-cat-owning friends receive postcards, which is $\frac{1}{15}$ of the friends. Together, this accounts for $\frac{1}{2} + \frac{1}{15} = \frac{17}{30}$ of the friends receiving postcards.

More formally, our state space is the set of friends, where each friend F has two attributes (owning/not owning a cat and receiving/not receiving a postcard). There are two random variables that naturally arise from this situation,

$$C(F) = \begin{cases} 1 & F \text{ owns a cat,} \\ 0 & F \text{ owns no cats,} \end{cases} \quad \text{and} \quad R(F) = \begin{cases} 1 & F \text{ receives a postcard,} \\ 0 & F \text{ receives no postcard.} \end{cases}$$

We want to know $P(R = 1)$, the probability that a friend receives a postcard. We can view this as the probability that a friend receives a postcard and owns a cat plus the probability that a friend receives a postcard and owns no cats.

In notation, we write the fact that $\frac{2}{3}$ of our friends own cats as $P(C = 1) = \frac{2}{3}$. What about the fact that $\frac{3}{4}$ of our cat-owning friends receive postcards? Ah, this is conditional probability: we know for sure that these friends own cats, and given this, we measure the probability that they get postcards. We see that $P(R = 1 | C = 1)P(C = 1) = \frac{1}{2}$. Similarly, we have $P(R = 1 | C = 0)P(C = 0) = \frac{1}{15}$. Finally, we compute $P(R = 1) = \frac{1}{2} + \frac{1}{15} = \frac{17}{30}$.

Now: if a postcard arrives, what is the probability that the recipient owns a cat? In other words, we want to measure the probability that a friend owns a cat given that ze has received a postcard, or $P(C = 1 | R = 1)$. By definition, this is $\frac{P(C=1 \text{ and } R=1)}{P(R=1)}$. We also know that $P(C = 1 \text{ and } R = 1) = P(R = 1 \text{ and } C = 1) = P(R = 1 | C = 1)P(C = 1) = \frac{1}{2}$. Thus, $P(C = 1 | R = 1) = \frac{1/2}{17/30} = \frac{15}{17}$.

The idea of conditional probability leads to another frequently used concept.

Definition 14.5.5. Two events A and B are *independent* if $P(A|B) = P(A)$ and $P(B|A) = P(B)$. In other words, if the probability that A happens is the same whether B happens or not, and if the probability that B happens is the same whether A happens or not, then events A and B are independent of each other.

> Interesting fact.   If A and B are independent, then $P(A|B) = \frac{P(A \text{ and } B)}{P(B)}$
> $= P(A)$, so that $P(A \text{ and } B) = P(A)P(B)$, and likewise, $P(B|A) =$
> $\frac{P(B \text{ and } A)}{P(A)} = P(B)$, with the same result. This gives us a tool for computing
> probabilities of multiple independent events as well as a criterion for deter-
> mining whether events are independent.

Example 14.5.6. Are the two events *receiving a postcard* and *owning a cat* from
Example 14.5.4 independent?   If they are, then $P(C = 1 \text{ and } R = 1) =$
$P(C = 1)P(R = 1)$. We have already determined their values; does $\frac{1}{2}$ equal
$\frac{2}{3}\frac{17}{30} = \frac{17}{45}$? No, so these events are not independent. We could also see this by
noticing that $P(C = 1) \neq P(C = 1|R = 1)$ (because $\frac{1}{2} \neq \frac{15}{17}$) and that $P(R = 1) \neq$
$P(R = 1|C = 1)$ (because $\frac{17}{30} \neq \frac{3}{4}$).

Independence as a concept means that two events have nothing to do with each
other; each occurs or does not occur independent of whether the other occurs (or
does not occur). But we need to have a mathematical definition of independence
as well, so that we can test to make sure that our assignments of probabilities to
events are consistent. Notice that if $P(A|B) = P(A)$, then $P(A \text{ and } B) = P(A)P(B)$,
from which it follows that $\frac{P(A \text{ and } B)}{P(A)} = P(B) = \frac{P(B \text{ and } A)}{P(A)}$, i.e., $P(B|A) = P(B)$.

> An independence criterion.   From the previous paragraph, $P(A|B) = P(A)$
> $\iff P(B|A) = P(B)$; this means we only need to check one of the two con-
> ditions in practice.

14.5.1   The Helpfulness of PIE in the Real World of Probability

An insurance company wants to offer a bare-bones policy for health insurance. It
will cover visits to the emergency room (ER), and it will cover hospitalizations.
(A hospitalization means that someone was checked into the hospital.) What is
the expected annual cost of issuing such a policy? The company needs to charge
more than the expected annual cost in order to break even (because there are also
administrative costs of running a business).

We will compute the expected annual cost incurred by a holder of an ER-and-
hospitalizations policy in Massachusetts in the 2008–2010 time frame. During
these years, there were about 28 outpatient ER visits per 100 people, 7 ER visits
that resulted in hospitalization per 100 people, and 5 hospitalizations from sources
other than the ER per 100 people (and all frequencies are given per year). If we let
EO represent the event of an outpatient ER visit, *EH* represent the event of an ER

visit that results in hospitalization, and *HN* represent the event of a hospitalization not occurring as the result of an ER visit, then we have (on average) that $P(EO) = .28$, $P(EH) = .07$, and $P(HN) = .05$ per person per year.

However, there is some overlap because people who have outpatient ER visits *and* hospitalizations not from ER visits (also known as the event *EO* and *HN*) are counted in the 28 outpatient ER visits per 100 people *and* in the 5 non-ER hospitalizations per 100 people. In order to compute the expected annual cost incurred by all possible ER-and-hospital healthcare events, we will need to find $P(EO$ and $HN)$ (and, for that matter, $P(EO$ and $EH)$, $P(EH$ and $HN)$, and $P(EO$ and EH and $HN)$).

Remember the principle of inclusion-exclusion from Sections 7.3 and 7.5? In its simplest form for sets A, B, PIE says that $|A \cup B| = |A| + |B| - |A \cap B|$.

> ProbabiliPIE.   PIE applies to probabilities of events E_1, E_2, so that $P(E_1$ or $E_2)$ $= P(E_1) + P(E_2) - P(E_1$ and $E_2)$. (Of course, PIE generalizes to probabilities of many events in the same way that PIE generalized for sets.)

We have to assume that *EO*, *EH*, and *HN* are independent of each other, because no data are readily available to calculate otherwise. We can now find

$$P(EO \text{ and } HN) = (.28) \cdot (.07) = .0196,$$
$$P(EO \text{ and } EH) = (.28) \cdot (.05) = .014,$$
$$P(EH \text{ and } HN) = (.07) \cdot (.05) = .0035,$$
$$P(EO \text{ and } EH \text{ and } HN) = (.28) \cdot (.07) \cdot (.05) \approx .001.$$

Notice now that the event *EO* measures whether or not an outpatient ER visit happens—it does not exclude the possibility of some sort of hospitalization happening as well. So we need to also compute, using PIE,

$P(EO$ only)

$$= P(EO) - P(EO \text{ and } EH) - P(EO \text{ and } HN) + P(EO \text{ and } EH \text{ and } HN)$$
$$= 0.28 - .0196 - .014 + .001 = .2474.$$

Similarly, $P(EH$ only$) = .07 - .0196 - .0035 + .001 = .0479$, and $P(HN$ only$) = .05 - .014 - .0035 + .001 = .0335$. Additionally, we must compute

$$P((EO \text{ and } EH) \text{ only}) = P(EO \text{ and } EH) - P(EO \text{ and } EH \text{ and } HN)$$
$$= .0196 - .001 = .0186,$$

and similarly, $P((EO$ and $HN)$ only$) = .014 - .001 = .013$ and $P((EH$ and $HN)$ only$) = .0035 - .001 = .0025$.

Even though we have broken down the possible healthcare events into a collection of exclusive events, we cannot yet compute the expected annual cost incurred! We have to define a random variable for which we can compute an expected value. Let $C(event)$ be the average cost of that event, whereby healthcare data shows that $C(EO) = \$700$ and $C(EH) = C(HN) = \$12{,}073$. This tells us that

$$C(EO \text{ and } EH) = \$700 + \$12{,}073 = \$12{,}773,$$

$$C(EO \text{ and } HN) = \$700 + \$12{,}073 = \$12{,}773,$$

$$C(EH \text{ and } HN) = \$12{,}073 + \$12{,}073 = \$24{,}146,$$

$$C(EO \text{ and } EH \text{ and } HN) = \$700 + \$12{,}073 + \$12{,}073 = \$24{,}846.$$

Now we can calculate the expected annual cost using the Lemma 14.3.10 expression $\mathbb{E}[C] = \sum_{s \in S} C(s)P(s)$, which in this case produces $\mathbb{E}[C] = \$700 \cdot (.2474) + \$12{,}073 \cdot (.0479) + \$12{,}073 \cdot (.0335) + \$12{,}773 \cdot (.0186) + \$12{,}773 \cdot (.0025) + \$24{,}146 \cdot (.013) + \$24{,}846 \cdot (.001) = \$1{,}764$.

For comparison, in this same time period, an individual healthcare policy in Massachusetts that covers ER visits and hospitalizations costs \$400/month, for a total annual revenue of \$4,800 for the insurance company. (This policy also covered routine physician visits and prescription drugs and had a \$1,000 deductible, but we could consider those to roughly cancel out over a year.)

14.5.2 Independence versus Exclusivity

We have earlier noted that two events being independent means that one event has no bearing on whether the other happens or not. We know that two events being exclusive means they cannot happen at the same time. So they are dependent on each other! If E_1 and E_2 are exclusive, when E_1 happens then $P(E_1) = 1$ and this implies $P(E_2) = 0$. That is, $P(E_1 \cap E_2) = 0$. In other words, independence and exclusivity are in some sense opposite ends of a spectrum.

Example 14.5.7. Let us again consider a fair die with six faces and the random variable Y as defined at the start of the section. Certainly the probability of rolling an even number is equal to the probability of rolling an odd number, so $P(Y = 1) = P(Y = 0) = \frac{1}{2}$. And the events of rolling an even number and of rolling an odd number are exclusive (and commensurately, Y cannot equal both 1 and 0) so that $P(Y = 1 \text{ and } Y = 0) = 0$. Now, notice that $P(Y = 1)P(Y = 0) = \frac{1}{2} \cdot \frac{1}{2} = \frac{1}{4} \neq 0$. So these exclusive events are *not* independent! Again, we can see that PIE works here, as $P(Y = 1 \text{ or } Y = 0) = P(Y = 1) + P(Y = 0) - P(Y = 1 \text{ and } Y = 0)$ or in other words, $1 = \frac{1}{2} + \frac{1}{2} - 0$.

Example 14.5.8. Here is an example of PIE with whipped probability topping. Suppose that we know that $P(E_1) = \frac{3}{5}$ and that $P(E_2) = \frac{1}{3}$. How large and how small might $P(E_1 \text{ and } E_2)$ and $P(E_1 \text{ or } E_2)$ be?

There are three possibilities for the behavior of E_1 and E_2: E_1 and E_2 are independent; E_1 and E_2 are exclusive; or, E_1 and E_2 are neither independent nor exclusive.

If E_1 and E_2 are exclusive, then from the probability axioms (or using PIE) we have $P(E_1 \text{ or } E_2) = P(E_1) + P(E_2) = \frac{3}{5} + \frac{1}{3} = \frac{14}{15}$. This is the maximum value $P(E_1 \text{ or } E_2)$ can have, as if E_1 and E_2 are not exclusive, then $P(E_1 \text{ and } E_2) > 0$. The minimum value of $P(E_1 \text{ or } E_2) = \max\{P(E_1), P(E_2)\} = P(E_1) = \frac{3}{5}$.

If E_1 and E_2 are independent, then $P(E_1 \text{ and } E_2) = P(E_1)P(E_2) = \frac{3}{5} \cdot \frac{1}{3} = \frac{1}{5}$. Notice that this is *not* the maximum value possible for $P(E_1 \text{ and } E_2)$! Suppose, for example, that whenever event E_2 happened, event E_1 happened as well. Then $P(E_1 \text{ and } E_2) = \min\{P(E_1), P(E_2)\} = \frac{1}{3}$. Another way to see this is to use PIE; $P(E_1 \text{ and } E_2) = P(E_1) + P(E_2) - P(E_1 \text{ or } E_2) \le \frac{3}{5} + \frac{1}{3} - \frac{3}{5} = \frac{1}{3}$. We also know that the minimum value for $P(E_1 \text{ and } E_2) = 0$ in the case that E_1, E_2 are exclusive. Additional information about either E_1 or E_2 informs our understanding of the probabilities of both.

Check Yourself

Verify your understanding by attempting these problems.

1. In Example 14.5.4, what if only $\frac{1}{10}$ of the friends own cats?

2. Are the events $W = 0$ and $Z = 0$ (as defined above) independent?

3. Consider the random variable Y as defined at the start of the section and a new random variable

    $$V(roll\ k) = \begin{cases} 1 & k \le 3, \\ 0 & k > 3. \end{cases}$$

    Are the events $Y = 1$ and $V = 1$ independent?

4. Examine Problem 3 of Section 14.4, and redo the problem using your new knowledge of independence and PIE.

5. State a version of PIE for the probabilities of three events.

14.6 Try This! …, Probably, Under Certain Conditions

Experiment with conditional probability with your peers, and do not be surprised if some of the later questions take you quite a while to resolve.

1. The Minbari have three castes (worker, religious, and warrior), and some Minbari enjoy eating the ceremonial food flarn whereas others do not. Translate the following questions into the language of random variables and conditional probability—no computations are desired.

   (a) What is the probability that a Minbari is religious caste and likes to eat flarn?

   (b) What is the probability that a religious-caste Minbari likes to eat flarn?

   (c) If a Minbari is religious caste, what is the probability that ze likes to eat flarn?

   (d) If a Minbari likes to eat flarn, what is the probability that ze is religious caste?

   (e) What is the probability that a Minbari who likes to eat flarn is religious caste?

2. The game *SET* is played with a deck of cards. Each card has one, two, or three symbols; each symbol is red, green, or purple; each symbol is shaded, open, or filled; and each symbol is a diamond, oval, or squiggle. There is exactly one card with each possible combination of attributes.

   (a) How many *SET* cards are there?

   (b) Draw a card from the *SET* deck. What is the probability that …

      (i) … the card has three symbols?

      (ii) … the card is shaded?

      (iii) … the card is shaded and has three symbols?

      (iv) … the card is shaded or has three symbols?

   (c) Are three-ness and shaded-ness independent events or not?

   (d) Draw two cards. What is the probability that both cards have filled symbols?

   (e) Draw two cards. What is the probability that at least one card is green?

   (f) Draw two cards. What is the probability that at least one card has two diamonds?

(g) Draw two cards. Use conditional probability to determine whether or not the first card being green is independent of the second card having two diamonds. (Think carefully. Does it matter whether the second card is green? Whether the first card has two diamonds?)

(h) Draw two cards. Use conditional probability to determine whether or not both cards having filled symbols is independent of at least one card being green.

(i) If you have never before played *SET*, go find a deck and learn how to play it. *SET* is way fun!

3. You enter the local doughnut shop. The pickings are slim:

> one maple-frosted raised doughnut,
>
> two chocolate-frosted chocolate cake doughnuts with sprinkles,
>
> one chocolate-frosted raised doughnut,
>
> three glazed raised doughnuts,
>
> one white-frosted raised doughnut with sprinkles,
>
> one blueberry cake doughnut,
>
> one chocolate-frosted cake doughnut,
>
> one glazed chocolate cake doughnut.

Sighing, you ask the counter clerk to choose three doughnuts for you arbitrarily. Assuming that hir choices *are* arbitrary, what is the probability that you will get…

(a) … three raised doughnuts?

(b) … exactly two raised doughnuts?

(c) … at least one doughnut with sprinkles?

14.7   Higher Expectations

In practice, the following theorem will make computing expected values much easier.

> **Theorem 14.7.1.** *For a bunch of random variables $X_1, X_2, \ldots, X_n$ on a state space S, we have that $\mathbb{E}_S[X_1 + X_2] = \mathbb{E}_S[X_1] + \mathbb{E}_S[X_2]$, and, in fact, $\mathbb{E}_S[\sum_{j=1}^{n} X_j] = \sum_{j=1}^{n} \mathbb{E}_S[X_j]$.*

Before we try to prove this statement, we need to know what it means! Basically, it says that if we want to compute the expected value of a random variable R, and we happen to be able to write R as a sum of other random variables (that is, $R = \sum_{j=1}^{n} X_j$), then we can compute the expected values for the individual X_j and add them up. Usually this is *much* easier than trying to figure out $\mathbb{E}[R]$ on its own.

We also need to know what it means to take the sum of some random variables! We will think of this in the sense of adding functions, because random variables *are* functions. Thus, $X_1 + X_2 : S \to N$, defined by $(X_1 + X_2)(s) = X_1(s) + X_2(s)$.

Proof: We will show that $\mathbb{E}_S[X_1] + \mathbb{E}_S[X_2] = \mathbb{E}_S[X_1 + X_2]$. Then the statement can be shown by induction on n.

By definition, $\mathbb{E}_S[X] = \sum_{k \in N} kP(X = k)$. Thus, $\mathbb{E}_S[X_1] + \mathbb{E}_S[X_2] = \sum_{k \in N} k \cdot P(X_1 = k) + \sum_{k \in N} kP(X_2 = k)$. By Lemma 14.3.10, this expression is equal to $\sum_{s \in S} X_1(s)P(s) + \sum_{s \in S} X_2(s)P(s)$. Combining the sums, $\sum_{s \in S} X_1(s)P(s) + \sum_{s \in S} X_2(s)P(s) = \sum_{s \in S} X_1(s)P(s) + X_2(s)P(s)$; collecting the terms, this is equal to $\sum_{s \in S}(X_1(s) + X_2(s))P(s)$, which by definition is equal to $\sum_{s \in S}(X_1 + X_2)(s)P(s)$. Again by Lemma 14.3.10, this expression equals $\sum_{k \in N} kP((X_1 + X_2) = k) = \mathbb{E}_S[X_1 + X_2]$.

By induction, the statement is also true for n random variables. (The induction portion of the proof is assigned as Problem 22 in Section 14.12.) □

Notice that we did not require any of the events involved in these random variables to be independent or exclusive! That makes the result somewhat surprising. Theorem 14.7.1 turns out to be incredibly useful because it reduces complex problems to collections of simpler problems.

Example 14.7.2. There are generally a dozen cat toys in rotation at any one time in the author's house. Every week when the vacuuming is done, these are rounded up and put in a pile, and every week the cats redistribute toys through the house. On any given day, what is the expected number of toys removed from the pile?

The random variable T describes the number of toys in the pile. We will write $T = \sum_{i=1}^{12} T_i$, where

$$T_i(\text{day } d) = \begin{cases} 1 & \text{toy } T_i \text{ is grabbed,} \\ 0 & \text{toy } T_i \text{ is in the pile} \end{cases}$$

because it is easier to compute with the T_i.

Note that $P(T_i = 1) = \frac{1}{7}$ because it is equally likely that a cat will first grab a toy on any day, so $\mathbb{E}[T_i] = 1P(T_i = 1) + 0P(T_i = 0) = \frac{1}{7}$.

In turn, by Theorem 14.7.1, $\mathbb{E}[T] = \sum_{i=1}^{12} \mathbb{E}[T_i] = \sum_{k=1}^{12} \frac{1}{7} = \frac{12}{7}$. Thus, we expect there to be almost two toys removed from the pile on any given day.

Example 14.7.3. Let's generalize our earlier example of two ebullient children playing near chairs. Imagine n staid adults, each of whom has a chair. When a musical triangle sounds (*ding!*), the adults tiredly arise and shuffle about, sitting again only on a second ding of the triangle. But wow, is that boring. Forget about boring adults! Instead, consider n ducks, each of whom has a paddlepool. The ducks are $d_1, \ldots, d_n$ and the paddlepools are $p_1, \ldots, p_n$. Every so often, a human walks by and the ducks become alarmed, quacking and beating their wings and waddling and flying about. After the human leaves, the ducks settle down again to their one-duck-per-pool state. Question: How many ducks do we expect will be swimming in paddlepools of the same label?

We have a natural random variable X that counts the number of ducks paddling in their matching pools. If it is equally likely that any arrangement of ducks in paddlepools happens, then the ith duck paddles in hir namesake pool about $\frac{1}{n}$ of the time. Now, if we let X_i be the indicator random variable with value 1 when the ith duck paddles in hir same-label pool and value 0 when the ith duck paddles elsewhere, then we can describe X as follows. The number of ducks paddling in their matching pools is the number of the X_i with value 1. So adding the X_i as functions, we have that $\sum_{i=1}^{n} X_i = X$.

Now we can answer, "How many ducks do we expect will be swimming in paddlepools of the same label?" by computing $\mathbb{E}[X]$. We know that $\mathbb{E}[X] = \mathbb{E}[\sum_{i=1}^{n} X_i]$ and can use Theorem 14.7.1 to see that this is equal to $\sum_{i=1}^{n} \mathbb{E}[X_i]$. Now we just need to compute $\mathbb{E}[X_i]$ and we should be all set. Each X_i only takes on the values 0 and 1, so $\mathbb{E}[X_i] = 0 \cdot P(X_i = 0) + 1 \cdot P(X_i = 1) = 0 + P(X_i = 1) = \frac{1}{n}$. Finally, $\sum_{i=1}^{n} \mathbb{E}[X_i] = \sum_{i=1}^{n} \frac{1}{n} = n\frac{1}{n} = 1$. So, no matter how many ducks are involved in swapping paddlepools, we expect that just one will be in its own pool!

14.7.1   That's Wild! (A Hint at the Probabilistic Method)

Some people are invited to a silly party. As they arrive, some of the guests shake hands, for a total of m handshakes. The host also hands out party hats, some of which are teal and the rest of which are emblazoned with ducks. Is there a way for the host to hand out hats so that no more than $\frac{m}{2}$ of the handshakes take place between duck-emblazoned-hat-wearing and teal-hat-wearing party-goers? The answer is "yes," as follows.

Theorem 14.7.4. *Suppose we have a simple connected graph with $m > 3$ edges and at least four vertices. Then there exists a set of at most $\frac{m}{2}$ edges whose removal disconnects the graph.*

At first glance, and probably even at sixth glance, this theorem has nothing to do with probability or expectation. Yet we would not mention it here if there were no connection! So hang on and bear with us.

It suffices to show that there exists a way of assigning the letters D (for duck-emblazoned) and T (for teal) to the vertices (party-goers) of the graph such that there are no more than $\frac{m}{2}$ edges (handshakes) with D on one vertex and T on the other. Why? Because if we ask all of the D vertices (people) to collect in one place and all of the T vertices to collect in another place, and remove all of the $D - T$ (or $T - D$) edges, then there will be no way to get from the D collective to the T collective, and therefore, the graph will not be connected.

Let's set up a state space, probability measure, and random variables. And let us label the vertices of the graph at random. Each edge could have its ends labeled as $D - D$, $D - T$, $T - D$, or $T - T$. The state space of a single edge is $S_e = \{D - D, D - T, T - D, T - T\}$. Because there are m edges, a state of the system can be represented by an m-tuple of edge labelings. We could think of a state as an element in the Cartesian product (m times) of the state space of a single edge.

It is equally likely that a vertex would be labeled D or T, and the vertex labels are independent of each other, so the probabilities for a single edge e are $P_e(D - D) = \frac{1}{4}$, $P_e(D - T) = \frac{1}{4}$, $P_e(T - D) = \frac{1}{4}$, and $P_e(T - T) = \frac{1}{4}$. There is a random variable associated with each single edge e,

$$X_e(S_e) = \begin{cases} 1 & e \text{ has a mixed label (two different vertex labels),} \\ 0 & e \text{ has the same label on both vertices.} \end{cases}$$

Note that $P(X_e = 1) = \frac{1}{2} = P(X_e = 0)$. Then we may define a random variable X as the number of edges with mixed labels, and $X = \sum_e X_e$.

Now we can reveal the crux of the proof. Oh, wait, we haven't even announced we are doing a proof!

Proof of Theorem 14.7.4: Using the notation introduced in the previous paragraphs, let us compute $\mathbb{E}[X] = \mathbb{E}[\sum_e X_e]$. By Theorem 14.7.1, $\mathbb{E}[\sum_e X_e] = \sum_e \mathbb{E}[X_e] = \sum_e (1 \cdot P(X_e = 1) + 0 \cdot P(X_e = 0)) = \sum_e \frac{1}{2}$. There are m edges, so $\sum_e \frac{1}{2} = m \cdot \frac{1}{2} = \frac{m}{2}$. So, the expected value of the number of edges with mixed labels is $\frac{m}{2}$. We may interpret the expected value as calculating the average number of edges with mixed labels; if on average there are $\frac{m}{2}$ edges with mixed labels, then there must exist some labeling with fewer than (or equal to) $\frac{m}{2}$ edges with mixed labels because not all labelings can have more mixed-label edges than the average. For this labeling, we put all of the D-labeled vertices in one pile and all of the T-labeled vertices in another pile. Then, removing the mixed-label edges disconnects the D vertices from the T vertices; we have thus produced an edge cut of size no more than $\frac{m}{2}$.   □

This is a hint at the type of proof discrete probability is used for most often. The *probabilistic method* is an approach to proofs whereby one proves existence by showing that the probability of the desired object occurring is greater than zero, or in this case, by showing that the expected value is greater than zero. Another example is given in Bonus Section 14.10.

Check Yourself

Make sure you can answer at least one of these problems.

1. Suppose five fair coins are flipped and consider the random variable H, defined as the number of heads revealed. Rewrite H as a sum of simpler random variables.

2. For the situation of rolling an eight-sided die, consider the random variables $E = 1$ when the result is even and $E = 0$ otherwise, and $G = 1$ if the result is greater than or equal to 5 and $G = 0$ if the result is four or less. Compute $(E + G)(roll\ 2)$, $(E + G)(roll\ 3)$, $(E + G)(roll\ 5)$, and $(E + G)(roll\ 6)$.

3. Consider a deck of four cards, labeled $1\triangle$, $1\bigcirc$, $2\triangle$, and $2\bigcirc$. Draw a card, put it back, shuffle the deck, and draw a second card. Compute the expected value of $Z = X + Y$, where X is the numerical value of the first card and Y is the numerical value of the second card.

14.8   Where to Go from Here

There are entire courses on probability; most require calculus as a prerequisite, as they focus on not-necessarily-discrete probability. Some famous mathematics problems, such as the Monty Hall problem, fall under the purview of probability. For an elementary and yet comprehensive introduction to the Monty Hall problem, check out *The Monty Hall Problem: The Remarkable Story of Math's Most Contentious Brain Teaser* by Jason Rosenhouse.

For a bit more on the uses and misuses of conditional probability in medicine and politics, see Steve Strogatz's *New York Times* column on the subject at https://opinionator.blogs.nytimes.com/2010/04/25/chances-are/?_r=0.

The probabilistic method is the primary use of probability in research in discrete mathematics, and the reference of choice (which is both challenging and advanced) is *The Probabilistic Method* by Noga Alon and Joel Spencer. Their first example (described there as "simple") is that given in Section 14.10.

Credit where credit is due: Thanks incredibly to Adam Marcus for teaching me the material exposed in this chapter. Much of my writing is based on notes taken from his introductory lectures in 2007. Some is enhanced by my reading of Ed Scheinerman's book *Mathematics: A Discrete Introduction*. Thanks also to Dylan Shepardson for suggesting the addition of sensitivity/specificity via Steve Strogatz's approach.

The Yarn Harlot mentioned in Problem 3 of Section 14.4 is real; her name is Stephanie Pearl-McPhee, and she is a *New York Times* best-selling author who writes about knitting. A brief description of her living-room-became-a-mitten-factory episode is at http://www. yarnharlot.ca/blog/archives/2011/12/05/resisting_only_makes_it_take_longer.html and http://www.yarnharlot.ca/blog/archives/2011/12/02/i_am_sure_it_will_pass.html.

Data for the healthcare example of Section 14.5.1 came from the sources "The Effect of Insurance on Emergency Room Visits: An Analysis of the 2006 Massachusetts Health Reform," by Sarah Miller, November 2011, downloadable from https://pdfs.semantic scholar.org/0b32/6cd7426ef7514c60d9bff2f7796ebd73f81b.pdf; "Potentially Preventable Hospitalizations in Massachusetts: Fiscal Years 2004 to 2008," July 2010, Publication Number: 10-200-HCF-01, downloadable from http://archives.lib.state.ma.us/handle/ 2452/50109; "Massachusetts QuickFacts from the US Census Bureau" at https://www. census.gov/quickfacts/MA; "Harvard Pilgrim Health Care – Cost of Services – MA" at https://www.harvardpilgrim.org/ (specific page no longer available); and "Health Connector: Health Insurance for Massachusetts Residents" at https://www.mahealthconnector. org/.

The Minbari in Problem 1 of Section 14.6 are from the television series *Babylon 5* by J. Michael Straczynski. Bonus Check-Yourself Problem 3 was suggested by Tom Hull. In Section 14.11, the Blue Sea Deck from Problem 6 can be found at https://www. thegamecrafter.com/games/the-blue-sea-deck. In Section 14.12, Problem 8 was inspired by Woot!.com's Bag of Crap, and data for Problems 29–32 came from the CDC at https:// www.cdc.gov/cancer/colorectal/statistics/index.htm and https://www.cdc.gov/cancer/ colorectal/statistics/age.htm; from "Multitarget Stool DNA Testing for Colorectal-Cancer Screening" in the *New England Journal of Medicine* (2014), available at https://www. nejm.org/doi/full/10.1056/NEJMoa1311194; and from "Emerging stool-based and blood-based non-invasive DNA tests for colorectal cancer screening" in *Abdominal Radiology* (2016), available at https://www.ncbi.nlm.nih.gov/pmc/articles/PMC4974132/.

14.9 Chapter 14 Definitions

probability: The likelihood that some given event will occur.

state: A configuration that a system might have.

state space: The set of different possible states.

sample space: A state space.

event: A collection of states; a subset of the state space.

exclusive: Two events that cannot happen at the same time. States are always exclusive; some events are exclusive and some are not.

probability axioms: Conditions that all probability functions must satisfy in order to actually measure probability, as follows; details are on page 447.

Let $P : \mathscr{P}(S) \to [0,1]$ be a function from all subsets of a state space to the unit interval $\{x \mid 0 \le x \le 1\}$:

- 🐤 $P(S) = 1$.
- 🐤 $P(\emptyset) = 0$.
- 🐤 For any element $s \in S$, $P(s) \ge 0$ and likewise for any subset (event) $E \subset S$, $P(E) \ge 0$.
- 🐤 Whenever two events E_1, E_2 are exclusive, $P(E_1 \text{ or } E_2) = P(E_1) + P(E_2)$.

random variable: A function $X : S \to N$ from a state space S to a finite set of real numbers N.

probability distribution: A display of all possible values for a random variable X with the corresponding probabilities.

expected value: The weighted average of the probability distribution for a random variable X, denoted $\mathbb{E}[X]$, and defined as $\mathbb{E}[X] = \sum_{k \in N} kP(X = k)$.

conditional probability: The probability that event E_1 happens, given that E_2 definitely occurs. Denoted $P(E_1 | E_2)$ and measured by $\frac{P(E_1 \text{ and } E_2)}{P(E_2)}$.

independent: Two events A and B are independent if $P(A|B) = P(A)$ and $P(B|A) = P(B)$. In other words, if the probability that A happens is the same whether B happens or not, and if the probability that B happens is the same whether A happens or not, then events A and B are independent of each other.

sensitivity: The measure, for a disease test, of someone having a disease testing positive for that disease.

specificity: The measure, for a disease test, of someone *not* having a disease testing *negative* for that disease.

probabilistic method: A way to prove existence by showing that the probability of the desired situation occurring is greater than zero.

14.10   Bonus: Ramsey Numbers and the Probabilistic Method

Recall from Section 3.9 that the Ramsey number $R(k,k)$ is the smallest number n such that a 2-edge-colored complete graph K_n must contain a monochromatic K_k. We would like to show that $R(k,k) \ge 2^{\lfloor \frac{k}{2} \rfloor}$. Yes, really. The plan will be to compute the expected number of monochromatic K_ks in a K_n and use this to obtain a lower bound on $R(k,k)$.

Consider a complete graph K_n. We want to choose a K_k in this K_n, which can be done by choosing k of the n vertices and then including the edges that connect those k vertices. There are $\binom{n}{k}$ ways for us to choose such a K_k.

We will color the edges of the chosen K_k with two colors (azure and lavender), so that $P_e(\text{azure}) = \frac{1}{2} = P_e(\text{lavender})$. The color of each edge is independent of the color of each other edge, so the probability that all edges are azure is $\frac{1}{2}^{\binom{k}{2}}$ because K_k has $\binom{k}{2}$ edges. Similarly, the probability that all edges are lavender is $\frac{1}{2}^{\binom{k}{2}}$. Therefore, the probability that all edges of the chosen K_k are azure *or* all edges of the chosen K_k are lavender is $2 \cdot \frac{1}{2}^{\binom{k}{2}} = \frac{1}{2}^{\binom{k}{2}-1}$.

Now, consider the state space of all possible 2-edge colorings of K_n. We shall define a random variable on this state space. Let M be the number of monochromatic K_ks in a K_n. We can write $M = \sum_{j=1}^{\binom{n}{k}} M_j$, where for the $j^{\text{th}} K_k$ we have

$$M_j(K_k) = \begin{cases} 1 & K_k \text{ is monochromatic,} \\ 0 & K_k \text{ has edges of two colors.} \end{cases}$$

We will compute the expected number of monochromatic K_ks in K_n by computing $\mathbb{E}[M] = \mathbb{E}[\sum_{j=1}^{\binom{n}{k}} M_j] = \sum_{j=1}^{\binom{n}{k}} \mathbb{E}[M_j] = \sum_{j=1}^{\binom{n}{k}} (1 \cdot P(M_j = 1) + 0 \cdot P(M_j = 0))$. We know that $P(M_j = 1) = \frac{1}{2}^{\binom{k}{2}-1}$ and that there are $\binom{n}{k}$ terms in the sum, so $\mathbb{E}[M] = \binom{n}{k} \frac{1}{2}^{\binom{k}{2}-1}$.

Now, what do we do with this computation? Remember that we want to find a lower bound on $R(k,k)$. That means we need to find a lower bound for n such that K_n definitely contains a monochromatic K_k. If we have an n for which K_n has no monochromatic K_k, we know that we need a larger n so that there *could* be a monochromatic K_k. Thus, such an n will provide a lower bound (though perhaps a terrible lower bound).

If $\mathbb{E}[M] < 1$, then we expect K_n to have less than one monochromatic K_k, which means we actually expect it to have no monochromatic K_k. This, in turn, means that it is possible for K_n to have no monochromatic K_k, and so if $\mathbb{E}[M] < 1$, then we have found an n such that $R(k,k) > n$.

Therefore, we want to figure out what conditions on n make

$$\mathbb{E}[M] = \binom{n}{k} \frac{1}{2}^{\binom{k}{2}-1} < 1.$$

This will require a close-to-ridiculous amount of algebra, so get ready and hang on tight to your mental seat!

First, let us show that $\binom{n}{k} < \left(\frac{n}{2}\right)^k$. By Section 6.7,

$$\binom{n}{k} = \frac{n \cdot (n-1) \cdot (n-2) \cdot \cdots \cdot (n-(k-1))}{1 \cdot 2 \cdot 3 \cdot 4 \cdot \cdots \cdot k}.$$

There are k terms in the numerator and k terms in the denominator. Observe that $n \cdot (n-1) \cdot (n-2) \cdot \cdots \cdot (n-(k-1)) < n \cdot \cdots \cdot n$ (k times) $= n^k$. So, $\binom{n}{k} < n^k/(1 \cdot 2 \cdot 3 \cdot 4 \cdot \cdots \cdot k)$. Also, $1 \cdot 2 \cdot 3 \cdot 4 \cdot \cdots \cdot k = 24 \cdot 5 \cdot \cdots \cdot k > 16 \cdot 5 \cdot \cdots \cdot k > 16 \cdot 2 \cdot \cdots \cdot 2 = 2^k$. Therefore, $1/(1 \cdot 2 \cdot 3 \cdot 4 \cdot \cdots \cdot k) < \frac{1}{2^k}$. This tells us that

$$\binom{n}{k} < \frac{n^k}{1 \cdot 2 \cdot 3 \cdot 4 \cdot \cdots \cdot k} < \frac{n^k}{2^k} = \left(\frac{n}{2}\right)^k.$$

Now, this means that

$$\binom{n}{k} \frac{1}{2}^{\binom{k}{2}-1} < \left(\frac{n}{2}\right)^k \frac{1}{2}^{\binom{k}{2}-1} = \frac{n^k}{2^k 2^{\binom{k}{2}-1}}.$$

Because $\binom{k}{2} = \frac{k(k-1)}{2}$, we know that

$$k + \binom{k}{2} - 1 = \frac{2k + k^2 - k - 2}{2} = \frac{k^2 + k + 2}{2}.$$

Thus,

$$\frac{n^k}{2^k 2^{\binom{k}{2}-1}} = \frac{n^k}{2^{\frac{2k+k^2-k-2}{2}}}.$$

If it were to be true that $n = 2^{something}$, then

$$\frac{n^k}{2^{\frac{2k+k^2-k-2}{2}}} = \frac{(2^{something})^k}{2^{\frac{2k+k^2-k-2}{2}}} = \frac{2^{k \cdot something}}{2^{\frac{2k+k^2-k-2}{2}}} = 2^{k \cdot something - \frac{2k+k^2-k-2}{2}}.$$

If by chance

$$k \cdot something - \frac{2k + k^2 - k - 2}{2} < 0,$$

then

$$2^{k \cdot something - \frac{2k+k^2-k-2}{2}} < 1.$$

In turn, this would mean

$$\binom{n}{k} \frac{1}{2}^{\binom{k}{2}-1} < 1,$$

which is what we desire. So let's see what *something* needs to be to get this to work out. If

$$k \cdot something - \frac{2k + k^2 - k - 2}{2} < 0,$$

then

$$k \cdot something < \frac{2k + k^2 - k - 2}{2} \quad \text{or} \quad something < \frac{2k + k^2 - k - 2}{2k}.$$

Certainly

$$\left\lfloor \frac{k}{2} \right\rfloor < \frac{k}{2} < \frac{k + (1 - \frac{1}{k})}{2} = \frac{2k + k^2 - k - 2}{2k}.$$

So what is our conclusion? It's that if $n = 2^{\lfloor \frac{k}{2} \rfloor}$, then $\binom{n}{k} \frac{1}{2}^{\binom{k}{2}-1} < 1$, which in turn means that $\mathbb{E}[M] < 1$ and so $R(k,k) > n = 2^{\lfloor \frac{k}{2} \rfloor}$.

14.11 Bonus Check-Yourself Problems

Solutions to these problems appear starting on page 628. Those solutions that model a formal write-up (such as one might hand in for homework) are to Problems 2, 5, and 6.

1. In Lucy Worsley's *If Walls Could Talk: An Intimate History of the Home*, the author says, "The medieval death rate was one in every fifty pregnancies. Considering that it wasn't unusual for a woman to give birth a dozen times, the odds quickly mounted up for reproductive wives."
So... what are these odds? Compute the probability of dying while pregnant for each of 1, 4, 6, and 12 pregnancies. What is the probability of dying during some one of 12 theoretical medieval pregnancies?

2. Suppose you have a box of colored pens (fuchsia, cinnamon, tangerine, gold, lime, forest, teal, cobalt, plum) and three pencils (mechanical, yellow No. 2, printed with cupcakes).

(a) Describe the state space of grabbing a pen and a pencil. What is the probability of each individual state?

(b) What is the probability of grabbing a pen whose color begins with "f" and a mechanical pencil?

(c) What is the probability of grabbing a pen whose color is greenish and a non-mechanical pencil?

(d) What is the probability of (grabbing a pen whose color begins with "f" and a non-mechanical pencil) or (grabbing a pen whose color is greenish and a non-mechanical pencil)?

3. A computer lab has 20 computers in it. On any given day, the probability that a given computer is not working

is p. How many computers do you expect will be functioning when you enter the lab today? Answer the question for $p = .001, p = .05$.

4. Chips of the World come in lots of flavors. In the sale bin are

2 bags of bacon ranch pita chips,
1 bag of salt and vinegar potato chips,
3 bags of hot-sauce cheese corn chips,
5 bags of crab potato chips, and
2 bags of peppercorn salsa pita chips.

If you close your eyes and grab three bags of chips (one at a time, so you know you have three), what is the probability you will get…

🦆 … all three bags of potato chips?

🦆 … exactly two bags of spicy chips?

🦆 … at least one bag of pita chips?

How many bags of corn chips do you expect to find in your three bags?

5. Shoes 'R' Us has a lot of different kinds of shoes in their display case, one of each kind they sell. A shoe can be brown, black, silver, or green; it can be a low shoe, a boot, or an athletic shoe; and, it can have laces or be a slip-on.

(a) How many different kinds of shoes does Shoes 'R' Us have in its display case?

(b) What is the probability that a Shoes 'R' Us display shoe is brown and slip-on?

(c) What is the probability that a Shoes 'R' Us display shoe is silver or a boot?

(d) Given that a Shoes 'R' Us display shoe is silver, what is the probability that it is a boot?

(e) Given that a Shoes 'R' Us display shoe is a green athletic shoe, what is the probability that it has laces?

(f) Are the properties *silver* and *boot* independent?

6. Consider a deck of cards that is standard, except for having six suits—the two additional suits are stars and squids. (This deck exists: it is the Blue Sea Deck.) Draw a card.

(a) What is the probability that the card is a queen or a squid?

(b) What is the expected value of the number on the card? (Here, Ace = 1, King = 13.)

7. The game *Elder Sign* has unusual dice. There are six six-sided green dice, each of which has three sides showing magnifying glasses, one side with a tentacle, one side with a skull, and one side with a scroll. There is also a six-sided yellow die with four sides showing magnifying glasses, one side with a skull, and one side with a scroll. Finally, there is a six-sided red die with three sides showing magnifying glasses, one side with a Wild sign, one side with a skull, and one side with a scroll.

(a) If you roll the six green dice, what is the expected number of magnifying glasses you'll see?

(b) If you roll seven of the dice, what is the probability that you will roll exactly one skull?

(c) If you roll seven of the dice, what is the probability that you will roll at least one scroll?

8. The game of *Qwirkle* uses a bag of tiles. Each black tile has a shape on it (circle, diamond, square, crisscross, starburst,

clover) that is colored (red, orange, yellow, green, blue, purple). There are three copies of each kind of tile.

(a) How many tiles are in a *Qwirkle* bag?

(b) What is the probability that a tile drawn is red?

(c) What is the probability that a tile drawn is a sunburst?

(d) What is the probability that a tile drawn is a red sunburst?

(e) What is the probability that a tile drawn is red or a sunburst?

(f) Is red-ness independent of sunburst-ness?

9. Another *Qwirkle* qwestion: pull two tiles from the bag.

(a) What is the probability that both are blue?

(b) What is the probability that the *second* tile is blue?

(c) What is the probability that at least one tile is blue?

10. What's the expected number of fixed points (items that do not move) in a permutation of n items?

14.12   Expect Problems, Probably

1. Above your mirror in the dressing room is a bank of four lightbulbs. Each lightbulb has a probability of .02 of being out when you flip the switch in your dressing room, and the workingness of each lightbulb is independent of the others. Describe the state space of the bank of lightbulbs and list the probability of each state.

2. Someone flips seven fair coins. Describe the state space for this situation. Define a random variable corresponding to the number of heads that show when the coins land. What is the probability that this random variable has value 3?

3. You and a friend are asked to each choose a number 1 to 8. Describe the state space for this situation. What is the probability that you each choose the same number? What is the probability that your numbers are one apart?

4. A gumball machine requests 25 cents and, in return, sends a colorful gum-

ball down a chute into your waiting hands. The gumballs in this machine come in white, red, yellow, blue, pink, green, and orange. The colors are not evenly distributed; $\frac{1}{10}$ of the gumballs are white, $\frac{11}{40}$ of the gumballs are orange, and each remaining color represents $\frac{1}{8}$ of the gumballs.

(a) Let the random variable $R = 1$ when a red gumball arrives, $R = \frac{1}{2}$ when a pink or orange gumball arrives, and $R = 0$ otherwise. (R measures the redness of a gumball.) Compute $\mathbb{E}[R]$.

(b) Let the random variable $Y = 1$ when a yellow gumball arrives, $Y = \frac{1}{2}$ when a green or orange gumball arrives, and $Y = 0$ otherwise. (Y measures the yellowness of a gumball.) Compute $\mathbb{E}[Y]$.

(c) Compute the expected value of $R + Y$.

5. Some of the Fairly Fun Folks own pets (but only ferrets and fish; no frogs). Of these Folks, $\frac{2}{5}$ own fish, $\frac{3}{20}$ own both ferrets and fish, and $\frac{11}{20}$ have no pets at all. The Fairly Fun Folks also have a variety of hair types: $\frac{1}{10}$ have wavy hair, $\frac{2}{5}$ have curly hair, $\frac{1}{5}$ have kinky hair, $\frac{1}{5}$ have straight hair, and $\frac{1}{10}$ have no hair.

   (a) What fraction of the Folks own ferrets?

   (b) What is the probability that a Folk has curly hair and owns a ferret?

   (c) What is the probability that a curly-haired Folk owns a ferret?

   (d) If a Folk has curly hair, what is the probability that ze owns a ferret?

   (e) If a Folk owns a ferret, what is the probability that ze has curly hair?

   (f) What is the probability that a ferret owner has curly hair?

   (g) Explain your answers to the previous questions using the language of random variables and conditional probability (if you have not already done so).

   (h) What is the probability that one of the Folk with no hair owns fish?

   (i) What is the probability that someone who owns both fish and ferrets has kinky hair?

   (j) What is the probability that someone has straight hair and no pets?

6. Explain why computing the average value of a random variable is the same as computing the expected value of a random variable in discrete probability.

7. Show that the Cartesian product $C = S_1 \times S_2$ of any two independent two-element state spaces has $P(C) = 1$.

8. The website Noot!.com sells one aquatic thing (like newts) each day. Every so often, the item for sale is a Bag of Carp, recognizable by its iconic chartreuse question mark on a plain brown lunchbag. A Bag of Carp contains three random carp-related items from the Noot! overstock. The possibilities are preserved and salted dried carp snacks (for people to eat), living koi, photos of carp in the wild, origami carp, crocheted carp, bead-knitted carp, aquatic insect pieces (for feeding carp), books on caring for carp, or a CD of carp music (made by carp for carp). Each item in a Bag of Carp is equally likely to be any of these possibilities.

   (a) Let R be a random variable that indicates whether the Bag of Carp contains a representation of carp ($R = 1$) or not ($R = 0$). Compute $\mathbb{E}[R]$.

   (b) Let C be a random variable that indicates whether the Bag of Carp contains an item that can be used by a pet carp ($C = 1$) or not ($C = 0$). Compute $\mathbb{E}[C]$.

   (c) What is $P(R = 1 \text{ and } C = 1)$? How about $P(R = 1 \text{ or } C = 1)$?

9. Suppose that we know that $P(E_1) = \frac{3}{8}$ and that $P(E_2) = \frac{7}{9}$. How large and how small might $P(E_1 \text{ and } E_2)$ and $P(E_1 \text{ or } E_2)$ be?

10. You have Xs and Os (nine of each) in a bag near a tic-tac-toe board. You arbitrarily select an X or an O, and then you place it arbitrarily on the board. Toss your letter back in the bag, shake the bag, and again arbitrarily select an X or an O and place it arbitrarily on the board.

(a) Consider the event in which you select the same letter twice; define a random variable corresponding to this event and compute its expected value.

(b) Consider the event of placing both letters in the same row; define a random variable corresponding to this event and compute its expected value.

(c) Consider the event of placing both letters in the same column; define a random variable corresponding to this event and compute its expected value.

(d) What is the probability that both letters are placed in the same column *and* in the same row?

(e) What is the probability that the same letter is selected twice *and* both are placed in the same row?

(f) Are any two of these three events independent?

11. You have Xs and Os (nine of each) in a bag near a tic-tac-toe board. You arbitrarily select an X or an O, and then you place it arbitrarily on the board. Shake the bag, and again arbitrarily select an X or an O and place it arbitrarily on the board (in an open spot).

(a) Consider the event in which you select the same letter twice; define a random variable corresponding to this event and compute its expected value.

(b) What is the probability that both letters are placed in the same row?

12. There is a throwing accuracy game in which a velcro board is hung on the wall (see Figure 14.2) and players throw small prickly balls that stick on the board. Points are awarded to players depending on the region in which each prickly ball lands. The probability that a prickly ball lands in any particular one of the outer teal regions is $\frac{1}{16}$, the probability that a prickly ball lands in any particular one of the white regions is $\frac{5}{48}$, and the probability that a prickly ball lands in the center oval is $\frac{1}{12}$.

(a) What assumption has been made in assigning these probabilities?

(b) What is the probability of hitting the upper-left-most region, given that you hit a region that is not white?

(c) You are awarded one point if your prickly ball lands on an outer teal region, three points if your prickly ball lands in a white region, and five points if your prickly ball lands within the oval. What is your expected number of points?

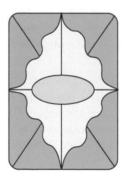

Figure 14.2. A velcro board.

13. (Also about the game using the board in Figure 14.2) Suppose that each time you throw a prickly ball, there is a probability of $\frac{1}{3}$ that you will miss the board entirely and a probability of $\frac{1}{5}$ that if the

prickly ball hits the board, it will not stick and will drop to the ground. Now what is your expected number of points?

14. Suppose you flip two identical coins simultaneously. Show that assigning $P(two\ heads) = P(two\ tails) = P(one\ head,\ one\ tail) = \frac{1}{3}$ violates the probability axioms for each individual coin.

15. **Challenge**: Show that there is no way to assign probabilities to heads and tails for two not-necessarily-identical coins so that

   🐦 the probability axioms are satisfied for both individual coins' state spaces and

   🐦 $P(two\ heads) = P(two\ tails) = P(one\ head,\ one\ tail) = \frac{1}{3}$.

16. Suppose you play the venerable game rock-paper-scissors. If you pick paper and your opponent chooses arbitrarily, what is the probability that you will win? If each of you pick arbitrarily, what is the probability that you will win?

17. Suppose you play rock-paper-scissors and there is a draw (that is, both you and your opponent choose the same object). You decide to try to break the tie by playing again. What is the probability that you win in the second game, given that you reached a draw in the first game? What is the probability that you draw in the first game and win the second game?

18. Consider flipping a fair coin while rolling an eight-sided die. Compute the expected value of $B = H + F$, where H is the number of heads revealed and

$F = 1$ when the roll is a multiple of four and $F = 0$ if not. Are the events $H = 1$ and $F = 1$ independent?

19. In the United States, mail is delivered six days each week. Your BFF (best friend forever) is on a trip and has sent you ten packages, all of which are scheduled to arrive this week. However, you have no idea on what day any of the packages will arrive. What is the expected number of packages that you receive on a particular weekday?

20. Consider a deck of four cards, labeled $1\triangle$, $1\bigcirc$, $2\triangle$, and $2\bigcirc$. Draw two of the cards. Compute the expected value of $Z = X + Y$, where X is the numerical value of the first card and Y is the numerical value of the second card.

21. (Also about the deck of four cards) Is $X = 1$ independent of $Y = 1$?

22. (Finishing the proof of Theorem 14.7.1) Given that $\mathbb{E}_S[X_1 + X_2] = \mathbb{E}_S[X_1] + \mathbb{E}_S[X_2]$, prove by induction that $\mathbb{E}_S[\sum_{j=1}^{n} X_j] = \sum_{j=1}^{n} \mathbb{E}_S[X_j]$.

23. (About the doughnut shop in Problem 3 of Section 14.6) Before the counter clerk has a chance to move, you change your mind and decide you are only in the mood for raised doughnuts, so you request that the three arbitrarily chosen doughnuts all be raised.

   (a) What is the probability that all three doughnuts will be glazed?

   (b) What is the probability that you get the maple doughnut?

24. (More doughnut shop adventures) Suddenly you remember that your best friend desperately wants a cake doughnut, so you ask for two arbitrarily chosen cake doughnuts instead.

(a) What is the probability that you get both double-chocolate doughnuts?

(b) What is the probability that you receive the blueberry doughnut?

25. (Still about the same doughnut shop) Supposing you are getting three arbitrary doughnuts... is *getting at least one raised doughnut* independent of *getting at least one doughnut with sprinkles*?

26. In the game *Neko Atsume*, cats visit a yard, eat food, engage with objects, and leave metal fish as gifts. Identify each of the following as a state, an event, both, or neither.

(a) Joe DiMeowgio plays with a baseball.

(b) The only cat present is Snowball, who sleeps on a sheepskin.

(c) Tubbs has eaten the sashimi.

(d) Pickles plays with a squishy ball and Bolt kicks a green fish; the rest of the yard is empty.

(e) Mr. Meowgi meditates on the pink silk pillow.

27. Suppose we roll an eight-sided die. What is the probability that we roll less than 5 given that the roll is even? Are the events *roll less than 5* and *roll even* independent?

28. Suppose we flip a coin and roll an eight-sided die. What is the probability that we get heads or roll a prime number?

29. The population incidence of colorectal cancer (CRC) is about 0.2% for average-risk individuals. (About 3% of 50-year-olds will get CRC by age 80, and CRC is the third most common cancer in both men and women.) For average-risk individuals ages 50–75, Cologuard (a multi-target stool DNA test) has 92.3% sensitivity and 86.6% specificity. (These terms are defined in Example 14.5.3.) Given that a person of average risk has a positive Cologuard test, what is the probability that ze has CRC?

30. (Also about Cologuard) Given that a person of average risk has a negative Cologuard test, what is the probability that ze has CRC?

31. (Also about CRC screening tests) For average-risk individuals ages 50–75, a fecal immunochemical test (FIT) has 73.8% sensitivity and 94.9% specificity. Given that a person of average risk has a positive FIT test, what is the probability that ze has CRC?

32. (Yet more on CRC screening) For average-risk individuals ages 50–75, the Epi proColon blood test has 49.2% sensitivity and 91.5% specificity. Given that a person of average risk has a positive Epi proColon test, what is the probability that ze has CRC? Given that a person of average risk has a negative Epi proColon test, what is the probability that ze has CRC? (What can we conclude?)

33. Consider the Blue Sea card deck from Bonus Check-Yourself Problem 6. Draw a card. What is the probability that the card is squid suit or star suit given that it is a face card (king, queen, jack)?

34. Give the probability distribution for the random variable *Sk* that measures the number of skulls rolled when rolling the red die and the yellow die in the game *Elder Sign* described in Bonus Check-Yourself Problem 7. What is the sum

of the values in the distribution? (What should it be?)

35. Give the probability distribution for the random variable M that measures the number of magnifying glasses rolled when rolling all eight dice in the game *Elder Sign* described in Bonus Check-Yourself Problem 7.

14.13   Instructor Notes

Often in discrete mathematics texts, the big ideas of probability are hidden and the science of probability has been reduced to combinatorics. Such combinatorial approaches to probability (particularly those introduced in high school) measure ratios of $\frac{\text{things we care about}}{\text{total things}}$ and concentrate on counting techniques to do such computations. There is a strong effort here to reveal the big ideas of probability instead.

Even though probability is widely used in mathematics, and in discrete mathematics via the probabilistic method, it is not central to an introductory discrete mathematics course (partly because the probabilistic method is so challenging). That's why this chapter has been placed outside the stream of the main course. Should you include this chapter in your course, it has some prerequisites, namely graphs, some elementary counting, and the principle of inclusion/exclusion. Therefore, it would most naturally fit after the major combinatorics chapters.

This material is reasonably challenging for students, so it is not presented in a discovery-based manner. For the first class day, assign students to read Sections 14.1, 14.2, and 14.3 as preparation. Review this material in a short lecture (and take questions!) before asking students to work in groups on Section 14.4. It will likely take at least two-thirds of a class period for most students to complete. Repeat this technique for the second day, asking students to read Section 14.5 before class, reviewing the material at the start of class, and then asking students to work in groups on Section 14.6. For the third day, ask students to read Section 14.7 as preparation and then spend the bulk of class going over the example in Section 14.7.1 in detail. With any remaining time, have students revisit the problems in Sections 14.4 and 14.6.

Chapter 15

Fun with Cardinality

15.1 Introduction and Summary

When it comes to infinity, just how big is it? And is that even a well-posed question—could there be more than one size for infinity to have? Addressing this mountainous question is the focus of our chapter. The vehicle for exploring the sizes of infinite sets is a play (that's Section 15.2) in which the characters encounter and discuss ways of relating infinite sets to each other.

This is followed by (or interleaved with, depending on how you decide to read the chapter) a lot of exploratory questions in Section 15.4. Some of the more challenging ideas are reviewed in Section 15.5, and then of course there are problems to think about. Along the way, we will need to introduce new ideas, new notations, and new sets… and we will even leave the realm of discrete mathematics, though just for a minute or two.

15.2 Read This! Parasitology, the Play

Perhaps a dramatic reading is in order. You will want ten willing readers (though you can make do with five), and a chalkboard or whiteboard would be handy for investigational pauses. Particularly good pausing moments are marked with $[\star_k]$ and have matching exploratory questions in Sections 15.4.1–15.4.4.

Characters:

CATALOGUER	NARRATOR
LAB ASSISTANT	PROTAGONIST
LAB ASSISTANT 2 (LA2)	RECEPTIONIST
LAB ASSISTANT 3 (LA3)	STORAGE COORDINATOR
LAB ASSISTANT 4 (LA4)	TAXONOMIST

15.2.1   Scene 1: The Storage Coordinator

Characters: LAB ASSISTANT, NARRATOR, PROTAGONIST, RECEPTIONIST, STORAGE COORDINATOR.

Enter RECEPTIONIST, *who stands as at a desk, and* NARRATOR. PROTAGONIST *enters as* NARRATOR *speaks.*

NARRATOR: (*in a declarative tone*) Recently when visiting the friendly neighborhood ducks, our protagonist noticed something odd about the fecal matter in the duck yard and, having recently read Carl Zimmer's *Parasite Rex*, wondered whether a parasite might be to blame. Our protagonist did a quick search and found that not only is there a local parasitology lab, but that this lab is a national center that accepts specimens for examination *and* among its specialities is parasites of waterfowl! We enter the scene as our protagonist arrives at the lab.

RECEPTIONIST: Welcome to Parasite Central! How may I help you?

PROTAGONIST: Hello. I understand that you accept specimens from the public for examination and identification?

RECEPTIONIST: Indeed we do. Have you brought a specimen with you? And if so, what kind of specimen is it, and what is your reason for bringing the specimen to us?

PROTAGONIST: Yes, I have brought a sample of duck fecal matter that I suspect contains parasites of some kind. I was hoping you could tell me what's in there.

RECEPTIONIST: Certainly. Let me page a lab assistant and we'll get that sample taken care of. (*The* RECEPTIONIST *presses a button.*)

Enter LAB ASSISTANT.

RECEPTIONIST: (*to* LAB ASSISTANT) We have a member of the public here who has an avian waterfowl sample to submit.

LAB ASSISTANT: (*to* PROTAGONIST) Excellent. Please accompany me to the sample analysis area.

Exit LAB ASSISTANT, *followed by* PROTAGONIST; *exit* RECEPTIONIST *in a different direction. Soon after, enter* LAB ASSISTANT, *followed by* PROTAGONIST.

LAB ASSISTANT: Let me quickly check your sample under a microscope. (*takes the sample from* PROTAGONIST *and quickly mimes creating a slide and examining it under the microscope*) That's interesting. There are definitely parasites in here, but I don't recognize them as a known species. (*looks up*) We will definitely want to store this sample for further study, and I'll need to consult with one of the senior scientists to find out whether the specimen warrants cataloguing as a probably new species. Would you like to accompany me on these tasks? I could give you a quick tour of part of the lab while I'm doing them.

PROTAGONIST: That would be great. (*They start to walk off.*) Do you get new samples all the time? If so, how do you have room for them all? Or do you not store that many of the samples?

LAB ASSISTANT: Oh, yes, we get new samples all the time. Many of them don't have parasites, but many do and many are new. Luckily, we have infinitely many sample drawers and so storage is not a problem.

Exit LAB ASSISTANT, *followed by* PROTAGONIST; *soon after, re-enter* LAB ASSISTANT, *followed by* PROTAGONIST, *whilst* STORAGE COORDINATOR *enters from another direction.*

LAB ASSISTANT: (*to* PROTAGONIST) This is our storage area.

STORAGE COORDINATOR: Are you bringing me a sample? I must warn you, all of the drawers are full at the moment.

PROTAGONIST: But I thought you had infinitely many sample drawers …?

STORAGE COORDINATOR: We do. But as samples get analyzed, it's often realized that we don't need them anymore and so drawers free up regularly. At the moment, all of them are full.

LAB ASSISTANT: Well, we need to put this sample *somewhere*. Should I just leave it on the desk here until a drawer is vacated?

STORAGE COORDINATOR: Oh, no, we can't do that. What if more samples come in before drawers are available? They could pile up, and imagine what a mess we would have—both literally and logistically. No, hang on and let me think for a moment… do you have a favorite number? We can move all the samples from that number onwards up a drawer, and then we can put your sample in the drawer with your favorite number.

PROTAGONIST: (*staring*) Seriously? Isn't moving all those samples going to create havoc with your system? And what about the last sample? Where does it go?

STORAGE COORDINATOR: You know how some library stacks have wheels so you can compress a bunch of aisles and just leave an opening for the aisle you want to go into? We have a system like that, except it moves drawers over and relabels them. It just takes a while to go into effect. And there is no last

[⋆1]   sample, because there are infinitely many.

What number drawer would you like?

PROTAGONIST: Umm… okay… drawer 59?

STORAGE COORDINATOR: Certainly. (*reaches to take sample*)

NARRATOR: Brrrrring! (*pause*) Brrrrrrring!

STORAGE COORDINATOR: (*answering the phone*) Yes? Storage facility…

RECEPTIONIST: (*from offstage*) I thought I should let you know that a field biologist has arrived with another eight samples, preverified as previously unknown. Are you able to handle them?

STORAGE COORDINATOR: Sure. Send them over with a lab assistant. (*sets down phone, speaks to* PROTAGONIST) I can't guarantee that your sample will remain

[⋆2]   in drawer 59. It seems that I'll have to make room for more samples.

NARRATOR: Brrrrring! (*pause*) Brrrrrrring!

STORAGE COORDINATOR: (*answering the phone*) Yes? Storage facility…

RECEPTIONIST: (*from offstage*) I'm sorry to bother you again, but another field biologist has arrived, and this one has infinitely many samples. What should I say? Can we take that many samples today?

STORAGE COORDINATOR: Oh, I think so. Send the field biologist here with a lab assistant and we'll work something out. (*setting down the phone and speaking to* PROTAGONIST *and* LAB ASSISTANT) Infinitely many samples have arrived. I really cannot guarantee the number of the drawer in which your sample will be stored. Will that be all right?

PROTAGONIST: Of course. But how will you store infinitely many *more* samples? You can't just move them all up by infinitely many places—that would take an infinite amount of time.

STORAGE COORDINATOR: It would take less time than you think. But no, there's a simpler way. Every sample is in a numbered drawer. I'll just move every sample to the drawer with twice the current number, and then all the odd-numbered drawers will be free. And there are infinitely many of those, so it should all be fine.

[★3]

LAB ASSISTANT: Thank you. We do need to get the slide from this sample checked by a senior scientist, so we'll be going now. Thanks again for your time.

STORAGE COORDINATOR: Wait, you can't go yet. I will need to write you a receipt for the sample, so that you can deduct it from your taxes as a charitable donation, and so that other scientists can find it if they need further observations of the sample.

NARRATOR: Brrrrring! (*pause*) Brrrrrrrring!

STORAGE COORDINATOR: I'm sorry for the delay caused by all these interruptions. (*answering the phone*) Yes? Storage facility...

RECEPTIONIST: (*from offstage*) Oh, dear! I'm afraid we're being deluged—infinitely many field biologists are arriving now, each with infinitely many specimens to store. Should I turn them away? Let some through? Which are most important?! Help!

PROTAGONIST: (*aside, to* LAB ASSISTANT) The receptionist must be frantic. I can hear the conversation through the phone!

LAB ASSISTANT: (*aside, to* PROTAGONIST) Yes, but it will be all right. You'll see.

STORAGE COORDINATOR: (*into the phone*) Well, I have a sample at the desk to store, and then there are the eight new samples you've sent over, and the infinitely many samples with the first field biologist, so... if you can delay the other infinitely many field biologists by about 15 minutes and then send them here with a coterie of lab assistants, I should be able to take them all. But hopefully that's all we will have to deal with today!

PROTAGONIST: (*aside, to* LAB ASSISTANT) Infinitely many field biologists will be a lot for a few lab assistants to handle. Aren't you going to need to help with managing these people? Maybe I should just go...

LAB ASSISTANT: (*aside, to* PROTAGONIST) It's not a problem. We have infinitely many lab assistants here.

STORAGE COORDINATOR: (*sets down the phone*) Whew. Perhaps you overheard that we have a bit of a situation. Let me get you on your way.

NARRATOR: The storage coordinator grabs a slip of paper and a pair of linked tags.

PROTAGONIST: I don't mean to delay things further, but how will you manage all infinitely many field biologists at once?

STORAGE COORDINATOR: It'll be a bit complicated, but totally doable. I'll leave the first specimen in the first drawer. Then I'll put two specimens from the first field biologist in the second and third drawers. I'll move the specimen that was in the second drawer to the fourth drawer, and the specimen from the third drawer to the ninth drawer. Then two more specimens from the first field biologist will go in drawers 5 and 6, and two specimens from the second field biologist will go in drawers 7 and 8. The specimens from those drawers will get moved down further… are you with me so far?

PROTAGONIST: I think so. You've moved the old specimens up by some number of drawers, and you've started placing specimens from the first and second field biologists. So far you've filled the first nine drawers. But you'll have to work in specimens from more field biologists, right?

STORAGE COORDINATOR: Yes, and that happens in the next round of placements. Two more specimens from the first field biologist will go in drawers 10 and 11, two more specimens from the second field biologist will go in drawers 12 and 13, and two specimens from the third field biologist will go in drawers 14 and 15. And so forth and so on, including specimens from an additional field biologist in each round of placements, until every specimen has been placed in a drawer or moved to a new drawer.

[⋆4]

Anyway, (*making some notes on the form and attaching one of the two tags*) here is your receipt and the tag that identifies your specimen. The matching tag stays with the specimen, so you can be sure upon retrieval that you got the specimen you submitted. Have a lovely day and enjoy your time at Parasite Central.

All characters exit.

15.2.2   Scene 2: The Taxonomist

Characters: LAB ASSISTANT, PROTAGONIST, TAXONOMIST.

LAB ASSISTANT *enters, followed by* PROTAGONIST; *from another direction,* TAXONOMIST *enters.*

Taxonomist: Greetings! Do you have some slides for me?

Lab Assistant: Yes, I have *a* slide for you. I am also showing the collector of the specimen around part of our lab.

Taxonomist: Lovely—(*to* Protagonist) Let me tell you a little bit about what we do here. (*gestures to the room*) I examine slides of waterfowl parasites and note their characteristics. With this information, I classify the parasites; if they are rare, or if they are different enough to be considered a new species, I store the slides in this array. The bulk of my job is deciding where in the "family tree" of waterfowl parasites new species should go.

(*to* Lab Assistant) The slide, please?

Lab Assistant *mimes handing a slide to* Taxonomist, *who mimes examining it under a microscope.*

Taxonomist: That's interesting… hmmm… (*looks up*) I'm going to need to do a little bit more work with this one before I can tell you whether it needs to be taken to Cataloguing. Could you take our guest to get some coffee or something for a few minutes? I should have an answer for you relatively soon. Oh, but I'll need your receipt for now, so I can make notes on it for potential cataloguing purposes.

Protagonist: I'm confused. You need the receipt to take notes? I thought it was for tax purposes…

Taxonomist: Oh, it *is*, as far as you are concerned. But before you leave with it, we will have preliminary notes recorded on it and a copy will be stored either here or with Cataloguing, depending on what the eventual fate of the sample and slide are.

Protagonist: Okay. Here's the receipt. Do you need the tag as well? … (*drifts off*) … that's odd; it says 134. But I thought we were told it was going to be number 59. Oh, well.   [⋆5]

Taxonomist: Thanks. The tag isn't necessary at this point. If I've already finished with your slide when you return, I'll be storing it in holder $(147, 2032)$ in the array. That one's empty at the moment.

Protagonist: As in the hundred-and-forty-seventh holder in the two-thousand-and-thirty-second row? You must have a lot of slides here!

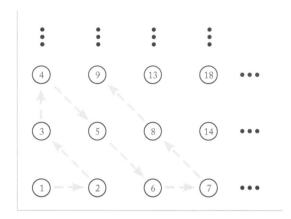

Figure 15.1. The old numbering of slide holders in Taxonomy.

TAXONOMIST: It's more like the two-thousand-and-thirty-second holder in the hundred-and-forty-seventh aisle. These coordinates are just like (x, y) coordinates of a point in the plane. And yes, we have a lot of slides here. Infinitely many, in fact.

PROTAGONIST: With two coordinates to count with, you must have room for infinitely many more than the Storage area has!

TAXONOMIST: Actually, I have exactly the same amount of room as Storage has. The only reason I use two coordinates is because my laboratory area is a different shape than the Storage area—when we first moved in, I numbered my holders in the array to match the drawers in Storage. It was like this, as shown in Figure 15.1: (*gestures*) holder 1 was in the corner here, holder 2 to the right of it, and then diagonally up and left was holder 3, holder 4 was just above it, and then I kept numbering by going diagonally back until I was at the right of holder 2, and so forth. But then over time, I got tired of zig-zagging back and forth from the corner to find a slide and decided to use array numbering instead. Besides, when we started, there were the same number of slides as samples, and now there are *way* more samples than slides—not everything we need to store is studied by me, but everything I study has a corresponding sample in Storage.

[⋆6]

LAB ASSISTANT: Before we go, I should warn you that a large influx of specimens arrived earlier today.

TAXONOMIST: How large are we talking?

LAB ASSISTANT: Well, over in Storage we heard that infinitely many field biologists had arrived with infinitely many samples each.

TAXONOMIST: Oh. Great.

LAB ASSISTANT: Will this be a problem? I mean, should we take a loooong coffee break, or ...?

TAXONOMIST: Don't worry. It's not like Storage will process all of them at once, and besides, there will probably not be very many that get slides passed to me; I just get the avian waterfowl slides, and even with infinitely many samples, there could be finitely many of those. And it's not like I could analyze more than one slide at a time anyway, so whoever comes in while I'm working can just wait. Or... I could pretend I'm on break and then no one will disturb me... just don't worry about it.

LAB ASSISTANT: Okay. We'll see you later.

PROTAGONIST: (*to* TAXONOMIST) Thanks for explaining your slide storage system to me! It was lovely viewing your area and hearing about your work.

All exit, LAB ASSISTANT *and* PROTAGONIST *in a different direction from* TAXONOMIST.

15.2.3   Scene 3: The Café

Characters: LAB ASSISTANT, LAB ASSISTANT 2 (LA2), LAB ASSISTANT 3 (LA3), NARRATOR, PROTAGONIST.

Enter LAB ASSISTANT *followed by* PROTAGONIST; *from another direction, enter* LAB ASSISTANT 2 (LA2) *and* LAB ASSISTANT 3 (LA3), *who sit down.*

NARRATOR: There is a counter with stools that stretches across the room. All the stools are occupied.

LAB ASSISTANT: Can I get you some coffee?

PROTAGONIST: (*looking around, staring*) ... Sure. ... There are a *lot* of people here. And I don't see anywhere to sit... is there anywhere besides the counter?

LAB ASSISTANT: Well, this is a common break time for lab assistants, so it's not surprising that there are a lot of people here. (*gazes towards where the counter would be, where* LA2 *and* LA3 *are sitting*) I'm pretty sure that while there are infinitely many people at the counter, there are also infinitely many lab assistants who aren't here. I don't see anyone whose name begins with "J," for example, and there are infinitely many of those. Anyway, we can sit at the counter. (*to* LA2 *and* LA3) Could you make room for two? I have a guest here who brought in a specimen.

LA2: Sure! (*to invisible people seated on the left*) Could you move over, please? We have a new arrival over here.

LA3: (*to invisible people seated to the right*) Please move down a seat; pass on the message.

LAB ASSISTANT: (*to* LA2, *while sitting down just to* LA2*'s right*) Has coffee been poured recently?

LA2: No, I think it will be soon, though. I heard some mugs clinking a moment ago.

NARRATOR: A server comes by, setting down steaming mugs of coffee while walking past.

LAB ASSISTANT: (*to* LA2) I usually like to have two cups. You?

LA2: Not usually, but today I'm very tired. (*turns to the left*) Could you please give me your coffee? And the coffee given to the person on your left?

LAB ASSISTANT: (*to* PROTAGONIST) You didn't want two cups of coffee, did you?

PROTAGONIST: No… but I would like cream for my coffee. And is it seriously okay to make those people to the left give up their coffee for you?

LAB ASSISTANT: Cream should be along in a minute; also, sugar. And the people to our left will get coffee from the people on their left. We're used to passing the mugs around before drinking.

[⋆7]

NARRATOR: A server comes by, setting down dishes of creamer and sugar cubes while walking past.

LA3: (*to* PROTAGONIST) Could you pass me the dish of creamer and sugar cubes? We don't have enough over here.

LA2: (*overhearing*) Extra creamer is being passed up this way, hang on a moment... here.

LA2 *hands a dish of creamer to* LAB ASSISTANT, *who hands it to* PROTAGONIST, *who takes two creamers before passing it to* LA3.

PROTAGONIST: A snack would be lovely... can we order anything? ... from anyone?

NARRATOR: A server walks by with an armload of cookies but doesn't set any down.

LAB ASSISTANT: We'll have to ask around. No one stops long enough for us to order drinks or snacks. (*to LA2*) Any cookies down that way? (*to LA3*) Do you see any cookies?

LA2: I'll ask. I'm pretty hungry myself.

LA3: Let me see....

LA2: (*to invisible people seated on the left*) Do you have any cookies?

LA3: (*to invisible people seated on the right*) Could we... Hey! You down there, stop hoarding all the cookies! (*pauses to listen*) It doesn't matter if you want infinitely many cookies, you can still send a few our way! (*turning to* PROTAGONIST *and shaking head*) Some people...   [⋆8]

LA2: (*to* LAB ASSISTANT) What group meeting do you have today?

LAB ASSISTANT: That's a good question. I'll have to check the schedule; it changes so often. You?

LA2: I'm lucky—for today I'm assigned to a group that's discussing my primary project!

LAB ASSISTANT: Wow, *nice*. Is it a big group?

LA2: Not too big. I think there are under 50 people.

LAB ASSISTANT: That's good. Last week I was in a group with 30,000,000 people, and it was tough for anyone to hear each other during the meeting.

PROTAGONIST: How often do you get to work on your primary project?

LAB ASSISTANT: We work on our primary projects whenever we have no other specific duties—dealing with sample intake, assisting senior scientists, and so on. But it's rare to have group meetings about our primary projects, because we are but lowly lab assistants.

PROTAGONIST: Oh, so most of the group meetings are about senior scientists' projects?

LAB ASSISTANT: No... they're about other lab assistants' primary projects. It's part of a workflow experiment. Every lab assistant has hir own primary project, but until recently we worked in relative isolation on our projects. One of the heads of the lab thought that some collaboration would increase research productivity on lab assistant projects and so asked the lab's operations researcher to figure out what would be best. But it's not a straightforward problem— What size discussion group is best? Should discussion groups be formed of lab assistants with close expertise to the project topic, or lab assistants who know very little and so bring fresh thoughts, or a mixture of the two? So the operations researcher decided we would try having every possible collection of lab assistants meet and have everyone take notes on the meetings, and then see what works best.

PROTAGONIST: Wow. That's a lot of meetings.

LAB ASSISTANT: Yeah, we have at least one in the schedule every day. And it's kind of random because the operations researcher is still trying to figure out the whole schedule, and so we don't know very far in advance when we're meeting about what with whom. But to answer your original question, some lab assistants don't get group meetings about their primary projects. They do get the notes from the meetings about their projects, though. I suspect that mostly meetings like that are a waste of time, but we'll see.

LA3: I heard that the operations researcher ran into a massive scheduling problem recently, and it might derail a chunk of the experiment.

LAB ASSISTANT: Really? What's going on?

LA3: The operations researcher made a list of all possible groups of lab assistants, right? And every group has to meet about someone's primary project because they have to have something to talk about, and it would be silly to have them talk about something that has nothing to do with the lab. But rumor has it that there is a group with no project, and this is throwing the operations researcher for a loop.

PROTAGONIST: Can't the operations researcher just pick a project no one has met about yet?

LA3: No, it's super-weird. It has to do with the particular group. The operations researcher came up with every possible group, and one of those is the group of lab assistants who are not in the group that discusses their primary project. And apparently there's no project for that group to discuss.

PROTAGONIST: I still don't see why the operations researcher can't just pick a project.

LAB ASSISTANT: Let's pretend that this particular group is going to discuss my primary project. I would love to be there, not that I'm bitter about not having had a group talk about my project yet, or anything…. Anyway, I want to be in the group. But every lab assistant in this group is *not* in the group that discusses hir primary project. So I can't be in the group. And *that* means I am not in the group that discusses my primary project, which means I *am* in the special group, which is discussing my primary project. It just goes 'round and 'round.   [⋆9]

PROTAGONIST: … Uhhh….

LA3: Well, I also heard that the operations researcher is considering expanding the focus of these groups; the problematic group might be told to discuss the functioning of this café, for example. That's not a lab assistant's primary project. But there might be other issues that come up.

LAB ASSISTANT: (*to* PROTAGONIST) Did you ever get the cookie you wanted? We should be getting back to Taxonomy to check on your slide. You still have your tag?

PROTAGONIST: Yeah, somewhere in there I got a cookie. The tag is in my pocket… (*pulls tag out of pocket*) Here. Wait a minute! The tag says 17,956, and I'm *sure* it was a much smaller number before. Wasn't it supposed to be 59?

LAB ASSISTANT: Oh, the tags are quantum entangled. Remember how it was originally one of a pair? They're linked, so when the Storage Coordinator moves a sample to a new drawer, the storage tag gets relabeled and that relabels the tag you have here.   [⋆10]

PROTAGONIST: (*to* LA2 *and* LA3) It was nice to meet you. Good luck with your meetings and projects!

LA2 *and* LA3 *wave as* LAB ASSISTANT *and* PROTAGONIST *exit, then exit themselves.*

15.2.4   Scene 4: Cataloguing

Characters: CATALOGUER, LAB ASSISTANT, LAB ASSISTANT 4 (LA4), NARRATOR, PROTAGONIST, TAXONOMIST.

LAB ASSISTANT *and* PROTAGONIST *enter from one direction and* TAXONOMIST *enters from another direction.*

TAXONOMIST: Ah! You're back! I have news—the avian waterfowl parasites on your slide are likely an unknown species, which means that they need to be catalogued. That means you will take the receipt and the sample tag down to Cataloguing, where a catalogue number will be assigned, a copy will be made of the receipt, and the tag will be stored with the catalogue information. Here. (*hands them the receipt*)

PROTAGONIST: Thanks!

All exit. Then LAB ASSISTANT *and* PROTAGONIST *enter from one direction and* LAB ASSISTANT 4 (LA4) *enters from another direction.*

LA4: Whoa! You can't go through here.

LAB ASSISTANT: Why not? We need to take some information to Cataloguing.

LA4: All infinitely many specimens in the Liver Fluke Lab need to be collected for transport and we are about to start. So you will need to wait.

PROTAGONIST: … Are we going to be waiting forever?

LA4: No. Just hang on a minute. (*backs away*)

NARRATOR: As PROTAGONIST and LAB ASSISTANT watch, a dolly stacked with boxes is brought across the hall and then brought back empty. This process takes about 30 seconds. In the next 15 seconds, the dolly passes across the hall full and returns empty. Again a load of boxes passes across, but in one-eighth of a minute. Continuing to pass back and forth twice as fast with each trip, the dolly becomes a blur, and about five seconds later the hall is still.

[⋆₁₁]   LA4: (*coming forwards again*) Okay, all clear. You can pass through.

LAB ASSISTANT *and* PROTAGONIST *cross and exit. Then* LA4 *exits. After a short pause,* LAB ASSISTANT *and* PROTAGONIST *enter from one direction and* CATALOGUER *enters from another direction.*

CATALOGUER: Hello. How may I help you?

LAB ASSISTANT: We've been informed by the avian waterfowl parasites taxonomist that the specimen brought in by this guest represents a new species, and so the specimen needs to be assigned a number in the catalogue.

CATALOGUER: Excellent! That should be no problem. May I see the receipt? I'll need the sample tag as well.

PROTAGONIST: Here you go. (*hands over the receipt and sample tag*) I've noticed that every department at Parasite Central seems to have its own way of organizing information. How do you do it here?

CATALOGUER: It's very simple. We create a catalogue entry for every identified species. The entries are numbered and have all the taxonomic and descriptive information reported by other departments. We can look up species by entry number or we can search the catalog by key words, though that's less useful as sometimes there are infinitely many results for a search.

PROTAGONIST: So when you get a large number of new species, you renumber?

CATALOGUER: (*runs the receipt through a scanner*) Oh, no. We'd never be able to find anything again! Besides, we don't just have the species identified at Parasite Central in our catalogue. We have every parasite species ever identified anywhere in our catalogue. One way we know a species is new is if its description doesn't match any of the entries in the catalogue.

PROTAGONIST: I'm confused. How do you avoid the renumbering? Do you have finitely many catalogue entries? I thought you said there were infinitely many search results....

CATALOGUER: We have infinitely many entries. But we don't renumber; we just find a number that hasn't been used yet. Here, let me show you how this works with the species you brought in. First, I'll pull up a few catalogue entries for you to see. (*types and then rotates the computer monitor so* PROTAGONIST *can see it*) Notice that at the top of the entry is the entry number. You can't see the whole thing on screen because it has more digits than will fit. And then below it is the description.

PROTAGONIST: (*squinting a bit*) Wait, is that a decimal point in front of the entry number? Or is it just a symbol in front of the number?

CATALOGUER: Yes, that's a decimal point—every entry number has a value between 0 and 1. Anyway, let's say I have a list of entry numbers and I need a new one. I just write down a decimal point, and then I go through the list. I write down a digit that is different from the first digit of the first number, and then I write down a digit that is different from the second digit of the second number, and I continue in this fashion. When I've gotten through the list, I know that I have a new number because it is different in at least one place from each number on the list. Easy peasy.

PROTAGONIST: Wouldn't it be even easier to use regular numbers instead of the numbers between 0 and 1?

CATALOGUER: No, because there wouldn't be enough of them unless I renumbered all the time. Imagine this: I number all of my current catalogue entries using regular counting numbers. The new entry I made a moment ago can't get a regular counting number because all of those are already in use. The decimals are much better. Anyway, here's your receipt back. The tag will be attached to [★12]   a copy of the receipt.

PROTAGONIST: Thanks so much. It's exciting to have a description of a new species!

LAB ASSISTANT: Come on, I'll walk you back to the reception area. Or did you take a shuttle to the building? It might be closer to leave through the café.

All characters exit. Soon LAB ASSISTANT *and* PROTAGONIST *enter from one direction and* LA2 *and* LA4 *enter from another direction and sit as at a café counter.*

NARRATOR: Again, there is a counter with stools that stretches across the room. And again, all the stools are occupied.

PROTAGONIST: It's still crowded in here. Is the café always full?

LAB ASSISTANT: Not always. It's a popular break time, so most of the lab assistants will be here now. Let me see…(*looks around*) In fact, I think that all but finitely many of the lab assistants are here. That must mean that Storage is caught up [★13]   with all the specimens brought in by the infinitely many field biologists.

PROTAGONIST: Thanks so much for the tour, and for letting me shadow your work with my sample. I'll have to tell my friendly neighborhood ducks that they've contributed to science.

LAB ASSISTANT: You're welcome! All in a day's work. Well, it's not *every* day that I get to handle a new species. So thank you, too. The door to the outside is this way…

Both characters exit.

NARRATOR: The End.

15.3   How Big Is Infinite?

Way back in Section 1.3, we gave the notation $|A|$ as the number of elements in a set A, and in Section 2.2, we explained that this is the cardinality of A. That all makes sense for finite sets, but what do we do with infinite sets? Do we just say $|A| = \infty$ and leave the matter at that? And how do we know whether two infinite sets have the same cardinality, or have the same infinite size?

We answered this question in Chapter 3—long before we even asked it! Remember the Facts of Section 3.2? The pertinent one here is...

> Fact 3.   If there is a bijective map from A to B, then $|A| = |B|$.

Well, there you have it. We never said that this was only true for finite sets, and so it gives us a way to determine when two infinite sets have the same size: we consider two sets to be the same size if we can put them in one-to-one correspondence.

15.4   Try This: Investigating the Play

There is a lot of mathematics embedded in Section 15.2, and our purpose here is to find, expose, and figure out that mathematics. The questions that follow correspond to the $[\star_k]$ markers scattered throughout Section 15.2.

15.4.1   Questions about Sample Storage

$[\star_1]$. "... do you have a favorite number? We can move all the samples from that number onwards up a drawer, and then we can put your sample in the drawer with your favorite number ... And there is no last sample, because there are infinitely many."

(a) To what set does the drawer numbering system correspond?

(b) Moving the samples can be described by a function. What is it? Make sure to specify the domain and target of the function as well as the defining rule.

(c) Is the function you just defined a bijection? If so, prove it; if not, explain why not.

(d) Consider the following statements: $\infty = \infty + 1$; $\infty < \infty + 1$. Argue that exactly one of these statements could be true.

[⋆₂]. Eight additional samples arrive and so the Storage Coordinator cannot guarantee that the Protagonist's sample will remain in drawer 59. What two sets seem to have the same cardinality here? Define a function to verify that their cardinality is, in fact, the same.

[⋆₃]. "Every sample is in a numbered drawer. I'll just move every sample to the drawer with twice the current number, and then all the odd-numbered drawers will be free."

    (a) What is the sample-moving function proposed by the storage coordinator?

    (b) Make and prove a conjecture about cardinalities of sets. Is there a bijection hiding here?

[⋆₄]. Infinitely many field biologists arrive, each with infinitely many specimens. "I'll leave the first specimen in the first drawer. Then I'll put two specimens from the first field biologist in the second and third drawers. I'll move the specimen that was in the second drawer to the fourth drawer, and the specimen from the third drawer to the ninth drawer. Then two more specimens from the first field biologist will go in drawers 5 and 6, and two specimens from the second field biologist will go in drawers 7 and 8. The specimens from those drawers will get moved down further ... two more specimens from the first field biologist will go in drawers 10 and 11, two more specimens from the second field biologist will go in drawers 12 and 13, and two specimens from the third field biologist will go in drawers 14 and 15. And so forth and so on, including specimens from an additional field biologist in each round of placements, until every specimen has been placed in a drawer or moved to a new drawer."

    (a) What is the sample-moving function described in this quotation?

    (b) What is the algorithm the Storage Coordinator is using to place new samples?

    (c) Will there actually be enough room for all of the new samples? Explain.

    (d) What monstrous cardinality fact is hiding here? Make and justify (but don't attempt to prove) a conjecture.

    (e) Did you expect that the set of drawers and the set of new samples would have the same cardinality? Why or why not?

15.4.2   More Questions about Sample Storage

[⋆5].   "… that's odd; [the sample tag] says 134. But I thought we were told it was
going to be number 59."

Assuming the sample was originally going to get the number 59, what does
this tell us about how many of the constantly arriving specimens the storage
coordinator has processed?

[⋆6].   "… holder 1 was in the corner here, holder 2 to the right of it, and then
diagonally up and left was holder 3, holder 4 was just above it, and then I
kept numbering by going diagonally back until I was at the right of holder
2, and so forth. But then over time, I got tired of zig-zagging back and forth
from the corner to find a slide and decided to use array numbering instead.
Besides, when we started, there were the same number of slides as samples,
and now there are *way* more samples than slides—not everything we need
to store is studied by me, but everything I study has a corresponding sample
in Storage."

(a) Add array numbering to Figure 15.1. What bijection does this suggest
(in particular, what sets are involved)? And what interesting cardinality fact can you thereby deduce?

(b) Find a surjection from array numbering to the rational numbers $\mathbb{Q}$.
(Why is this not an injective map?) What surprising and interesting
cardinality fact does this suggest?

(c) The taxonomist says that there are more samples than slides, but you
know that there are $|\mathbb{N}|$ samples (because the sample drawers are num-
bered) and $|\mathbb{N}|$ slides (because these slide holders used to be num-
bered). What gives? (How can both statements be true?)

15.4.3   Questions about Café Conversations

[⋆7].   Lab Assistant and LA2 each want to have two cups of coffee. LA2 turns to
the left and asks, "Could you please give me your coffee? And the coffee
given to the person on your left?" Lab Assistant explains that "… the people
to our left will get coffee from the people on their left."

(a) The stools in the café stretch all the way across the room, left and right.
What set would be convenient to use in numbering the stools?

(b) What map on coffee cups is described (that will give Lab Assistant and
LA2 two cups of coffee each)?

[⋆8]. "It doesn't matter if you want infinitely many cookies, you can still send a few our way!"

How is it possible for one person to have infinitely many cookies and still send some down?

[⋆9]. "LA3: The operations researcher made a list of all possible groups of lab assistants, right? And every group has to meet about someone's primary project …. The operations researcher came up with every possible group, and one of those is the group of lab assistants who are not in the group that discusses their primary project. And apparently there's no project for that group to discuss. …

"LAB ASSISTANT: Let's pretend that this particular group is going to discuss my primary project … every lab assistant in this group is *not* in the group that discusses hir primary project. So I can't be in the group. And *that* means I am not in the group that discusses my primary project, which means I *am* in the special group, which is discussing my primary project. It just goes 'round and 'round."

(a) LA3 and Lab Assistant's dialogue suggests a function from the set of groups of lab assistants to the set of projects. What is that function? Describe/define it verbally and using function-and-set-theoretic notation.

(b) In turn, this suggests a function from the set of groups of lab assistants to the set of lab assistants. What is *that* function? (Please call it f.)

(c) Consider the group of lab assistants who are not in the group that discusses their primary project. Define this group g using set-theoretic notation.

(d) What is $f(g)$?

(e) Is $f(g) \in g$? Explain.

(f) Explain why there cannot be a bijection between the set of groups of lab assistants and the set of lab assistants.

(g) What is the relationship between the groups of lab assistants and the lab assistants themselves?

(h) What cardinality conclusion does this present?

[⋆10]. "Wait a minute! The tag says 17,956, and I'm *sure* it was a much smaller number before. Wasn't it supposed to be 59?"

Assuming again that the sample was originally going to get the number 59, what does this tell us about how many of the constantly arriving specimens the storage coordinator has processed?

15.4.4   Indiscrete Questions

[⋆11]. The infinitely many specimens in the Liver Fluke Lab are collected for transport: "… a dolly stacked with boxes is brought across the hall and then brought back empty. This process takes about 30 seconds. In the next 15 seconds, the dolly passes across the hall full and returns empty. Again a load of boxes passes across, but in one-eighth of a minute. Continuing to pass back and forth twice as fast with each trip, the dolly becomes a blur, and about five seconds later the hall is still."

(a) How long does it take for all of the boxes to be taken across the hall?

(b) How many trips does the dolly make?

(c) Write the duration of each trip as a fraction of the total time, and add these up. Rewrite the result using summation notation to conclude an interesting fact.

[⋆12]. The Cataloguer avoids relabeling as follows: "… every entry number has a value between 0 and 1 … Let's say I have a list of entry numbers and I need a new one. I just write down a decimal point, and then I go through the list. I write down a digit that is different from the first digit of the first number, and then I write down a digit that is different from the second digit of the second number, and I continue in this fashion. When I've gotten through the list, I know that I have a new number because it is different in at least one place from each number on the list."

(a) What is the set of entry numbers?

(b) How does its cardinality compare to that of $\mathbb{N}$? Explain.

[⋆13]. "In fact, I think that all but finitely many of the lab assistants are here. That must mean that Storage is caught up with all the specimens brought in by the infinitely many field biologists."

Are there more lab assistants in the café now than there were earlier?

15.5   How High Can We Count?

Hey! You! Don't read this unless you have worked through the problems in Section 15.4. I mean it!

We can only count at a finite pace (see Section 15.6 for other possibilities), so the literal answer to the question raised in the title of this section is "to some large finite number." But considering the question more figuratively, we have a pretty good idea of how many natural numbers there are, and thus $|\mathbb{N}|$ is a yardstick against which we can measure the cardinalities of other sets. However, as we observed in Sections 15.2 and 15.4, not all infinite sets have the same cardinality, so it is not useful to say $|\mathbb{N}| = \infty$ (even though that statement is true). Instead, we say $|\mathbb{N}| = \aleph_0$ (pronounced *aleph-naught* or *aleph-null*) so that we have a name for this particular size of infinity. We also say, because $\mathbb{N}$ is the set of counting numbers, that any set with size $\aleph_0$ is *countable*. (Indeed, any set with size larger than $\aleph_0$ is *uncountable*.)

   With this terminology, $\mathbb{N}$, $\mathbb{Z}$, and $\mathbb{Q}$ are countable, whereas $\mathbb{R}$ is uncountable. An indirect way to see that $\mathbb{R}$ is uncountable is to rewrite every real number in binary; for any number, there are only finitely many places before the point, but there may be infinitely many places after the point. So the number of possible real numbers (in binary) is $2^{\aleph_0}$. This is the size of the power set of $\mathbb{N}$, which we saw in Problem 9 of Section 15.4.3 is larger than $\aleph_0$. A different indirect way to see that $\mathbb{R}$ is uncountable is to note that $\mathbb{R} \supset [0, 1]$, and Problem 12 of Section 15.4.4 showed that $[0, 1]$ is uncountable.

Example 15.5.1. You are reading an element of an uncountable set! That is, you are reading words that form a subset of a page, which in turn is a subset of the plane, and the plane is $\mathbb{R}^2$. Therefore, the text of this book is an element of $\mathscr{P}(\mathbb{R}^2)$. We know that $|\mathscr{P}(\mathbb{R}^2)| \geq |\mathbb{R}^2| > |\mathbb{N}| = \aleph_0$.

   The proofs that $[0, 1]$ and $\mathscr{P}(\mathbb{N})$ are uncountable have the same structure: they both show by contradiction that there cannot be a one-to-one correspondence with $\mathbb{N}$. For $[0, 1]$, we use the naturals to list all elements of $[0, 1]$ (in any order!), which is a one-to-one correspondence with $\mathbb{N}$, and then produce an element of $[0, 1]$ that does not appear on this list. Thus, the bijection produced by the supposed one-to-one correspondence turns out to not be onto $[0, 1]$. For $\mathscr{P}(\mathbb{N})$, we assume there is a bijection between $\mathscr{P}(\mathbb{N})$ and $\mathbb{N}$ and produce an element of $\mathscr{P}(\mathbb{N})$ that cannot correspond to any element of $\mathbb{N}$.

Figure 15.2. Cardinality is mysterious, perhaps because the iconic cardinal wears a mask.

> Summary of approaches. Our three favorite sets $\mathbb{N}$, $\mathbb{Z}$, and $\mathbb{Q}$ are countable. We show a set is countable by exhibiting a bijection with $\mathbb{N}$.
>
> On the other hand, $\mathbb{R}$, $[0,1]$ and $\mathscr{P}(\mathbb{N})$ are uncountable. To show a set is uncountable, we show by contradiction that there cannot be a one-to-one correspondence with $\mathbb{N}$.

These proofs are conceptually challenging. Because it tangles with infinity, the study of cardinality is inherently mysterious (see Figure 15.2), and in fact, brain-twisting aspects of infinity arise even in relatively simple situations. For example, the sets $\mathbb{N}$ and $2\mathbb{N}$ have the same cardinality, but $\mathbb{N}$ has infinitely more elements than $2\mathbb{N}$ has. (The extra elements are the odd counting numbers.) This shows the need for precision in language—we could colloquially say that $\mathbb{N}$ and $2\mathbb{N}$ are the same size (they have the same cardinality) and also that $\mathbb{N}$ is larger than $2\mathbb{N}$ ($\mathbb{N} \supset 2\mathbb{N}$ and $\mathbb{N}$ has *way* more elements).

In order to show that two sets have the same cardinality, we do not necessarily need to give a single function that we can prove is a bijection. We can instead show that a set of functions together provide an injection and a surjection.

Example 15.5.2. Let us show that $|\mathbb{N} \cup \mathbb{N} \cup \mathbb{N} \cup \mathbb{N} \cup \mathbb{N}| = |\mathbb{N}|$.

First, some setup. We will denote an element $t \in \mathbb{N} \cup \mathbb{N} \cup \mathbb{N} \cup \mathbb{N} \cup \mathbb{N}$ by $t = (n, p)$ where $n \in \mathbb{N}$ and $p \in \{1, 2, 3, 4, 5\}$ to indicate which copy of $\mathbb{N}$ we're looking at. Next, we partition $\mathbb{N}$ into five subsets, $\{n \mid n \equiv 0 \pmod{5}\}$, $\{n \mid n \equiv 1 \pmod{5}\}$, $\{n \mid n \equiv 2 \pmod{5}\}$, $\{n \mid n \equiv 3 \pmod{5}\}$, $\{n \mid n \equiv 4 \pmod{5}\}$.

Now we will define our bijection by $f(t) = f((n, p)) = 5(n-1) + p$.

It is an injection: Suppose $f(t_1) = f(t_2)$. Then we have $5(n_1 - 1) + p_1 = 5(n_2 - 1) + p_2$. Examining this equation mod 5, we see that $p_1 \equiv p_2 \pmod{5}$.

Because $p_1, p_2 \in \{1,2,3,4,5\}$, we know $p_1 = p_2$. Then $5(n_1 - 1) = 5(n_2 - 1)$ implies $n_1 = n_2$. Therefore, $(n_1, p_1) = (n_2, p_2)$.

It is a surjection: For any $m \in \mathbb{N}$, we note that $m = 5k + r$ for some $k \in \mathbb{W}$ and $r \in \{0, 1, 2, 3, 4\}$, so $f((k+1, r)) = 5(k + 1 - 1) + r = m$.

As another example, consider showing that infinitely many copies of $\mathbb{N}$ have the same cardinality as $\mathbb{N}$ (as in Problem 4 of Section 15.4.1). Wait! What do we mean by "infinitely many copies"? That phrase could indicate countably many copies or uncountably many copies. Before we go further, let us give some notation for describing each of these sets. Countably many copies of $\mathbb{N}$ can be denoted $\bigcup_{k \in \mathbb{N}} \mathbb{N}_k$. (If you are disturbed by the notion that we are distinguishing between copies of $\mathbb{N}$, imagine that the kth copy of $\mathbb{N}$ is $\{k, k\,k, k\,k\,k, k\,k\,k\,k, \dots\}$.) Uncountably many copies of $\mathbb{N}$ can be denoted $\bigcup_{\alpha \in I} \mathbb{N}_\alpha$, where I is an uncountable index set (such as $[0, 1]$).

Now, when we say we want to show that infinitely many copies of $\mathbb{N}$ have the same cardinality as $\mathbb{N}$, we must mean to consider countably many copies of $\mathbb{N}$. After all, if we consider uncountably many copies, the number of copies of $\mathbb{N}$ is already of larger cardinality than $\mathbb{N}$ itself!

Here is one way to show that infinitely many copies of $\mathbb{N}$ have the same cardinality as $\mathbb{N}$, without using a lone function. We partition $\mathbb{N}$ into subsets that each have cardinality $\aleph_0$ (for example, $2\mathbb{N}$ is one such subset) and then put each of these subsets into one-to-one correspondence with copies of $\mathbb{N}$. The challenge is that we must partition $\mathbb{N}$ into $\aleph_0$ subsets!

We start with noting that $\mathbb{N}$ is in one-to-one correspondence with the square numbers $\{1, 4, 9, 16, \dots\}$ via the map $f(n) = n^2$. This is our first of the $\aleph_0$ copies of $\mathbb{N}$. Actually, let's call it the zeroth copy of $\mathbb{N}$ for ease in indexing and because its map differs from the remaining maps. We will deal the elements of the kth copy of $\mathbb{N}$ out like a deck of cards; element 1 will be sent to element $k^2 + 2(k - 1) + 1$ and element 2 will be sent to element $k^2 + 2(k - 1) + 2$. (This is exactly the map described in Problem 4 of Section 15.4.1.) Element 3 will be sent to element $(k + 1)^2 + 2(k - 1) + 1$, element 4 will be sent to $(k + 1)^2 + 2(k - 1) + 2$, and elements $2n - 1$ and $2n$ will be sent to elements $(k + n - 1)^2 + 2(k - 1) + 1$ and $(k + n - 1)^2 + 2(k - 1) + 2$, respectively. How can we see that this set of one-to-one correspondences partitions $\mathbb{N}$? Notice that we are filling in the numbers between k^2 and $(k + 1)^2 = k^2 + 2k + 1$ with pairs of elements from $\mathbb{N}_1, \dots, \mathbb{N}_k$. Because there are $2k$ numbers between k^2 and $(k + 1)^2$, we have at each stage exactly the same number of "cards" dealt as places to deal them.

Further, here is a simpler observation that follows from the fact that $\aleph_0$ copies of $\mathbb{N}$ have the same cardinality as $\mathbb{N}$. We can view $\bigcup_{k \in \mathbb{N}} \mathbb{N}_k$ as $|\mathbb{N}|$ columns, each containing a copy of $\mathbb{N}$. This forms the set $\mathbb{N} \times \mathbb{N}$ but still has the same cardinality as $\bigcup_{k \in \mathbb{N}} \mathbb{N}_k$; therefore, $\left| \bigcup_{k \in \mathbb{N}} \mathbb{N}_k \right| = |\mathbb{N}| = |\mathbb{N} \times \mathbb{N}|$.

15.5.1   The Continuum Hypothesis

We have shown that $2^{\aleph_0} > \aleph_0$. But is there some number *weird* so that $2^{\aleph_0} >$ *weird* $> \aleph_0$? Let us define $\aleph_1$ as the smallest number that is bigger than $\aleph_0$. Is $2^{\aleph_0} > \aleph_1$? Or is $2^{\aleph_0} = \aleph_1$?

No one knows.

Unless we change the foundations of mathematics, no one will ever know.

Seriously. The deal is that underlying the mathematics we study is a set of axioms, collectively called the *Zermelo-Fraenkel axioms*. Most people use one additional axiom (the axiom of choice) and together, our axiom system is referred to as *ZFC*. It has been proven that under ZFC, if you assume that $2^{\aleph_0} = \aleph_1$, no contradiction arises. And also, if you assume that $2^{\aleph_0} \neq \aleph_1$, no contradiction arises! (By the way, the statement that $2^{\aleph_0} = \aleph_1$ is known as the *continuum hypothesis* because it says that the cardinality of the real numbers, or continuum, is $\aleph_1$.)

On the other hand, Hugh Woodin proposed (in 2001) adding an axiom to ZFC that would make the continuum hypothesis false. In other words, with his axiom, assuming that $2^{\aleph_0} = \aleph_1$ would lead to a contradiction.

Cardinality is quite strange (see Figure 15.2).

Check Yourself ───

1. What is the cardinality of the sample drawers (in the Storage facility)?

2. What is the cardinality of the stools in the coffee area?

15.6   Where to Go from Here

The topic of cardinality is generally classified as part of set theory and part of the foundations of mathematics. (To see where to go with these areas, consult Section 2.8.) However, understanding cardinality at a basic level is necessary for most of advanced mathematics and particularly for real analysis (which is the branch of mathematics that contains calculus).

If you enjoyed this chapter, you might like the story *Surreal Numbers* by Donald Knuth. Its mathematical content is set-theoretic in nature. For one mathematician's interpretation of what it would mean to be able to count to $\aleph_0$ and how that differs from being able to count to $\aleph_1$, see the excellent novel *White Light* by Rudy Rucker.

Credit where credit is due: The story in Section 15.2 was based on the folklore known as Hilbert's Hotel, of which there are many, many versions available. The best one is "Hilbert's Hotel," by Ian Stewart. It appeared in *New Scientist*, 19/26 December 1998–2 January 1999, pages 59–61. The style of Section 15.2 was inspired by Imre Lakatos's *Proofs and Refutations*, though, of course, this play does not measure up to his. Bonus Check-Yourself Problem 10 was suggested by Tom Hull. Problems 26–31 of Section 15.10 were donated by Heather Ames Lewis.

15.7   Chapter 15 Definitions

cardinality: The number of elements in a set.

same size: We consider two sets to be the same size if we can put them in one-to-one correspondence.

aleph-naught: $\aleph_0$, also pronounced *aleph-null*, the cardinality of the natural numbers.

countable: Any set with size $\aleph_0$ (because we can count the natural numbers).

uncountable: Any set with size larger than $\aleph_0$.

Zermelo-Fraenkel axioms: The set of axioms underlying ordinary mathematics. Generally used with the additional axiom of choice and then called ZFC.

continuum hypothesis: The statement that $2^{\aleph_0} = \aleph_1$ is known as the continuum hypothesis because it says that the cardinality of the real numbers, or continuum, is $\aleph_1$.

15.8   Bonus: The Schröder–Bernstein Theorem

Back in the mists of time and pages that were Chapter 3, we presented Theorem 3.2.9, which said that if $f : A \to B$ is a function on finite sets A, B and $|A| = |B|$, then f is one-to-one if and only if f is onto. We gave an informal proof and promised that later we would reveal a similar theorem for infinite sets. That time has come.

Theorem 15.8.1. *Let A, B be infinite sets. Then if $f : A \to B$ and $g : B \to A$ are both one-to-one functions, there exists a bijection $h : A \to B$.*

This is known as the Schröder–Bernstein theorem because Schröder and Bernstein proved it independently in the 1890s. We will give a sketch of the proof of Theorem 15.8.1 because a full proof is beyond the scope of this text.

Sketch of proof. First, if it turns out that f is onto, then we're done because f is the bijection we seek. So we only need to consider the case that f is not onto. That means there is some element $b_\alpha \in B$ such that there is no $a \in A$ with $f(a) = b_\alpha$. Check out $g(b_\alpha) = a_\alpha$. We can use this to construct part of h. For each such b_α that is not in the image of f, define $h(a_\alpha) = b_\alpha$.

Now consider the set *Notonto* $= \{a_\alpha \in A \mid h(a_\alpha) = b_\alpha \notin Range(f)\}$. These are the elements of A we have reserved for defining h. (We have defined $h :$ *Notonto* $\to B$ so far.) And look at the set $A \setminus Notonto$. That's the rest of A, so for $a_\beta \in (A \setminus Notonto)$, we will define $h(a_\beta) = f(a_\beta)$.

At first, this looks like the h we want: for $a \in A \setminus Notonto$, $h(a) = f(a)$ and for $a \in Notonto$, $h(a) = b$, where we got b by reversing a bit of g. This map h is one-to-one; the f parts are one-to-one, the non-f parts are one-to-one, and because these two parts are defined using disjoint subsets of B, the whole thing is one-to-one. And it should be onto, because we made sure to start by filling in the gaps in B left by f. However, there are some $a \in A$ for which $h(a) \neq f(a)$. For those a, there were some $b \in B$ that were mapped to by f ... and now, for those b, there are no $a \in A$ such that $h(a) = b$. This means that h is, unfortunately, not onto.

What we *want* to do is look at the new set of not-mapped-to $b \in B$, use g to find $a \in A$ so that new-$h(a) = b$, and use new-$h(a) = $ old-$h(a)$ for the rest of the $a \in A$. But this has the same problem as before!

It turns out the fix is as follows: Think about following each element $a \in A$ through a sequence of applications of the functions. So we start with a_α and then follow it across f to get $f(a_\alpha) = b_\alpha$, and follow this across g to get $g(b_\alpha) = a_\beta$, and follow this across f to get $f(a_\beta) = b_\beta$ and so forth. And if there is some b_{nice} such that $g(b_{nice}) = a_\alpha$, stick that onto the front of the sequence for a_α ... and continue this process as well. In the end, you'll have some sequences that start in A and some sequences that start in B and some sequences that have no start at all. In this view, the elements of B to which f does not map elements of A (the elements that got missed by f) have become the elements of B that start these sequences of applications of f and g.

Now define real-h as follows. If the sequence corresponding to a starts in A or has no start, use real-$h(a) = f(a)$. If the sequence corresponding to a starts in B, then there is a $b_{nice} \in B$ such that $g(b_{nice}) = a$, so use real-$h(a) = b_{nice}$.

As mentioned before, the full proof is beyond the scope of this text, so we will not attempt to explain why real-h is a bijection. If you are familiar with composition of functions and inverses of functions, and if you are very persistent, you may find it rewarding to research full proofs of the Schröder–Bernstein theorem. A compact one can be found at http://www.artofproblemsolving.com/wiki/index.php/Schroeder-Bernstein_Theorem.

15.9   Bonus Check-Yourself Problems

Solutions to these problems appear starting on page 634. Those solutions that model a formal write-up (such as one might hand in for homework) are to Problems 4, 5, and 8.

1. Show that $|\mathbb{Z}| = |\mathbb{Z}| + 72$.

2. Show that $\mathbb{Z}$ has the same cardinality as $4\mathbb{N}$.

3. Show that $\mathbb{Z}$ has the same cardinality as $\mathbb{N} \times \mathbb{N}$.

4. Prove that $|\mathscr{P}(\mathbb{Q})| > |\mathbb{Q}|$.

5. Show that $\mathbb{W}$ has the same cardinality as $\mathbb{Z}$.

6. What is the cardinality of the set $\{\frac{p}{q} \mid p \in \mathbb{W}, q \in \mathbb{Z}\}$?

7. What is $(\aleph_0)^3$? How about $(\aleph_0)^8$? Or $(\aleph_0)^{\aleph_0}$? Explain.

8. Consider the set $\mathscr{F}$ of all functions from $\mathbb{N}$ to $\mathbb{N}$. Is $\mathscr{F}$ countable or uncountable?

9. Is the total number of steps in an algorithm that does not terminate countable or uncountable?

10. Consider the set H of length-$\frac{1}{2}$ intervals that are contained in the interval $[0, 1]$. What is $|H|$?

15.10   Infinitely Large Problems

1. Show that $|\mathbb{N}| = |\mathbb{N}| + 1$.

2. Show that $|\mathbb{N}| = |\mathbb{N}| + 100{,}000{,}000$.

3. Show that $\mathbb{N}$ has the same cardinality as $3\mathbb{N}$.

4. Show that $\mathbb{N}$ has the same cardinality as $\mathbb{W}$.

5. Show that $\mathbb{N}$ has the same cardinality as $\mathbb{Z}$.

6. Show that $\mathbb{Z}$ has the same cardinality as $2\mathbb{Z}$.

7. Show that $\mathbb{N}$ has the same cardinality as $\mathbb{Z} \times \mathbb{Z}$.

8. Suppose that the additional eight samples brought into Storage just after the Protagonist leaves the sample are placed into drawers with numbers higher than 59. What would happen to the Protagonist's tag number later in the play?

9. Suppose two field biologists come in, each with infinitely many samples. Devise a simple way to store them all. (Be sure to use a clearly defined map.)

10. Is the map described in Problem 7 of Section 15.4.3 a bijection?

11. What if everyone wanted two cups of coffee—what map would accomplish this goal?

12. Prove that $[1,2)$ is uncountable.

13. Prove that $|\mathscr{P}(\mathbb{Z})| > |\mathbb{Z}|$.

14. Explain why $\aleph_0 + 1 = \aleph_0$. In fact, prove it.

15. Show that $2\aleph_0 = \aleph_0$.

16. Show that $\aleph_0 \cdot \aleph_0 = \aleph_0$.

17. What should $\aleph_2$ mean?

18. What is the cardinality of the subset of $[0,1]$ consisting of infinite decimal expansions with only the digits 2 and 5?

19. What is the cardinality of the set $\{\frac{1}{k} \mid k \in \mathbb{N}\}$?

20. What is the cardinality of the set $\{\frac{p}{q} \mid p, q \in \mathbb{N}\}$?

21. Consider the set W of all words in the English alphabet (sensical or otherwise). What is $|W|$?

22. Consider the set L of all lines in the plane that pass through the origin. Is L countable or uncountable, and why?

23. Is the set of all finite graphs countable or uncountable?

24. In this problem we will count polynomials.

    (a) Consider polynomials of the form ax, where $a \in \mathbb{N}$. How many such polynomials are there?

    (b) Now consider polynomials of the form $ax + b$, where $a, b \in \mathbb{N}$. How many such polynomials are there?

    (c) Continue considering, and this time examine polynomials of the form $ax^2 + bx + c$, where $a, b, c \in \mathbb{N}$. How many such polynomials are there?

    (d) Finally, consider polynomials of the form $a_n x^n + a_{n-1} x^{n-1} + \cdots + a_1 x + a_0$. How many such polynomials are there?

25. **Challenge:** Is the set of polynomials with integer coefficients countable or uncountable?

26. Prove that...

    (a) ... $f : \mathbb{Z} \to 5\mathbb{Z}$ given by $f(x) = 5x$ is a bijection.

    (b) ... $g : \mathbb{R} \to \mathbb{R}$ given by $g(x) = 10x - 7$ is a bijection.

    (c) ... $h : \mathbb{Z} \to \mathbb{Z}$ given by $h(x) = 10x - 7$ is *not* a bijection.

27. A temporary excursion to the land of finity:

    (a) If $|A| = 4$ and $|B| = 3$, does there exist a one-to-one map $A \to B$? What about an onto map? Why or why not?

    (b) If $|A| = 4$ and $|B| = 3$, does there exist an injection $B \to A$? What about a surjection? Why or why not?

    (c) In general, if there exists an injective function between finite sets S_1, S_2, which set has larger cardinality?

    (d) In general, if there exists a surjective map between finite sets S_1, S_2, which set has larger cardinality?

28. In what drawer is the Protagonist's sample after all samples that arrive during Scene 1 are stored? Explain.

29. After the infinitely many field biologists arrive and their samples are stored, what samples are in the first 20 drawers?

30. After the infinitely many field biologists arrive and their samples are stored, in what drawer is the fifth sample of the tenth field biologist? In what drawer is

the tenth sample of the fifth field biologist?

31. In Scene 4, when the Liver Fluke Lab specimens are brought across the hallway, for how many seconds is the dolly full of boxes? For how many seconds is it empty?

32. Prove that there are uncountably many disjoint copies of $\mathbb{Z}$ in $\mathbb{R}$.

33. Prove that $|\bigcup_{i=1}^{10} \mathbb{N}_i| = |\mathbb{N} \times \mathbb{N}|$.

34. What is the cardinality of the set $\{p \cdot q \mid p, q \in \mathbb{Z}\}$?

35. Give five examples of sets with cardinality $2^{\aleph_0}$.

15.11   Instructor Notes

This chapter is designed to convey the basics of cardinality in a discovery-based way. Working through Sections 15.2 and 15.4 will take at least most of a week! However, there are a couple of different ways of approaching Sections 15.2 and 15.4, depending on your particular class and your teaching style.

You may wish to have your class (or a subset of your class) read Section 15.2 aloud, with pauses for discussion of the mathematics. To facilitate this presentation of the material, there are periodically $[\star_k]$ notations in the script to cue you and the students as to where such pauses might most profitably fall. On the other hand, you may wish to have the class read Section 15.2 aloud straight through and discuss the mathematics afterwards. (A direct reading takes about 30 minutes.) In this case, Section 15.4 contains the questions that correspond to the $[\star_k]$ notations from Section 15.2.

Even though there are ten characters, no more than five are on stage at once. The Narrator, Protagonist, and Lab Assistant are regularly used at the same time. Two of the other three Lab Assistants (2, 3, 4) are on stage together; and, at no time do any two of the Receptionist, Storage Coordinator, Taxonomist, and Cataloguer interact.

Of course it is possible, though far less fun, to simply assign students to read Section 15.2.

No matter how you choose to treat Section 15.2, students can be assigned to read Section 15.3 after the first day of class (and to read or reread Section 15.2), as it was specifically designed to be independent of how far into the previous material students have gotten.

Section 15.5 adds some very necessary terminology but also reviews ideas that arise in Sections 15.2 and 15.4. It may be best to have students read Section 15.5 after the second class day spent on this material, but whether this is effective timing depends on what discussions your class has already had.

Chapter 16

Number Theory

16.1   Introduction and Summary

We have already encountered some very basic number theory in Chapter 5. There we do some modular arithmetic and equivalence relations (as well as a bunch of basic cryptography), and that material is prerequisite to the material in this here number theory chapter.

This chapter is no substitute for a course in number theory! It only discusses greatest common divisors, congruence equations, Euler's phi function, and mediants... and there is much, much more to the topic. Please look at Silverman's *A Friendly Introduction to Number Theory* [23] for a deeper look at this and related material.

16.2   Try This! Divisors and Congruences

Before we start exploring, a quick...

Definition 16.2.1. Two numbers are *relatively prime* if they have no divisors (other than 1) in common. For example, 15 and 8 are relatively prime because $\{1,3,5,15\} \cap \{1,2,4,8\} = \{1\}$.

16.2.1   Phi, Phi, Pho, Phum

For this exploration, we also want a second...

Definition 16.2.2. The *Euler phi function*, $\varphi(n)$, counts the natural numbers that are relatively prime to and less than n. For example, $\varphi(15) = 8$ because 1, 2, 4, 7, 8, 11, 13, and 14 are relatively prime to 15.

1. Compute $\varphi(3)$, $\varphi(5)$, $\varphi(7)$, and $\varphi(11)$. What do you suspect $\varphi(p)$ is when p is a prime number? Prove it.

2. Compute $\varphi(4)$, $\varphi(9)$, $\varphi(25)$, and $\varphi(49)$. What do you suspect $\varphi(p^2)$ is when p is a prime number? Prove it. Can you extend your reasoning to determine $\varphi(p^k)$ for $k \geq 2$?

3. Compute $\varphi(10)$, $\varphi(14)$, $\varphi(15)$, $\varphi(21)$, $\varphi(33)$, $\varphi(35)$, and $\varphi(55)$. What form do these numbers have, and what do you suspect $\varphi(n)$ is when n is of this form?

16.2.2   Getting More out of Prime Factorizations

Think all the way back to Chapter 1, where an example theorem was given in Example 1.4.6: every natural number greater than 1 has a unique factorization into prime numbers. (You can prove this later in Problem 1 of Section 16.12.) We will write this factorization as $n = p_1^{k_1} \cdot \cdots \cdot p_m^{k_m}$, so that we are all using the same language. Now, here is some terminology that helps us discuss numbers that are not relatively prime:

Definition 16.2.3. The *greatest common divisor* of two numbers a and b, denoted by $\text{GCD}(a, b)$, is the largest natural number that divides a and divides b. For example, $\text{GCD}(6, 9) = 3$. By convention, we order the numbers so that $a \leq b$. The *least common multiple* of two numbers a and b, denoted by $\text{LCM}(a, b)$, is the smallest natural number that is divisible by a and divisible by b. For example, $\text{LCM}(6, 9) = 18$.

1. Compute $\text{GCD}(5, 7)$, $\text{GCD}(6, 12)$, $\text{GCD}(3, 10)$, and $\text{GCD}(20, 30)$.

2. Compute $\text{LCM}(5, 7)$, $\text{LCM}(6, 12)$, $\text{LCM}(3, 10)$, and $\text{LCM}(20, 30)$.

3. What is the relationship between $\text{GCD}(a, b)$ and $\text{LCM}(a, b)$? Make and prove a conjecture.

4. Prove that $\text{GCD}(a, b)$ divides $a - kb$ for any integer k.

5. Recall from Example 5.3.2 that $b \pmod{a}$ is shorthand for "the smallest nonnegative integer r such that $b \equiv r \pmod{a}$." Show that $\text{GCD}(a, b) = \text{GCD}(b \pmod{a}, a)$.

6. Use this result to calculate $\text{GCD}(20, 30)$.

7. Use this result to calculate $\text{GCD}(230, 382)$.

16.2.3   Back to φ: More Phi, Less Pho and Phum

Now that you've had a little bit of a GCD rest, let's think about $\varphi(n)$ again.

1. Compute $\varphi(20)$. You've already computed $\varphi(2)$, $\varphi(4)$, $\varphi(5)$, and $\varphi(10)$. Which of these, if any, might be used to compute $\varphi(20)$?

2. Compute $\varphi(12)$, $\varphi(18)$, $\varphi(24)$, $\varphi(30)$, and $\varphi(36)$. Now examine the prime factorizations of 12, 18, 24, 30, and 36. Do any of the factorizations suggest other ways to calculate φ?

3. Consider any natural n. Conjecture a way to calculate $\varphi(n)$ in terms of the prime factorization of n

4. What needs to be proved in order for this conjecture to be shown true? (What parts or aspects of the conjecture have already been proved?)

5. Write a precise statement for each remaining-to-be-proved part of the conjecture. Can you use a common proof technique to reduce any of these statements to simpler statements?

6. Your should have at least one simpler statement that is in the form of an equation involving $\varphi(n)$. Write each side of that equation for $\varphi(20)$ in terms of (sizes of) sets. What does this suggest as a possible proof technique?

7. The phrase "write each side of that equation in terms of sets" is a bit ambiguous. It could mean "for each set involved, write out a list of elements" or it could mean "for each set involved, write the definition as $\{m \in S \mid$ condition(s)$\}$." Whichever meaning you chose before, rewrite the equation using the other meaning here.

16.3   Computing the GCD

Hey! You! Don't read this unless you have worked through the problems in Section 16.2.2. I mean it!

The most straightforward way to compute $\text{GCD}(a,b)$ is to look at the prime factorizations of a and b and determine which prime factors they have in common. We can also observe that $\text{GCD}(a,b) = \text{GCD}(b \pmod{a}, a)$: every divisor of both a and b also divides $ak + b$, and every divisor of $ak + b$ and a also divides b. Therefore, the greatest divisor of a and b divides $ak + b$ and a, and vice-versa, so the two GCDs must be equal.

Indeed, this equation gives an algorithm (see Chapter 5), often called the *Euclidean algorithm*, for computing $GCD(a,b)$, where we assume that $a \leq b$ (that's okay, because no matter what order we write them in, a and b have the same set of common divisors):

1. Does a divide b? If so, $GCD(a,b) = a$. If not, continue.

2. Compute $b \pmod{a}$; replace a with $b \pmod{a}$ and replace b with a.

3. Go to step 1.

We know this algorithm terminates because if $a < b$ does not divide b, then $b \pmod{a} < a$ and we are dealing with smaller numbers on the next round; this can only continue for a finite number of iterations (at most a of them).

The Euclidean algorithm can be used to prove a...

> **Theorem 16.3.1.** *There exist integers k, ℓ such that $GCD(a,b) = ka + \ell b$. In particular, if a and b are relatively prime, then there exist integers k, ℓ such that $ka + \ell b = 1$.*

Proof: We will use induction on a. As a base case consider $a = 1$. Then $1 \cdot 1 + 0 \cdot b = 1$ and we have produced $k = 1, \ell = 0$ with $k \cdot 1 + \ell b = 1$. Our inductive hypothesis is that if $s < a$, then there exist integers i, j such that $GCD(s,b) = ia + jb$.

We will use the relationship $GCD(a,b) = GCD(b \pmod{a}, a)$ for the inductive step. We know that $b \pmod{a} < a$ and so the inductive hypothesis applies. Thus, there exist i, j such that $GCD(b \pmod{a}, a) = i(b \pmod{a}) + ja$. It's hard to know what to do with this because of the awkward presence of $b \pmod{a}$ in the equation. So, consider the relationship between a and b: because $b > a$, we can write $b = qa + r$ where $r < a$. Indeed, $r = b - qa = b \bmod a$. Aha! Therefore $GCD(a,b) = i(b - qa) + ja = (j - iq)a + ib$. Setting $k = j - iq$ and $\ell = i$, we have $GCD(a,b) = ka + \ell b$. (If a, b are relatively prime, then $GCD(a,b) = 1$ and we have just produced integers k, ℓ such that $ka + \ell b = 1$.) □

Actually, we can prove a stronger property:

> **Theorem 16.3.2.** *Given two positive integers a and b, there exist integers k, ℓ such that $ka + \ell b = 1$ if and only if a and b are relatively prime.*

Proof: We know that if a and b are relatively prime, then there exist integers k, ℓ such that $ka + \ell b = 1$ by Theorem 16.3.1.

Now suppose that there exist integers k, ℓ such that $ka + \ell b = 1$, and also suppose that a and b are *not* relatively prime. Pick some common divisor of a and b

(greater than 1) and name it d. Then we know that $a = di$ and $b = dj$ for some positive integers i, j. Now we can write $ka + \ell b = 1$ as $kdi + \ell dj = d(ki + \ell j) = 1$. But both d and $ki + \ell j$ are integers, and $d > 1$, and we know that no product of an integer greater than 1 with another integer will produce 1. Contradiction! It must be that a and b are relatively prime. □

Check Yourself

1. Use the Euclidean algorithm to compute $\text{GCD}(8, 12)$.

2. Use the Euclidean algorithm to compute $\text{GCD}(1233, 1234)$.

3. Find integers k, ℓ such that $\text{GCD}(8, 12) = k8 + \ell 12$.

4. Find integers k, ℓ such that $\text{GCD}(1233, 1234) = k1233 + \ell 1234$.

5. **Challenge:** Pick two natural numbers a, b where a does not divide b, and find their GCD using the Euclidean algorithm. Then find integers k, ℓ such that $\text{GCD}(a, b) = ka + \ell b$.

16.4   Try This! Congruence Experiments

Recall the notation $x \equiv s \pmod{a}$ from Section 5.3, and recall that this means that $x \in \{\ldots, s - 2a, s - a, s, s + a, s + 2a, s + 3a, \ldots\}$. We could also say that this set is all solutions to the equation $x = ka + s$. Note that there is exactly one $x \in \mathbb{Z}_a = \{0, \ldots, a - 1\}$ that satisfies $x \equiv s \pmod{a}$. (All of this meaning in a simple equivalence statement!)

1. Which numbers x satisfy $x + 9 \equiv 2 \pmod{4}$?

2. Which numbers x satisfy $x + r \equiv s \pmod{a}$?

3. Which numbers x satisfy $2x \equiv 1 \pmod{4}$?

4. Which numbers x satisfy $3x \equiv 0 \pmod{6}$?

5. Which numbers x satisfy $2x \equiv 1 \pmod{3}$?

6. It would be really convenient if when trying to solve $kx \equiv s \pmod{a}$, we could just cancel out the k somehow, to get $x \equiv \textit{thing} \pmod{a}$. In regular arithmetic, we would divide, but we know from Example 5.3.5 that this does not always work in modular arithmetic. The next best thing we can try is to

get rid of the k using multiplication—that is, we hope for a w with $wk \equiv 1$ (mod a), so that $kx \equiv s$ (mod a) becomes $wkx \equiv ws$ (mod a) or $x \equiv ws$ (mod a). Let's experiment. (A related exploration took place in Problem 19 of Section 5.11.)

(a) Consider the equivalence $2x \equiv 1$ (mod 3). We would like to get rid of that 2. What number w has the property that $w \cdot 2 \equiv 1$ (mod 3)?

(b) Multiply $2x \equiv 1$ (mod 3) through by the w you just found. What equivalence relation do you get? What are its solutions?

(c) Consider the equivalence $3x \equiv 2$ (mod 5). Is there a number w by which we can multiply this equivalence in order to ditch the coefficient 3? Explain.

(d) What about $3x \equiv 0$ (mod 6)? Does an appropriate w exist?

(e) How about $3x \equiv 2$ (mod 10)? ... $5x \equiv 4$ (mod 6)? ... $6x \equiv 7$ (mod 9)?

(f) Make a conjecture as to when the w we seek exists.

7. Which numbers x are both $x \equiv 0$ (mod 2) and $x \equiv 0$ (mod 3)?

8. Which numbers x are both $x \equiv 1$ (mod 2) and $x \equiv 2$ (mod 3)?

9. Which numbers x are both $x \equiv 1$ (mod 2) and $x \equiv 3$ (mod 4)?

10. What techniques did you use to solve the previous simultaneous equations? Do any of them generalize to help solve equations $x \equiv s$ (mod a) and $x \equiv t$ (mod b)?

16.5   Counting with Congruence Equations

Hey! You! Don't read this unless you have worked through the problems in Sections 16.2.1 and 16.4. I mean it!

As you will have deduced from Section 16.2.3, we could compute $\varphi(n)$ for any n if we only could prove a certain conjecture. Our goal in this section will be to prove that when a and b are relatively prime, then $\varphi(ab) = \varphi(a)(b)$. Interestingly, our proof will involve linear congruences.

16.5.1 Just One

Suppose we have the congruence $kx + r \equiv s$ (mod a). We can convert this to $kx \equiv s - r$ (mod a) and reduce the number of variables running around by rewriting to $kx \equiv t$ (mod a). But now what? We know from our experiments in Section 16.4 that there is not always a solution to this equation. An additional example: consider $4x \equiv 3$ (mod 6). Here 4 does not divide 6, but $4x$ is always even and any number of the form $6m + 3 = 3(2m + 1)$ is odd because it is the product of two odd numbers. Therefore there are no solutions to $4x \equiv 3$ (mod 6). You most likely conjectured that as long as k and a are relatively prime, the congruence $kx \equiv t$ (mod a) has a solution. Let's prove that this is true. Actually, let's prove a slightly stronger statement:

> **Theorem 16.5.1.** *The congruence $kx \equiv t$ (mod a) has exactly one solution $0 \leq x < a$ if k and a are relatively prime.*

Proof: Consider the set $\mathbb{Z}_a = \{0, 1, \ldots, a - 1\}$. It has n elements. Now multiply all of these elements by k, so we have $\{0 \pmod{a}, k \pmod{a}, \ldots, k(a - 1) \pmod{a}\}$. How many elements does this set have? At most there are n elements. Could there be fewer? Maybe two (or more) of them have the same value. Suppose two of them are equal, so that we have $ku \pmod{a} = kv \pmod{a}$. They are both equal to the same number $0 \leq r < a$, which means $ku = q_1 a + r$ and $kv = q_2 a + r$. Now $r = ku - q_1 a = kv - q_2 a$, or $k(u - v) = a(q_1 - q_2)$. Therefore a must divide $k(u - v)$... but if k and a are relatively prime, that means a must divide $u - v$. However, $u, v < a$ so $u - v < a$, which means that the only way for a to divide $u - v$ is for $u - v = 0$. Thus $u = v$, which is a contradiction.

What does all this have to do with our theorem? Well, the set $\{0 \pmod{a}, k \pmod{a}, \ldots, k(a - 1) \pmod{a}\}$ has a different elements, all of which are non-negative and less than a, so one of them must be 1. And that means that there exists some $w \in \mathbb{Z}_a$ such that $wk \equiv 1$ (mod a). Therefore $wkx \equiv x \equiv wt$ (mod a). And while there are infinitely many $x \equiv wt$ (mod a), there is exactly one with $0 \leq x < a$. $\square$

> How to use Theorem 16.5.1. So what does this tell us? If we want to solve the congruence $kx \equiv t$ (mod a), and k and a are relatively prime, then we know there is some whole number $w < a$ such that $wk = 1$. We can find this by trial-and-error (there are at most $a - 1$ different values to check). Then we know that $x \equiv wt$ (mod a) is a solution, and in fact all numbers $x = wt + da$, where d is any integer, will satisfy the congruence equation.

16.5.2   Two at Once

Suppose we have two congruences, $x \equiv s \pmod{a}$ and $x \equiv t \pmod{b}$. We know that the first congruence is satisfied by all numbers $x \in \{\ldots, s - 2a, s - a, s, s + a, s + 2a, s + 3a, \ldots\}$, which is also all solutions to $x = ja + s$. Ah, but having this equation means that we can use it to substitute: $x = ja + s \equiv t \pmod{b}$. We know from Section 16.5.1 how to deal with this. First say $ja \equiv t - s \pmod{b}$. This is a congruence equation with variable j (remember that a, b, t, and s were fixed in the original problem). If a and b are relatively prime, we are guaranteed that there is a solution to the congruence equation. Indeed, we have $waj \equiv j \equiv w(t - s) \pmod{b}$ and $0 \leq w(t - s) \pmod{b} < b$.

Plugging back in, we have $x = ja + s = (w(t - s) \pmod{b})a + s$. Notice that this gives us only *one* value for x, but of course there should be infinitely many. We can write $j \equiv w(t - s) \pmod{b}$ as $j = w(t - s) + qb$ for some—really, any—integer q, and then $x = (w(t - s) + qb)a + s = aw(t - s) + s + qab$. In turn, that means $x \equiv aw(t - s) \pmod{ab}$, and now we can see $x \in \{\ldots, aw(t - s) - 2ab, aw(t - s) - ab, aw(t - s), aw(t - s) + ab, \ldots\}$.

16.5.3   Computing $\varphi(n)$

Back in Section 16.2.1 we experimented with computing $\varphi(n)$ for various special n. For example, $\varphi(p) = p - 1$ because every natural number less than p is relatively prime to p. Similarly, $\varphi(p^k) = p^k - p^{k-1}$ for $k \geq 2$ because the multiples of p less than or equal to p^k are $\{p, 2p, \ldots, (p-1)p, p^2, (p+1)p, \ldots, (p^{k-1} - 1)p, p^{k-1}p\}$; there are p^{k-1} of these, and all other natural numbers less than p^k are relatively prime to p^k. Looking at some examples, we suspect that if $n = pq$ is the product of two primes, then $\varphi(n) = \varphi(p)\varphi(q)$. What if there are more than two primes involved, or higher powers of these primes? Let $n = p_1^{k_1} \cdots p_m^{k_m}$. Experiments are consistent with the conjecture that $\varphi(n) = \varphi(p_1^{k_1}) \cdots \varphi(p_m^{k_m}) = (p_1^{k_1} - p_1^{k_1 - 1}) \cdots (p_m^{k_m} - p_m^{k_m - 1})$.

How can we prove this conjecture? We know how to calculate $\varphi(p^k)$, so if we could calculate $\varphi(p_1^{k_1} p_2^{k_2})$, then we could extend to $\varphi(n)$ by induction. And more generally, what seems to be true is that $\varphi(ab) = \varphi(a)\varphi(b)$ when a and b are relatively prime. Let's prove it.

> **Theorem 16.5.2.** *If a and b are relatively prime, then $\varphi(ab) = \varphi(a)\varphi(b)$.*

Proof: The Euler phi function $\varphi(n)$ counts the natural numbers that are relatively prime to and less than n. Literally, $\varphi(n)$ is the size of a set. So let us look at what the definition says: $\varphi(ab) = |\{r \in \mathbb{N} \mid r < ab, \ r \text{ is relatively prime to } ab\}|$;

$\varphi(a) = |\{s \in \mathbb{N} \mid s < a,\ s \text{ is relatively prime to } a\}|$; and $\varphi(b) = |\{t \in \mathbb{N} \mid t < b,\ t$ is relatively prime to $b\}|$. We will now name these sets: $R = \{r \in \mathbb{N} \mid r < ab,\ r$ is relatively prime to $ab\}$, $S = \{s \in \mathbb{N} \mid s < a,\ s$ is relatively prime to $a\}$, and $T = \{t \in \mathbb{M} \mid t < b,\ t$ is relatively prime to $b\}$.

We would like to show that $|R| = |S| \cdot |T|$, and we will do this by showing $|R| = |S \times T|$. This suggests we look for a bijection between the single set and the Cartesian product of sets. It may seem a bit strange at first, but as we work through the proof, it will start to make sense.

An element of R has the form r and an element of $S \times T$ has the form (s,t). First, notice that $R = \{r \in \mathbb{N} \mid r < ab$ is relatively prime to $ab\}$ is a subset of $\mathbb{Z}_{ab} = \{r \in \mathbb{W} \mid r < ab\}$, and likewise $S \subset \mathbb{Z}_a, T \subset \mathbb{Z}_b$. So we will create a map $f : \mathbb{Z}_{ab} \to \mathbb{Z}_a \times \mathbb{Z}_b$ and show that when restricted to $R \to S \times T$, it is a bijection.

We define $f(r) = (r \pmod{a}, r \pmod{b})$.

First, we will show that this map is injective. (See Chapter 3 for a refresher on injectivity proofs.) Suppose $f(r_1) = f(r_2)$. Then $(r_1 \pmod{a}, r_1 \pmod{b}) = (r_2 \pmod{a}, r_2 \pmod{b})$. This means that $r_1 \pmod{a} = r_2 \pmod{a}$ and $r_1 \pmod{b} = r_2 \pmod{b}$. Reworded, that says that $r_1 \equiv r_2 \pmod{a}$ and $r_1 \equiv r_2 \pmod{b}$. By definition, we have that $r_1 = ak + r_2$ and $r_1 = b\ell + r_2$, or $r_1 - r_2 = ak = b\ell$. That means $r_1 - r_2$ is divisible by a and also divisible by b, and because a and b are relatively prime and so have no prime factors in common, $r_1 - r_2$ must be divisible by ab. Therefore $r_1 - r_2 = abw$, or $r_1 = r_2 \pmod{ab}$. But r_1 and r_2 are both natural numbers less than ab, so $r_1 = r_2$. This completes the proof of injectivity.

Now, what about surjectivity? Consider an element from $S \times T$. It is (s,t), where s is relatively prime to a and t is relatively prime to b. We seek an element r relatively prime to ab such that $r \pmod{a} = s$ and $r \pmod{b} = t$. Rewritten slightly, those equations become $r \equiv s \pmod{a}$ and $r \equiv t \pmod{b}$.

And that's exactly the situation we were considering in Section 16.5.2! We know that the solution is $aj + s$, where $j < b$ and $aj + s \equiv t \pmod{b}$ (from our original substitution).

But there's one question left—is $aj + s$ relatively prime to ab? The answer is *yes*, and here's why. We know that a and b are relatively prime to each other, so each is composed of distinct primes. That means that if $ay + s$ is relatively prime to a, and also relatively prime to b, then it is relatively prime to ab. (Great. Now we have *two* questions left.)

So, is $aj + s$ relatively prime to a? If there were a common divisor d, then $a = k_1 d$ and $aj + s = k_2 d = k_1 d j + s$, so that $d(k_2 - k_1 j) = s$ and therefore d is a divisor of s as well. But we know that s is relatively prime to a, so we must have

$d = 1$ and so $aj + s$ is relatively prime to a. Similarly, if $aj + s$ and b had a common divisor v, then $aj + s = z_1 v$ and $b = z_2 v$. Remember that $aj + s \equiv t \pmod{b}$, so $aj + s = qb + t$ for some integer q. Now we have $z_1 v = aj + s = qb + t = z_2 vq + t$, so that $v(z_2 - z_1 q) = t$. But we know that t is relatively prime to b, so we must have $v = 1$ and so $qb + t = aj + s$ is relatively prime to b.

We're done!   □

Check Yourself

Do at least one of the first two, one of the next four, and the seventh problem, to make sure you're comfortable with the computations this section enabled.

1. Compute $\varphi(210)$.

2. Compute $\varphi(3200)$.

3. Find w such that $w \cdot 2 \equiv 1 \pmod{5}$ and use this to find all x that satisfy $2x \equiv 3 \pmod{5}$.

4. Find w such that $w \cdot 3 \equiv 1 \pmod{4}$ and use this to find all x that satisfy $3x \equiv 2 \pmod{4}$.

5. Find w such that $w \cdot 3 \equiv 1 \pmod{7}$ and use this to find all x that satisfy $3x \equiv 6 \pmod{7}$.

6. Find w such that $w \cdot 5 \equiv 1 \pmod{7}$ and use this to find all x that satisfy $5x \equiv 3 \pmod{7}$.

7. Re-solve the first two congruence pairs from Section 16.4 using the techniques given in Section 16.5.2:

    (a) Which numbers x are both $x \equiv 0 \pmod{2}$ and $x \equiv 0 \pmod{3}$?

    (b) Which numbers x are both $x \equiv 1 \pmod{2}$ and $x \equiv 2 \pmod{3}$?

8. **Challenge:** Invent your own congruence equation $kx \equiv t \pmod{a}$ and find its solutions.

9. **Challenge:** Invent your own pair of congruence equations $x \equiv s \pmod{a}$ and $x \equiv t \pmod{b}$ (but make sure that a and b are relatively prime!) and find their common solutions.

16.6   Try This! Investigate Freaky Fraction Lists

Here is the first fraction list:

$$\frac{0}{1} \quad \frac{1}{1}$$

We will add the numerators to make a new numerator, add the denominators to make a new denominator, and slap the resulting fraction in the middle, to get a second fraction list:

$$\frac{0}{1} \quad \frac{1}{2} \quad \frac{1}{1}$$

Now we'll do the same thing for each pair of adjacent fractions to create a third fraction list:

$$\frac{0}{1} \quad \frac{1}{3} \quad \frac{1}{2} \quad \frac{2}{3} \quad \frac{1}{1}$$

In general, to make the next fraction list we look at two fractions that are adjacent, say $\frac{a}{b}$ and $\frac{c}{d}$, and stick the new fraction $\frac{a+c}{b+d}$ between them.

1. Make the next three fraction lists.

2. What's freaky about these fraction lists? Make and list some observations and conjectures (both freaky and non-freaky).

3. Before trying to prove your conjectures, prove this: for any two adjacent fractions $\frac{a}{b}$ and $\frac{c}{d}$, $bc - ad = 1$.

4. Show that when creating and placing new fractions, you never have to reduce— $\frac{a+c}{b+d}$ is already in lowest terms.

5. Show that new fractions have values that are between the old fractions that lie on either side. That is, show that $\frac{a}{b} < \frac{a+c}{b+d} < \frac{c}{d}$.

6. Can a fraction appear more than once in a freaky fraction list? Explain.

7. Which rationals (between 0 and 1) appear in the freaky fraction lists? Conjecture a characterization for these fractions, and try to prove your conjecture.

16.7   Mysterious Mediants

Hey! You! Don't read this unless you have worked through the problems in Section 16.6. I mean it!

Given two fractions $\frac{a}{b}$ and $\frac{c}{d}$, their *mediant* is $\frac{a+c}{b+d}$. The freaky fraction lists (or FFLs for short) are made from mediants—to create the next FFL, we insert all possible mediants into the previous FFL. In order to refer to the FFLs more precisely, let us use FFL_n to denote the nth FFL, where FFL_1 is $\frac{0}{1}$ $\frac{1}{1}$.

It turns out that the crucial fact about mediants is...

> Fact.   For any two adjacent fractions $\frac{a}{b}$ and $\frac{c}{d}$ in an FFL, $bc - ad = 1$.

Proof: We can prove this by induction. The base case is our first FFL: $1 \cdot 1 - 0 \cdot 1 = 1$. The inductive hypothesis is that for $n < k$, all adjacent pairs of fractions $\frac{a}{b}$ and $\frac{c}{d}$ in FFL_n have the property $bc - ad = 1$. For the inductive step, consider FFL_k. Adjacent fractions either have the form $\frac{a}{b}$ and $\frac{a+c}{b+d}$, or the form $\frac{a+c}{b+d}$ and $\frac{c}{d}$, where $\frac{a}{b}$ and $\frac{c}{d}$ are from FFL_{k-1}. For this reason, we know that $bc - ad = 1$. In order to complete the proof, we need to show that $b(a+c) - a(b+d) = 1$ and that $(b+d)c - (a+c)d = 1$.

Now, $b(a+c) - a(b+d) = ab + bc - ab - ad = bc - ad = 1$, and $(b+d)c - (a+c)d = bc + cd - ad - cd = bc - ad = 1$. Done!   □

Surely you noticed that for every mediant $\frac{a+c}{b+d}$, $GCD(a+c, b+d) = 1$. (That's a little bit freaky.) Check out this slickness: Our freshly proven fact $bc - ad = 1$ can be rewritten as $ab + bc - ab - ad = b(a+c) + (-a)(b+d) = 1$. Therefore, we have found $k = b$ and $\ell = -a$ such that $k(a+c) + \ell(b+d) = 1$, which means by Theorem 16.3.2 that $GCD(a+c, b+d) = 1$.

Also, every FFL has the fractions automatically listed in increasing order. To see this, it is enough to show that $\frac{a}{b} < \frac{a+c}{b+d} < \frac{c}{d}$. Again, we use that $bc - ad = 1$. (See why it was useful to show that first?) It follows that $bc = 1 + ad$, so $bc > ad$. Now a little bit of algebra gives

$$bc + ab > ad + ab \qquad\qquad bc + cd > ad + cd$$

$$b(a+c) > a(b+d) \qquad\qquad c(b+d) > d(a+c)$$

$$\frac{a+c}{b+d} > \frac{a}{b}; \qquad\qquad\qquad \frac{c}{d} > \frac{a+c}{b+d}.$$

Because each FFL is in increasing order, no fraction can appear more than once. If a fraction *did* appear more than once, it would have to appear next to itself. This is not possible because $\frac{a}{b} < \frac{a+c}{b+d}$, so no two adjacent fractions can be equal. (Or, $\frac{a}{b} = \frac{a+c}{b+d}$ implies that $bc = ad$, which is false.)

Now, here's the really interesting thing about all the FFLs.

> **Theorem 16.7.1.** *Every rational number $0 \leq \frac{m}{n} \leq 1$ appears in some freaky fraction list.*

Proof: Suppose the fraction $\frac{m}{n}$ does not appear on any FFL. Pick some FFL (any one will do, it turns out), say FFL_k, and find the two nearest consecutive fractions, so $\frac{a}{b} < \frac{m}{n} < \frac{c}{d}$. Now, $\frac{a}{b} < \frac{m}{n} \Rightarrow an < bm \Rightarrow bm - an > 0$, and $\frac{m}{n} < \frac{c}{d} \Rightarrow md < cn \Rightarrow cn - md > 0$. Because a, b, c, d, m, n are all integers, this actually means that $bm - an \geq 1$ and $cn - md \geq 1$. Now, using these two inequalities, we have that $(a + b)(cn - md) + (c + d)(bm - an) \geq a + b + c + d$. Check out the left-hand side: it expands to $acn - amd + bcn - bmd + bcm - acn + bdm - adn$, which then reduces to and factors as $bc(m + n) - ad(m + n) = (bc - ad)(m + n)$. Using our old reliable relationship $bc - ad = 1$, the inequality becomes $m + n \geq a + b + c + d$.

Now, if we look at FFL_{k+1}, we have $\frac{a}{b} < \frac{a+c}{b+d} < \frac{c}{d}$, with either $\frac{a}{b} < \frac{m}{n} < \frac{a+c}{b+d} < \frac{c}{d}$ or $\frac{a}{b} < \frac{a+c}{b+d} < \frac{m}{n} < \frac{c}{d}$. (One of these must be true or else $\frac{m}{n}$ is in FFL_{k+1}.) Repeating the argument from the previous paragraph gives us $m + n \geq a + b + (a + c) + (b + d) = 2a + 2b + c + d$ or $m + n \geq (a + c) + (b + d) + c + d = a + b + 2c + 2d$. That is, with every new FFL we examine, we increase the right-hand side of the inequality. So after looking at (no more than) $m + n$ more FFLs, so at worst FFL_{k+m+n}, we get a contradiction as the right-hand side will now be larger than the left-hand side!

Therefore $\frac{m}{n}$ appears in some FFL. □

Check Yourself ─────────────────────────────

Do all of these quick problems!

1. Pick three pairs of adjacent fractions in FFL_5, and verify that for each pair, $bc - ad = 1$.

2. Show that not every mediant $\frac{a+c}{b+d}$ must be in lowest terms by finding two fractions $\frac{a}{b}$ and $\frac{c}{d}$ that are *not* adjacent in an FFL and whose mediant is not in lowest terms.

3. In which FFL does $\frac{1}{9}$ appear?

4. In which FFL does $\frac{2}{11}$ appear?

16.8   Where to Go from Here

Number theory is properly its own course (which has been stated at least k times in this book already) and is a subfield of mathematics popular all the way to the research level. The best textbook on this material, in my opinion, is *A Friendly Introduction to Number Theory* by Joseph Silverman [23]. It is eminently readable and contains many approachable explorations. You can find more about the Euler phi function, about the GCD, about solving single congruence equations, about solving sets of congruence equations, and about the RSA cryptosystem there. It's great. There are also connections to material from Chapters 6 and 8 from Part II of the book you are reading now. If instead of a textbook you want a reference book, try *Elementary Number Theory and Its Applications* by Kenneth Rosen or *Elementary Number Theory* by David Burton.

As for those freaky fraction lists, they are lists of node-labels in the Stern-Brocot tree. Stern was a math person, and Brocot was a clockmaker. To find out how Brocot used these fractions in clock manufacture and to see a totally different take on the same mathematics we encountered in Section 16.7, check out David Austin's column at http://www.ams.org/publicoutreach/feature-column/fcarc-stern-brocot.

Credit where credit is due: Thanks to Tamara Veenstra for consultation on Sections 16.2 and 16.4. Much of the exposition in this chapter was inspired by both [10] and [23]; the latter also inspired some of the problems in Section 16.12. The exposition in Section 16.10 was inspired by an Art of Problem Solving lesson. Several of the problems in Section 16.12 were adapted from problems or text in [20]. Thanks to Ollie Levy for telling me which social networking website ducks use.

16.9   Chapter 16 Definitions

relatively prime: Two numbers that have no divisors (other than 1) in common.

Euler phi function: $\varphi(n)$ counts the natural numbers that are relatively prime to and less than n.

greatest common divisor: The largest natural number that divides a and divides b.

least common multiple: The smallest natural number that is divisible by a and is divisible by b.

Euclidean algorithm: A way of computing $\text{GCD}(a, b)$, where we assume that $a \le b$:

1. Does a divide b? If so, $\text{GCD}(a, b) = a$. If not, continue.

2. Compute $b \pmod{a}$; re-place a with $b \pmod{a}$ and replace b with a.

3. Go to step 1.

mediant: Given two fractions $\frac{a}{b}$ and $\frac{c}{d}$, their mediant is $\frac{a+c}{b+d}$.

16.10   Bonus: The RSA Cryptosystem

The RSA cryptosystem is named after Ron Rivest, Adi Shamir, and Leonard Adelman, who collaborated to invent it in the late 1970s. RSA is used widely to encrypt internet transactions and is based on the idea that it is difficult to factor large numbers. In particular, this encryption method uses a publicly-published product P of two many-digit primes (approximately-300-digit primes). Even knowing that $P = p_1 p_2$, it's impractically difficult to factor. As an example, consider the publisher's webpage from which you can buy this very book. The security information shows that it uses RSA Encryption (1.2.840.113549.1.1.11, to be precise) and that its public key P is 256 bytes long: A3 89 57 45 7B C6 AF 98 92 CD 73 C3 4C 1B D9 66 46 3F E6 9C 7F 93 1B E4 8B 60 98 5D 36 8F 2D 99 C9 4E 83 FA B4 9B 7F E6 78 E8 8F D6 D1 11 59 7E 1D B1 E3 46 15 03 69 5F BD 3F 3C CD 84 1C 30 97 55 BC 50 E1 1C 00 20 66 FA 58 0F 8F D6 2E 20 00 49 09 4A 06 A5 EE 61 E1 DC A4 AB A4 9C 50 32 47 0E 8A F0 27 47 01 EE 33 E6 41 2B F7 FF BD BB 6A FE 90 95 A5 23 FA 67 85 5F 1A 47 ED 21 67 3B 76 02 22 B6 D6 FC 93 D2 07 BD 25 7E 86 B4 C0 7E 67 FF E8 DD B9 F5 EE 9F E1 47 2E 9C 0F EA FE 76 74 E4 0C 5E 33 B8 D5 01 BE 6A 3C F3 79 5F D3 88 2A 1A D1 2C 68 F4 67 D2 22 4A 06 A2 89 40 3D 46 1D 7D 63 35 89 29 D6 3E 8A 4B 43 8D 9F BB 3A 09 96 30 4C 08 9F 2C D4 FA E9 9B 9C C4 4D 14 6E 66 9A 83 A6 97 F6 A4 C6 83 C6 AB 4E 9F 6F 9F E7 C5 26 32 0B C4 5F 64 4D 6B 37 FA 0D D4 0E 68 79 02 6F (and that's valid through April 2019). We're also told that the Exponent is 65537. What is this all about? Why are there letters all over the place? That first question will take a bit to answer. But the second is easy: P is written in hexadecimal (base 16).

Okay, let's look at the big picture of RSA. It's a type of public key cryptography. Back in Chapter 5, we dealt with cryptographic systems that were based on the communicating parties both knowing the encryption method and decryption key in advance. For a shift cipher, the recipient of the message needs to know the shift length. For a Vigenère cipher, the recipient of the message needs to know a key word. But with public key cryptography, there is no communication in advance and thus no way to hand over a shift length or any kind of key. So how does this work?

We'll use the case of Grey Duck and White Duck to illustrate the process. Grey Duck has flown to Oklahoma for the winter in order to be warm, while White Duck went to Toronto to spend a few months at a luxury spa. In order to plan their summer duck espionage, they must communicate securely, and like sensible modern ducks they decide to use RSA.

Grey Duck posts hir public key G (on hir WebFoot page) and White Duck posts hir public key W. Grey Duck starts a conversation by first using W to encrypt a message and then sending the encrypted text. White Duck knows that $W = w_1 w_2$ and uses w_1, w_2 to decrypt the message. Then White Duck composes a response, uses G to encrypt it, and sends the encrypted response on to Grey Duck, who uses g_1, g_2 to see what it says.

The mathematical details of RSA depend on modular arithmetic. The ducks use a common exponent e—humans usually use $e = 65537$, but ducks prefer $e = 5$. Basically, Grey Duck computes $(message)^5 \pmod{W}$ to encrypt and produce *encrypted*, and White Duck solves $5 \cdot d \equiv 1 \pmod{(w_1 - 1)(w_2 - 1)}$ for d and then computes $(encrypted)^d \pmod{W}$ to decrypt and get *message* back.

To start their communication, the ducks choose their public keys and post them on WebFoot as $W = 95, G = 39$. (You probably already figured out that $95 = 5 \cdot 19, 39 = 3 \cdot 13, \ldots$ but each duck only needs to keep the factorization secret from other ducks, who really don't know how to factor at all, so it's okay that you, the reader, can determine their primes.)

Of course, Grey Duck wants to start with the message *hi*, or 78 in the standard conversion of Section 5.4. Grey duck uses W to encrypt and finds $78^5 \pmod{95} = 13$, and sends this to White Duck over their shared Quack channel. White Duck knows that *bd* is probably not what Grey Duck wanted to convey and so decides to decrypt the message. A while ago, after selecting $W = 5 \cdot 19$, White Duck solved the equivalence $5 \cdot d \equiv 1 \pmod{(5-1)(19-1)} \equiv 1 \pmod{4 \cdot 18} \equiv 1 \pmod{72}$. After some trial and error, we notice that $5 \cdot 29 = 145 = 2 \cdot 72 + 1$, so $d = 29$ works. Finally, White Duck computes $13^{29} \pmod{95} = 78$ and knows that Grey Duck was just saying "Hi." (Of course, if things are working properly, after encrypting and then decrypting, we should always get back the message we started with.)

White Duck decides to respond with the message *ah*, or 07. White duck uses G to encrypt and finds $07^5 \pmod{39} = 37$ and sends this to Grey Duck. Grey Duck is certain that White Duck wouldn't say *di* and therefore decrypts. Long ago, Grey Duck solved the equivalence $5 \cdot d \equiv 1 \pmod{(3-1)(13-1)} \equiv 1 \pmod{2 \cdot 12} \equiv 1 \pmod{24}$. Remembering that $5 \cdot 5 = 25 = 24 + 1$, we see that $d = 5$. Grey Duck computes $37^5 \pmod{39} = 07$ and receives White Duck's ambivalent message "Ah."

Let's look at the process more generally: Someone who wants to receive secret messages publicizes P and e, generally with $e = 65537$. A sender computes
$(message)^e \pmod{P}$ to encrypt and produce *encrypted* and sends *encrypted* to the
recipient. The recipient knows that $P = p_1 p_2$ and solves $e \cdot d \equiv 1 \pmod{(p_1 - 1)}$
$(p_2 - 1))$ for d. Then the recipient computes $(encrypted)^d \pmod{P} = message$.
How do we know this process works?

🦆 *How do we know that $(encrypted)^d \pmod{P} = message$ instead of $=$*
wacktext?

     ◉ First, we break up our messages into blocks so that each block
$messageblock_i < P$. In practice a message is broken into blocks that are
about the same length as the public key P, so that after taking powers
and computing the result modulo P, the length of $encryptedblock_i$ will
be the same as the length of $messageblock_i$. This is how the recipient
computer knows how to break up the encrypted message correctly into
blocks.

     ◉ Then if $(encrypted)^d \pmod{P} = message \pmod{P}$, or equivalently
$(encrypted)^d \equiv message \pmod{P}$, that's good enough because
$message < P$. Using some advanced number theory (that we won't
discuss here), it can be shown that $(encrypted)^d \equiv message \pmod{p_1}$
and $(encrypted)^d \equiv message \pmod{p_2}$. Knowing that $P = p_1 p_2$, we
can use the ideas from Section 16.5.2 to conclude that $(encrypted)^d \equiv$
$message \pmod{P}$: because $(encrypted)^d \equiv message \pmod{p_1}$, we
know $(encrypted)^d = jp_1 + message$. Substituting into $(encrypted)^d \equiv$
$message \pmod{p_2}$, we have that $jp_1 + message \equiv message \pmod{p_2}$
or $jp_1 \equiv 0 \pmod{p_2}$ or $jp_1 = kp_2$. Therefore, because p_1 and p_2 are
relatively prime, $j = qp_2$ for some integer q. We can then rewrite
$(encrypted)^d = jp_1 + message$ as $(encrypted)^d = qp_2 p_1 + message$
and conclude $(encrypted)^d \equiv message \pmod{p_1 p_2}$.

🦆 *How do we know that the equivalence $e \cdot d \equiv 1 \pmod{(p_1 - 1)(p_2 - 1)}$ can*
be solved for d? As we know from Theorem 16.5.1, as long as e is relatively
prime to $(p_1 - 1)(p_2 - 1)$, there is always a solution to the equivalence. So
we pick e to be a prime number. But notice that because p_1 and p_2 are prime,
$p_1 - 1$ and $p_2 - 1$ are even and therefore very much not prime. It could be
that e divides $p_1 - 1$ or $p_2 - 1$. For example, if $e = 3, p_1 = 5, p_2 = 13$, then
because $12 = 3 \cdot 4$ we have that e divides $p_2 - 1$.

Therefore, in practice we need to check to make sure that e is relatively prime to $p_1 - 1, p_2 - 1$. We do this using what? The Euclidean algorithm. We compute $\mathrm{GCD}(e, (p_1 - 1)(p_2 - 1))$ and make sure it's 1. (If not, we need to pick at least one different prime!)

🐦 *Why does this process keep messages secret?* It's because in order to compute $p_1 - 1$ and $p_2 - 1$ we need to know what p_1 and p_2 are. Only P is published, so in order for someone to figure out p_1 and p_2, ze would have to factor P. But P is huge, and it's very hard to factor large numbers… both in theory and in practice. Computers can't do it quickly enough to decrypt messages for unintended recipients.

By the way, you too can use public key cryptography! There is a protocol called Pretty Good Privacy that is widely used—if you want to find a plugin for software, go to https://www.openpgp.org/, or if you want your own public key to publish, there are lots of websites that will give you a public/private key pair (google "generate a pgp key") to find them.

RSA exercises: For the purpose of sending messages, we will use the same text-to-numbers conversion as in Section 5.4, where $a \to 0$ and $b \to 1$, except that we will assume that each letter uses 2 digits so that we can mash the letter-equivalents in a message together into one massive number (instead of encrypting one letter at a time). Thus, $a \to 00, b \to 01$, and so forth.

Also, to do modular arithmetic calculations with large numbers, use your favorite computer algebra system (or Wolfram|Alpha, if you don't have one— it understands if you type in (4343654)∧(34) mod 45313).

1. You prepare your public key with $p_1 = 557, p_2 = 1453$. If you use $e = 5$, will that work? What about $e = 11$?

2. Your fiend Molls, who is allergic to cats, posts $P = 3314221, e = 13$. Encode the message *peacock feathers make great cat toys* to send hir. What is a good block size to use when breaking up the numerical version of the message?

3. An encrypted message arrives for you. You've previously chosen $p_1 = 1867, p_2 = 2017, e = 23$. Here's what you receive:
1248178   2903440   3185267   549167   3006642   17453   3620483
3234036   1284537   730584   487841   3135256   1134082
Decode it!

4. Here's another message to decode:

861322 770240 24514 1625142 916196 1465653 1609436
311552 318157 1275548 1688431 97178 311552 263089
139683 1202933 1260030 743663 1625990 1231643 936225

Oh, wait, that was for a different triple of primes: $p_1 = 919, p_2 = 1907, e = 62989$.

16.11 Bonus Check-Yourself Problems

Solutions to these problems appear starting on page 636. Those solutions that model a formal write-up (such as one might hand in for homework) are to Problems 1, 3, and 8.

1. Given any n integers $k_1, k_2, \ldots, k_n$, show that there exist k_i, k_j such that $k_i \equiv k_j \pmod{n-1}$.

2. Find a set of 13 natural numbers, each of which has a different value modulo 13 and all of which are multiples of 5.

3. Consider the equation $5x + 2y = 3$. Why can it not have any solutions with both x and y whole numbers? For which k does the equation $5x + 2y = k$ have solutions with both x and y whole numbers?

4. Compute $\varphi(5^k 4)$, for any positive k.

5. For which n is $\varphi(n) = 4$?

6. Prove that the number of natural numbers relatively prime to n and $\leq mn$ is $m\varphi(n)$.

7. Use the Euclidean algorithm to compute $\mathrm{GCD}(1234, 12345)$. Now find integers k, ℓ such that $\mathrm{GCD}(1234, 12345) = k1234 + \ell12345$.

8. Prove that if $\mathrm{GCD}(a, b) = 1$, then for any integer c, there is always a solution to $ax + by = c$, where x and y are integers. Use this fact to find a solution to the equation $2x + 3y = 4$.

9. Find all x that satisfy $2x \equiv 2 \pmod 7$.

10. Which x satisfy both $x \equiv 1 \pmod 3$ and $x \equiv 3 \pmod 4$?

16.12 Problems on Number Theory Topics

1. Prove that every natural number $n > 1$ has a unique factorization into prime numbers. Specifically:

   (a) Prove that every natural number $n > 1$ can be factored into prime numbers.

   (b) Prove that this factorization is unique, perhaps by assuming that

   there are two different factorizations and deriving a contradiction.

2. Find a condition on n that determines whether or not n is squarefree, in terms of n's prime factorization. (A *squarefree* number is not divisible by k^2 for any natural $k \geq 2$.)

3. In a ball pit (literally, a space the size of a room filled knee-deep with plastic balls), all the balls are numbered with natural numbers. You grab eight of them to take home. Show that there must be two of them that have numbers that differ by some multiple of seven.

4. Prove that if n is odd, then the n natural numbers $2, 4, \ldots, 2(n-1), 2n$ have distinct values $\bmod n$.

5. A friend of yours comes by with a really large bag that ze says contains 175 balls from the ball pit in Problem 3. You exclaim, "Ha! I bet I can choose 18 of the balls such that all of them have the same tens digit!" Why should your friend believe you? (What does this have to do with number theory?)

6. For which k does the equation $9x + 6y = k$ have solutions with both x and y whole numbers? Why?

7. For which k does the equation $3x + 7y = k$ have solutions with both x and y whole numbers? Why?

8. A followup to the previous two problems: Make a conjecture about what conditions on a, b produce what results for which k the equation $ax + by = k$ has solutions with both x and y whole numbers. Explain why this makes sense.

9. An n-digit number with digits $a_{n-1}, \ldots, a_0$ can be written as $10^{n-1}a_{n-1} + \cdots + 10a_1 + a_0$. Use this fact along with some modular arithmetic to show that a number m is divisible by 3 if and only if the sum of its digits is divisible by 3.

10. Compute $\varphi(10^k)$, for any positive k.

11. Compute $\varphi(2^k 3)$, for any positive k.

12. For which n is $\varphi(n)$ even? Why?

13. For which n is $\varphi(n)$ divisible by 4? Why?

14. Suppose $\varphi(n) = 20$. Which primes might divide n? Find two different numbers n with $\varphi(n) = 20$. Could there be more n with $\varphi(n) = 20$?

15. Suppose $\varphi(n) = 125$. Which primes might divide n? How many possible n could there be with $\varphi(n) = 125$?

16. List as many different n as you can that have $\varphi(n) = 24$. Are you certain that you have listed all possible n?

17. Under what conditions on n and m is $\varphi(n) = \varphi(mn)$?

18. **Challenge:**

   (a) For each of the numbers $5, 6, 8, 9, 11, 12$, consider the set of divisors $\{d_i\}$, and compute $\sum_i \varphi(d_i)$. (We say that $\varphi(1) = 1$.)

   (b) Make a conjecture: for any natural number n, what is $\sum_i \varphi(d_i)$?

   (c) Prove your conjecture using the prime factorization of n and the formula for $\varphi(n)$.

   (d) Prove your conjecture using a combinatorial proof involving the sets $S_{d_i} = \{k \mid \mathrm{GCD}(k, n) = d_i\}$ and $\varphi(\frac{n}{d_i})$.

19. Use the Euclidean algorithm to compute $\mathrm{GCD}(4620, 7644)$. Now find integers k, ℓ such that $\mathrm{GCD}(4620, 7644) = k4620 + \ell7644$.

20. Prove that
   $\mathrm{GCD}(km, kn) = k\mathrm{GCD}(m, n)$.

21. Find all x that satisfy $3x \equiv 4 \pmod 5$.

22. Find all x that satisfy $4x \equiv 2 \pmod{11}$.

23. Find all x that satisfy $13x \equiv 4 \pmod{16}$.

24. Find all x that satisfy
    $9x \equiv 8 \pmod{22}$.

25. You saw in Section 16.4 that the solutions to $2x \equiv 1 \pmod 3$ are $x \equiv 2 \pmod 3$. Which numbers x satisfy $2x \equiv 4 \pmod 6$? What is the relationship between these solution sets, and why?

26. Find the solution set for $x \equiv 3 \pmod 5$ and $x \equiv 6 \pmod 7$.

27. Which x satisfy both $x \equiv 2 \pmod 3$ and $x \equiv 4 \pmod 5$?

28. Which x satisfy both $x \equiv 2 \pmod 5$ and $x \equiv 4 \pmod 6$?

29. Which x satisfy both $x \equiv 3 \pmod{11}$ and $x \equiv 12 \pmod{13}$?

30. **Challenge:** Find all x that satisfy all of $x \equiv 1 \pmod 3$ and $x \equiv 3 \pmod 5$ and $x \equiv 5 \pmod 7$.

31. **Challenge:** Prove that if $x \equiv s \pmod a$ and $x \equiv s \pmod b$ and $\mathrm{GCD}(a,b) = 1$, the common solutions are $x \equiv s \pmod{ab}$. Use this to show that there are no solutions to the pair of equations $x \equiv 5 \pmod{12}$ and $x \equiv 38 \pmod{86}$.

32. The *Farey sequence* $\mathscr{F}_n$ is the increasing list of all rational numbers $\frac{a}{b}$ with $a \le b \le n$. Write out $\mathscr{F}_2$ through $\mathscr{F}_5$. (You will notice that $\mathscr{F}_n$ is much shorter than FFL_n.) How many fractions do you add to $\mathscr{F}_{n-1}$ to get $\mathscr{F}_n$? How many fractions are in $\mathscr{F}_n$?

33. What is the sum of the fractions in $\mathscr{F}_n$?

34. (Knowledge of Chapter 8 may be useful here.) How many fractions are in FFL_k?

35. (Knowledge of Chapter 8 may be useful here.) What is the sum of the numerators in FFL_k? How about the sum of the denominators?

16.13   Instructor Notes

The intent of this chapter is to give a sampler of ideas that are both accessible and not the standard beginning of a number theory course. It is outside the main text because not only is number theory its own course, but as a mathematical research subfield it is not considered part of discrete mathematics. As with other supplemental chapters, the prerequisites for this chapter are Chapters 1 through 5 (the last of which introduces very basic modular arithmetic).

The material is structured to allow instructors to use the chapter flexibly: the chapter can be done in its entirety in a week (see below for suggested lesson plans), or some sections can be used interstitially between other chapters, or just the start of the chapter can be used to supplement Chapter 5. Note that Sections 16.4, 16.5, and 16.6 depend on Sections 16.2.2 and 16.3.

An instructor might want to place this chapter right after Chapter 5, or after Chapter 8, or at the end of the course. If you are using the whole chapter, begin the week by diving right into the Try This! Section 16.2 on investigating $\varphi(n)$ and GCD calculations, which will take an entire class period. Make sure to leave time for students to share their (preliminary) results. Ask students to read Section 16.3 on the Euclidian algorithm as preparation for the second day, when you might discuss the Euclidean algorithm; then work in groups to do congruence equation explorations via Section 16.4; and then summarize strategies for solving congruence equations. Then Section 16.5 can be assigned as reading to prove that $\varphi(ab)$ is multiplicative for a, b relatively prime. Finally, do and discuss Section 16.6 on the third day. Time can be carved out for discussion of Section 16.5.

Chapter 17

Computational Complexity

17.1 Introduction and Summary

In this chapter our goal will be to measure and compare the relative efficiency of algorithms. The same algorithm, when implemented on different computers or using different programming languages, will take different amounts of time to run. However, if we say that one algorithm runs much faster than another algorithm, then what we mean is that for most similar implementations (similar computers, similar programming languages) the first algorithm will take less time to execute than the second algorithm. Because these measurements take into account how involved the procedures of an algorithm are, we say that we are measuring the complexity of the algorithm.

This topic is very important in practice: paying attention to the computational complexity of an algorithm as it's being designed can make the difference between a program that won't finish before the universe ends and a program that finishes in an hour or two.

17.2 Try This! Count the Operations

As we saw in several earlier chapters, there are often multiple algorithms that will produce the same output or result. These often vary widely in efficiency and complexity. Our task here will be quantifying the complexity of an algorithm by counting the number of operations it takes to execute.

How do we know how to count these operations? We want to count the actions taken. So for example, storing a value, loading a value from memory, adding two numbers, comparing two numbers, or taking the ceiling of a number each take one operation to accomplish. We can't simply count the number of lines of code, because some lines may contain a lot more action than others.

Example 5.2.13 on page 135 gives two different algorithms that produce the same result. We know that one algorithm does more work than the other, but not *how much* more work. Let's try to measure that now.

1. How many operations does the first algorithm take with input 4? The second algorithm?

2. How many operations does the first algorithm take with input 7? The second algorithm?

3. Now generalize—How many operations does the first algorithm take with input n? The second algorithm?

Here are two (of the many that exist!) sorting algorithms; we will analyze them shortly.

Insertion sort.   This algorithm takes a list of length n as input. We will call that input *list*.

1. Let $i = 2$.

2. Let $x =$ the ith entry of *list*.

3. Let $j = i - 1$.

   (a) If x is less than the jth entry of *list*, put the jth entry of *list* in the $(j+1)$st slot of *list*. Otherwise, put x in the $(j+1)$st slot of *list* and go to step 4.

   (b) If $j = 1$, put x in the first slot of *list* and go to step 4. Otherwise, replace j with $j - 1$ and go to step (a).

4. If $i < n$, replace i with $i + 1$ and go to step 2; otherwise, continue.

5. Output *list*.

Selection sort.   This algorithm also takes a list as input, which we will also call *list*.

1. Let $i = 1$, let $j = 2$, let $k = 1$, and let x be the first entry of *list*.

2. If the jth entry of *list* is less than x, then replace x with the jth entry of *list*, set $k = j$, and continue; otherwise, continue.

3. If $j < n$, replace j with $j + 1$ and go to the previous step; otherwise, continue.

4. Put the ith entry of *list* into the kth spot of *list*.

5. Put x into the ith spot of *list*.

6. If $i < n - 1$, replace j with $i + 2$, replace k with $i + 1$, replace x with the $(i + 1)$st entry of *list*, replace i with $i + 1$, and go to step 2; otherwise, continue.

7. Output *list*.

If you would like to see examples of these algorithms in action, check out https://www.cs.usfca.edu/~galles/visualization/ComparisonSort.html (and choose the relevant options from the menu).

Let us compare these algorithms.

1. Run each of the sorting algorithms on the input $\{6, 3, 7, 2, 5, 4, 1\}$.

2. Now run them on $\{1, 2, 3, 4, 5, 6, 7\}$. How many operations does each take? As you count, keep track of how many operations are used for each specific value of i.

3. Now run them on $\{7, 6, 5, 4, 3, 2, 1\}$. How many operations does each take? As you count, keep track of how many operations are used for each specific value of i.

4. The list $\{1, 2, 3, 4, 5, 6, 7\}$ is a best-case input because it's already sorted. The list $\{7, 6, 5, 4, 3, 2, 1\}$ is a worst-case input because it's anti-sorted. How do insertion sort and selection sort compare in terms of how many operations they take on these two inputs?

5. Use your notes from Problem 2 to determine how many operations it takes to run each of the two algorithms on $\{1, 2, \ldots, n\}$.

6. Use your notes from Problem 3 to determine how many operations it takes to run insertion sort on $\{n, n - 1, \ldots, 1\}$, and estimate how many operations it takes to run selection sort on $\{n, n - 1, \ldots, 1\}$.

7. Now, make a general comparison between the number of operations insertion sort and selection sort each need for sorted input and anti-sorted input.

17.3   Computation of Runtime Functions

In Section 17.2, you started by counting the number of operations an algorithm takes to execute given a specific input and moved on to counting the number of operations an algorithm takes to execute given more general inputs. In each case, the expression you obtained was a function of the size of the input. Such functions let us compare how quickly (or slowly) different algorithms run and are thus called *runtime functions*; we will define this term carefully in Section 17.3.3.

17.3.1   Challenges in Measuring Complexity of Algorithms

In trying to measure the complexity of algorithms in Section 17.2, you probably came up against a bunch of issues.

- The number of operations it takes to execute an algorithm depends on the size of the input. But how do we measure that? Different algorithms have different kinds of inputs—numbers or sets or lists, for example—and so how the input is measured depends on what kind of input we have. For the algorithms from Example 5.2.13 that added integers, the input size was a number. For the sorting algorithms, the input size was a list length.

- The number of operations it takes to execute an algorithm also depends on qualities of the particular input. As you saw (and as you would expect anyway), it takes fewer operations to sort an already sorted list than it does to sort an unsorted list!

- It's not at all obvious how to count the number of operations it takes to execute an algorithm. Which commands "take time," so to speak, and which don't?

- Counting the number of operations an algorithm takes is really tedious and can be quite confusing.

And there's more—what we really want to know (in practice) is how long it takes for an algorithm to run, but not all operations take exactly the same amount of time to execute (even multiplication and addition don't take *exactly* the same amount of time). Worse yet, the same operation is likely to take different amounts of time on different computers, or when implemented in different programming languages. To account for this variation, we pretend that the amount of time every operation takes is the same, and say that this amount of time is large (within the range of times) so that algorithms might execute more quickly than we expect, but are unlikely to execute more slowly than we expect.

Another issue is that two different programmers might implement the same algorithm in the same programming language but do it subtly differently, so that their two programs need slightly different numbers of operations. It seems hopeless to count the number of operations an algorithm takes in any precise way! So, we will not even try to be precise—we will approximate. (Are you noticing a theme yet?) As long as we are consistent in our guidelines for approximating, then we can compare the overall complexity of one algorithm to that of another.

We will choose to always compare performance on worst-case inputs. This is not a flawless choice! It's entirely possible that Algorithm A performs really badly in the worst case but really well on average cases, whereas Algorithm B performs poorly both on average cases and in the worst case. This would mean that our analysis says that Algorithm B is better, but in practice Algorithm A is better. (This actually happens—the famous classical simplex algorithm, used in solving linear programming problems, is The Worst for worst case scenarios and Really Fast for almost all scenarios. It turns out that the worst case just doesn't happen in practice.) However, examining worst-case inputs makes the task of estimating algorithm complexity simpler. What would it really mean to examine average-case inputs? We could calculate the runtime for all possible inputs of a given size and take the average. (That seems like a *lot* of computation.) Or we could take a random sample of inputs and average their runtimes. Or we could calculate the runtimes for the kinds of inputs that arise most often in actual use—but how would we know what those are? In comparison, determining the runtime for the worst case is straightforward.

Example 17.3.1. Consider an algorithm that seeks a yellow marble from a line of marbles.

1. Pick up a marble.
2. If the marble is yellow, go to step 5; otherwise, toss the marble aside and continue.
3. If marbles remain, pick up the next marble in the line, and go to the previous step; otherwise, continue.
4. State dejectedly, "There are no yellow marbles," and take a nap.
5. Announce, "I have a yellow marble!" and take a nap.

The number of operations this algorithm takes to execute depends on the number of marbles in the line as well as where (if anywhere) in the line the first yellow marble is placed. A worst-case scenario is that there are no yellow marbles in the line. We now count the number of operations used in this instance: step 1 takes one operation, step 2 takes two operations for each marble, step 3 takes two operations for each marble except the last one, step 4 takes one operation, and step 5 is never reached so takes zero operations. Therefore, for a line of n non-yellow marbles, we use $1 + 2n + 2(n - 1) + 1 = 4n$ operations.

17.3.2 Why This Matters

Before we *really* get going on analyzing algorithm complexity... what is the point of doing so?

Suppose you want to know how many lists of a certain length n that have integer entries in the range k to ℓ also have entries that add up to s. Here is a naïve pseudo-algorithm for doing so: In an orderly fashion, make every list of length n with integer elements in the range k to ℓ. As each list is made, compute the sum of the entries. If the sum is s, then add 1 to a variable *total*.

Now, we know how to vaguely estimate how long this will take to run. There are $\ell - k + 1$ choices for each list entry, so there are $(\ell - k + 1)^n$ possible lists. For each of those lists, we have to write it down (that takes n operations to put in the entries) and compute its sum (this takes another $n - 1$ operations for all the additions) and check its sum against s (there's another operation). This all means there are at least $(2n)(\ell - k + 1)^n$ operations involved. (We will return to this estimate in Example 17.5.1.)

Imagine that you've coded this algorithm on your computer. In this imaginary world, you casually run some small examples on your computer: using entries ranging from 1 to 10, and lists of length up to 10, it takes about a second to run—you barely notice this. (Estimates are that as of 2017, a decent home computer will do 10^8 to 10^{14} instructions per second.) It occurs to you that maybe you'll get sums of s more often if you use negative values, so you change your range to -10 to 10, and then—why not?—to -50 to 50. As you're waiting for the computer to finish the calculation, you surf the web and run across an article saying that the universe is more than 10 billion years old and has at least $\approx 10^{80}$ atoms in it. The clickbait on the page has articles on "How do you know if you're at risk for becoming a mummy?" and "12 weird signs that you're getting old." Sighing, you decide to figure out how long this calculation might take. It has about 10^{21} operations involved, so might take about 10^7 seconds, which is... about four months. Oops.

Good thing that imaginary you didn't change the length of the lists (with range of -10 to 10) to length 50—that would have taken 10^{46} years to finish! And if you'd used a range of -50 to 50 with those length-50 lists, it would have been more like 10^{80} years... about as many years as there are atoms in the universe.

For a problem like this, it's not just that we need an algorithm that is better than using brute force—we need an algorithm that is a *lot* better than using brute force. Even a new generation of computers won't be sufficiently faster—consider that over the last decade, standard computer chips have only gotten a few times faster (maybe 10 times, or one order of magnitude). Even a hundredfold speed increase (two orders of magnitude) only gets us to 10^{44} years. And it saves a *lot* of both human and computer time if one thinks through the speed of an algorithm and how long it will take to execute, instead of naïvely coding the first thing that comes to mind and discovering mid-calculation that one must start over. (If only the author could heed her own words instead of repeatedly wasting time in this fashion....)

If we are going to have faster algorithms that are sufficiently fast to run on large inputs in small amounts of time, we need to know how to determine the speed of an algorithm *and* how to compare those speeds. This is where we're going: in the following sections we will compute runtime functions, and in Section 17.5 we will discuss the necessary comparisons.

17.3.3   What Is a Runtime Function?

A function, as we know from Section 3.2, outputs a unique output for each input. A *runtime function* for an algorithm has as input the size n of whatever data we feed into the algorithm and outputs the worst-case number of operations it takes to execute the algorithm on an input of size n. If we use $\mathbb{R}^{\geq 0}$ to denote the nonnegative real numbers, then formally a runtime function is denoted $r : \mathbb{N} \to \mathbb{R}^{\geq 0}$. The name *runtime* comes from the idea that this function roughly measures the amount of time it will take for this algorithm to run.

The most common runtime function types are *linear* (of the form an), *quadratic* (of the form $bn^2 + an$), *polynomial* (of the form $cn^k +$ terms of lower degree), *logarithmic* (of the form $d\log(n)$), and *exponential* (of the form $d2^n$ or dk^n). For example, an algorithm investigated earlier in this chapter has runtime function $r(n) = 3n + 2$.

Two different people who attempt to count the number of operations it takes for a given algorithm to execute may get slightly different results. Because of all the issues raised in Section 17.3.1, this is to be expected—and in particular, because different implementations of an algorithm give different runtimes, there is no point in trying to produce very precise runtime functions. (This is fortunate because trying to count operations precisely is frustratingly tedious.) A small tweak to the code on a particular algorithm implementation could make it run twice as fast!

Therefore, when we compute runtime functions, we don't worry too much about the coefficients.

In fact, how much detail we worry about in computing runtime functions depends on the purpose for which we're creating those functions. We might start with two algorithms and determine the types of their runtime functions. If they're of the same function type, we might then try to be more precise to understand the difference in their runtimes. There is, of course, a significant difference between the runtime functions $0.003n^2$ and $426n^2$, so we do try to keep track of each coefficient to within an order of magnitude (closest power of 10).

Example 17.3.2. Section 6.12, starting on page 194, approximates the runtime function for the bubble sort algorithm.

What does this all mean about how we compute runtime functions in practice? We give some general guidelines, followed by one algorithm (ha!) for computing runtime functions.

First, we have to be cautious as to how we look at steps: A step that includes several sub-steps is not really just one operation. For example, a step with a *while* statement includes an entire loop so it definitely cannot be counted as a single operation. And in general, a step that commands us to evaluate a function or do several mathematical operations is hiding many smaller operations.

We also need to decide which actions count as single operations. Most of the time we consider addition, subtraction, multiplication, division, calculation of x (mod n), executing and/or/not, comparison (implementing the condition of an *if* statement), and assignment of variables to values to be single operations. (The reason to say "most of the time" is that sometimes in the course of execution an algorithm will internally generate numbers disproportionately large compared to the input, in which case basic arithmetic takes longer because of the intermediate calculations.) In contrast, we cannot consider taking exponents or factorials to be single operations because they are repeated application of multiplication (which is a single operation).

How to compute a runtime function for an algorithm:

1. Familiarize yourself with the algorithm under consideration. Run it by hand on a couple of random inputs to see how it works.

2. Using this experience, determine what constitutes a worst-case input for this algorithm. Create a sample generic worst-case input *sagewci* (possibly pronounced "so juicy").

3. How should the input size of *sagewci* be measured? That is, what would make the most sense as an input size, given your understanding of the algorithm?

4. For each step of the algorithm, determine how many operations are used on *sagewci*.

5. Also determine how many times each step is executed for *sagewci*. Try to write this amount as a function of the input size you determined earlier.

6. Add it all up, as in $\sum_{i=1}^{steps}$ (number of operations in step$_i$)(number of times step$_i$ is executed).

There are of course sensible variations on this procedure—sometimes it's better to group steps together that are always executed in sequence; sometimes it makes sense to try to write the number of operations as a function of the iteration number. This is just one possible way to proceed.

Check Yourself ————————————————————————————

Try counting the numbers of operations for these simple algorithms.

1. Consider the counting algorithm with input a natural number n.

   1. Let $k = 1$.
   2. Output k.
   3. Replace k with $k + 1$.
   4. If $k = n$, output k, and stop; otherwise, return to step 2.

   How many operations does this algorithm take to execute?

2. Consider the laundry sorting algorithm that works on a pile of n pieces of clothing:

   1. Pick up a piece of clothing.
   2. If the item is a dark color, toss it to the right. If the item is a pale color, toss it to the left. If the item is a bright color, toss it behind you.
   3. If there is more clothing in the pile, go to step 1; otherwise, be done.

   What are best- and worst-case inputs for this algorithm? How many operations does it take to execute?

3. Write a runtime function for the algorithm given in Example 5.2.7.

4. Write a runtime function for the algorithm given in Problem 10 of Section 5.11.

17.4   Try This! Compute Runtime Functions

1. Review your work on the sorting algorithm comparison problems in Section 17.2 (or do the problems for the first time!) and use this to produce runtime functions for insertion sort and selection sort. What type of function is the insertion sort runtime function? How about the selection sort runtime function?

2. **List length:** Depending on the programming language, a list may be ended with an end-of-list character such as $'\backslash 0'$ or may have its length specified as part of the list object itself.

    (a) Suppose list length is specified as part of the list object. What is the runtime function type of a command that has input *list* and outputs the length of the list?

    (b) Suppose list length is determined by counting entries until an end-of-list character is encountered. What is the runtime function type of a command that has input *list* and outputs the length of the list?

    (c) Suppose again that list length is specified as part of the list object. Consider the following algorithm for inserting a 4 between the second and third list entries of *list* with length n. What is its runtime function? What function type is that?

        1. Let $i = 1$ and let *newlist* $= \{\}$ with length(*newlist*) $= 0$.
        2. If $i = 3$, append 4 to *newlist* and replace length(*newlist*) with length(*newlist*) $+ 1$; otherwise, continue.
        3. Append *list$_i$* to *newlist* and replace length(*newlist*) with length(*newlist*) $+ 1$.
        4. If $i = n$, output *newlist*; otherwise, replace i with $i + 1$ and go to step 2.

    (d) Suppose again that list length is determined via an end-of-list character. Consider the following algorithm for inserting a 4 between the second and third list entries of *list* with length n. What is its runtime function? What function type is that?

        1. Let $i = 1$ and let *newlist* $= \{\}$.
        2. If $i = 3$, append 4 to *newlist*; otherwise, continue.
        3. Append *list$_i$* to *newlist*.
        4. If *list$_i$* $= '\backslash 0'$, output *newlist*; otherwise, replace i with $i + 1$ and go to step 2.

3. Consider the following vaguely described algorithm, which counts the number of nonempty subsets of a set of integers that sum to zero. The input is a length-n list of integers *ints*.

   1. Let $num = 0$.
   2. Construct a new subset *subs* of *ints*.
   3. If $\sum_k subs_k = 0$, replace *num* with $num + 1$; otherwise, continue.
   4. If there are any subsets left to construct, go to step 2; otherwise, output *num*.

   (a) What is a worst-case input here?

   (b) Approximate the runtime function.

   (c) Is your approximation (aside from the constants involved) more likely an underestimate or more likely an overestimate of the actual runtime function?

4. Our *Dictionary of 1000 Duck-Related Words* consists of a length-1000 list named *duckdict*, where each entry is a pair $\{word, definition\}$. The *word*s are in alphabetical order. Here is an algorithm for finding the definition of *drwid*, an arbitrary duck-related word in the dictionary.

   1. Let $i = 500$ and let $last = temp = 0$.
   2. If $drwid = duckdict_{i,1}$, output $duckdict_{i,2}$; otherwise, continue.
   3. Replace *temp* with i.
   4. If *drwid* is earlier in the alphabet than $duckdict_{i,1}$, then replace i with $i - \lceil \frac{|i-last|}{2} \rceil$; otherwise, replace i with $i + \lceil \frac{|i-last|}{2} \rceil$.
   5. Replace *last* with *temp*.
   6. Go to step 2.

   Approximate the runtime of this algorithm. What is the input size? How does the algorithm generalize? Approximate the runtime for the generalized algorithm.

17.5 Comparing Runtime Functions

By now we're so deep into the ideas of runtime functions and how to compute them that it's easy to lose sight of our overall goal: we need to compare the complexity of algorithms, and we especially need to compare the complexity of different

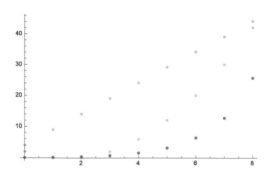

Figure 17.1. Runtime functions $r_1(n) = 5n + 4$ (teal), $r_2(n) = n^2 - 3n + 2$ (pale grey), and $r_3(n) = \frac{2^n}{10}$ (dark grey) shown for small input sizes.

algorithms that accomplish the same task. We will do this by comparing runtime functions, and in particular by comparing the growth types of runtime functions.

To be clear, for this purpose we are completely abandoning any sense of measuring what an algorithm will do in practice, meaning that in addition to ignoring the specifics of a way an algorithm is implemented, we are not looking at how long it takes an algorithm to run (despite basing our analysis on runtime functions). Instead we will examine how the number of operations scales as the input size gets larger, so that we can see which algorithms are better (or worse) for large inputs. For this purpose, we look at the runtime function type of the algorithm. This is called the *complexity* of an algorithm.

One caveat is that we can only compare runtime functions that have the same type of input size—it doesn't make sense to compare a runtime function $r_1(n)$ where n is the length of a list with $r_2(n)$ where n is the magnitude of a number. In the case where two algorithms that we wish to compare have different natural ways to measure the input size, we must find a way to convert (or at least approximate) one input size in terms of the other. Of course, how that conversion could be done depends entirely on the details of the inputs themselves.

So, what does it mean to compare the growth types of runtime functions? And why do growth types matter? Let us consider three sample runtime functions, $r_1(n) = 5n + 4$, $r_2(n) = n^2 - 3n + 2$, and $r_3(n) = \frac{2^n}{10}$. We plot these functions for input sizes up through 8 in Figure 17.1. Notice that because we have only natural-number inputs, the plots only show values at natural-number x-values. The x-axis shows the input size and the y-axis shows the runtime for that input size. From this figure, it appears that $r_1(n)$ is the slowest, using the largest number of operations for a given input, with r_2 a bit faster and $r_3(n)$ faster yet. What happens when we

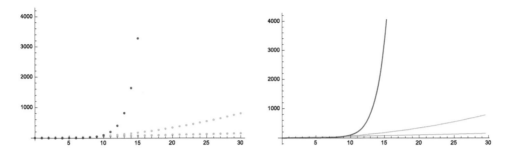

Figure 17.2. Runtime functions $r_1(n) = 5n + 4$ (teal), $r_2(n) = n^2 - 3n + 2$ (pale grey), and $r_3(n) = \frac{2^n}{10}$ (dark grey) shown for moderate input sizes with discrete and continuous inputs.

consider larger input size, say, up through input size 30? The results are shown in Figure 17.2, along with the same graph with continuous input in case the spacing between the dots makes it too hard to see what's happening. Gosh! The situation changes dramatically when we examine larger input sizes! Now $r_1(n)$ seems like the most efficient, using far fewer operations for a given input, with $r_2(n)$ slower and $r_3(n)$ using a whole honking pile of operations. In this view, the three functions seem to be very similar for small input sizes but to differ substantially as the input size grows.

These three functions are representatives of different growth types of functions: $r_1(n)$ is a linear function, $r_2(n)$ is a quadratic function, and $r_3(n)$ is an exponential function. Let us also look at two more quadratic functions, $r_4(n) = 2n^2$ and $r_5(n) = \frac{1}{2}n^2$. In Figure 17.3 we plot these functions together with $r_2(n)$ for two ranges of input sizes. Despite the fact that $r_2(n)$ is not of the simple form

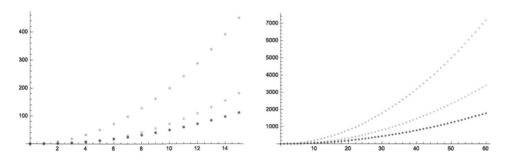

Figure 17.3. Runtime functions $r_4(n) = 2n^2$ (teal), $r_2(n) = n^2 - 3n + 2$ (pale grey), and $r_5(n) = \frac{1}{2}n^2$ (dark grey) shown for small and moderate input sizes.

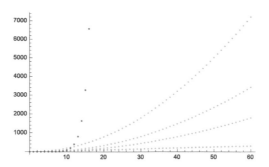

Figure 17.4. Runtime functions $r_1(n) = 5n + 4$ (teal), $r_2(n) = n^2 - 3n + 2$, $r_4(n) = 2n^2$, $r_5(n) = \frac{1}{2}n^2$ (all pale grey), and $r_3(n) = \frac{2^n}{10}$ (dark grey) shown for moderate input sizes.

bn^2, the linear term $-3n$ does not change its behavior; it is still solidly between the quadratic functions with higher and lower coefficients. And as we see in Figure 17.4, the three quadratic functions have behavior that is similar to each other and different from linear and exponential functions. Notice that when we're looking at the behavior of these functions, we are looking at what happens as the inputs get larger—this is what is meant by comparing the growth types of the functions.

In fact, as long as n is large enough (in this case, larger than 10), we have $r_1(n) < r_5(n) < r_2(n) < r_4(n) < r_3(n)$. If we also ignore leading coefficients, we can say that as long as n is large enough, then $linear(n) < quadratic(n) < exponential(n)$. We give common technical refinements on this idea in Section 17.5.1. Linear, quadratic, and exponential are common complexity classes for runtime functions. Given a new algorithm, determining or estimating the complexity class is more important than determining the runtime function—after all, if the complexity class indicates that the new algorithm will be on the whole less efficient than older algorithms, perhaps no one will bother to implement it!

Example 17.5.1. In Section 17.3.2, we counted the number of lists of length n that have integer entries in a given range (k to ℓ) that add up to s. Our vaguely described algorithm created every list that met our criteria, computed the sum of its entries, and compared the sum to s.

We noted that because there are $\ell - k + 1$ choices for each list entry, there are $(\ell - k + 1)^n$ possible lists. For each of the possible lists, we said there are approximately $2n$ operations—n operations to determine the n entries, $n - 1$ operations to add them up, and one operation to check whether the sum is s.

Depending on how exactly we write that algorithm, there are likely to be a few more operations per list. That's probably a small constant, or maybe somehow a constant multiple of n. However, there won't be n^2 more operations per list as that would require additional algorithmic structure (such as an extra loop executed for each list). So we can feel confident that for some constant d, our runtime function is $r(n) = (dn)(\ell - k + 1)^n$.

What kind of runtime function is this, though? It looks like there are three different variables, k, ℓ, and n! Notice that what matters is not the values of ℓ or k, but the value of $\ell - k$, so we can let $m = \ell - k + 1$ and rewrite the function in two variables as $r(n, m) = dn \cdot m^n$. If we had a fixed list length, this would be a polynomial function of the value range m. If instead we had a fixed value range, this would be an exponential function of the list length n.

As a multivariable function, this is discouragingly large. Furthermore, the troublesome exponential factor is present because we are checking the sum for each and every one of our exponentially many lists. Before throwing up our hands and deciding we might as well not even try to answer our original question, mathematics comes in: is there a way to determine whether some lists have sum s without actually calculating their sums? Noting that two lists that differ only in one entry have similar sums, we see that if we find a list with sum s, we know that neighboring lists will *not* have sum s... and we suspect that backtracking (see Section 10.8) could be used to reduce the infeasibility of solving the problem. (Unfortunately, there's no simple way of estimating the complexity of a backtracking algorithm, but certainly using backtracking must improve the runtime!)

Complexity analysis has its limits, though—if all known algorithms to solve Problem Type X are exponential runtime, and if we still need to solve Problem Y (which is of Problem Type X) within the next two weeks, then we have to look at details of implementation. What input sizes do we have for Problem Y? What are the runtime coefficients for various algorithms that solve Problem Type X? Is there an available algorithm that we can code so it will run in under two weeks to solve Problem Y, or do we need to seek tweaks (fast!)?

Conversely, the famous classical simplex problem has exponential runtime, but in practice it runs in polynomial time. This is because the worst-case input does not happen in practice.

17.5.1 Big-O (and Ω and Θ) Notation

We have already grouped runtime functions by type—linear, quadratic, logarithmic, exponential—and now we will combine worst-case scenarios with these types to make new groupings of runtime functions. First we formalize the process we have already been using to classify runtime functions:

Definition 17.5.2. Suppose *type* refers to an example of a function type, so that n^2 would be an example of the quadratic function type, or 3^n would be an example of the exponential function type. We use $\Theta(type)$ to denote the set of all runtime functions $r(n)$ such that

- there are positive real constants c_1 and c_2 and

- there is a natural number k

so that whenever $n \geq k$, $c_1 \cdot type \leq r(n) \leq c_2 \cdot type$.

This is a technically complicated definition, but not as difficult to use as it seems at first. It says that aside from small inputs (that's the $n \geq k$ part), we group together runtime functions that can be sandwiched between constant multiples of the function type we're interested in. In conversation, the statement "$f(n) \in \Theta(type)$" is pronounced "$f(n)$ is in big theta of *type*."

Example 17.5.3. When we noted that generally $r_5(n) < r_2(n) < r_4(n)$, we were also saying that $r_2(n) \in \Theta(n^2)$. If we let $c_1 = \frac{1}{2}$, $c_2 = 2$, and $k = 6$, then whenever $n \geq 6$, $\frac{1}{2}n^2 \leq r_2(n) \leq 2n^2$.

Our next grouping is perhaps the most common one used by computer scientists.

Definition 17.5.4. We use $O(type)$ to denote the set of all runtime functions $r(n)$ such that

- there is a positive real constant c and

- there is a natural number k

so that whenever $n \geq k$, $r(n) \leq c \cdot type$.

The set $O(type)$ collects runtime functions that are less than some constant multiple (maybe a big one) of the function type we're talking about. Why would big O (yes, that's how it's pronounced) be preferable to Θ? It's an upper bound on how slowly an algorithm might run—sort of a worst case among worst cases. We might not be able to determine a runtime function $r(n)$ precisely enough to show $r(n) \in \Theta(type)$, but we might be able to show that $r(n) \in O(type)$. The process of proving that a given runtime function is, or is not, in $O(type)$ is similar to using integral convergence comparison tests in calculus.

Be careful in using big-O notation in conversation, and be cautious in interpreting other people's claims involving $O(type)$. People frequently misuse $r(n) \in O(type)$ to mean "I think this is the worst-case runtime" rather than the actual "this is possibly much worse than the worst-case runtime." Sometimes people misuse $O(type)$ as a single function rather than a set of functions. It's tricky to use $O(type)$ correctly!

Notice that $\Theta(type) \subset O(type)$, but $O(type) \not\subset \Theta(type)$. For example, $3n + 2 \in O(n^2)$, but $3n + 2 \notin \Theta(n^2)$. Given any $c_1 > 0$, we note that $3n + 2 < c_1 n^2$ if $n > \frac{3 + \sqrt{9 - 8c_1}}{2}$. Thus there is no k that will satisfy the definition of $\Theta(n^2)$.

Finally, we have a lower-bound grouping:

Definition 17.5.5. We use $\Omega(type)$ to denote the set of all runtime functions $r(n)$ such that

♣ there is a positive real constant c and

♣ there is a natural number k

so that whenever $n \geq k$, $r(n) \geq c \cdot type$.

Again, a bit of caution is needed in using $\Omega(type)$. If $r(n) \in \Omega(type)$, that doesn't mean that the algorithm always runs slower than *type*—it means that in worst cases, the algorithm runs slower than *type*.

It follows directly from these definitions that $\Theta(type) = \Omega(type) \cap O(type)$.

Example 17.5.6. We reinterpret earlier examples from this chapter in terms of big-O/Θ/Ω notation.

Example 17.3.1 computes the runtime function $y(n) = 4n$. This is an exact runtime function, essentially independent of implementation, so $y(n) \in \Theta(n)$. Additionally, $y(n) \in O(n^{53})$, though it's useless to say so, and $y(n) \in \Omega(1)$... which is true of every algorithm, so it's even more useless as a measure.

Problems 1–4 of the Check Yourself in Section 17.3 have runtime functions $cy_1(n)$, $cy_2(n)$, $cy_3(n)$, and $cy_4(n)$. We can say $cy_1(n), cy_2(n) \in \Theta(n)$ and $cy_3(n)$, $cy_4(n) \in \Theta(1)$.

Example 17.5.1 discussed the runtime function $r(n, m) = dn \cdot m^n$. The original algorithm was vague and perhaps oversimplified, so our estimate of the runtime is likely to be an underestimate. Thus we know $r(n, m) \in \Omega(m^n)$.

If you find it difficult to remember the definitions of and differences between $\Theta(type)$, $O(type)$, and $\Omega(type)$, just don't use that notation! Say what you mean explicitly: $r(n) \leq c \cdot type$ is clear.

Check Yourself ————————————————————————————

Try these problems for practice in the details of runtime function comparison.

1. We claimed that as long as n is larger than 10, we have $r_1(n) < r_5(n) < r_2(n) < r_4(n) < r_3(n)$. Verify that this is not true if $n = 10$.

2. How small can n be and still have $r_5(n) < r_2(n) < r_4(n) < r_3(n)$ be true?

3. True or false (and explain): $n^3 \in O(n^2)$.

4. True or false (and explain): $3^n \in O(3^{n-1})$.

5. True or false (and explain): $r(n) \in O(r(n))$.

——

17.6   Try This! Determine Complexity Classes

1. For each problem you did in Sections 17.2 and 17.4, rewrite your conclusions using big-O, Θ, or Ω notation. Can you use more than one type of notation for any of these conclusions? Does it make sense to compare any of these runtime functions to each other?

2. Example 5.2.2 on page 128 describes an algorithm for multiplying multidigit integers.

   (a) Suppose A has m digits. (We are given that B has n digits.) How might we measure the size of the input?

   (b) For each step of the algorithm, compute the number of operations.

   (c) Write a runtime function for this algorithm.

   (d) What is the complexity of the algorithm?

   (e) When doing multiplication by hand, we make an assumption about the relationship between m and n. What is it?

   (f) How is the runtime function affected if we swap the roles of A and B in the algorithm? What is the effect on the complexity of the algorithm?

3. The following algorithm computes the nth Fibonacci number. What is its complexity?

   1. Let $fib(n) = 0$, let $k = n$, let $fib(1) = 1$, and let $fib(2) = 1$.
   2. If $k = 1$, go to step 6; otherwise, continue.

3. If $k = 2$, go to step 6; otherwise, continue.

4. Replace $fib(k)$ with $fib(k-1) + fib(k-2)$.

5. Replace k with $k-1$ and go to step 2.

6. Output $fib(n)$.

(You may wish to compare this to other algorithms for computing the Fibonacci numbers; see Problems 13 and 14 in Section 17.10. Note also that it would be difficult to analyze the complexity of an algorithm that uses the Binet formula—either we use a computer algebra system, which has varying complexity associated with each of its commands, or we need to understand how the computer stores irrational numbers.)

17.7   Where to Go from Here

There are whole courses on computational complexity, often titled Analysis of Algorithms, in computer science departments. In this chapter, we only examined algorithms with simple runtime functions. More complicated algorithms have much harder-to-determine runtime functions—and one of the reasons we use big O is because that might be easier to figure out than Θ. An excellent source to learn more about efficiency and complexity of algorithms is [5], particularly Chapters 1, 2, and 10. Anything you want to know is in there somewhere, though be warned: it's written clearly but is still challenging to read.

For a more computer-science-y perspective on some of the same material in this chapter, see Prof. Mary K. Vernon's page at http://pages.cs.wisc.edu/~vernon/cs367/notes/3.COMPLEXITY.html. Some technical notes on how to prove functions are (or are not) in $O(type)$ may be found at http://www.math.uvic.ca/faculty/gmacgill/guide/big-O.pdf.

Study of the growth rates of functions in general is part of calculus, so if you are interested in learning more about that topic, go take (or take more) calculus!

Credit where credit is due: Much of my understanding of computational complexity stems from [5], which also inspired chunks of the material in this chapter.

Gary Lewandowski pointed out to me why we need to think beyond the complexity of algorithms—after all, we still need to accomplish the most computationally difficult tasks, so there is value in finding smaller-coefficient algorithms; just because a problem is computationally complex doesn't mean we don't need to solve it.

Mirella Damian (Villanova University), Robin Flatland (Siena College), Kyle Burke (University of New Hampshire), and Thomas Hull (Western New England University) shared suggestions of some algorithm and problem types while at a conference in Barbados (at which I was not present).

The algorithms for computing the Fibonacci numbers given in Problem 3 of Section 17.6 and in Problems 13 and 14 of Section 17.10 were generated in Hannah Alpert's 2017 MathILy Branch class.

Problem 23 in Section 17.10 has exhortations by Blaartholomew and Tod, respectively, from the *Heavenly Nostrils* universe.

The author determined that...

🐦 ... the number of atoms in the universe is $\approx 10^{80}$ and definitely not more than 10^{85} from *Universe Today*, at https://www.universetoday.com/36302/atoms-in-the-universe/.

🐦 ... the age of the universe is $14 \times 10^9 \approx 10^{10}$ years old from NASA, at https://map.gsfc.nasa.gov/universe/uni_age.html.

🐦 ... a desktop computer does 100 million (which is 10^8) instructions per second from http://computer.howstuffworks.com/question54.htm, and calculations from claims made in the wikipedia page https://en.wikipedia.org/wiki/Instructions_per_second and related links produce a range of 10^{11} to 10^{14} instructions per second.

17.8   Chapter 17 Definitions

runtime function: A function that has as input the size n of whatever data we feed into the algorithm and outputs the worst-case number of operations it takes to execute the algorithm on an input of size n; formally, $r : \mathbb{N} \to \mathbb{R}^{\geq 0}$.

linear function: A function of the form an.

quadratic function: A function of the form $bn^2 + an$.

polynomial function: A function of the form $cn^k +$ terms of lower degree.

logarithmic function: A function of the form $d \log(n)$.

exponential function: A function of the form $d2^n$ or dk^n.

complexity: The runtime function type of an algorithm.

17.9   Bonus Check-Yourself Problems

Solutions to these problems appear starting on page 637. Those solutions that model a formal write-up (such as one might hand in for homework) are to Problems 3 and 4.

1. Write a runtime function for the marble-sorting algorithm given in Example 5.2.8.

2. Consider the following algorithm that has input *list* with length n. What does the algorithm do? What is a worst-case

input? What is its runtime function? What function type or complexity class is that function?

1. Let $i = 1$.
2. If $list_i \geq list_{i+1}$, say "nope" and exit; otherwise, continue.
3. If $i = n - 1$, say "yup!"; otherwise, replace i with $i + 1$ and go to step 2.

3. Prove that $r(n) = n^2 - 16$ is in $O(n^3)$.

4. Estimate the complexity of the following pointless algorithm that takes as input a natural number n.

1. Let $i = 1$.
2. If it's Tuesday, then continue; otherwise, go to step 5.
3. Let $j = 2^{n+1}$ and let $m = i(2 - j)$.
4. If $i = n$, output "Ha ha ha!"; otherwise, replace i with $i + 1$ and go to step 2.
5. Let $j = 2i$.
6. If $i = n$, output "Ha ha ha!"; otherwise, replace i with $i + 1$ and go to step 2.

5. Consider the following algorithm with input *list* of length n with integer elements. What does it do? What is its complexity?

1. Let $i = 1$ and let *newlist* $= \{\}$.
2. Append $2 \cdot list_i$ to *newlist*.
3. If $i < n$, replace i with $i + 1$ and go to the previous step; otherwise, output *newlist*.

6. Determine the complexity class of the bubble sort algorithm, whose runtime function is approximated in Section 6.12.

7. True or false: $\frac{1}{2}n^2 + n\log(n) \in O(n\log(n))$. Explain, and if the statement is false, make a corrected statement.

8. Write an algorithm that inputs a number n and outputs whether the number is even, odd, or neither (not an integer). What is its complexity?

9. Consider the following algorithm. What does it do? Estimate its complexity.

1. Input n.
2. Let $i = 0$ and let $j = n$.
3. If $\lfloor \frac{j}{2} \rfloor = 0$, output i; otherwise, continue.
4. Replace j with $\lfloor \frac{j}{2} \rfloor$, replace i with $i + 1$, and go to the previous step.

10. True or false: $\sqrt{n} + n \in \Omega(n)$. Explain, and if the statement is false, make a corrected statement.

17.10 Computation Problems

1. Example 5.2.5 gives four algorithms, three of which terminate. Write runtime functions for these three algorithms.

2. Consider the following algorithm for removing all instances of the entry 61 in *list* with length n. What is a worst-case

input? What is its runtime function? What function type or complexity class is that function?

1. Let $i = 1$ and let *newlist* = {} with length(*newlist*) = 0.
2. If *list$_i$* $\neq$ 61, append *list$_i$* to *newlist* and replace length(*newlist*) with length(*newlist*) + 1; otherwise, continue.
3. If $i = n$, output *newlist*; otherwise, replace i with $i + 1$ and go to step 2.

3. Consider the following algorithm that has input *list* with length n. What does the algorithm do? What is a worst-case input? What is its runtime function? What function type or complexity class is that function?

1. Let $i = 1$.
2. If *list$_i$* = *banana*, say "yes" and exit; otherwise, continue.
3. If $i = n$, say "no"; otherwise, replace i with $i + 1$ and go to step 2.

4. In Section 10.8, we examined a problem of placing coins on a square grid so that no two are in the same row or column or diagonal, and we proposed three non-backtracking procedures for finding solutions. Give a lower bound for the complexity of each of these procedures for an $n \times n$ grid.

5. Plot the functions you derived in Problem 4 on the same axes, and adjust the x-axis range so that you can see how they compare. How do each of the functions compare to 2^n? ... to x^2? Based

on your observations, make a statement about the complexity class of each procedure.

6. What does the following algorithm do with input *list* of length n? Describe the complexity of its runtime function.

1. Let $i = 1$ and let $j = 2$.
2. If *list$_i$* = *list$_j$*, output "Yup." Otherwise, continue.
3. If $j < n$, replace j with $j + 1$ and go to step 2; otherwise, continue.
4. If $i < n - 1$, replace i with $i + 1$, replace j with $i + 1$, and go to step 2; otherwise, continue.
5. Output "Nope."

7. **Challenge:** Example 5.2.11 gives a Russian-style algorithm for multiplying multidigit integers. Write a runtime function for this algorithm. (It will be useful to first do Problem 24 in Section 5.11.)

8. **Challenge:** Compare the runtime functions for US-style and Russian-style multiplication.

9. Write a runtime function for the algorithm in Example 5.2.14.

10. Write a runtime function for the algorithm given in Problem 12 in Section 5.11. What's the complexity class?

11. Section 5.8 describes a depth-first algorithm for searching a graph (by visiting each node in the graph). Consider the case of a rooted tree. How does depth-first search proceed? What should the input size of the algorithm be? What is the complexity class of depth-first search for a tree?

12. Section 5.8 also describes a breadth-first algorithm for searching a graph. Consider the case of a rooted tree. How does breadth-first search proceed? What should the input size of the algorithm be? What is the complexity class of breadth-first search for a tree?

13. The following algorithm computes the nth Fibonacci number. What is its complexity?

    1. Let $place = 1$, let $fib = 1$, and let $i = 1$.

    2. Replace fib with $place + fib$.

    3. Replace $place$ with $fib - place$.

    4. If $i < n$, replace i with $i + 1$ and go to step 2; otherwise, output fib.

14. Here is yet another algorithm that computes the nth Fibonacci number. What is its complexity?

    1. Let $fiblist = \{1,1\}$ and let $i = 2$.

    2. If $i = n$, output $fiblist_{n-1}$; otherwise, append $fiblist_i + fiblist_{i-1}$ to $fiblist$.

    3. Replace i with $i + 1$ and go to step 2.

15. **Challenge:** In Problems 13 and 14 the algorithms compute Fibonacci numbers by adding previous Fibonacci numbers together. Fibonacci numbers get large very quickly (the 25th Fibonacci number is 75,025, for example) and so addition gets slower as the algorithm progresses——it takes more time to add two 30-digit numbers than it does to add two 2-digit numbers. Suppose that it takes approximately n operations to add two n-digit numbers. Also suppose that the number of digits in the nth Fibonacci number is at most kn, some constant multiple of n. How does taking this into account affect the complexity of these two Fibonacci-number-computing algorithms?

16. Compare the complexity classes of the algorithms given in Problems 13 and 14 above and Problem 3 of Section 17.6.

17. Prove that $r(n) = n^2$ is not in $O(n)$.

18. Prove that $r(n) = 45n^5 - 89n^4 + 40{,}789n^3 - 67n^2 + 57{,}840{,}578n - 2$ is in $O(n^5)$.

19. Prove that if $r_1(n) \in O(r_2(n))$ and $r_2(n) \in O(r_3(n))$, then $r_1(n) \in O(r_3(n))$.

20. Prove that $r(n) = 4{,}782{,}947{,}384 \log(n) \in \Theta(\log(n))$.

21. Write an algorithm that detects whether an input list is anti-sorted, and give a runtime function for this algorithm.

22. Consider the following algorithm. What does it do? Estimate its complexity.

    1. Input $string$.

    2. Let $i = 1$.

    3. If $string_i = \,'m'$, continue; otherwise, go to step 10.

    4. If $string_{i+1} = \,'a'$, continue; otherwise, go to step 10.

    5. If $string_{i+2} = \,'t'$, continue; otherwise, go to step 10.

    6. If $string_{i+3} = \,'c'$, continue; otherwise, go to step 10.

    7. If $string_{i+4} = \,'h'$, continue; otherwise, go to step 10.

8. If $string_{i+5} = 'a'$, continue; otherwise, go to step 10.

9. Output "Let's drink some tea!"

10. If $i < n - 6$, replace i with $i + 1$ and to go step 3; otherwise, output "This is not the tea you're looking for."

23. Estimate the complexity of the following pointless algorithm that takes as input a natural number n.

    1. Let $i = 1$.
    2. If it's before noon, then continue; otherwise, go to step 10.
    3. Let $j = 1$.
    4. If $j = i$ then eat a cookie; otherwise, replace j with $j + 5$.
    5. Let $list = \{i, j, i - 1, j + 1\}$.
    6. Append $'m'$ to $list$.
    7. Append $'cookie'$ to $list$.
    8. If $list_5 = 3$, then append j to $list$; otherwise, continue.
    9. If $i = n$, output "Blaaaart!"; otherwise, replace i with $i + 1$ and go to step 2.
    10. Let $j = 2$ and replace j with 3.
    11. If $i = n$, output "Rar!"; otherwise, replace i with $i + 1$ and go to step 2.

24. Estimate the complexity of the following pointless algorithm that takes as input a natural number n.

    1. Let $i = 1$, let $j = i + 1$, and let $k = 1$.
    2. If $i = n$, output "Pointless! Totally pointless. No points

made. Worth no points. Not even a dot." Otherwise, continue.

3. If $j = n + 1$, then go to step 8; otherwise, continue.

4. If $k = j + 1$, then go to step 7; otherwise, continue.

5. Output $3 + 4$.

6. Replace k with $k + 1$ and go to step 2.

7. Replace j with $j + 1$ and go to step 2.

8. Replace i with $i + 1$ and go to step 2.

25. Consider the following algorithm with input $list$ of length n with integer elements. What does it do? What is its complexity?

    1. Let $i = 1$, let $j = 1$, let $sum = 0$, and let $lesser = \{\}$.
    2. Replace sum with $sum + list_i$.
    3. If $i < n$, replace i with $i + 1$ and go to the previous step; otherwise, continue.
    4. Let $ave = \frac{sum}{n}$.
    5. If $list_j < ave$, append $list_j$ to $lesser$; otherwise, continue.
    6. If $j < n$, replace j with $j + 1$ and go to the previous step; otherwise, output $lesser$.

26. Consider the following algorithm with input $list$ of length n with integer elements. What does it do? What is its complexity?

    1. Let $i = 1$, let $j = 2$, let $diff = 0$, let $imax = 0$, and let $jmax = 0$.

2. If $|list_i - list_j| > diff$, replace *imax* with *i*, replace *jmax* with *j*, replace *diff* with $|list_i - list_j|$, and continue; otherwise, continue.

3. If $j < n$, replace *j* with $j + 1$ and go to the previous step.

4. If $i < n$, replace *i* with $i + 1$ and go to step 2; otherwise, output $\{list_{imax}, list_{jmax}, diff\}$.

27. Consider the functions $r(n) = 236n + 5\log(n)$ and $t(n) = n + (\log(n))^2$. Plot $r(n)$ and appropriate multiples of $t(n)$ on the same axes, and adjust the *x*-axis range so that you can see how they compare. Is $r(n) \in O(t(n))$? Is $r(n) \in \Omega(t(n))$? Is $r(n) \in \Theta(t(n))$? Explain.

28. Consider the functions $r(n) = 99n2^n$ and $t(n) = 3^n$. Plot $r(n)$ and $t(n)$ on the same axes, and adjust the *x*-axis range so that you can see how they compare. Is $r(n) \in O(t(n))$? Is $r(n) \in \Omega(t(n))$? Is $r(n) \in \Theta(t(n))$? Explain.

29. Suppose that we wish to compute $k^{(2^m)}$. We could use Algorithm A:

    1. Let $i = 1$, let $p = 2^m$, and let *pow* $= k$.
    2. If $i < p$, replace *pow* with *pow* $\cdot k$ and continue; otherwise, output *pow*.
    3. Replace *i* with $i + 1$ and go to the previous step.

    Or, we could use Algorithm B:

    1. Let $i = 1$ and let *pow* $= k$.
    2. If $i < m + 1$, replace *pow* with *pow* $\cdot$ *pow* and continue; otherwise, output *pow*.

3. Replace *i* with $i + 1$ and go to the previous step.

Compare the complexity of these two algorithms.

30. True or false: $n\log n \in \Theta(n)$. Explain, and if the statement is false, make a corrected statement.

31. True or false: $\sqrt{n} + \log(n) \in O(n)$. Explain, and if the statement is false, make a corrected statement.

32. Consider the problem of deleting a directory structure—a folder cannot be deleted until all files it contains have been deleted. (And a folder may contain other folders!) Write an algorithm that deletes a directory structure. What should the input size be? Estimate the complexity of the algorithm.

33. Consider the problem of copying a directory structure—files cannot be copied until the folder that should contain them exists. Write an algorithm that copies a directory structure from location A to location B. What should the input size be? Estimate the complexity of the algorithm.

34. Write an algorithm that inputs a list of *n* integers and outputs the list element of greatest value. What is its complexity?

35. **Challenge:** Consider a list *catprice* of pairs $\{cost, item\}$ of *m* items in a catalog, no two of which have the same price, ordered by increasing cost. You have a $50 gift certificate and know that you want to buy the special duck-head-capped fountain pen (in teal, of course), which costs $d. Write an algorithm that will determine which sets of distinct catalog items you can buy to exactly cash in your gift certificate. What is its complexity?

17.11   Instructor Notes

This chapter is intended to develop the idea of complexity classes from scratch, so that students will understand how the abstraction involved arises and the importance of the nuances in the definitions. It is placed outside the main text because the topic is more part of the computer science curriculum than the mathematics curriculum, and because further study leads into continuous mathematics rather than discrete mathematics. The prerequisites for this chapter are Chapters 5 and 8. An instructor might choose to place this chapter after Chapter 10 in order to continue the algorithmic theme introduced there, or choose to place it at the end of the course.

Before introducing the chapter, have students reread Section 5.2.2 as preparation for the first class. Some students may also benefit from reviewing Chapter 8, as some exercises require students to find closed forms for simple summations. You may want to start class with students working in groups on Section 17.2. It's easy for people to get things wrong in the initial problems of the Try This!, particularly because students will slightly miscount the number of increment operations that are taken, or count an if-then as only one operation instead of two (one for compare and one for the then/else). And not everyone agrees on the definition of "action"! So check in with students to make sure that everyone is in the same place after this problem—or at least close enough to the same place for the purposes of this chapter. It will help to leave 10 minutes at the end of class to give a summarizing wrap-up that sets their work in context and prefaces the reading.

Students should read Section 17.3 in preparation for the second class day, which can be devoted to the topic of computing runtime functions. After a short review of the reading, much of the time can be spent on Section 17.4. These questions are designed to transition students from counting operations individually through counting them systematically to approximating Θ. Instructors may want to encourage students in this direction. Problem 3(c) is especially likely to reveal student confusion over what runtime functions are measuring—they may be likely to say that their approximation is an overestimate of the runtime function on the grounds that they are considering worst-case inputs, when that is in fact part of an accurate runtime function. Instead, the approximation is a Ω.

In preparation for the third class day, students should read Section 17.5. The definitions of big O, Θ, and Ω are often seen as confusing, so a lecture on asymptotic behavior of runtime functions that includes attention to these definitions is recommended. Section 17.6 is intended for students to practice implementing the ideas of measuring the growth type of runtime functions and using the asymptotic notations, and it can be used in the class time remaining after the lecture.

Appendix A 🦆

Solutions to Check Yourself Problems

Please do not look at these solutions until *after* you have sincerely tried to do the problems!

Section 1.3
page 9

1. Gelly Roll pens come in 6 colors of fine point and 11 of medium point, 10 moonlight colors, 10 shadow colors, 12 stardust colors, and 14 metallic colors. (Not kidding.) How many different Gelly Roll pens are there?

   By the sum principle, $6 + 11 + 10 + 10 + 12 + 14 = 63$.

2. When redeeming a prize coupon, you may choose one of six charms and *either* one of three carabiners *or* one of two bracelets. How many different prize choices could you make?

   By the product and sum principles, $6 \cdot 3 + 6 \cdot 2 = 18 + 12 = 30$.

3. Challenge: Invent your own problem that uses both the sum principle and the product principle.

   Answers will vary.

Section 1.4
page 14

1. Prove that if n is even, then n^2 is even.

   Because n is even, we may write $n = 2k$, and $n = 2k \Rightarrow n^2 = 4k^2 = 2(2k^2)$.

2. Prove that if n is odd, then $n^2 + 5n - 3$ is also odd.

   Because n is odd, we may write $n = 2k + 1$, and $n = 2k + 1 \Rightarrow n^2 + 5n - 3 = (2k+1)^2 + 5(2k+1) - 3 = 4k^2 + 4k + 1 + 10k + 5 - 3 = 2(2k^2 + 2k + 5k + 3) - 3 = 2(2k^2 + 2k + 5k + 1) + 1 = \text{(odd)}$.

3. Challenge: Invent your own false proposition and accompany it with a counterexample.

   Answers will vary.

Section 1.5   page 19

1. List all the subsets of {*egg*, *duck*, *goose*}. How many are there? How many of them contain *egg*? ... *duck*? ... *goose*?

   There are eight subsets: ∅, {*egg*}, {*duck*}, {*goose*}, {*egg*, *goose*}, {*egg*, *duck*}, {*duck*, *goose*}, and {*egg*, *duck*, *goose*}, four of which each contain *egg*, *duck*, and *goose*, respectively.

2. Consider a standard deck of cards with suits hearts (♡), spades (♠), clubs (♣), and diamonds (◇), and values 2–10, jack, queen, king, and ace. How many cards must you deal out before being assured that two will have the same suit? How many must you deal out before being assured that two will have the same value?

   There are four suits, so we need an additional card there, and $5 = 4 + 1$. There are 13 different card values, so we need an additional card there, and $14 = 13 + 1$.

3. Challenge: Invent your own counting question that can be answered using the pigeonhole principle.

   Answers will vary.

Section 2.2   page 36

1. List the elements of $\{z \in \mathbb{Z} \mid -10 \leq z < 10\}$.

   $\{-10, -9, -8, -7, -6, -5, -4, -3, -2, -1, 0, 1, 2, 3, 4, 5, 6, 7, 8, 9\}$.

2. Write the set $\{2, 4, 6, 8, 10\}$ as a set of elements subject to a condition.

   $\{z \in 2\mathbb{Z} \mid 2 \leq z \leq 10\}$.

3. What is the cardinality of the set $\{duck, ∅, \{duck, egg\}, \{duck, \{duck, egg, ∅\}\}\}$?

   It has cardinality 4.

4. Is $\{3, 6, 13, 67\} \subset \{67, 4, 53, 5, 13, 6\}$?   No, because $3 \notin \{67, 4, 53, 5, 13, 6\}$.

5. List the elements of $\mathscr{P}(\{-1, 5, 20\})$.

   $\{∅, \{-1\}, \{5\}, \{20\}, \{-1, 5\}, \{-1, 20\}, \{5, 20\}, \{-1, 5, 20\}\}$.

6. Let $A = \{5, 6, 7, 8, 9, 23\}$, $B = \{6, 7, 9, 456, 3.142\}$, and $C = \{7, 4, 8, 2, 3, \pi, 6\}$. List the elements of...

   (a) ... $A \cup B$. $\{5, 6, 7, 8, 9, 23, 456, 3.142\}$.
   (b) ... $B \cap C$. $\{6, 7\}$.
   (c) ... $A \setminus C$. $\{5, 9, 23\}$.

7. Let $D = \{6.53, 42, 1, hat\}$ and $F = \{0, -2\}$. List the elements of...

   (a) ... $D \times F$.

   $\{(6.53, 0), (42, 0), (1, 0), (hat, 0), (6.53, -2), (42, -2), (1, -2), (hat, -2)\}$.

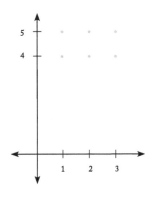

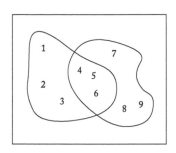

Figure A.1. **A visual representation of the set** $\{1,2,3\} \times \{4,5\}$.

Figure A.2. **A Venn diagram that represents** $\{1,2,3,4,5,6\} \cap \{4,5,6,7,8,9\}$.

(b) ... $F \times D$.

$\{(0,6.53),(-2,6.53),(0,42),(-2,42),(0,1),(-2,1),(0,hat),(-2,hat)\}$.

(c) ... $D \times D$.

$\{(6.53,6.53),(6.53,42),(6.53,1),(6.53,hat),(42,6.53),(42,42),(42,1),$
$(42,hat),(1,6.53),(1,42),(1,1),(1,hat),(hat,6.53),(hat,42),(hat,1),$
$(hat,hat)\}$.

(d) ... $\emptyset \times F$. $\emptyset$.

8. Draw a visual representation of the set $\{1,2,3\} \times \{4,5\}$.   **See Figure A.1.**

9. Make a Venn diagram that represents $\{1,2,3,4,5,6\} \cap \{4,5,6,7,8,9\}$.

   See Figure A.2.

10. Challenge:

   (a) Invent three sets of your own.

   (b) Find a different way to write each of the sets (for example, list the elements, or describe what the elements have in common using set notation).

   (c) Make a Venn diagram showing the relationships between your three sets.

   Answers will vary.

Section 2.3   page 46

1. Let P represent the statement *Ximena is pretty*, Q represent *Ximena is quizzical*, and R represent *Ximena is a rugby player*. Write $(P \vee Q) \wedge R$ as an English sentence.

   Ximena is pretty or quizzical, and Ximena is a rugby player.

2. Write *Miyuki does not like kumquats, but ze likes pickles or daikon* in logic notation.

   $\neg K \wedge (P \vee D)$.

3. Rewrite *every cat drinks beer* as an implication.

   If there is a cat, then the cat drinks beer.

4. Challenge: Come up with two examples of mathematical statements and two examples of mathematical non-statements.

   Answers will vary, but here is an example of a mathematical statement ($7 = 5 - 14$, which is false) and a mathematical non-statement (12).

5. Using truth tables, verify that the converse of a statement is not logically equivalent to the original statement. (Suggestion: make the columns P, Q, $P \Rightarrow Q$, and $Q \Rightarrow P$, and compare the last two columns.)

P	Q	$P \Rightarrow Q$	$Q \Rightarrow P$
T	T	T	T
T	F	F	T
F	T	T	F
F	F	T	T

   The last two columns are different, so the converse of a statement and the original statement are not logically equivalent.

6. Write the contrapositive of the statement *if the maple tree is orange, then the scissors are closed.*

   If the scissors are open, then the maple tree is not orange.

7. Using truth tables, verify that the statement *if I am at the combination Pizza Hut and Taco Bell, then I am at the Pizza Hut* is always true.

P	T	$P \wedge T$	$P \wedge T \Rightarrow T$
T	T	T	T
T	F	F	T
F	T	F	T
F	F	F	T

   The last column has all true values, so the statement P is always true.

8. Negate the statement *there exists an even number n such that n < 10.*

   $\neg(\exists$ even n such that $n < 10)$ becomes $\forall$ even n, $n \geq 10$.

Section 2.5 page 49

1. Prove that if n^2 is odd, then n is odd. (Suggestion: try proving the contrapositive.)

   Use the contrapositive; suppose n is even. Then $n = 2k$, and $n^2 = (2k)^2 = 2(2k^2)$, which is even.

2. Prove that if there are ten ducks paddling in four ponds, then some pond must contain at least three paddling ducks. (Suggestion: try contradiction.)

   Suppose not. Then no pond contains more than two ducks. The total number of ducks in four ponds is no more than eight.

3. Challenge: Develop your own statement that can be proved by contradiction.

   Answers will vary.

Section 3.2 page 71

1. Here are some gipos that have domain $\mathbb{N}$. For each gipo, determine whether it is a function, whether the target space is also $\mathbb{N}$, and whether it is one-to-one.

   (a) $f(n) = \frac{n}{3} + 1$.

   It is a function; note that $a = b \Rightarrow \frac{a}{3} = \frac{b}{3} \Rightarrow \frac{a}{3} + 1 = \frac{b}{3} + 1 \Rightarrow f(a) = f(b)$. The target is not $\mathbb{N}$ because $1 \mapsto \frac{4}{3}$; it is one-to-one.

   (b) $f(n) = n$. It is trivially a one-to-one function with target $\mathbb{N}$.

   (c) $f(n) = n - 1$.

   It is a one-to-one function with target $\mathbb{W}$. Note that $a = b \Leftrightarrow a - 1 = b - 1 \Leftrightarrow f(a) = f(b)$.

   (d) $f(n) = n^2 - 1$.

   It is a one-to-one function (because there are no negative inputs) with target $\mathbb{W}$. Note that $a = b \Leftrightarrow a^2 = b^2 \Leftrightarrow a^2 - 1 = b^2 - 1 \Leftrightarrow f(a) = f(b)$.

2. Here are some functions that have domain $\mathbb{Z}$ and target space $\mathbb{W}$. For each function, determine whether it is one-to-one or onto.

   (a) $f(k) = 0$.

   It is not onto, as $1 \in \mathbb{W}$ has nothing mapping to it. It is not one-to-one as $1, -6$ both map to 0. The range is 0.

   (b) $f(k) = |\lfloor \frac{k}{2} \rfloor|$. (The notation $\lfloor x \rfloor$ is known as the *floor* function, as it returns the integer equal to or just less than the input. Thus, $\lfloor \frac{k}{2} \rfloor$ returns $\frac{k}{2}$ if k is even and $\frac{k-1}{2}$ if k is odd.) (Oh, and there is a matching *ceiling* function, which returns the integer equal to or just greater than the input.)

   It is not one-to-one, as 1 and 0 both map to 0.

   Here is a proof that it is onto: given $w \in \mathbb{W}$, consider $2w \in \mathbb{Z}$. Then, $f(2w) = |\lfloor \frac{2w}{2} \rfloor| = |\lfloor w \rfloor| = |w| = w$.

(c) $f(k) = k^2 + 2$.

This function is not one-to-one as $-4, 4$ both map to 18. It is not onto as nothing maps to $0 \in \mathbb{W}$.

For those functions that are not onto, what is the range? Are any of the functions bijections?

The range of $f(k) = 0$ is 0.

The range of $f(k) = k^2 + 2$ is $\{2, 3, 6, 11, 27, \dots\} = \{w \in \mathbb{W} \mid w = k^2 + 2, k \in \mathbb{W}\}$.

None is a bijection.

3. Challenge: Write out proofs for Problems 1 and 2: that is, prove that the relevant gipos are well defined, one-to-one, and onto, and for those that are not, give counterexamples.

Proofs are given in the solutions to the previous problems.

Section 3.5   page 79

1. Find the degree sequences of the graphs in Figure 3.5.

They are $(3, 3, 3, 3, 3, 3, 3, 3)$ and $(0, 1, 1, 1, 2, 3, 4)$.

2. Look through the graphs pictured so far; identify one that is simple and one that is not simple.

All are simple except for the left-hand graph in Figure 3.3 and the right-hand graph in Figure 3.5.

3. For each graph in Figures 3.3 and 3.5, decide whether or not the graph is connected. Is any of the graphs a tree? A forest?

Connected? Yes, yes, yes, yes, no. Yes, the middle graph of Figure 3.3 is a tree and therefore also a forest.

4. Find the longest possible path in the middle graph of Figure 3.3 and in the left-hand graph of Figure 3.5.

Length 4; length 7.

5. What is the largest cycle in any graph shown in Figures 3.3 and 3.5? How about the smallest?

The longest are a cycle of length 7 in the left graph of Figure 3.3 and a cycle of length 8 in the left graph of Figure 3.5. The smallest cycle is of length 1, as there's a loop.

6. There is at least one bipartite graph pictured in Section 3.3. Identify one; is it complete?

It is the left-hand graph in Figure 3.5; no, it is not complete.

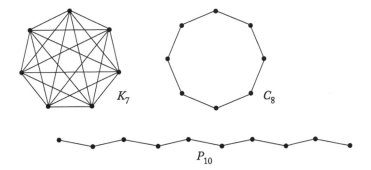

Figure A.3. Graphs K_7, C_8, and P_{10}.

7. Draw K_7, C_8, and P_{10}.   See Figure A.3.

8. Draw two 2-regular graphs on ten vertices, one of which is connected and one of which has two components.

   Multiple answers are possible; one is given in Figure A.4.

9. What is the length of a smallest cycle in the Petersen graph?   5.

10. Draw a bipartite graph with nine vertices.   See Figure A.5.

Section 3.6   page 82

1. Pick a graph from Figures 3.3 and 3.5 and draw it so that it looks different but is, in fact, the same graph.

   Answers will vary.

2. List all nonisomorphic subgraphs of C_4.

   Here they are: one vertex, two vertices, three vertices, four vertices, one edge, one edge and a vertex, one edge and two vertices, two edges, P_3, P_3 and a vertex, P_4.

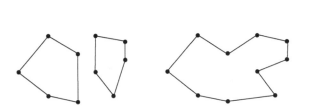

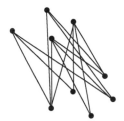

Figure A.4. Two 2-regular graphs on ten vertices, one of which is connected (right) and one of which has two components (left).

Figure A.5. A bipartite graph with nine vertices.

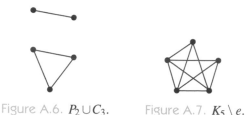

Figure A.6. $P_2 \cup C_3$.   Figure A.7. $K_5 \setminus e$.

3. Label the vertices of the graphs in Figure 3.8 and define a function between them that shows the graphs are isomorphic.

Number the vertices across the top of the left-hand graph as $1, 2, 3, 4$ and across the bottom of the left-hand graph as $5, 6, 7, 8$; similarly, letter the vertices clockwise around the right-hand graph (starting at the top teal vertex) as a, b, c, d, e, f and across the middle of the right-hand graph as g, h. Then the map $1 \mapsto f, 2 \mapsto h, 3 \mapsto d, 4 \mapsto b, 5 \mapsto e, 6 \mapsto c, 7 \mapsto g, 8 \mapsto a$ is a one-to-one and onto function. You can check edge by edge that edges of the left-hand graph correspond to those of the right-hand graph; for example, $\{1, 5\}$ is mapped to the edge $\{e, f\}$.

Section 3.7 page 86

1. Draw $P_2 \cup C_3$.   See Figure A.6.

2. What are $K_5 \setminus v$, $K_5 \setminus e$, and $\overline{K_5}$? (Note that the symmetry of K_5 means that it doesn't matter which vertex is chosen to be v or which edge is chosen to be e.)

$K_5 \setminus v$ is K_4, $K_5 \setminus e$ is shown in Figure A.7, and $\overline{K_5}$ is five vertices with no edges.

3. Choose one of the graphs pictured in this chapter (other than the one in Figure 3.19) and encode it using vertex/edge lists, as an adjacency matrix, and using vertex/adjacency lists.

Answers will vary.

Section 3.9 page 89

1. What is $R(2, 2)$?

It's 2 because starting with two vertices, there will be a K_2 in whichever color moves first.

2. Given three particular numbers k, m, N, what are the two ways you could show that $R(k, m) \neq N$?

We could exhibit a K_N with no monochromatic K_k and no monochromatic K_m; this would show $R(k, m) > N$. Or we could show that $R(k, m) < N$ by proving that for some $M < N$, either a monochromatic K_k or a monochromatic K_m is forced.

Section 4.2   page 111

1. If the statement you want to prove is made in terms of n, should your inductive step
    be done using n or using k (or some other variable)?

    It should use k; if n is used, nothing is proven because the inductive hypothesis
    restricts n to be less than or equal to k.

2. Prove by induction that the path graph P_n has $n - 1$ edges.

    For a base case, note that P_2 has one edge. Suppose that for $n \le k$, P_n has $n - 1$
    edges, and consider P_{k+1}. Remove one of the leaves to get P_k, which has $k - 1$
    edges. Replacing the leaf, we get one more edge, so P_{k+1} has k edges. That's it!

3. Write $2! + 4! + 6! + 8! + 10!$ in summation notation. (Knowing what $2!$, $4!$, $6!$, etc.
    means is not necessary for completing this problem.)

    $$2! + 4! + 6! + 8! + 10! = \sum_{j=1}^{5} (2j)!.$$

4. Write $\sum_{j=0}^{6} \dfrac{3j-1}{2}$ out in full. $\sum_{j=0}^{6} \dfrac{3j-1}{2} = \dfrac{-1}{2} + 1 + \dfrac{5}{2} + 4 + \dfrac{11}{2} + 7 + \dfrac{17}{2}.$

5. How is $\sum_{j=1}^{5} j^2 - j$ related to $\sum_{j=1}^{4} j^2 - j$? Try writing $\sum_{j=1}^{5} j^2 - j$ in terms of
    $\sum_{j=1}^{4} j^2 - j$.

    $$\sum_{j=1}^{5} j^2 - j = \sum_{j=1}^{4} j^2 - j + (5^2 - 5).$$

    More generally, how is $\sum_{j=1}^{5} q(j)$ related to $\sum_{j=1}^{4} q(j)$? $\sum_{j=1}^{5} q(j) = \sum_{j=1}^{4} q(j) + q(5).$

    And even more generally, how is $\sum_{j=1}^{k+1} q(j)$ related to $\sum_{j=1}^{k} q(j)$?

    $$\sum_{j=1}^{k+1} q(j) = \sum_{j=1}^{k} q(j) + q(k+1).$$

Section 4.4   page 114

1. Use direct proof to show that $2^n \le 2^{n+2} + 5$.

    Because $1 < 4$, we multiply both sides by 2^n and get $2^n \le 2^{n+2}$; then, $0 < 5$, so we
    can add 5 to the right-hand side of the previous inequality to obtain $2^n \le 2^{n+2} + 5$.

2. Show, by induction, that a polygon formed by n arbitrarily chosen points on a circle
    has exactly n edges.

    Three points form a triangle, which has exactly three edges. Assume that for $n \le k$,
    a polygon formed by n arbitrarily chosen points on a circle has exactly n edges.
    Consider $k + 1$ points on a circle chosen arbitrarily; cover one of them and consider

the polygon formed by the remaining k points. It has k edges by the inductive hypothesis. The remaining point forms a triangle with the two points on either side of it on the circle. One edge of that triangle coincides with a side of the k-sided polygon, and the other two coincide with uncounted sides of the $(k+1)$-sided polygon. So the number of edges of the $(k+1)$-sided polygon is $k+2-1$ or $k+3-2$ (depending on how you count the coincided edge) $= k+1$.

Section 4.5   page 116

1. Go through the inducktive step of the proof for the case $n = 5$ ducks to see how the subsets of ducks interact.

   Take five ducks and set one in the pond, leaving four ducks. By the inducktive hypothesis, these four ducks are grey. Now set one grey duck in the pond, leaving three grey ducks, and take the wetter uncolored duck and set it with those three grey ducks. Now there are four ducks, which are all grey by the inducktive hypothesis. Return the slightly wet grey duck to the yard to obtain five grey ducks.

2. Rewrite this proof for the statement *all owls are teal*, noting that whereas ducks swim about, owls fly and perch in trees.

   We just rewrite the induction step. Consider a set of $k+1$ owls. Choose one arbitrarily and perch it in a tree behind the barn. The remaining k owls must all be teal. Perch one of these k teal owls in a different tree, leaving $k-1$ teal owls, and hoot for the uncolored owl perched in a tree behind the barn. When it returns, you see it is teal because it is part of a set of $k-1+1 = k$ owls that are therefore teal (after all, $k-1$ of them are teal and all are the same color). Now call back the lone teal owl to have k teal owls. Thus, all owls are teal.

3. Do you believe that all ducks are grey? Many students claim that they have seen white ducks, but Section 4.5 proves that all ducks are grey. (A "white" duck is very pale grey.) Remember, a correct proof compels assent—so either you believe a correct proof or you believe that the given proof is problematic. Try to find an error in the proof, or justify completely that all ducks are grey.

   Do *not* try looking this up (e.g., on the internet). That would spoil your fun! Instead, think through the details of the proof. Does the base case make sense? Is the inducktive hypothesis correctly stated? How does the inducktive step hold up under scrutiny?

   See the text of Section 4.7.

Section 4.7   page 117

1. Prove that $3j^2 < 2j^3$. Be sure to use a base case of $j = 1$.

   Ha! Setting $j = 1$ creates a false statement. But $j = 2$ is a lovely base case that produces $12 < 16$. So assume that for $2 \leq j \leq k$, $3j^2 < 2j^3$. Now consider

$3(k+1)^2 = 3k^2 + 6k + 3$; by the inductive hypothesis, this is $< 2k^3 + 6k + 3$. Now, because $1 < 2k^2$, we have $3 < 6k^2 < 6k^2 + 2$ and so $3(k+1)^2 < 2k^3 + 6k + 6k^2 + 2 = 2(k+1)^3$. Thus, we have proven the revised statement that when $j \geq 2, 3j^2 < 2j^3$.

Section 5.2   page 136

1. Try performing the $3n+1$ algorithm given in Example 5.2.12 for $n = 3, n = 4, n = 7$, $n = 8$, and $n = 13$. How many iterations are required for each of these numbers? Do any of the sequences generated appear within any of the others (and if so, which)?

   3: $10, 5, 16, 8, 4, 2, 1$ (7 iterations).
   4: $2, 1$ (2 iterations).
   7: $22, 11, 34, 17, 52, 26, 13, 40, 20, 10, 5, 16, 8, 4, 2, 1$ (16 iterations).
   8: $4, 2, 1$ (3 iterations).
   13: $40, 20, 10, 5, 16, 8, 4, 2, 1$ (9 iterations).
   The sequence for 4 appears within the sequence for 8, which appears within the sequence for 3, which appears within the sequence for 13, which appears within the sequence for 7.

2. Translate the instruction *replace t with t/2 while t is even* into plain English.

   Start with t and keep dividing by two until you obtain an odd number.

3. What does this list of instructions do? Comment on whether it forms an algorithm, and if so, whether it terminates and/or is correct.

   1. Let $n = 2$.
   2. Replace n with $n + 4$.
   3. If n is even, go to step 2; otherwise, continue.
   4. Output n.

   The instructions are an algorithm that does not terminate; it counts indefinitely by 4s, starting with 2.

Section 5.3   page 142

1. True or false:

   (a) $2 \equiv 10 \pmod{12}$.
       False. The remainder when 2 is divided by 12 is 2; the remainder when 10 is divided by 12 is 10.
   (b) $2 \equiv -10 \pmod{12}$.   True. $-10 + 12 = 2$.
   (c) $22 \equiv 10 \pmod{12}$.   True. $10 + 12 = 22$.
   (d) $-2 \equiv 10 \pmod{12}$.   True. $-2 + 12 = 10$.

2. What is the set $[2]$ if we are working modulo 3?   It is $\{\ldots, -4, -1, 2, 5, 8, 11, \ldots\}$.

3. Show that $=$ is an equivalence relation.

   We check: $a = a$ so we have reflexivity; if $a = b$ then $b = a$ as well, so we have symmetry; and $a = b$ and $b = c$ implies $a = c$, so we have transitivity.

4. Is $\{1,2\}, \{2,3\}, \{3,4\}, \{4,5,6\}$ a partition of $\{1,2,3,4,5,6\}$?

   No, because $\{1,2\} \cap \{2,3\} = \{2\} \neq \emptyset$.

5. Create a partition of $\{1,4,2,7,9,14,89,246\}$.

   Answers will vary, but one is $\{1, 246\}, \{4, 2, 7, 9\}, \{14, 89\}$.

6. **Challenge**: We know that $=$ has the property that if $a = b$, then $ac = bc$; Theorem 5.3.3 says that this property also holds for $\equiv$ (mod n). Think of another property that holds for $=$ in ordinary arithmetic, and test to see whether that property holds for $\equiv$ (mod n).

   Answers will vary; some properties are represented in the exercises; others may be quite unprovable at this level.

Section 5.4   page 148

1. Encrypt the message *lemon drops* using a Caesar cipher.   *mfnpo espqt.*

2. Decrypt the message *pvaanzba ohaf* using ROT13.   *cinnamon buns.*

3. Encrypt the message *quilt blocks* using a shift cipher with shift 7.   *xbpsa isvjrz.*

4. Decrypt the message *bdpja lxxtrnb*, which was encrypted using a shift of 9.

   sugar cookies.

5. Encrypt the message *lions tigers and bears oh my* using a Vigenère cipher and key word *zoo*.

   kwcmg hhusqg omr pdofr cv lm.

6. Decrypt the message *wwrfw aiw wowl*, which was encrypted with a standard Vigenère cipher using key word *ears*.

   swans are soft.

Section 6.2   page 178

1. Write the solutions to the questions that begin this section in choice notation.

   $\binom{5}{1} + 2\binom{5}{2} + \binom{5}{3} = 5 + 20 + 10 = 35$, then just $\binom{5}{3} = 10$, then $\binom{4}{1} + 2\binom{4}{2} + \binom{4}{3} = 4 + 12 + 4 = 20$ and $\binom{4}{3} = 4$.

2. Compute $\binom{4}{2}$ using the basic choice notation identity (and a little bit of exhaustive listing).

   $\binom{4}{2} = \binom{3}{1} + \binom{3}{2}$; $\binom{3}{1} = 3$ because the possibilities are $1, 2, 3$, and $\binom{3}{2} = 3$ because the possibilities are $(1,2), (1,3), (2,3)$.

3. Compute $\binom{5}{2}$ using the basic choice notation identity and the previous problem.

$\binom{5}{2} = \binom{4}{1} + \binom{4}{2}$; $\binom{4}{1} = 4$ because the possibilities are $1, 2, 3, 4$, and $\binom{4}{2} = 6$ by the previous problem.

Section 6.4   page 181

1. Quickly compute $\binom{8}{3}$.   By Pascal's triangle, 56.

2. Quickly compute $\binom{7}{5}$.   By Pascal's triangle, 21.

3. How many ways are there to choose two ducks out of a raft of nine ducks?

There are $\binom{9}{2}$ or 36 ways.

4. How many ways are there for two of the author's three cats to sit on her bed?

There are $\binom{3}{2}$ or three ways.

5. Invent a question to which the answer is $\binom{13}{4}$.

Answers will vary. One could be, "How many ways are there to choose 4 pillows from a pile of 13 pillows?"

Section 6.5   page 186

1. You receive a shipment of 36 legs for stools to go with the stock of mass-manu-factured stool seats you already have. How many stools can you complete?

Twelve three-legged stools, or nine four-legged stools.

2. Suppose we wanted to place all nine different numbers onto a Sudoku board without reusing rows or columns—how many ways would there be to do it?

There are $(9!)^2 = 131{,}681{,}894{,}400$ ways.

3. On the other hand, what if we wanted to place nine 4s onto a Sudoku board without reusing rows or columns? (Again, we will ignore the fact that on a Sudoku board, a player also cannot have two of the same number appear in the same 3×3 block.) How many different ways would there be to make that placement?

There are $\frac{(9!)^2}{9!} = 9! = 362{,}880$ different ways.

4. How many orderings *are* there of $a, b, c, d, e, f, g, h, i, j, k, l, m, n, o, p$?

There are 16 letters, so $16! = 20{,}922{,}789{,}888{,}000$.

Section 6.7
page 189

1. Find the coefficient of $x^5 y^5$ in $(x+y)^{10}$. $\binom{10}{5} = 252$.

2. Find the coefficient of the monomial containing c^3 in $(5b^2 - 4c)^4$.

   $\binom{4}{3} \cdot (-4)^3 \cdot 5 = -4^4 \cdot 5 = -256 \cdot 5 = -1{,}280$.

3. Compute, by hand, a numerical value for $\binom{36}{32}$.

   $$\frac{36!}{32!(36-32)!} = \frac{36 \cdot 35 \cdot 34 \cdot 33}{4!} = 3 \cdot 35 \cdot 17 \cdot 33 = 58{,}905.$$

4. Challenge: Create a binomial $(x+y)^n$ (with n greater than one) with neither x nor y a constant, such that when expanded it will have a constant term.

   Answers will vary; $(x + \frac{1}{x})^2$ is one possibility.

Section 6.8
page 191

1. Let $m \le n$. What might $\binom{n}{3}\binom{n}{m-3}$ be counting?

   Answers will vary, but here's one: the number of ways to choose three items from a pile of n items and then choose $m - 3$ items from a different pile of n items.

2. What might $\binom{n}{6}\binom{6}{k}$ be counting?

   Answers will vary, but here's one: the number of ways to first choose six items from n items and then pick k of those six items.

3. What might $\binom{n}{k}2^k$ be counting?

   Answers will vary, but here's one: the number of ways to first choose k items from n items and then color each of those items teal or purple.

4. Challenge: Create a situation similar to the lemon-and-chrysanthemum-drops situation, and use this to write down a new binomial identity.

   Answers will vary. One possibility is to have n lemon drops and $2n$ chrysanthemum drops and to grab m drops total:

   $$\sum_{k=0}^{n} \binom{2n}{k}\binom{n}{m-k} = \binom{3n}{m} \text{ for } m \le n.$$

Section 7.2
page 213

1. How many ways are there to give four snacks to six puppies, with no more than one snack going to each puppy?

   How many ways are there to place four unlabeled balls into six labeled boxes, with at most one ball per box?

2. How many ways are there to give four snacks to six puppies, with gluttony and cruelty allowed?

How many ways are there to place four unlabeled balls into six labeled boxes?

3. How many ways are there to deal five ace cards and four queen cards to nine card players?

This is like asking how many ways there are to line up five aces and four queens, so it is the same as asking how many ways there are to distribute four unlabeled balls into six labeled boxes, or five unlabeled balls into five boxes.

4. How many ways are there to feed 12 spinach stems (of different lengths) to four ducks such that the grey duck gets five spinach stems, the white duck gets five spinach stems, and the pale-grey and black ducks get one spinach stem each?

How many ways are there to put 12 labeled balls into 4 labeled boxes, with the boxes getting 5, 5, 1, and 1 balls, respectively?

5. How many ways are there to arrange 5 monster figurines and 12 angel figurines in a line on a shelf?

This is either …

How many ways are there to distribute 5 unlabeled balls into 13 labeled boxes? or

How many ways are there to distribute 12 unlabeled balls into 6 labeled boxes?

6. How many ways are there to distribute three chocolates (one white, one milk, and one dark) to four classmates, at most one chocolate per classmate?

The chocolates are distinct, so this is the same as asking how many ways there are to place three labeled balls in four labeled boxes, at most one ball per box.

7. How many ways are there to put a star sticker (they come in gold, silver, red, green, and blue) on every student's paper in your class?

This will depend on how many students are in the class (let's say n). The students are the boxes, and each could get any of the five different star stickers, so this is the same as asking how many ways there are to place one ball with five possible labels into each of n boxes.

8. How many ways are there to give two different catnip toys to five cats, such that no cat gets more than one toy?

The catnip toys are distinct, so this is the same as asking how many ways there are to place two labeled balls in five labeled boxes, at most one ball per box.

Section 7.4

page 225

1. How many ways are there to give four snacks to six puppies, with no more than one snack going to each puppy?

Choose which four of the six puppies will get snacks: $\binom{6}{4}$ ways.

2. How many ways are there to give four snacks to six puppies, with gluttony and cruelty allowed?

   Oh, no, gluttony and cruelty are allowed! Aside from being unhealthy and unfair, this means we have four star snacks and five bars that divide the snacks into puppy piles: there are $\binom{4+5}{4} = \binom{4+5}{5}$ ways to do it.

3. How many ways are there to deal five aces and four queens to nine card players?

   We just need to know which players get aces (or which players get queens), so $\binom{9}{5} = \binom{9}{4}$.

4. How many ways are there to feed 12 spinach stems (of different lengths) to four ducks such that the grey duck gets five spinach stems, the white duck gets five spinach stems, and the pale-grey and black ducks get one spinach stem each?

   We can do $\binom{12}{5}\binom{12-5}{5}\binom{7-5}{1} = 33{,}264$.

5. How many ways are there to arrange 5 monster figurines and 12 angel figurines in a line on a shelf?

   This is $\binom{5+12}{5} = 6{,}188$.

6. How many ways are there to distribute three chocolates (one white, one milk, and one dark) to four classmates, at most one chocolate per classmate?

   The chocolates are distinct, so this is the same as asking how many ways there are to place three labeled balls in four labeled boxes, at most one ball per box. By the formula, that gives us $\frac{4!}{1!} = 24$.

   We could also say, an empty box is a fourth (invisible) kind of chocolate, so one ball per box and 4!. Or, we choose the three boxes and then take all possibilities for which balls go in which of those three boxes, so $\binom{4}{3}3! = 24$.

7. How many ways are there to put a star sticker (they come in gold, silver, red, green, and blue) on every student's paper in your class?

   This will depend on how many students are in the class (let's say n). The students are the boxes, and each could get any of the five different star stickers, so this is the same as asking how many ways there are to place one ball with five possible labels into each of n boxes. That's 5^n.

8. How many ways are there to give two different catnip toys to five cats, such that no cat gets more than one toy?

   The catnip toys are distinct, so this is the same as asking how many ways there are to place two labeled balls in five labeled boxes, at most one ball per box. So $\frac{5!}{(5-2)!} = 20$.

Section 7.5   page 229

1. Write out Theorem 7.5.4 for four sets A_1, A_2, A_3, A_4.

   $|A_1 \cup A_2 \cup A_3 \cup A_4| = |A_1| + |A_2| + |A_3| + |A_4| - |A_1 \cap A_2| - |A_1 \cap A_3| - |A_1 \cap A_4| - |A_2 \cap A_3| - |A_2 \cap A_4| - |A_3 \cap A_4| + |A_1 \cap A_2 \cap A_3| + |A_1 \cap A_2 \cap A_4| + |A_1 \cap A_3 \cap A_4| + |A_2 \cap A_3 \cap A_4| - |A_1 \cap A_2 \cap A_3 \cap A_4|$. Whew!

2. How many permutations of the numbers $0, 1, 2, 3, 4$ have substrings 03 or 21?

   Permutations containing 03: 4!. Permutations containing 21: 4!. Overlap: size 6, because we can consider arrangements of the three "digits" 03, 21, 4. So, $2 \cdot 4! - 3! = 42$ permutations of $0, 1, 2, 3, 4$ have substrings 03 or 21.

3. How many permutations of the numbers $5, 6, 7, 8, 9$ have the substrings 59 or 85?

   Permutations containing 59: 4!. Permutations containing 85: 4!. Overlap: permutations containing 859, of which there are 3!. So, $2 \cdot 4! - 3! = 42$ permutations of $5, 6, 7, 8, 9$ have the substrings 59 or 85.

4. Challenge: Write and solve your own three-set PIE problem (but not involving pies).

   Answers will vary.

Section 8.2   page 248

1. List the third, sixth, and seventh Fibonacci numbers.   2, 8, 13.

2. Write out the ten Fibonacci numbers after 55.

   89, 144, 233, 377, 610, 987, 1597, 2584, 4181, 6765.

3. Find a formula for the sum of the first n odd-index Fibonacci numbers (F_1, F_3, etc.).

   F_{2n}.

4. Find a formula for the sum of the first n even-index Fibonacci numbers (F_2, F_4, etc.).

   $F_{2n+1} - 1$.

Section 8.3   page 252

1. Find a closed form for each of these sequences:

   (a) $1, 2, 3, 4, 5, \ldots$ $a_n = n$.

   (b) $1, 2, 4, 8, 16, \ldots$ $a_n = 2^n$.

   (c) $1, 2, 6, 24, 120, \ldots$ $a_n = n!$.

   (d) $1, 4, 9, 16, 25, \ldots$ $a_n = n^2$.

   (e) $1, -4, 16, -64, \ldots$ $a_n = (-4)^{n-1}$.

2. Write out the first five terms of each of the sequences defined by these closed-form formulas:

(a) $a_n = n^2 - 3n + 4$. $2, 2, 4, 8, 14$.

(b) $a_n = n - n^2 + 5$. $5, 3, -1, -7, -15$.

(c) $a_n = 3F_n - 1$. $2, 2, 5, 8, 14$.

(d) $a_n = 3\binom{n}{2}$. $0, 3, 9, 18, 30$.

(e) $a_n = 2^n - 2$. $0, 2, 6, 14, 30$.

(f) $a_n = \binom{n+2}{3}$. $1, 4, 10, 20, 35$.

(g) $a_n = (n-2)^2$. $1, 0, 1, 4, 9$.

3. Give an example of a sequence that is not an integer sequence.

Answers will vary, but here is one example: $0.5, 0.75, 1, 1.25, 1.5, \ldots$.

4. Find a sequence different from $a_n = 2^n$ that satisfies the recurrence $a_n = 2a_{n-1}$.

Answers will vary, but one example is $3, 6, 12, 24, \ldots$ defined by $a_n = 3 \cdot 2^n$.

5. Find the recurrence that defines the sequence $1, 3, 6, 10, 15, \ldots$. $a_n = a_{n-1} + n$.

6. Find a description for the sequence $2, 3, 5, 7, 11, 13, \ldots$. Can you find a closed form or a recurrence that defines this sequence?

a_n is the nth prime. There is neither a closed form nor a recurrence that defines this sequence.

Section 8.5   page 255

Figure out closed forms for some of these recurrences and check (briefly) by induction that they're correct.

1. $a_0 = 1. a_n = 3a_{n-1}$.

Sequence is $1, 3, 9, 27, 81, \ldots$. Closed form is $a_n = 3^n$.

We will proceed, abbreviatedly, by induction. The base case is covered by prior calculations. Consider a_{k+1}. By the recurrence, this equals $3a_{k+1-1} = 3a_k$. This falls under the inductive hypothesis, so $= 3 \cdot 3^k = 3^{k+1}$ as desired.

2. $a_1 = 1. a_n = -2a_{n-1}$.

Sequence is $1, -2, 4, -8, 16, -32, \ldots$. Closed form is $a_n = (-2)^{n-1}$.

We will proceed, abbreviatedly, by induction. The base case is covered by prior calculations. Consider a_{k+1}. By the recurrence, this equals $-2a_{k+1-1} = -2a_k$. This falls under the inductive hypothesis, so $= (-2) \cdot (-2)^{k-1} = (-2)^k$ as desired.

3. $a_0 = 1, a_n = 2a_{n-1} - 1$.

   Sequence is $1, 1, 1, 1, \ldots$. Closed form is $a_n = 1$.

   We will proceed, abbreviatedly, by induction. The base case is covered by prior calculations. Consider a_{k+1}. By the recurrence, this equals $2a_{k+1-1} - 1 = 2a_k - 1$. This falls under the inductive hypothesis, so $= 2 \cdot 1 - 1 = 1$ as desired.

4. $a_1 = 0, a_2 = 1, a_n = a_{n-1}a_{n-2}$.

   Sequence is $0, 1, 0, 0, 0, 0, \ldots$. Closed form is $a_n = 0$ for $n > 2$.

   We will proceed, abbreviatedly, by induction. The base case is covered by prior calculations. Consider a_{k+1}. By the recurrence, this equals $a_{k+1-1}a_{k+1-2} = a_k a_{k-1}$. This falls under the inductive hypothesis, so $= 0 \cdot 0 = 0$ as desired.

5. $a_1 = a_2 = 1, a_n = a_{n-1}a_{n-2}$.

   Sequence is $1, 1, 1, 1, 1, \ldots$. Closed form is $a_n = 1$.

   We will proceed, abbreviatedly, by induction. The base case is covered by prior calculations. Consider a_{k+1}. By the recurrence, this equals $a_{k+1-1}a_{k+1-2} = a_k a_{k-1}$. This falls under the inductive hypothesis, so $= 1 \cdot 1 = 1$ as desired.

Section 8.6   page 258

1. Can a closed form for the recurrence $a_0 = 2, a_n = 3a_{n-1} - 7$ be found using the techniques of this section? Why or why not?

   No, because the coefficient of a_{n-1} is not 1.

2. Can a closed form for the recurrence $a_0 = 22, a_n = a_{n-1} + 9n^9 - 12n^7 + n^6 - 43n - 7$ be found using the techniques of this section? Why or why not?

   Yes, because the recurrence has the form $a_n = a_{n-1} + p_9(n)$.

3. Can a closed form for the recurrence $a_0 = 2, a_n = a_{n-1} + 3^{n-7}$ be found using the techniques of this section? Why or why not?

   No, because the added function is exponential, not polynomial.

4. Can a closed form for the recurrence $a_0 = 2, a_1 = 22, a_n = a_{n-1} + a_{n-2} - 22$ be found using the techniques of this section? Why or why not?

   No, because there are lower recursive terms in addition to a_{n-1}.

5. Try to use kth differences to find a closed form for the recurrence $a_0 = 1, a_n = 2a_{n-1}$. What happens?

   All first, second, ..., kth differences produce the original sequence.

6. Try to use kth differences to find a closed form for the recurrence $a_0 = 0, a_n = a_{n-1} + (n+1)(-1)^{n+1}$. What happens?

   Sequence is $0, 2, -1, 3, -2, 4, -3, \ldots$. First differences are $2, -3, 4, -5, 6, -7, \ldots$.

   Second differences are $-5, 7, -9, 11, -13, \ldots$.

   Third differences are $12, -16, 20, -24, \ldots$. They get larger and larger!

7. Find a closed form for $a_0 = 8, a_n = a_{n-1} - 4$ and check that your formula is correct.

   Sequence is $8, 4, 0, -4, -8, \ldots$. First differences are $-4, -4, -4, -4, \ldots$.

   Therefore, the closed form is $a_n = 8 - 4n$.

   Abbreviated inductive proof: The first five terms of the sequence form base cases. We consider $a_{k+1} = a_k - 4 = 8 - 4k - 4 = 8 - 4(k+1)$ as desired.

8. Find a closed form for $a_0 = 3, a_n = a_{n-1} + 2$ and check that your formula is correct.

   Sequence is $3, 5, 7, 9, 11, \ldots$. First differences are $2, 2, 2, 2, \ldots$.

   Therefore, the closed form is $a_n = 3 + 2n$.

   Abbreviated inductive proof: The first five terms of the sequence form base cases. We consider $a_{k+1} = a_k + 2 = 3 + 2k + 2 = 3 + 2(k+1)$ as desired.

9. **Challenge:** Create your own recurrence for which a closed form can be found using kth differences, and find that closed form.

   Answers will vary.

Section 8.8   page 265

1. Which parts of the definition of a linear homogeneous recurrence relation with constant coefficients do each of these recurrences violate?

   (a) $a_n = na_{n-3} + 6$. **This has a nonconstant coefficient and is not homogeneous.**

   (b) $a_n = a_{n-1}a_{n-43} + 23n$.   **This is not linear and is not homogeneous.**

   (c) $a_n = 5n^2 a_{n-3} - (-1)^n a_{n-6}$.   **This has variable coefficients.**

2. Find the characteristic equation for...

   (a) ... $a_n = 4a_{n-1}$. $x = 4$.

   (b) ... $a_n = a_{n-1} - 3a_{n-2}$. $x^2 = x - 3$.

   (c) ... $a_n = 2a_{n-2}$. $x^2 = 2$.

3. Determine a_0 for each of these recurrences.

   (a) $a_1 = 3, a_n = 4a_{n-1}$. $3 = a_1 = 4a_0$, so $a_0 = \frac{3}{4}$.

   (b) $a_1 = a_2 = 1, a_n = a_{n-1} - 3a_{n-2}$. $a_2 = a_1 - 3a_0$ or $1 = 1 - 3a_0$, so $a_0 = 0$.

   (c) $a_1 = 1, a_2 = 2, a_n = 2a_{n-2}$. $2 = a_2 = 2a_0$, so $a_0 = 1$.

4. Determine the characteristic equation for $a_n = 2a_{n-1}$. $x = 2$.

   What are its roots?   It has one root, $r_1 = 2$.

   Using this information and the initial condition $a_1 = 42$, determine a closed-form formula.

   $a_n = q_1 2^n$. So $42 = q_1 \cdot 2$ or $q_1 = 21$, so $a_n = 21(2^n)$.

5. For each of the following recurrences, decide whether one could use (i) kth differences, (ii) the characteristic equation, or (iii) neither in order to find a closed form for the recurrence.

(a) $a_n = 2a_{n-1} + 2$.

Neither, because this isn't homogeneous and doesn't have a coefficient of 1 on the a_{n-1} term.

(b) $a_n = a_{n-1} + 2a_{n-2} + 3a_{n-3} + 4a_{n-4}$.   The characteristic equation.

(c) $a_n = 3a_{n-3} + 3n$.

Neither, because this isn't homogeneous and doesn't fit the kth differences form.

(d) $a_n = a_{n-1} - 2n^2$. kth differences.

Section 9.3   page 282

1. Prove by induction that $p_n = p_{n-1} + n, p_0 = 1, p_1 = 2$ has closed form $\frac{n^2}{2} + \frac{n}{2} + 1$.

Base cases have been done empirically. Consider $p_{n+1} = p_n + n + 1$ and use the inductive hypothesis to see this $= \frac{n^2}{2} + \frac{n}{2} + 1 + n + 1 = \frac{n^2 + n + 2 + 2n + 2}{2} = \frac{n^2 + 3n + 4}{2} = \frac{n^2 + 2n + 1 + n + 1 + 2}{2} = \frac{(n+1)^2}{2} + \frac{n+1}{2} + 1$ as desired.

2. How many regions of the plane result from four lines passing through a point? What if n lines pass through the point?

Eight, and $2n$.

3. How many regions of the plane result from three lines passing through a point and a fourth line intersecting the others pairwise?

Six from the first three lines plus four from the additional pairwise intersections, for a total of ten.

4. Draw two parallel lines and then two additional lines that intersect one of the parallel lines at a point. Make two nudges to place these four lines in general position.

See Figure A.8.

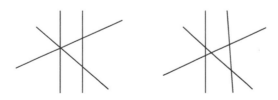

Figure A.8. Nudging lines to place them in general position.

1. Show that $\binom{n}{2} + n + 1 = \frac{n^2}{2} + \frac{n}{2} + 1 = \binom{n+1}{2} + 1$.

   $\binom{n}{2} = \frac{n!}{2!(n-2)!} = \frac{n(n-1)}{2}$. Thus, $\binom{n}{2} + n + 1 = \frac{n(n-1)}{2} + \frac{2n}{2} + \frac{2}{2} = \frac{n^2-n+2n+2}{2} = \frac{n^2}{2} + \frac{n}{2} + 1 = \frac{n(n+1)}{2} + 1 = \binom{n+1}{2} + 1$.

2. For what sequence is y_n the sequence of first differences? List the first five terms of that sequence.

   y_n is also $h_{n,3}$. It should be the sequence of first differences of $h_{n,4}$. And the first five terms should be $1, 2, 4, 8, 16, 31$.

3. What are the first seven terms of $h_{n,6}$?

   $1, 2, 4, 8, 16, 32, 64$ because six mutually perpendicular cuts can be inserted into a six-dimensional hyperbeet.

1. Draw all trees on five vertices.   See Figure A.9.

2. What must be true about the degree sequence of a tree?

   It must begin with at least two 1s.

3. Suppose a graph G has 364 vertices and 365 edges. Can it be a tree?

   No. It has two more edges than a tree should have.

4. Suppose a graph G has 28 vertices and 27 edges. Must it be a tree?

   Not necessarily. It might be nonconnected and have cycles.

5. Suppose a connected graph G with 432,894,789 vertices has no cycles. How many edges might it have?

   It must have 432,894,788 edges because it is a tree.

6. Draw a tree that has exactly two leaves.   Consider a path graph.

7. Draw a tree that has exactly three leaves.   See Figure A.10.

Figure A.9. All trees on five vertices.

Figure A.10. **A tree that has exactly three leaves.**

Section 10.4 page 324

1. An unweighted graph could be considered an edge-weighted graph with all edges of the same weight (perhaps 1). What happens if you run Prim's algorithm on it?

   You get the exact same result as the start-small algorithm on unweighted graphs.

2. What happens if you run one of the spanning tree algorithms for unweighted graphs on an edge-weighted graph?

   You get a spanning tree, but generally not a minimum-weight one.

3. If an edge-weighted graph has several edges of the same weight, there will be more than one way to order the edges while still having them in increasing order of weight. What difference do these orderings make to Kruskal's algorithm?

   You just get a different minimum-weight spanning tree from a different ordering. Sometimes. (Sometimes you get the same minimum-weight spanning tree.)

4. Prim's algorithm does not specify an edge of G with which to start. What would happen if you ran Prim's algorithm twice, but starting with different edges?

   You'd probably get different minimum-weight spanning trees.

5. In Example 10.4.3, would the same spanning tree have resulted if the labels were switched on edges e_4 and e_6?

   Nope. We would have picked e_5 instead of the edge to its left.

6. In Example 10.4.4, at what stage could one have made a choice of edge that would have resulted in a different spanning graph?

   Any time after the first three edges were chosen, one could have picked the leftmost weight-2 edge.

Section 10.5 page 330

1. Suppose that a binary decision tree for set membership is labeled consistently (i.e., "left" indicates an element is in the set and "right" indicates an element is not in the set). What subset will be assigned to the leftmost leaf? ... the rightmost leaf?

   The entire set; the empty set.

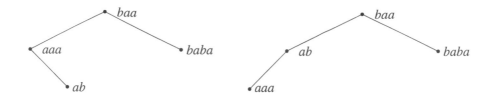

Figure A.11. Two binary search trees for the micro-dictionary $aaa, ab, baa, baba$.

2. Consider a language with only two letters (a and b), and a binary decision tree that encodes dictionary ordering for short words (no more than five letters long) in this language. What is the practical meaning indicated by the tree being incomplete?

   Not all sequences of letters are words in the language.

3. In Example 10.5.3, what principle allows us to conclude that two nodes must represent the same person?

   The pigeonhole principle!

4. Placing baa at the root, draw a binary search tree for the micro-dictionary $aaa, ab, baa, baba$.

   Figure A.11 shows two possibilities.

Section 10.7   page 333

1. What is the relationship between the number of edges in a matching and the number of vertices in that matching?

   There are twice as many vertices as edges.

2. Is it possible for a graph with an odd number of vertices to have a matching? **Yes.**

3. Is it possible for a graph with an odd number of vertices to have a perfect matching?

   No. Not all vertices can be included.

Section 10.8   page 335

1. We claimed that the column-position list $(3, 2, 5, 6, 2, 7, 1, 8)$ had two coins on the same diagonal. Which two and why?

   There are coins in column 1, row 3 and in column 2, row 2; the slope is 1 and so they are on the same diagonal. There are also coins in column 3, row 5 and in column 4, row 6; the slope is -1 and so they also are on the same diagonal.

2. Why is this approach called backtracking?

Because we trace a path through the tree of possible solutions, and when we find an invalid configuration, we backtrack along that path until we have a different choice to make.

Section 11.3 page 350

1. Go to http://planarity.net; enjoy. Answers will vary. ⌣

Section 11.5 page 354

1. Verify Euler's formula for K_4. (Be sure to draw K_4 without edges crossing.)
   $4 - 6 + 4 = 2$.

2. Draw K_3. Count the number of vertices, edges, and faces. How many edges must you remove to obtain a spanning tree? Do so. Count the number of vertices, edges, and faces of the spanning tree. Verify Euler's formula for K_3 and for the spanning tree you obtained.

   See Figure A.12 and note that we have three vertices, three edges and two faces; then, three vertices, two edges, and one face; $3 - 3 + 2 = 2$ and $3 - 2 + 1 = 2$.

3. Verify Euler's formula for W_6, the wheel with five spokes. $6 - 10 + 6 = 2$.

4. Explain why every planar drawing of a graph has the same number of faces.

   The number of vertices is constant across drawings, as is the number of edges. And, $|F(G)| = 2 - |V(G)| + |E(G)|$, so it is constant.

Section 11.6 page 357

1. Why is the constraint $|V(G)| \geq 3$ necessary in Theorem 11.6.1?

   If there are only two vertices, then $3|V(G)| - 6 = 0$, but we do have a planar graph with one edge and two vertices. Similarly, the null graph has more than -3 edges.

2. Draw a nonsimple graph that violates Theorem 11.6.6.

   Answers will vary, but see Figure A.13 for one example.

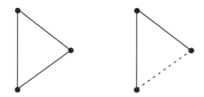

Figure A.12. K_3 and a spanning tree of K_3.

Figure A.13. A nonsimple graph that violates Theorem 11.6.6.

Figure A.14. A graph with a non-facial 3-cycle highlighted.

3. Theorem 11.6.3 requires that G have no 3-cycles. This requirement could be replaced with the constraint that G be drawn with no faces of size 3. Why is this a weaker constraint?

All faces of size 3 are 3-cycles, but there can be 3-cycles that are not faces. See Figure A.14 for an example.

Section 12.3   page 377

1. Does the graph in Figure 12.3 have an Euler traversal? ... an Euler circuit?

Yes; no. It has exactly two odd vertices (of degree 3).

2. Does K_5 have an Euler traversal? ... an Euler circuit?

Yes; yes. It has five vertices, each of degree 4.

3. Does K_6 have an Euler traversal? ... an Euler circuit?

No; no. It has six vertices, each of degree 5.

4. Which cycle graphs have Euler traversals? ... Euler circuits?

All of them have Euler circuits!

5. Challenge: Draw three graphs with the same number of vertices, that differ from each other by at most two edges, and where one graph has an Euler circuit, one has only an Euler traversal, and one has no Euler traversal.

Answers will vary, but one way to construct this triple is to make a graph with all even degrees, then delete an edge to get a second graph, and delete another edge to get a third graph.

Section 12.4   page 381

1. In the third subdiagram of Figure 12.5, why are there two labels $(4, d)$?

There are two edges of length 2 emanating from d, and d is tagged with weight-from-s 2 as well.

2. In the first list of distances computed by Dijkstra's algorithm, how many tags are present?

   As many as there are edges emanating from s—in other words, the degree of s.

3. In the third subdiagram of Figure 12.5, there are two labels $(5, f)$. Neither is placed on the graph in the following step... so why is only one of them left in the list in the fourth subdiagram?

   One of the $(5, f)$s pointed to b, which got tagged with $(3, d)$ and was no longer eligible to receive the $(5, f)$ tag.

Section 13.3   page 415

1. What is $\chi(K_n)$?   It is n, as every vertex is adjacent to every other.

2. Find a graph with $\chi(G) = 1$.   A single vertex—with no edges!

3. Find the chromatic number and index of a path of length 5.

   The chromatic number (index) is 2; color the first vertex (edge) red, the second vertex (edge) blue, and continue along the path alternating colors.

4. Find the chromatic number and index of a cycle of length 423.

   The chromatic number (index) is 3 by Example 13.3.4.

5. Challenge: Create a graph G for which $\chi(G) > \chi'(G)$. K_4 will do it.

Section 13.5   page 422

1. What is $\chi'(K_{578349})$?   578,349.

2. What is the shortest length a cycle can be in a bipartite graph?

   4. (A 2-cycle is a double edge, which is not generally permitted in a bipartite graph.)

3. What is $\chi'(K_{42,87})$?   87.

4. What proof method(s) is/are used in the proof of Theorem 13.5.2?

   Direct proof, twice (once for each conditional).

5. Let G be planar. When is the upper bound on $\chi(G)$ better from planarity than from $\Delta(G) + 1$?

   When $\Delta(G) \geq 6$.

6. Use Theorem 13.5.4 to determine $\chi'(K_{m,n})$.   It gives m or n, whichever is larger.

7. Challenge: Create a graph (other than the one in Example 13.5.6) on which a greedy algorithm produces a truly awful vertex coloring.

   Answers will vary, but one such graph is $K_{n,n}$ with all the vertical edges (of a standard drawing) removed and with the vertices addressed in a zig-zag order.

Section 14.2   page 448

1. What is the probability of rolling a 12 using a fair 20-sided die? $\frac{1}{20}$.

2. List the elements of the state space for flipping three fair coins at once.

   $\{head_1/head_2/head_3, head_1/tail_2/head_3, tail_1/head_2/head_3, tail_1/tail_2/head_3, head_1/head_2/tail_3, head_1/tail_2/tail_3, tail_1/head_2/tail_3, tail_1/tail_2/tail_3\}$.

3. List the elements of the state space for rolling an eight-sided die.

   $S_{D_8} = \{roll\ 1, roll\ 2, roll\ 3, roll\ 4, roll\ 5, roll\ 6, roll\ 7, roll\ 8\}$.

4. Determine the probability for each state in Example 14.2.5. What is the probability of getting one head and one tail? To what subset of the state space does this event correspond?

   Each state has probability $\frac{1}{4}$. Because any two states are exclusive, $P(head_1/tail_2$ or $tail_1/head_2) = P(head_1/tail_2) + P(tail_1/head_2) = \frac{1}{2}$. The subset is $\{head_1/tail_2, tail_1/head_2\}$.

5. Challenge: Invent your own situation and list the elements of the corresponding state space.

   Answers will vary.

Section 14.3   page 454

1. Compute the expected value for each of the random variables defined in Example 14.3.4, assuming that you always see all four ducks when you go visiting. Compute the probability distribution, use the definition of expected value, and then use Lemma 14.3.10.

   Because we always see all four ducks, $P(W=4)=1$ and $P(W\neq4)=0$. The probability distribution for W is (values, probabilities) $= \{(0,0), (1,0), (2,0), (3,0), (4,1)\}$. Then $\mathbb{E}[W] = 0\cdot0 + 1\cdot0 + 2\cdot0 + 3\cdot0 + 4\cdot1 = 4$ white ducks.

   Using the lemma, we note that for a subset A of the ducks, $W(A) = |A|$, and we have only one state: we see all four of the ducks. This state occurs with probability 1. Thus, $\mathbb{E}[W] = 4\cdot1 = 4$ white ducks.

   $P(WH=1) = 1$ and $P(WH\neq1)=0$, producing the probability distribution (values, probabilities) $= \{(0,0), (1,1)\}$, so $\mathbb{E}[WH] = 0\cdot0 + 1\cdot1 = 1$ all-white duck.

   Using the Lemma, we note that $WH(A) = 1$ because we see all of the ducks; seeing all four ducks is our only possible state, with probability 1, so $\mathbb{E}[WH] = 1\cdot1 = 1$ all-white duck.

2. Define a random variable B describing the number of black ducks you see when you visit the ducks from Example 14.3.4.

   $B(d) = 1$ if d is black and 0 otherwise.

(a) What is N? $N = \{0,1\}$ because only one duck has any black on it.

(b) Give $P(B = n)$ for each $n \in N$ (the probability distribution of B).

There are 16 different subsets of ducks. The black duck appears in 8 of these subsets. So $P(B = 1) = \frac{8}{16}$ and $P(B = 0) = \frac{8}{16}$.

(c) Compute $\mathbb{E}[B]$. $\mathbb{E}[B] = 0 \cdot \dfrac{8}{16} + 1 \cdot \dfrac{8}{16} = \dfrac{1}{2}$.

3. Define a random variable G, the number of grey ducks you see when you visit the ducks from Example 14.3.4, and compute its expected value.

Let $G : S \to \{0,1,2\}$ be defined by $G(d) = 1$ if the duck is grey and $G(d) = 0$ if the duck is not grey. ($G(E) = 2$ if we have an event in which we see two grey ducks.) We find that four subsets of ducks contain no grey ducks, eight subsets contain a single grey duck, and four subsets contain two grey ducks. Thus, the probability distribution of G is (values, probabilities) $= \{(0, \frac{1}{4}),(1, \frac{1}{2}),(2, \frac{1}{4})\}$ and $\mathbb{E}[G] = 0 \cdot \frac{1}{4} + 1 \cdot \frac{1}{2} + 2 \cdot \frac{1}{4} = 1$ grey duck.

4. Challenge: For the state space you created in the Check Yourself Challenge in Section 14.2, define at least one random variable.

Answers will vary.

Section 14.5   page 462

1. In Example 14.5.4, what if only $\frac{1}{10}$ of the friends own cats?

Indeed, just replace $\frac{2}{3}$ with $\frac{1}{10}$ to get $\frac{3}{4} \cdot \frac{1}{10} + \frac{1}{5} \cdot \frac{9}{10} = \frac{51}{200}$ and $\dfrac{\frac{3}{4} \cdot \frac{1}{10}}{\frac{51}{200}} = \frac{5}{17}$.

2. Are the events $W = 0$ and $Z = 0$ (as defined above) independent?

No, because their probabilities don't equal their conditional probabilities.

3. Consider the random variable Y as defined at the start of the section and a new random variable
$$V(roll\ k) = \begin{cases} 1 & k \leq 3, \\ 0 & k > 3. \end{cases}$$
Are the events $Y = 1$ and $V = 1$ independent?

Compute $P(Y = 1)P(V = 1) = \frac{1}{2} \cdot \frac{1}{2} = \frac{1}{4}$. Compare this to $P(Y = 1$ and $V = 1)$; note that the only state $Y = 1$ and $V = 1$ have in common is $k = 2$, so $P(Y = 1$ and $V = 1) = \frac{1}{6} \neq \frac{1}{4}$ and so these events are *not* independent.

4. Examine Problem 3 of Section 14.4, and redo the problem using your new knowledge of independence and PIE.

The relevant questions are, "What is $P($(flipping heads) or (rolling an even number))?" and "What is $P($(flipping heads) and (rolling an even number))?"

By PIE, $P((\text{flipping heads}) \text{ or } (\text{rolling an even number})) = P(\text{flipping heads}) + P(\text{rolling an even number}) - P((\text{flipping heads}) \text{ and } (\text{rolling an even number}))$. And, we know that flipping a coin is independent of rolling a die. Therefore, $P((\text{flipping heads}) \text{ and } (\text{rolling an even number})) = P(\text{flipping heads}) \cdot P(\text{rolling an even number}) = \frac{1}{2} \cdot \frac{1}{2} = \frac{1}{4}$. This in turn tells us that $P((\text{flipping heads}) \text{ or } (\text{rolling an even number})) = \frac{1}{2} + \frac{1}{2} - \frac{1}{4} = \frac{3}{4}$ as we computed before.

5. State a version of PIE for the probabilities of three events.

   Let's call the events A, B, and C. Then $P(A \text{ or } B \text{ or } C) = P(A) + P(B) + P(C) - P(A \text{ and } B) - P(A \text{ and } C) - P(B \text{ and } C) + P(A \text{ and } B \text{ and } C)$.

Section 14.7   page 468

1. Suppose five fair coins are flipped and consider the random variable H, the number of heads revealed. Rewrite H as a sum of simpler random variables.

   $H = H_1 + H_2 + H_3 + H_4 + H_5$, where $H_j = 1$ if the jth coin lands head-side-up and $H_j = 0$ otherwise.

2. For the situation of rolling an eight-sided die, consider the random variables $E = 1$ when the result is even and $E = 0$ otherwise, and $G = 1$ if the result is greater than or equal to 5 and $G = 0$ if the result is four or less. Compute $(E + G)(\text{roll } 2)$, $(E + G)(\text{roll } 3)$, $(E + G)(\text{roll } 5)$, and $(E + G)(\text{roll } 6)$.

   $(E + G)(\text{roll } 2) = E(\text{roll } 2) + G(\text{roll } 2) = 1 + 0 = 1, (E + G)(\text{roll } 3) = E(\text{roll } 3) + G(\text{roll } 3) = 0 + 0 = 0, (E + G)(\text{roll } 5) = E(\text{roll } 5) + G(\text{roll } 5) = 0 + 1 = 1$, and $(E + G)(\text{roll } 6) = E(\text{roll } 6) + G(\text{roll } 6) = 1 + 1 = 2$.

3. Consider a deck of four cards, labeled $1\triangle$, $1\bigcirc$, $2\triangle$, and $2\bigcirc$. Draw a card, put it back, shuffle the deck, and draw a second card. Compute the expected value of $Z = X + Y$, where X is the numerical value of the first card and Y is the numerical value of the second card.

   By Theorem 14.7.1, we just need to compute $\mathbb{E}[Z] = \mathbb{E}[X] + \mathbb{E}[Y]$. $\mathbb{E}[X] = 1P(X = 1) + 2P(X = 2) = 1 \cdot \frac{1}{2} + 2 \cdot \frac{1}{2} = \frac{3}{2}$. Likewise, $\mathbb{E}[Y] = 1P(Y = 1) + 2P(Y = 2) = 1 \cdot \frac{1}{2} + 2 \cdot \frac{1}{2} = \frac{3}{2}$. Thus, $\mathbb{E}[Z] = 3$.

Section 15.5   page 505

1. What is the cardinality of the sample drawers (in the Storage facility)?

   The drawers are numbered as $\mathbb{N}$, so the cardinality is $\aleph_0$.

2. What is the cardinality of the stools in the coffee area?

   The stools are lined up and could be numbered as $\mathbb{Z}$, so the cardinality is $\aleph_0$.

Section 16.3
page 515

1. Use the Euclidean algorithm to compute GCD(8, 12).

   12 (mod 8) is 4, so we have $\text{GCD}(8, 12) = \text{GCD}(4, 8) = 4$.

2. Use the Euclidean algorithm to compute GCD(1233, 1234).

   1234 (mod 1233) is 1, so we have $\text{GCD}(1233, 1234) = \text{GCD}(1, 1233) = 1$.

3. Find integers k, ℓ such that $\text{GCD}(8, 12) = k8 + \ell 12$.

   $4 = 1 \cdot 4 + 0 \cdot 8$ so using the inductive thing in the proof we have $k = 0 - 1 \cdot 1$ and $\ell = 1$. Oh, wait, $4 = (-1) \cdot 8 + 1 \cdot 12$ by inspection anyway.

4. Find integers k, ℓ such that $\text{GCD}(1233, 1234) = k1233 + \ell 1234$.

   Let's do $k = -1$ and $\ell = 1$ again. Yup.

5. **Challenge:** Pick two natural numbers a, b where a does not divide b, and find their GCD using the Euclidean algorithm. Then find integers k, ℓ such that $\text{GCD}(a, b) = ka + \ell b$.

   Answers will vary.

Section 16.5
page 520

1. Compute $\varphi(210)$.

   $210 = 2 \cdot 3 \cdot 5 \cdot 7$ so $\varphi(210) = \varphi(2)\varphi(3)\varphi(5)\varphi(7) = 1 \cdot 2 \cdot 4 \cdot 6 = 48$.

2. Compute $\varphi(3200)$.

   $3200 = 2^7 \cdot 5^2$, so $\varphi(3200) = \varphi(2^7)\varphi(5^2) = (2^7 - 2^6)(5^2 - 5^1) = 64 \cdot 20 = 1280$.

3. Find w such that $w \cdot 2 \equiv 1 \pmod 5$ and use this to find all x that satisfy $2x \equiv 3 \pmod 5$.

   $3 \cdot 2 = 6 \equiv 1 \pmod 5$; the equivalence becomes $x \equiv 3 \cdot 3 \equiv 4 \pmod 5$, so the desired set is all $x \equiv 4 \pmod 5$.

4. Find w such that $w \cdot 3 \equiv 1 \pmod 4$ and use this to find all x that satisfy $3x \equiv 2 \pmod 4$.

   $3 \cdot 3 = 9 = 8 + 1 \equiv 1 \pmod 4$; the equivalence becomes $x \equiv 3 \cdot 2 \equiv 2 \pmod 4$, so the desired set is all $x \equiv 2 \pmod 4$.

5. Find w such that $w \cdot 3 \equiv 1 \pmod 7$ and use this to find all x that satisfy $3x \equiv 6 \pmod 7$.

   $5 \cdot 3 = 15 \equiv 1 \pmod 7$; the equivalence becomes $x \equiv 5 \cdot 6 \equiv 2 \pmod 7$, so the desired set is all $x \equiv 2 \pmod 7$.

6. Find w such that $w \cdot 5 \equiv 1$ (mod 7) and use this to find all x that satisfy $5x \equiv 3$ (mod 7).

   $3 \cdot 5 = 15 \equiv 1$ (mod 7); the equivalence becomes $x \equiv 3 \cdot 3 \equiv 2$ (mod 7), so the desired set is all $x \equiv 2$ (mod 7).

7. Re-solve the first two congruence pairs from Section 16.4 using the techniques given in Section 16.5.2:

   (a) Which numbers x are both $x \equiv 0$ (mod 2) and $x \equiv 0$ (mod 3)?

   $x = 2j \equiv 0$ (mod 3), and $2 \cdot 2 \equiv 1$ (mod 3), so $j \equiv 0$ (mod 3). Then $j = 3q$ and $x = 6q$. Therefore $x \equiv 0$ (mod 6).

   (b) Which numbers x are both $x \equiv 1$ (mod 2) and $x \equiv 2$ (mod 3)?

   We have that $x = 2j + 1$ and plugging in gives $2j + 1 \equiv 2$ (mod 3) or $2j \equiv 1$ (mod 3), which we solved earlier to give us $j \equiv 2$ (mod 3). Now $j = 3q + 2$ so $x = 2(3q + 2) + 1 = 5 + 6q$ and $x \equiv 5$ (mod 6).

8. **Challenge:** Invent your own congruence equation $kx \equiv t$ (mod a) and find its solutions.

   Answers will vary.

9. **Challenge:** Invent your own pair of congruence equations $x \equiv s$ (mod a) and $x \equiv t$ (mod b) (but make sure that a and b are relatively prime!) and find their common solutions.

   Answers will vary.

Section 16.7   page 523

1. Pick three pairs of adjacent fractions in FFL_5, and verify that for each pair, $bc - ad = 1$.

   Answers will vary, but here are four: for $\frac{1}{4}, \frac{2}{7}, 4 \cdot 2 - 1 \cdot 7 = 1$; for $\frac{4}{7}, \frac{3}{5}, 7 \cdot 3 - 4 \cdot 5 = 1$; for $\frac{5}{8}, \frac{2}{3}, 8 \cdot 2 - 5 \cdot 3 = 1$; for $\frac{3}{4}, \frac{4}{5}, 4 \cdot 4 - 3 \cdot 5 = 1$.

2. Show that not every mediant $\frac{a+c}{b+d}$ must be in lowest terms by finding two fractions $\frac{a}{b}$ and $\frac{c}{d}$ that are *not* adjacent in an FFL and whose mediant is not in lowest terms.

   Consider $\frac{2}{3}$ and $\frac{2}{5}$, whose mediant is $\frac{4}{8}$.

3. In which FFL does $\frac{1}{9}$ appear? FFL_9, as it's the second fraction there.

4. In which FFL does $\frac{2}{11}$ appear? FFL_7, where it's the fourth fraction.

Section 17.3 page 541

page 541

1. Consider the counting algorithm with input a natural number n.

    1. Let $k = 1$.
    2. Output k.
    3. Replace k with $k + 1$.
    4. If $k = n$, output k, and stop; otherwise, return to step 2.

   How many operations does this algorithm take to execute?

   Each of the first three steps involves a single operation. The fourth step also involves a single operation if $k < n$ and two operations when $k = n$ (we don't count "stop" or "return" as operations). Step 1 happens once, for 1 operation; steps 2 and 3 happen once for each value of $k \in \{1, \ldots, n-1\}$, for $2(n-1)$ operations; step 4 happens once for each value of $k \in \{2, \ldots, n\}$ for $(n-2) + 2$ operations. In total, there are $1 + 2n - 2 + n - 2 + 2 = 3n - 1$ operations.

2. Consider the laundry sorting algorithm that works on a pile of n pieces of clothing:

    1. Pick up a piece of clothing.
    2. If the item is a dark color, toss it to the right. If the item is a pale color, toss it to the left. If the item is a bright color, toss it behind you.
    3. If there is more clothing in the pile, go to step 1; otherwise, be done.

   What are best- and worst-case inputs for this algorithm? How many operations does it take to execute?

   All inputs are the same (in the best/worst sense) for this algorithm because it treats them identically. The algorithm takes 3 operations per piece of clothing (the three listed above!) so the runtime is $3n$.

3. Write a runtime function for the algorithm given in Example 5.2.7.

   There are only two steps, the second of which contains at most four operations in the worst case (the marble is neither red nor green), so the function is... 4.

4. Write a runtime function for the algorithm given in Problem 10 of Section 5.11.

   There are only 3 operations before it terminates! $r(n) = 3$.

Section 17.5 page 550

page 550

1. We claimed that as long as n is larger than 10, we have $r_1(n) < r_5(n) < r_2(n) < r_4(n) < r_3(n)$. Verify that this is not true if $n = 10$.

   $r_1(10) = 54$, $r_5(10) = 50$, $r_2(10) = 72$, $r_4(10) = 200$, $r_3(10) = 102.4$.

2. How small can n be and still have $r_5(n) < r_2(n) < r_4(n) < r_3(n)$ be true?

   $n \geq 12$. However, for $n = 11$ we have $r_5(n) < r_2(n) < r_3(n) \leq r_4(n)$!

3. True or false (and explain): $n^3 \in O(n^2)$.

   False, because for any $c > 0$, when $n > c$ we have $n^3 > cn^2$.

4. True or false (and explain): $3^n \in O(3^{n-1})$.

   True, because $3^n = 3 \cdot 3^{n-1}$, so we may choose $c = 4$ and have $3^n < 4 \cdot 3^{n-1}$ for all natural n.

5. True or false (and explain): $r(n) \in O(r(n))$.

   True, because for $c = 1$ and $n > 1$, $r(n) \leq r(n)$.

Appendix B 🐤🐤

Solutions to Bonus Check-Yourself Problems

1. A Timbuk2 custom messenger bag comes in four sizes, has 46 options for the left-panel and center-panel and right-panel fabrics, 18 different binding options, 27 logo colors, 11 liner colors, three options for pocket style, two handednesses, and 47 different options for the strap pad. (Really, not kidding—these numbers came from the Timbuk2 website in October 2014.) How many different custom messenger bags could one order?

   This is a total mix-and-match situation, so the product principle applies and we multiply together all the numbers of options. There are $4 \cdot 46 \cdot 46 \cdot 46 \cdot 18 \cdot 27 \cdot 11 \cdot 3 \cdot 2 \cdot 47 = 586{,}964{,}112{,}768$ ways to order a Timbuk2 custom messenger bag.

2. Prove that the product of any three odd numbers is also odd.

   We first name the three odd numbers n_1, n_2, n_3. Because they are odd, each can be written in the form $2k + 1$—but the k is likely different for each, so we have $n_1 = 2k_1 + 1$, $n_2 = 2k_2 + 1$, and $n_3 = 2k_3 + 1$. The product of the numbers is $n_1 n_2 n_3 = (2k_1 + 1)(2k_2 + 1)(2k_3 + 1)$. Expanding this expression, we get $8n_1 n_2 n_3 + 4n_1 n_2 + 4n_1 n_3 + 4n_2 n_3 + 2n_1 + 2n_2 + 2n_3 + 1$, which can be rewritten as $2(n_1 n_2 n_3 + 2n_1 n_2 + 2n_1 n_3 + 2n_2 n_3 + n_1 + n_2 + n_3) + 1 = 2q + 1$ for some integer q. Therefore, the product of any three odd numbers is also odd.

3. Takeo, a paper store in Tokyo, has walls lined with coded drawers. Each code designates a type of paper. One such drawer is 2Q08. If the first entry has to be 1, 2, or 3 (there are only three walls with drawers), the second is a letter, and the last two are numbers, then how many drawers could Takeo have?

   We think of filling slots: the first slot has 3 possibilities, the second 26, and the third and fourth slots have 10 possibilities each. The product principle says there are $3 \cdot 26 \cdot 10 \cdot 10 = 7{,}800$ paper drawers. (It turns out that at Takeo there are no drawers that end in 00, so this is certainly an overestimate.)

4. You want to buy an electric car. The Chevy Volt comes in eight colors (red, brown, grey, pale blue, two blacks, two whites), offers three kinds of wheels, and has five kinds of interiors (two cloth, three leather). The Tesla comes in nine colors (black, two whites, two greys, brown, red, green, blue), and gives a choice of three roof styles (one is glass), four wheel styles, four seat colors, four dashboard prints, and three door-trim colors. There are three versions of the Nissan Leaf (S, SV, SL), each of which comes in seven colors (two whites, two greys, red, blue, black). How many different choices of car do you have?

For the Chevy Volt, there are $8 \cdot 3 \cdot 5 = 120$ ways to specify the car because we can have any combination of exterior, interior, and wheels. The Tesla has a ridiculous number of options: $9 \cdot 3 \cdot 4 \cdot 4 \cdot 4 \cdot 3 = 5{,}184$. In contrast, the $3 \cdot 7 = 21$ types of Nissan Leaf seem understated. Still, we add these three numbers together because we're only buying one car: $120 + 5{,}184 + 21 = 5{,}325$ choices of electric car.

5. Prove, or find a counterexample: the sum of two consecutive perfect cubes is odd.

Here are two proofs:
(1) First, observe that a number and its cube have the same parity. We do two cases: An even number may be written as $2k$, and $(2k)^3 = 8k^3 = 2(4k^3) = 2q$, which is even. An odd number may be written as $2k + 1$, and $(2k+1)^3 = 8k^3 + 12k^2 + 6k + 1 = 2(4k^3 + 6k^2 + 3k) + 1 = 2r + 1$, which is odd.

Thus, consecutive perfect cubes have the property that one is odd and the other even. The sum of an odd number and an even number is odd, so the sum of two consecutive perfect cubes is odd.

(2) Either the consecutive perfect cubes are $(2k)^3$ and $(2k+1)^3$, or $(2k-1)^3$ and $(2k)^3$, depending on which of the numbers is even. In the first case, we have $(2k)^3 + (2k+1)^3 = 8k^3 + 8k^3 + 12k^2 + 6k + 1 = 2(4k^3 + 4k^3 + 6k^2 + 3k) + 1$, which is odd. In the second case, we have $(2k-1)^3 + (2k)^3 = 8k^3 - 12k^2 + 6k - 1 + 8k^3 = 2(4k^3 + 4k^3 - 6k^2 + 3k - 1) + 2 - 1 = 2(4k^3 + 4k^3 - 6k^2 + 3k - 1) + 1$, which is odd.

6. How many four-digit phone extensions have no 0s and begin with 3?

There is only one choice for the first digit (3) and 9 choices for each of the other three digits, so there are $1 \cdot 9 \cdot 9 \cdot 9 = 729$ such extensions.

7. In 2016, there were 3,945,875 live births in the US. (Source: http://www.cdc.gov/nchs/fastats/births.htm.) Did there have to be two of these births within the same second?

2016 was a leap year, so it had 366 days. Each of those days had 24 hours, each of which had 60 minutes, each of which had 60 seconds. Thus there were $366 \cdot 24 \cdot 60 \cdot 60 = 31{,}622{,}400$ seconds in 2016. There are more seconds than live births, so each birth could happen in a different second.

8. How many length-8 binary strings have no 0s in the fourth place?

2^7, because the fourth place must be a 1 and there are two choices for each of the other 7 places.

9. You receive a choose-your-own-adventure certificate for a jewelry store! The deal is that you get to pick one of eight precious gems, and either a ring or a bracelet to put it in. There are three possible ring styles and six possible bracelet styles.

   (a) How many possible prizes are there?   72.

   (b) How did you answer the previous question? If you used the product principle first, re-answer the question using the sum principle first. (And if you used the sum principle first, re-answer the problem using the product principle first.)

   Product principle then sum principle: $8 \cdot 3 + 8 \cdot 6$.
   Sum principle then product principle: $8 \cdot (3 + 6)$.

   (c) On closer look, you realize that neither the ruby nor the emerald would look good on the bracelet. How many prizes are still possible?

   Here, you have to use the product principle first: $8 \cdot 3 + 6 \cdot 6 = 60$.

10. I have a lot of stuff in my stuff-holder: six ball-point pens, a silver star wand, three teal signature pens, a bronze-yellow colored pencil, five liquid ink pens, three mechanical pencils, a highlighter, six permanent markers, seven gel pens, a Hello Kitty lollipop, two markers, three wooden pencils, a 3-inch-long pen, a calligraphy marker, a pen shaped like a cat, and a pair of left-handed office scissors.

   How many writing utensils do I have in the stuff-holder?

   Good grief, that seems like a lot. But I just need to add up the numbers of writing utensils: $6 + 3 + 1 + 5 + 3 + 1 + 6 + 7 + 2 + 3 + 1 + 1 + 1 = 40$. Seriously?

Chapter 2, Section 2.11   page 59

1. On an October 2014 visit to the CVS Minute Clinic, the check-in kiosk asked the question, "If you have a copay for today's visit, will you be paying for it with a credit or debit card?"

   (a) Identify the formal logic quantifiers and structure in this question.

   If $\exists$ (copay for today's procedure), then is the statement *I will pay with a credit or debit card* true? So, we have an $\exists$ and a $\Rightarrow$.

   (b) The visit in question was for a flu vaccine, which does not require a copay. The kiosk gave options of *Yes* and *No*. How should the visitor have answered?

   Because we have an implication of the form $P \Rightarrow Q$ with P false, we know from the truth table that $P \Rightarrow Q$ is true. But this doesn't say anything about whether we should answer *Yes* or *No*, because the question is about payment methods, not the truth of the statement. It shouldn't matter which answer is selected.

(c) Can you find a simpler way to word the question clearly? (In other words, what *should* the kiosk question ask?)

A better question might be, "If there are charges for today's visit, will you be paying for them with a credit or debit card?" That way it's not dependent on what is already known (there is no copay), but what might happen (additional services rendered, for example).

2. There was a recent campaign slogan heard on the radio: *Not just Blue Cross Blue Shield of Massachusetts, but Blue Cross Blue Shield ... of you.* Why is this mathematically nonsensical for residents of Massachusetts?

You are in Massachusetts, so $\{you\} \subset$ Massachusetts and therefore you were already included in the described set. The slogan implies that Massachusetts $\subsetneq$ $\{you\} \cup$ Massachusetts.

3. Consider the Venn diagram in Figure 2.14.

(a) Express the shaded area as a set using unions, intersections, and/or complements of the sets Q, R, and S.

$(Q \setminus (R \cup S)) \cup (S \setminus (Q \cup R))$ is one way of expressing that area.

(b) Let $Q = \{k \in \mathbb{Z} \mid |k| \leq 10\}$, $R = $ even numbers, and $S = \{n \in \mathbb{N} \mid n$ is a perfect square$\}$. List the elements of the shaded area.

The elements are $\{-9, -7, -5, -3, -1, 3, 5, 7, 25, 49, 81, \dots\}$

4. Let $A = $ multiples of 4, and $B = $ multiples of 6. Write $A \cap B$ as a set in the form $\{$ sets $\mid$ conditions $\}$. $\{k \in \mathbb{Z} \mid k$ is a multiple of 12$\}$.

5. Negate the statement $\forall n \in \mathbb{Z}, \exists y \in 2\mathbb{N}$ such that $n = y \cdot k$ for some $k \in \mathbb{Z}$. Is either the statement or its negation true?

$\neg(\forall n \in \mathbb{Z}, \exists y \in 2\mathbb{N}$ such that $n = y \cdot k$ for some $k \in \mathbb{Z})$.
$\exists n \in \mathbb{Z}, \neg(\exists y \in 2\mathbb{N}$ such that $n = y \cdot k$ for some $k \in \mathbb{Z})$.
$\exists n \in \mathbb{Z}$, such that $\forall y \in 2\mathbb{N}, \neg(n = y \cdot k$ for some $k \in \mathbb{Z})$.
$\exists n \in \mathbb{Z}$, such that $\forall y \in 2\mathbb{N}$, there is no $k \in \mathbb{Z}$ such that $n = y \cdot k$.

The statement is false; a counterexample is 1. The negation is true; consider any odd number.

6. Prove that $k \in \mathbb{Z}$ is positive if and only if k^3 is positive.

This is a biconditional, so it has two parts:

($\Rightarrow$) Suppose $k \in \mathbb{Z}$ is positive. Then k^3 is positive because the product of positive numbers is positive.

($\Leftarrow$) The simplest way of doing this is a proof by contrapositive, i.e., to prove that *if k^3 is positive, then k is positive* we'll show that *if k is not positive, then k^3 is not positive*. (It's difficult to be convincing about cube roots in a discrete context.)

Suppose $k \in \mathbb{Z}$ is negative. Then k^3 is negative because the product of an odd number of negative numbers is negative. This neglects the non-positive case of $k = 0$, but of course $0^3 = 0$ is not positive, so we are done.

7. Make a truth table for $\neg(P \wedge Q) \wedge ((P \vee Q) \wedge R)$. Can you express this statement (henceforth referred to as *aaaaaa!*) more simply?

P	Q	$P \vee Q$	R	$(P \vee Q) \wedge R$	$P \wedge Q$	$\neg(P \wedge Q)$	*aaaaaa!*
T	T	T	T	T	T	F	F
T	F	T	T	T	F	T	T
F	T	T	T	T	F	T	T
F	F	F	T	F	F	T	F
T	T	T	F	F	T	F	F
T	F	T	F	F	F	T	F
F	T	T	F	F	F	T	F
F	F	F	F	F	F	T	F

Notice that when R is false, so is the *aaaaaa!* statement. When R is true, we have the pattern of P xor Q (and also the pattern of $\neg(P \Leftrightarrow Q)$), so *aaaaaa!* is equivalent to $(P$ xor $Q) \wedge R$ (and also equivalent to $\neg(P \Leftrightarrow Q) \wedge R$).

8. Let $A = \{0, 1, 2\}$ and $B = \{1, 3, 5, 7\}$.

    (a) List the elements of $(A \times B) \cap (B \times A)$.

    Any element that is in both $A \times B$ and $B \times A$ must have each component in A and in B. Because $A \cap B = \{1\}$, the only element that qualifies is $(1, 1)$.

    (b) List the elements of $(A \setminus B) \times (B \setminus A)$.

    $A \setminus B = \{0, 2\}; B \setminus A = \{3, 5, 7\}$.
    Thus $(A \setminus B) \times (B \setminus A) = \{(0, 3), (0, 5), (0, 7), (2, 3), (2, 5), (2, 7)\}$.

9. Show that $(A \times B) \cup (C \times B) = (A \cup C) \times B$.

    We proceed by double-inclusion.

    If $x \in (A \times B) \cup (C \times B)$, then there are two cases: $x \in A \times B$ or $x \in C \times B$. In each case, $x \in (A \cup C) \times B$. One direction is done.

    If $x \in (A \cup C) \times B$, then $x = (x_1, x_2)$ and $x_1 \in A$ or $x_1 \in C$. If $x_1 \in A$, then $x \in A \times B$; if $x_1 \in C$, then $x \in C \times B$. Thus, $x \in (A \times B) \cup (C \times B)$. Both directions are done.

10. Show that $\{2k \mid k \in \mathbb{N}\} \cup \{4k + 1 \mid k \in \mathbb{W}\} \cup \{4k + 3 \mid k \in \mathbb{W}\} = \mathbb{N}$.

    We proceed by double-inclusion.

    First, we show that $\{2k \mid k \in \mathbb{N}\} \cup \{4k + 1 \mid k \in \mathbb{W}\} \cup \{4k + 3 \mid k \in \mathbb{W}\} \subset \mathbb{N}$.

    Consider $x \in \{2k \mid k \in \mathbb{N}\}$; because $k \in \mathbb{N}$, then $2k \in \mathbb{N}$ so $x \in \mathbb{N}$. Now, let $y \in \{4k + 1 \mid k \in \mathbb{W}\}$; for any $k \neq 0$, $k \in \mathbb{N}$ so $4k + 1 \in \mathbb{N}$, and if $k = 0$, then $4k + 1 = 1 \in \mathbb{N}$. Thus $y \in \mathbb{N}$. Similarly, we see that for $z \in \{4k + 3 \mid k \in \mathbb{W}\}$, $z \in \mathbb{N}$.

    Second, we show that $\{2k \mid k \in \mathbb{N}\} \cup \{4k + 1 \mid k \in \mathbb{W}\} \cup \{4k + 3 \mid k \in \mathbb{W}\} \supset \mathbb{N}$.

Consider $n \in \mathbb{N}$. If n is even, then $n \in \{2k \mid k \in \mathbb{N}\}$. Otherwise, divide n by 4; the remainder must be 1 or 3 (as if it were 0 or 2, n would be even). If the remainder is 1, then $n \in \{4k+1 \mid k \in \mathbb{W}\}$ and if the remainder is 3, then $\ell \in \{4k+3 \mid k \in \mathbb{W}\}$. Therefore, $n \in \{2k \mid k \in \mathbb{N}\} \cup \{4k+1 \mid k \in \mathbb{W}\} \cup \{4k+3 \mid k \in \mathbb{W}\}$.

We conclude that $\{2k \mid k \in \mathbb{N}\} \cup \{4k+1 \mid k \in \mathbb{W}\} \cup \{4k+3 \mid k \in \mathbb{W}\} = \mathbb{N}$.

Chapter 3, Section 3.14   page 95

1. Let $S = \{s_1, s_2, \dots, s_n\}$. How many functions are there with domain $\mathbb{Z}_3$ and target S? Of those functions, how many are one-to-one? How many are onto?

   There are n^3 functions from $\mathbb{Z}_3$ to S. Of those, $n \cdot (n-1) \cdot (n-2)$ are one-to-one. Only if $|S| \leq 3$ are any of the functions onto. If $|S| = 3$, then there are six onto functions (for the six different ways of assigning the three elements of $\mathbb{Z}_3$ to the three elements of S); if $|S| = 2$, then there are still six onto functions (two choices of which element of S is hit by only one element of $\mathbb{Z}_3$, and for each of those three choices of which element of $\mathbb{Z}_3$ goes there); if $|S| = 1$, then there is exactly one onto function (send all the elements of $\mathbb{Z}_3$ to the only element of S).

2. Draw all connected 3-regular graphs with four vertices.

   Find them by careful casework, for example, by doing cases on the number of loops and the number of multiple edges in the graph. See Figure B.1 for the five graphs.

3. Are the two graphs in Figure 3.27 isomorphic? Justify your response.

   Yes, these are isomorphic. Use the map $\varphi(1) = a, \varphi(2) = b, \varphi(3) = f, \varphi(4) = c, \varphi(5) = d, \varphi(6) = e$. Many other isomorphisms are also possible.

4. Is the function $f : \mathbb{Z} \to \mathbb{Z}$ defined by $f(n) = \lfloor \sin(n) \rfloor$ a one-to-one function? Prove or disprove.

   The range of $\sin(x)$ is $[-1, 1]$. Thus, $\lfloor \sin(x) \rfloor$ can only take on the values $\{-1, 0, 1\}$. To show $f(n)$ is not one-to-one, we just have to find two values of n that both have $0 < \sin(n) < 1$. Trial and error gives $\sin(1) \approx 0.841471, \sin(2) \approx 0.909297$, so $f(1) = f(2)$; yet, $1 \neq 2$ and thus $f(n)$ is not one-to-one.

5. Is it possible to draw a graph with six vertices of degrees 2, 2, 3, 3, 4, and 4? If so, draw one. If not, explain why not.

   See Figure B.2.

Figure B.1. **All 3-regular graphs with four vertices.**

Figure B.2. I'm a sixy, sixy graph.

6. A *finger-finger* graph is denoted by $F_{m,n}$ and has m fingers, from each of which grows n fingers; see Figure 3.28. Conjecture and prove formulas for the number of vertices and the number of edges of a finger-finger graph.

A finger-finger graph $F_{m,n}$ has m edges for primary fingers, and from each primary finger n edges for secondary fingers. Thus by the sum and product principles, it has $m+mn$ edges. A finger-finger graph $F_{m,n}$ has one palm vertex, and one vertex from each of the m primary fingers, and then a vertex on the end of each secondary finger; there are n secondary fingers per primary finger, so by the sum and product principles, it has $1+m+mn$ vertices.

7. What can you say about the number of vertices of a 3-regular graph?

The number of vertices of a 3-regular graph must be even by the handshake lemma, as the total degree must be even and 3 is odd.

8. The following statement is true: *any cycle C_n with $n \geq k$ has complement $\overline{C_n}$ containing a triangle.* Determine k and prove the statement.

By trial and error, we find that $\overline{C_3}$ has no edges, neither $\overline{C_4}$ nor $\overline{C_5}$ has triangles, but $\overline{C_6}$ contains plenty of triangles. We suspect that $k = 6$. Now suppose $n \geq 6$, and number the vertices $1, 2, \ldots, n$. Vertices $1, 3, 5$ are not adjacent in C_n but form a triangle in $\overline{C_n}$.

9. Consider the map $g : (\mathbb{N} \times \mathbb{N}) \to \mathbb{N}$ defined by $g((a,b)) = ab$. Is this one-to-one? Onto? Give proofs.

g is onto because given any $n \in \mathbb{N}$, the element $(1,n) \in \mathbb{N} \times \mathbb{N}$ is such that $g((1,n)) = 1 \cdot n = n$. However, g is not injective because we have $g((1,n)) = n = g((n,1))$ but $(1,n) \neq (n,1)$.

10. Shown in Figure 3.29 are four infinite graphs in pairs A, B and C, D. One of these pairs is isomorphic and the other nonisomorphic. Which is which? Justify your response.

Graphs A and B are nonisomorphic, because A has vertices of degree 2 but B does not. Graphs C and D are isomorphic. Intuitively, we see that doing a vertical flip on one of each pair of diamonds will do the trick. We need to give a labeling to each graph and use this to define an isomorphism between the graphs. Figure B.3 shows such a labeling.

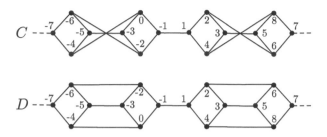

Figure B.3. Oh, *that's* who you are.

Our isomorphism is defined by $\varphi(j) = j$, and this is clearly a bijection. To show formally that it preserves operations, we have to check that the adjacencies match. In both C and D, we can see the following:

- If $j = 8k+1$ or $8k+5$ $(k \in \mathbb{Z})$, j is adjacent to $j-1, j+1, j+3$.
- If $j = 8k+2$ $(k \in \mathbb{Z})$, j is adjacent to $j-1, j+1, j+4$.
- If $j = 8k+3$ or $8k+7$ $(k \in \mathbb{Z})$, j is adjacent to $j-1, j+1, j+2$.
- If $j = 8k+4$ $(k \in \mathbb{Z})$, j is adjacent to $j-3, j-1, j+4$.
- If $j = 8k+6$ $(k \in \mathbb{Z})$, j is adjacent to $j-4, j-1, j+1$.
- If $j = 8k$ $(k \in \mathbb{Z})$, j is adjacent to $j-4, j-3, j-1$.

Chapter 4, Section 4.11   page 120

1. Prove that $\sum_{j=1}^{n} 3+5j = \frac{1}{2}(11n + 5n^2)$.

   When $n = 1$, we have $3+5 = 8$ and $\frac{1}{2}(11+5) = \frac{16}{2} = 8$. Check.

   $\sum_{j=1}^{k+1} 3+5j = \sum_{j=1}^{k} 3+5j+3+5(k+1)$. Using the inductive hypothesis, $= \frac{1}{2}(11k+5k^2) + 3 + 5(k+1) = \frac{1}{2}(11k+5k^2) + \frac{1}{2}(6+10(k+1)) = \frac{1}{2}(11k+5k^2+5+1+10k+10) = \frac{1}{2}(11(k+1)+5k^2+10k+5) = \frac{1}{2}(11(k+1)+5(k+1)^2)$ as desired.

2. Prove that $n^4 < 3 \cdot 8^n$.

   (Base case) When $n = 1$, we have $1^4 = 1 < 24 = 3 \cdot 8^1$.

   (Inductive hypothesis) For $n \le k$, $k^4 < 3 \cdot 8^k$.

   (Inductive step) $(k+1)^4 = k^4 + 4k^3 + 6k^2 + 4k + 1$. The inductive hypothesis applies to k^4, so we have $(k+1)^4 < 3 \cdot 8^k + 4k^3 + 6k^2 + 4k + 1$.

   Now, we want to show that $3 \cdot 8^k + 4k^3 + 6k^2 + 4k + 1 < 3 \cdot 8^{k+1} = 24 \cdot 8^k$. If we can show that $4k^3 + 6k^2 + 4k + 1 < 21 \cdot 8^k$, that will do the trick.

   We do know that $1 \le k^2$ and $4k \le 4k^2$, so $4k^3 + 6k^2 + 4k + 1 \le 4k^3 + 6k^2 + 4k^2 + k^2 = 4k^3 + 11k^2$ and $11k^2 \le 11k^3$, so $4k^3 + 11k^2 < 4k^3 + 11k^3$, and by the first example in this chapter we know that $15k^3 < 15 \cdot (2^k)^3 = 15 \cdot 2^{3k} = 15 \cdot 8^k < 21 \cdot 8^k$.

Therefore $4k^3 + 6k^2 + 4k + 1 < 21 \cdot 8^k$, which means (from above) that $(k+1)^4 < 3 \cdot 8^k + 4k^3 + 6k^2 + 4k + 1 < 3 \cdot 8^k + 21 \cdot 8^k = 3 \cdot 8^{k+1}$. And we're done.

3. Show that every convex polygon can be decomposed into triangles.

(Base cases) A triangle is already a triangle. Adding an edge joining opposite corners of a quadrilateral shows that the quadrilateral is composed of two triangles.
(Inductive hypothesis) Suppose any convex polygon with $n \le k$ sides can be decomposed into triangles.
(Inductive step) Consider a convex polygon with $k+1$ sides. Pick a vertex and travel along the edges of the polygon; skip the next vertex, but pick the one after that. Join these two vertices with an edge. On one side of the edge is a triangle made by the edge and the skipped vertex. On the other side of the edge is a convex polygon with $k-1$ sides. Therefore, by the inductive hypothesis it can be decomposed into triangles. Together with the triangle made by the edge and the skipped vertex, we have a decomposition of our $(k+1)$-sided polygon into triangles.

4. Show by induction that $K_{m,n}$ has mn edges.

(Base cases) Let's do three for good measure: $K_{1,1}$ has $1 = 1 \cdot 1$ edge, $K_{1,2}$ has $2 = 1 \cdot 2$ edges, and $K_{2,2}$ has $4 = 2 \cdot 2$ edges.
(Inductive hypothesis) For $m \le n$ and $n \le k$, $K_{m,n}$ has mn edges.
(Inductive step) Consider $K_{m,k+1}$, where $m \le k+1$. (We can do this because $K_{m,n} = K_{n,m}$.) If we remove one of the $k+1$ vertices, we are left with $K_{m,k}$. If $m \le k$, then the inductive hypothesis applies and we know $K_{m,k}$ has mk edges. Replacing the removed vertex, we also restore m edges for a total of $mk + m = m(k+1)$ edges as desired. If $m = k+1$, then $K_{m,k} = K_{k,k+1}$ and we may remove one of the $k+1$ vertices; this leaves us with $K_{k,k}$ which, by previous argument, has k^2 edges. Replacing the most recently removed vertex shows that $K_{k,k+1}$ has $k^2 + k = k(k+1)$ edges and replacing the first-removed vertex shows that $K_{k+1,k+1}$ has $k(k+1) + (k+1) = (k+1)^2$ edges, as desired.

5. Prove that $\sum_{j=0}^{n} (j+1)(j-2) = \frac{1}{3}(n-3)(n+1)(n+2)$.

When $n = 0$, we have $1(-2) = -2$ and $\frac{1}{3}(-3)(1)(2) = -2$. Check.

$\sum_{j=0}^{k+1}(j+1)(j-2) = \sum_{j=0}^{k}(j+1)(j-2) + (k+1+1)(k+1-2)$. Using the inductive hypothesis, $= \frac{1}{3}(k-3)(k+1)(k+2) + (k+2)(k-1) = \dots$ algebra $\dots$ $= \frac{1}{3}(k+3)(k+2)(k-2) = \frac{1}{3}(k+1-3)(k+1+1)(k+1+2)$ as desired.

6. Prove $(2(n!))^2 < 2^{(n!)^2}$ for sufficiently large values of n.

	n	$(2(n)!)^2$	$2^{(n!)^2}$	
What are these sufficiently large values? Let's do some experiments:	1	4	2	Looks like the statement holds for $n \ge 3$, and $n = 3$ is a base case.
	2	16	16	
	3	144	68,719,476,736	

Suppose that for $3 \leq n \leq k$, $(2(k!))^2 < 2^{(k!)^2}$. Consider $((2(k+1)!))^2$ and rewrite it as $(k+1)^2(2(k!))^2$. By the inductive hypothesis, $(k+1)^2(2(k!))^2 < (k+1)^2 2^{(k!)^2}$. Now, $2^{((k+1)!)^2} = 2^{(k+1)^2(k!)^2} = 2^{(k+1)^2} 2^{(k!)^2}$, so it remains to show that $(k+1)^2$ is less than $2^{(k+1)^2} = 2^{k^2+2k+1} = 2^{k^2} 2^{2k} 2$.

Because $2 < k$, we know $k+1 < k+k = 2k$, so $(k+1)^2 < (2k)^2 = 4k^2$. By the first example in this chapter, we know that $k < 2^k$, so $4k^2 < 4 \cdot 2^k 2^k = 4 \cdot 2^{2k}$ and certainly $2 < 2^{k^2}$, so we can conclude that $(k+1)^2 < 2^{k^2} 2^{2k} 2$. And we're done!

7. Use induction to prove the sum principle for n finite sets.

(Base case) If A_1 has a_1 elements, then it has... a_1 elements. Okay, that doesn't feel like we're saying anything. If $|A_1| = a_1$, $|A_2| = a_2$, and $A_1 \cap A_2 = \emptyset$, then $A_1 \cup A_2$ has $a_1 + a_2$ elements.

(Inductive hypothesis) Let the finite set A_i have a_i elements, and let sets A_i and A_j be disjoint. Then for $n \leq k$, $\bigcup_{i=1}^n A_i$ has $\sum_{i=1}^n a_i$ elements.

(Inductive step) Consider $\bigcup_{i=1}^{k+1} A_i$. We may rewrite this as $\left(\bigcup_{i=1}^k A_i\right) \cup A_{k+1}$. The inductive hypothesis applies to $\bigcup_{i=1}^k A_i$ so we know it has $\sum_{i=1}^k a_i$ elements. And, $\left(\bigcup_{i=1}^k A_i\right) \cup A_{k+1}$ is the union of two sets so the base case applies and we know it has $\left(\sum_{i=1}^k a_i\right) + a_{k+1}$ elements. That last expression is simply $\sum_{i=1}^{k+1} a_i$ as desired.

8. Take a piece of paper and fold it—not necessarily in half, but definitely with a single straight crease somewhere in the paper. Fold the (still folded) paper again. In fact, fold it n times, wherever you like. Now unfold it completely. Prove by induction that you can always color the paper with two colors (teal and purple) so that no fold line has the same color on both sides.

(Base case) The base case is $n = 1$ fold. There are two regions in the paper, and one can be colored teal and the other purple.

(Inductive hypothesis) Suppose that for any $n \leq k$, a piece of paper folded n times and unfolded can be colored teal and purple so that no piece of a fold line has the same color on both sides.

(Inductive step) Consider a piece of paper that has been folded $k+1$ times. Unfold it once, and mark the fold line (it's the $(k+1)st$) so that when you unfold it all the way you know which folds belong to that fold line. Unfold the paper completely. Now, ignore the $(k+1)st$ line's presence—what you have is a piece of paper that has been folded k times and unfolded. This can be colored teal and purple so that no piece of a fold line has the same color on both sides.

Examine the $(k+1)st$ fold line. It bisects some regions. Leave the portions of those regions to one side (the first side) of the fold line alone, and switch the colors of the portions of regions and whole regions on the other side (the second side) of the fold line.

We need to show that there is no piece of a fold line that has the same color on both sides. There are three kinds of fold line pieces: those on the first side of the $(k+1)st$ fold line, those on the second side of the $(k+1)st$ fold line, and those that are part of the $(k+1)st$ fold line. Those on the first side already had different colors; those on the second side had different colors, both of which switched, so they still have different colors; and those part of the $(k+1)st$ fold line had the same color but one has been switched so there are now different colors on the two sides. Thus, we've correctly colored the paper teal and purple.

9. For what values of n is $5^{n+2} < 6^n$? Prove it.

By trial and error we note that for $n = 17$, we have $5^{19} = 19{,}073{,}486{,}328{,}125 > 16{,}926{,}659{,}444{,}736 = 6^{17}$, but for $n = 18$ we have $5^{20} = 95{,}367{,}431{,}640{,}625 < 101{,}559{,}956{,}668{,}416 = 6^{18}$. (We generally expect that larger bases will produce larger functions in the long run.)

So we will use a base case of $n = 18$ and suppose that for $18 \leq n \leq k$, $5^{k+2} < 6^k$. Consider $5^{k+3} = 5 \cdot 5^{k+2}$. By the inductive hypothesis, we have $5 \cdot 5^{k+2} < 5 \cdot 6^k$. But, we also know $5 < 6$ so $5 \cdot 6^k < 6 \cdot 6^k$, and the result is that $5^{k+3} < 6^{k+1}$ as desired.

10. Prove that any natural number $n \geq 2$ can be written as the product of prime numbers.

The base case is clear; 2 is the product of 2 (itself), which is prime. Suppose that any $n \leq k$ can be written as the product of prime numbers, and consider $k + 1$. Either $k + 1$ is prime, or it is not. If $k + 1$ is prime, then it is the product of one prime (itself). If $k + 1$ is not prime, then it is the product of (at least) two smaller numbers, each of which can be written as a product of primes (by the inductive hypothesis). Thus, the product of those products of prime numbers is *also* a product of prime numbers, and we are done.

Chapter 5, Section 5.10   page 157

1. Find the smallest nonnegative integer x that satisfies the equation $3(x + 7) \equiv 4(9 - x) + 1 \pmod{5}$.

That simplifies to $3x + 21 \equiv 36 - 4x + 1 \pmod{5}$ and then $7x \equiv 1 \pmod{5}$. The positive multiples of 7 are $7, 14, 21, \ldots$ and $21 = 7 \cdot 3 \equiv 1 \pmod{5}$, so $x = 3$.

2. Encrypt this message from a supportive shark using a shift-by-10 cipher: YOU ARE SUPER GREAT AND FACES ARE HIGH IN PROTEIN

First we convert to numbers: $24, 14, 20, 0, 17, 4, 18, 20, 15, 4, 17, 6, 17,$ $4, 0, 19, 0, 13, 3, 5, 0, 2, 4, 18, 0, 17, 4, 7, 8, 6, 7, 8, 13, 15, 17, 14,$ $19, 4, 8, 13$.

Then we add 10 (mod 26): $8, 24, 4, 10, 1, 14, 2, 4, 25, 14, 1, 16, 1, 14, 10, 3, 10, 23,$ $13, 15, 10, 12, 14, 2, 10, 1, 14, 17, 18, 16, 17, 18, 23, 25, 1, 24, 3, 14, 18, 23$.

Then we convert back to letters: *iyekbocezobqbokdkxnpkmockborsqrsxzbydosx* and we're done.

3. Prove, using only the definition of congruence modulo n, that if $a \equiv b \pmod{n}$, then $a + c \equiv b + c \pmod{n}$.

   We know from the definition that $a - b = kn$. Therefore $a - b + c - c = kn$ or $a + c - (b + c) = kn$ so $a + c \equiv b + c \pmod{n}$.

4. While you are distraught over your latest discrete math exam, a passerby shoves a scrap of paper into your hand that reads *xvghdibhvivozz 21*. You suspect that this could be a shift cipher. What does the message say?

   First convert to numbers: $23, 21, 6, 7, 3, 8, 1, 7, 21, 8, 21, 14, 25, 25$.

   Then subtract 21 (mod 26) in the hopes that this was the shift: $2, 0, 11, 12, 8, 13, 6, 12, 0, 13, 0, 19, 4, 4$.

   In letters, this is *calmingmanatee*. How nice to be handed a calming manatee!

5. Here is an algorithm:

   1. Get a pot, a cover, a stove, and an egg.
   2. Put the egg in a pot.
   3. Fill the pot with enough water to cover the egg.
   4. Turn a burner to high heat.
   5. Set the pot on the burner.
   6. Put on a hat.
   7. Wait until the water boils.
   8. Wait for 3 minutes.
   9. Remove the pot from the heat and add a cover.
   10. Wait for 10 minutes.
   11. Crack the shell of the egg.
   12. Drain the water, replace with cold water, and let stand for 3 minutes.
   13. Put away the egg.

   What are the inputs? What are the outputs? Does the algorithm terminate? What does the algorithm do? Are there any problems with this algorithm?

   Inputs: a pot, a cover, a stove, an egg, and, later, water and a hat.
   Outputs: A boiled egg.
   Terminate: Yes.
   Action: The algorithm boils an egg.
   Problems: The hat and water (or a sink) are not listed as inputs. There is an excess hat. The stove burner doesn't get turned off.

6. Let $a \sim b$ exactly when ab^2 is even. Is $\sim$ an equivalence relation?

   Let's check: Is $a \sim a$? Yes, a is even if and only if $aa^2 = a^3$ is even, and if a^3 is divisible by 2 then so is a.

If $a \sim b$, then is $b \sim a$? Yes. If ab^2 is even, then so is ab and thus so is ba^2.
If $a \sim b$ and $b \sim c$, then is $a \sim c$? No. Suppose that b is even but a,c are odd; for
example, let $a = 3, b = 4, c = 5$. Then $3 \sim 4$ because $3 \cdot 16 = 48$ is even, and $4 \sim 5$
because $4 \cdot 25 = 100$ is even, but $3 \not\sim 5$ because $3 \cdot 25 = 75$ is odd.
Therefore, this $\sim$ is not an equivalence relation.

7. Write an algorithm that lists the first 10 negative multiples of 9.

   1. Set $k = 1$.
   2. Output $-9 \cdot k$.
   3. If $k = 10$, stop. Otherwise, continue.
   4 Replace k with $k + 1$.
   5. Go to step 2.

8. Encrypt *the foam shark visor is intended only for children* using the original Vigenère cipher with key word *pickles*.

   The message becomes $19, 7, 4, 5, 14, 0, 12, 18, 7, 0, 17, 10, 21, 8, 18, 14, 17, 8, 18,$
   $8, 13, 19, 4, 13, 3, 4, 3, 14, 13, 11, 24, 5, 14, 17, 2, 7, 8, 11, 3, 17, 4, 13$. That's 42 letters.
   The key word becomes $15, 8, 2, 10, 11, 4, 18$. That's 7 letters.
   How convenient! There are exactly 6 repetitions of the key word in the message.
   We add and get $8, 15, 6, 15, 25, 4, 4, 7, 15, 2, 1, 21, 25, 0, 7, 22, 19, 18, 3, \ 12, 5, 8, 12,$
   $15, 13, 15, 7, 6, 2, 19, 0, 15, 25, 21, 20, 22, 16, 13, \ 13, 2, 8, 5$, which comes out as
   ipg pzee hpcbv zahwt sd mfimpnph gcta pzv uwqnncif when translated to letters.

9. Let $\sim$ be defined so that $a \sim b$ exactly when $b - a \geq 2$. Is this an equivalence relation? If so, list the equivalence classes. If not, which of the three properties (reflexive, symmetric, transitive) does not hold?

   $a - a = 0 \not\geq 2$, so $\sim$ is not reflexive.
   Suppose $a = 2$ and $b = 16$. Then $b - a \geq 2$ but $a - b \not\geq 2$, so $\sim$ is not symmetric.
   If $a \sim b$ and $b \sim c$ then $b - a \geq 2$ and $c - b \geq 2$ so $c \geq 2 + b \geq 2 + 2 + a$ and thus
   $c \geq a + 4 > a + 2$, so $c - a \geq 2$ and $\sim$ is transitive.

10. Decrypt *xx ut e kcyrp nvavximtsfl ixoegwwpbggn* using a Vigenère cipher and the key word *pemberley*. Is this an original or a standard Vigenère cipher?

    In numbers, *xx ut e kcyrp nvavximtsfl ixoegwwpbggn* is $23, 23, 20, 19, 4, 10, 2, 24,$
    $17, 15, 13, 21, 0, 21, 23, 8, 12, 19, \ 18, 5, 11, 8, 23, 14, 4, 6, 22, 22, 15, 1, 6, 6, 13,$ and
    pemberley is $15, 4, 12, 1, 4, 17, 11, 4, 24$. That's 9 characters long, so we subtract
    this from the first 9 numbers of the message, and obtain $8, 19, 8, 18, 0, 19, 17, 20, 19$.
    In letters, this is *it is a trut*. (Hm. I wonder what a "trut" is?) In order to determine
    whether this is an original or a standard Vigenère cipher, we test both options: we
    look at the next 9 numbers of the message, namely $15, 13, 21, 0, 21, 23, 8, 12, 19$,
    subtract $15, 4, 12, 1, 4, 17, 11, 4, 24$ to get $0, 9, 9, 25, 17, 6, 23, 8, 21$, which translates

to *ajjzrgxiv*, and subtract $8, 19, 8, 18, 0, 19, 17, 20, 19$ to get $7, 20, 13, 8, 21, 4, 17,$ $18, 0$, which translates to *huniversa*. The second makes a lot more sense than the first, so we will suppose this is an original Vigenère cipher and proceed: Take the next 9 numbers, $18, 5, 11, 8, 23, 14, 4, 6, 22$ and subtract $7, 20, 13, 8, 21, 4, 17, 18, 0$ to get $11, 11, 24, 0, 2, 10, 13, 14, 22$, which translates to *llyacknow*. Only 6 numbers remain, $22, 15, 1, 6, 6, 13$, so we subtract $11, 11, 24, 0, 2, 10$ to get $11, 4, 3, 6, 4, 3$, which translates to *ledged*. Our final text is *it is a truth universally acknowledged...*, the opening line of a famous novel.

Chapter 6, Section 6.14   page 202

1. Find a combinatorial proof for the identity $\sum_{k=0}^{n} k \binom{n}{k} = n2^{n-1}$.

   Let's say we need to make a store display for a bin of n Pretty Rocks. One has to be "featured," meaning that it will be put on top of the Pretty Rocks sign.

   The left-hand side of the equation says that we pick k of the n Pretty Rocks, for $\binom{n}{k}$ ways, and then pick one of those to be featured, for $\binom{k}{1} = k$ ways. But we didn't determine in advance how many Pretty Rocks we were going to put in the display, so we need to sum over k because we could have picked any number.

   The right-hand side of the equation says we pick a Pretty Rock to be featured (n choices for this) and then for each other Pretty Rock, we decide whether or not it's going to be in the display (2 choices for each of the $n - 1$ remaining Pretty Rocks). This completes the proof.

2. Show that if n is even and k is odd, then $\binom{n}{k}$ is even.

   A direct proof might seem like a good idea here, but induction turns out to be *waaay* simpler. Informally, compute a few base cases, and assume that the statement holds for "numerators" less than n. (If we were being formal, we'd need to add a new variable because k already means something here.)

   Note that $\binom{n}{k} = \binom{n-1}{k} + \binom{n-1}{k-1}$. This doesn't help (yet) because $n - 1$ is odd and so the inductive hypothesis doesn't apply. So, we need to break it down further. That expression $= (\binom{n-2}{k} + \binom{n-2}{k-1}) + (\binom{n-2}{k-1} + \binom{n-2}{k-2}) = \binom{n-2}{k} + 2\binom{n-2}{k-1} + \binom{n-2}{k-2}$. That does it. The first and third terms are even because $n - 2$ is even and both k and $k - 2$ are odd, so the inductive hypothesis holds; the second term is clearly even because it's a multiple of 2. Done!

3. Evaluate $\sum_{r=0}^{2m} 3^r 2^{2m-r} \binom{2m}{r}$.   This is the binomial theorem for $(3+2)^{2m} = 5^{2m}$.

4. The four students Ariel, Bingwen, Clarissa, and Dwayne have albums they need to listen to for a music appreciation class: *Duck Rock* (by Malcolm McLaren), *Duck Stab* (by The Residents), *Quack* (by Duck Sauce), and *This Time* (by Galapagos Duck).

   (a) How many ways are there to match the students with the albums? $4!$.

(b) The library has two listening rooms, each of which has two listening stations. How many ways are there to pair the students in the rooms?

There are $\binom{4}{2}$ ways to put two students in one room, and then this determines who is in the other room.

(c) Suppose the students have to sign up in advance, so they have to specify which listening station each student is using. Now how many ways are there for the students to be distributed into the rooms?

There are six ways to assign students to rooms, and for each room there are two possible matchings of students to listening stations. So $6 \cdot 2 \cdot 2 = 24$. *Or*, notice that this is just matching each student to one of four listening stations, so $4! = 24$.

5. Give a combinatorial proof that $\binom{n}{4} = \frac{n!}{4!(n-4)!} = \frac{n(n-1)(n-2)(n-3)}{24}$.

The left-hand side is the number of ways to choose 4 items from n items. The right-hand side counts the number of ways to choose one item from n, then one item from the remaining $n-1$ items, then one item from the remaining $n-2$ items, and finally one item from the remaining $n-3$ items. If we don't care about the order in which these were picked, we need to divide by $4! = 24$. But this is the same as the number of ways of just picking 4 things from n, so we're done.

6. At the art museum, you are decorating a round spinny top with stickers. However, this is an anti-creative art museum, so there are only four equally spaced spots on the spinny top that are designated for receiving stickers, and there are only two colors of sticker available—gray and grey. How many ways are there to "decorate" the spinny top? (There are quotation marks because it is hard to envision the spinny top as actually being decorated....)

Let's see. For each sticker-spot, we could use grey (e) or gray (a), so there are $2^4 = 16$ possibilities. But the spinny top is round, so some of these are actually the same (e.g., *eaea* is the same as *aeae*). We suspect we can just account for our overcounting, but it's not straightforward. We have three cases, (i) all one color: *aaaa*, *eeee*; (ii) one color different: *aeee = eaee = eeae = eeea*, and another four patterns all equivalent to *eaaa*; (iii) two colors each: *eaea = aeae*, and *aaee = aeea = eeaa = eaae*. That's all 16 accounted for, but only 6 are different, so there are 6 ways to "decorate" the spinny top.

7. There are 18 students gathering to work on making a campus duck pond. They need to work in groups of three on various tasks. How many ways are there for the students to form groups?

We may place the first three students in a group in $\binom{18}{3}$ ways. Then we can choose the next group in $\binom{15}{3}$ ways, and so on for a total of $\binom{18}{3}\binom{15}{3}\binom{12}{3}\binom{9}{3}\binom{6}{3}\binom{3}{3}$ ways. However, the order in which we form the groups doesn't matter, so we need to divide by 6!. By the way, that final answer is 190,590,400.

8. Conjecture and prove a binomial identity for $\sum_{i=0}^{n} \binom{i}{5}$.

   We start by generating a few values—oops, this only makes sense for $n \geq 5$:

n	5	6	7	8	9
sum	1	7	28	84	210

   Let's see. These are all numbers from Pascal's triangle, so what might they be? There are lots of 1s, but only a couple of 7s in Pascal's triangle—they're $\binom{7}{1}, \binom{7}{6}$. Looking nearby, we see $28 = \binom{8}{2} = \binom{8}{6}$, $84 = \binom{9}{3} = \binom{9}{6}$, and $210 = \binom{10}{4} = \binom{10}{6}$. We conjecture that $\sum_{i=0}^{n} \binom{i}{5} = \binom{n+1}{6}$.

   Proof by induction: Base cases have already been checked, and we assume the statement holds for $n \leq k$. Here is the inductive step: $\sum_{i=0}^{k+1} \binom{i}{5} = \sum_{i=0}^{k} \binom{i}{5} + \binom{k+1}{5} = \binom{k+1}{6} + \binom{k+1}{5} = \binom{k+2}{6}$ as desired.

9. Find the coefficient of $x^4 y^6$ in $(5x^2 - 3y^3)^4$.

   First, let $a = 5x^2$ and let $b = -3y^3$. From Pascal's triangle, we know the coefficients of $(a+b)^4$ are $(1,4,6,4,1)$. The binomial theorem tells us we have $(5x^2)^4 + 4(5x^2)^3(-3y^3) + 6(5x^2)^2(-3y^3)^2 + 4(5x^2)(-3y^3)^3 + (-3y^3)^4$. The monomial with variables $x^4 y^6$ is $6(5x^2)^2(-3y^3)^2$, and it has coefficient $6 \cdot 5^2 \cdot (-3)^2 = 1{,}350$.

10. Prove that $\binom{2n}{2} = 2\binom{n}{2} + n^2$.

    We will do a combinatorial proof. $\binom{2n}{2}$ is the number of ways of choosing two books from a shelf with $2n$ books. On the other hand, we could split the shelf into two halves of n books each. We could choose two by picking one from each half $(\binom{n}{1}^2 = n^2)$ or by picking both from the same half $(\binom{n}{2}$ for each half).

Chapter 7, Section 7.10   page 237

1. Around Halloween, one can find bags of minipacks of SweeTarts. There are three SweeTarts in each pack, and the available color-flavors are orange, pink, purple, and blue.

   (a) How many different kinds of three-SweeTart minipacks are there?

       The three SweeTarts are unlabeled balls, and the four color-flavors are boxes; a box can get any number of balls (including 0). Thus we have Question D, which has solution $\binom{4+3-1}{3} = \binom{6}{3} = 20$. So 20 different kinds of minipacks.

   (b) Actually, if you open a pack reasonably (instead of ripping it completely apart), you get only one SweeTart out to eat at a time. How many different experiences of three-SweeTart minipacks are there? 4^3.

2. In a 300-home neighborhood of Batamji, there are four different kinds of trees (magnolias, cypress, willow, and river birch). Forty homes have just cypress trees; 32 homes have just willow trees; 9 homes have just river birch. Seventy homes have magnolia and willow; 47 homes have magnolia and cypress; 40 homes have cypress and river birch; 61 homes have magnolia and river birch; 44 homes have cypress and willow; 56 homes have willow and river birch. Twelve homes have magnolias, cypress, willow, and river birch; 38 homes have magnolias, cypress, and willow; 19 homes have magnolias, willow, and river birch; 28 homes have magnolias, cypress, and river birch; 29 homes have cypress, willow, and river birch. How many homes have just magnolia trees?

This is a little bit tricky because we're given some numbers of the form $|A \setminus (B \cup C \cup D)|$ ("just cypress") and others that are $|E \cap F|$ ("willow and river birch") ... that means we only have to deal with overlaps starting with the paired intersections. We have $300 = M + 40 + 32 + 9 + [(70 + 47 + 40 + 61 + 44 + 56) - (38 + 19 + 28 + 29) + 12] = M + 297$, so there are 3 homes with only magnolia trees.

3. A hungry ninja is making tacos with the following ingredients: beans, guacamole, cheese, tomatoes, scallions, salsa, and lettuce. How many ways can the ninja assemble tacos for different meals (breakfast, snack, lunch, tea, dinner), the first of which has three fillings, the next two of which have four fillings, and the final two of which have five fillings?

This initially looks like Question E because we have labeled ingredients (balls) and labeled tacos (boxes). However, we need repetition of the 7 ingredients to get 21 fillings! So we proceed by counting the number of ways there are to construct each taco. For the first taco, we have $\binom{7}{3}$ toppings, for the second two we have $\binom{7}{4}$ toppings, and for the last two we have $\binom{7}{5}$ toppings, for a total of $\binom{7}{3} \cdot \binom{7}{4} \cdot \binom{7}{4} \cdot \binom{7}{5} \cdot \binom{7}{5} = 18,907,875$ meal possibilities for the ninja's day.

4. The Edgy Ruck company uses length-10 serial numbers that mix letters (except Y) and numbers. How many serial numbers are there that have a 7 in the fourth slot and a consonant in the eighth slot, or have a letter in the fifth slot and a vowel in the ninth slot?

First, there are 25 letters, 5 of which are vowels and 20 of which are consonants, and 10 numbers. We will use slots and PIE for this problem.

Serial numbers with a 7 in the fourth slot and a consonant in the eighth slot: $35 \cdot 35 \cdot 35 \cdot 1 \cdot 35 \cdot 35 \cdot 35 \cdot 20 \cdot 35 \cdot 35 = 45,037,507,812,500$.

Serial numbers with a letter in the fifth slot and a vowel in the ninth slot: $35 \cdot 35 \cdot 35 \cdot 35 \cdot 25 \cdot 35 \cdot 35 \cdot 35 \cdot 5 \cdot 35 = 281,484,423,828,125$.

Serial numbers with a 7 in the fourth slot, a letter in the fifth slot, a consonant in the eighth slot, and a vowel in the ninth slot: $35 \cdot 35 \cdot 35 \cdot 1 \cdot 25 \cdot 35 \cdot 35 \cdot 20 \cdot 5 \cdot 35 = 4,595,664,062,500$.

PIE says we have $45{,}037{,}507{,}812{,}500 + 281{,}484{,}423{,}828{,}125 - 4{,}595{,}664{,}062{,}500 = 321{,}926{,}267{,}578{,}125$ serial numbers with 7 in the fourth slot and a consonant in the eighth slot, or have a letter in the fifth slot and a vowel in the ninth slot.

5. All that is left of your Hello Kitty Jelly Belly sampler is the 12 Very Cherry flavored Jelly Bellies (because you *hate* that flavor) and you have four friends who volunteer to eat them for you. How many ways are there to hand out the Jelly Bellies?

Question D, 12 unlabeled balls, 5 labeled boxes, and $\binom{12+4-1}{4} = \binom{16}{4} = 1{,}820$ ways.

6. You've made a pile of eight cute notes for your best friend to find. Ze has 12 folders, one for each of hir classes and activities. How many ways are there to tuck the notes into folders? (Of course, you will not put more than one note in a folder. That would be excessive.)

The notes are all different and so are the folders, so we have labeled balls and labeled boxes. There are 12 choices for the folder for the first note, then 11 for the second, and so forth—$12 \cdot 11 \cdot 10 \cdot 9 \cdot 8 \cdot 7 \cdot 6 \cdot 5 = 19{,}958{,}400$ is the count.

7. The computer print-out says it all: Your first student needs three Learning Modules inserted, your second student needs five Learning Modules inserted, and your third student needs 54 Learning Modules inserted from the bank of 62 new government-approved-topic Learning Modules. But wait… The computer print-out doesn't say which Learning Modules should go to which student. How many ways can you assign Learning Module topics to students?

This is classic Question E: $\binom{62}{3} \cdot \binom{62}{5} \cdot \binom{62}{54} = 827{,}467{,}389{,}801{,}458{,}843{,}800$. Hm. Guess the computer must not care much about which way you assign the topics.

8. Your spiky little plant has once again outgrown its pot, and when you split off all the small bits into different pots, you discover you have 23 spiky plant-spawn. You've promised eight people they can have baby spiky plants, but really you want to get rid of *all* of the spiky plant-spawn so they don't take over your house. How many ways are there to distribute the 23 baby spiky plants to the eight people?

Each person gets at least one plant, so Question D′ gives us $\binom{23-1}{8-1} = 170{,}544$.

9. How many anagrams are there of the word ENUMERATE?

ENUMERATE has 9 letters, of which 3 are Es, and so $9!/3! = 60{,}480$ anagrams.

10. How many ways are there to list the 50 U.S. states so that no two states beginning with "A" are next to each other?

Oh, good grief. What *are* the states that start with "A"? Alaska, Arkansas, Arizona, and Alabama. So we have 46 states that can be ordered any-which-way (for 46! ways), and for each of those orderings we then need to stick the 4 "A" states into the 47 places between and around those 46 states (for $\binom{47}{4}$ ways). Then for each of *those* placements, there are 4! ways we could have ordered the "A" states. In total we have

$46! \cdot \binom{47}{4} \cdot 4!$, which is a number with 65 digits, so I will not write it out. Okay, fine. It's 23,555,404,836,837,197,892,961,193,467,390,979,135,594,520,358,001,049, 600,000,000,000. (This problem can also be solved using PIE, but that solution is much more complicated.)

Chapter 8, Section 8.13   page 272

1. Dandelions reproduce very quickly, as anyone who maintains a lawn knows. In fact, did you know that on any given day, if you went to your lawn and counted the dandelions, then the next day twice as many *new* dandelions will have emerged from the ground? Luckily, dandelions die after two days, so that helps keep the numbers down. Still, if on day 0 you had 1 dandelion, then on day 1 you would have 3 dandelions, on day 2 you'd have 8 dandelions, and then on day 3 you'd have 22 dandelions.

   (a) Write a recurrence equation for d_n = the number of dandelions on day n.

   Let's see what happens day by day.
   Day 0: 1.
   Day 1: $1+2=3$ (one old, two new).
   Day 2: $3+6-1=8$ (the two-day-old one dies).
   Day 3: $8+16-2=22$ (the two two-day-old dandelions die).
   Day 4: $22+44-6=60$. It looks like we have $d_n = 3d_{n-1} - 2d_{n-3}$.

   (b) Find a closed-form formula for d_n.

   The characteristic polynomial is $x^n = 3x^{n-1} - 2x^{n-3}$, which becomes $x^3 = 3x^2 - 2$ and then $x^3 - 3x^2 + 2 = 0$, which has roots $x = 1, x = 1 + \sqrt{3}$. Now we have $d_n = q_1(1)^n + q_2(1+\sqrt{3})^n + q_3(1-\sqrt{3})^n$. For low values of n, this gives us
   $d_0 = 1 = q_1 + q_2 + q_3$,
   $d_1 = 3 = q_1 + q_2(1+\sqrt{3}) + q_3(1-\sqrt{3})$,
   $d_2 = 8 = q_1 + q_2(1+\sqrt{3})^2 + q_3(1-\sqrt{3})^2 = q_1 + q_2(4+2\sqrt{3}) + q_3(4-2\sqrt{3})$.
   Look at $-2d_1 + d_2$ to get $2 = -q_1 + 2q_2 + 2q_3$. Add $-2d_0$ to both sides to get $0 = -3q_1$ or $q_1 = 0$. The equations for d_0, d_1 now become $1 = q_2 + q_3$ and $3 = q_2(1+\sqrt{3}) + q_3(1-\sqrt{3})$. Substitute $q_2 = 1 - q_3$ to get $3 = (1-q_3)(1+\sqrt{3}) + q_3(1-\sqrt{3}) = 1+\sqrt{3} - 2\sqrt{3}q_3$, so that $q_3 = \frac{\sqrt{3}-2}{2\sqrt{3}}$ and $q_2 = 1 - \frac{\sqrt{3}-2}{2\sqrt{3}} = \frac{1}{2} + \frac{1}{\sqrt{3}}$.
   Our final formula is $d_n = (\frac{1}{2} + \frac{1}{\sqrt{3}})(1+\sqrt{3})^n + \frac{2-\sqrt{3}}{2\sqrt{3}}(1-\sqrt{3})^n$. Yuck!

2. Generate the first 30 terms of the sequence $a_n = a_{n-1} + a_{n-2} - a_{n-3}$, $a_0 = 0, a_1 = 1, a_2 = 1$.

   This is so much fun we'll generate 40 terms: $1, 1, 2, 2, 3, 3, 4, 4, 5, 5, 6, 6, 7, 7, 8, 8, 9, 9, 10, 10, 11, 11, 12, 12, 13, 13, 14, 14, 15, 15, 16, 16, 17, 17, 18, 18, 19, 19, 20, 20$. Ha!

3. Suppose that $a_n = (-4)^n$, $a_n = 1$, and $a_n = 2^n$ are all closed forms for the same recurrence. Find a recurrence that fits this criterion and verify that it really does work for all three closed forms.

This criterion suggests we have a geometric sequence whose characteristic polynomial has roots $r_1 = -4, r_2 = 1, r_3 = 2$. That characteristic polynomial would be $(x+4)(x-1)(x-2) = x^3 + x^2 - 10x + 8 = 0$, or $x^3 = -x^2 + 10x - 8$, which corresponds to a recurrence $a_n = -a_{n-1} + 10a_{n-2} - 8a_{n-3}$.

Can we find initial conditions that correspond to these three sequences? Check it out: If $a_n = (-4)^n$, then $a_0 = 1, a_1 = -4, a_2 = 16, a_3 = -64$, and $-64 = -16 + 10(-4) - 8$. If $a_n = 1$, then $a_0 = 1, a_1 = 1, a_2 = 1, a_3 = 1$, and $1 = -1 + 10(1) - 8$. If $a_n = 2^n$, then $a_0 = 1, a_1 = 2, a_2 = 4, a_3 = 8$, and $8 = -4 + 10(2) - 8$. Yes!

4. Consider the sequence $1, 3, 4, 7, 11, 18, 29, \ldots$.

   (a) Find a recurrence that L_n satisfies. $L_n = L_{n-1} + L_{n-2}$.

   (b) Prove that $L_n = F_{n-1} + F_{n+1}$.

   We'll do this by induction. Do we want $L_0 = 1$ or $L_1 = 1$? Note that $3 = 1 + 2$, so we have $L_2 = F_1 + F_3$. We'll check another couple of base cases to be sure: $L_3 = 4 = 1 + 3 = F_2 + F_4$ and $L_4 = 7 = 2 + 5 = F_3 + F_5$.

   Now assume that our statement holds for values of $n \leq k$. $L_{k+1} = L_k + L_{k-1}$ and the inductive hypothesis gives us $L_{k+1} = F_{k-1} + F_{k+1} + F_{k-2} + F_k = (F_{k-1} + F_{k-2}) + (F_{k+1} + F_k) = F_k + F_{k+2}$ as desired.

5. Find a closed-form formula for the sequence $a_0 = -1, a_n = a_{n-1} + 3n + 1$.

   This is an arithmetic sequence, so we use the technique of finite differences.
   Sequence: $-1, 3, 10, 20, 33, 49, 68, 90, 115, 143, 174, \ldots$.
   First differences: $4, 7, 10, 13, 16, 19, 22, 25, 28, 31, \ldots$.
   Second differences: $3, 3, 3, 3, 3, 3, 3, 3, 3, \ldots$.
   The closed form is generically $a_n = c + dn + fn^2$. When $n = 0$, we have $-1 = c$. When $n = 1$, we have $3 = c + d + f$ or $4 = d + f$ or $f = 4 - d$. When $n = 2$, we have $10 = c + 2d + 4f$ or $11 = 2d + 4(4 - d) = 16 - 2d$. Then $d = \frac{5}{2}$ and $f = \frac{3}{2}$, so $a_n = -1 + \frac{5}{2}n + \frac{3}{2}n^2$.

6. Consider the recurrence relation $a_n = 3a_{n-1} - a_{n-2}$ with $a_0 = 0, a_1 = 1$. Generate some terms, make a conjecture as to what sequence this is, try to find the closed form, and try to explain what is going on here.

   Terms: $0, 1, 3, 8, 21, 55, \ldots$. This looks like every second Fibonacci number. I wonder what kind of formula that will give us? Let's start with $F_n = F_{n-1} + F_{n-2}$. We want only the even ones, so F_{2k}. Those have the form $F_{2k} = F_{2k-1} + F_{2k-2}$. But $2k - 1$ is odd, so let's break that down further. (Uh-oh. This might never end.) $F_{2k} = (F_{2k-2} + F_{2k-3}) + F_{2k-2}$. But wait, $F_{2k-2} = F_{2k-3} + F_{2k-4}$ can be written as $F_{2k-3} = F_{2k-2} - F_{2k-4}$, which means we get $F_{2k} = (F_{2k-2} + F_{2k-2} - F_{2k-4}) +$

$F_{2k-2} = 3F_{2k-2} - F_{2k-4}$. And *that* can be rewritten as $F_{2(k)} = 3F_{2(k-1)} - F_{2(k-2)}$, or, ditching the 2s, as $a_n = 3a_{n-1} - a_{n-2}$. So that's what's going on—this recurrence is the same as for every second Fibonacci number.

Now we have to find a closed form. Okay. Characteristic equation: $x^n = 3x^{n-1} - x^{n-2}$ becomes $x^2 = 3x - 1$ or $x^2 - 3x + 1 = 0$. The roots of this equation are $\frac{3-\sqrt{5}}{2}, \frac{3+\sqrt{5}}{2}$. Therefore we have $a_n = q_1(\frac{3-\sqrt{5}}{2})^n + q_2(\frac{3+\sqrt{5}}{2})^n$. For low values of n, we have $a_0 = 0 = q_1 + q_2$ and $a_1 = 1 = q_1(\frac{3-\sqrt{5}}{2}) + q_2(\frac{3+\sqrt{5}}{2})$. Now $q_2 = -q_1$, so $1 = q_1(\frac{3-\sqrt{5}}{2}) - q_1(\frac{3+\sqrt{5}}{2}) = q_(-4\sqrt{5})$; so $q_1 = \frac{-1}{4\sqrt{5}}$ and $q_2 = \frac{1}{4\sqrt{5}}$ for a final closed-form formula of $a_n = (\frac{-1}{4\sqrt{5}})(\frac{3-\sqrt{5}}{2})^n + (\frac{1}{4\sqrt{5}})(\frac{3+\sqrt{5}}{2})^n$.

7. Consider the sequence $5, -3, 5, \ 3, 5, \ 3, 5, -3, 5, -3, 5, \ldots$. Find a recurrence for this sequence, and find two more (different) sequences that satisfy that recurrence.

    One recurrence for this sequence is $a_n = -a_{n-1} + 2$.
    Starting with $a_0 = 1$, we get $1, 1, 1, 1, 1, 1, 1, 1, 1, 1, 1, \ldots$.
    Starting with $a_0 = 0$, we get $0, 2, 0, 2, 0, 2, 0, 2, 0, 2, 0, \ldots$.

8. Find a closed form for the sequence defined by the recurrence $a_n = -a_{n-1}a_{n-2} + 2, a_0 = 1, a_1 = 1$. How do things change if $a_0 = 0, a_1 = 0$?

    The sequence is $1, 1, 1, 1, 1, 1, 1, 1, 1, 1, 1, \ldots$, so the closed form is $a_n = 1$.

    If we change the initial conditions, we get the sequence $0, 0, 2, 2, -2, 6, 14, -82, 1150, 94302, -108447298, \ldots$, which is (as of August 2018) not even in the OEIS.

9. Here is a characteristic equation: $x^5 + 4x^3 - 3x^2 - 1 = 0$. What is the associated recurrence?

    First rewrite as $x^5 = -4x^3 + 3x^2 + 1$, then multiply through by x^{n-5} to get $x^n = -4x^{n-2} + 3x^{n-3} + x^{n-5}$. Finally, translate to $a_n = -4a_{n-2} + 3a_{n-3} + a_{n-5}$.

10. Find a closed-form formula for the sequence $a_0 = 1, a_n = a_{n-1} + n^2 - 2n$.

    This is an arithmetic sequence, so we use the technique of finite differences.
    Sequence: $1, 0, 0, 3, 11, 26, 50, 85, 133, 196, 276, \ldots$.
    First differences: $-1, 0, 3, 8, 15, 24, 35, 48, 63, 80, \ldots$.
    Second differences: $1, 3, 5, 7, 9, 11, 13, 15, 17, \ldots$.
    Third differences: $2, 2, 2, 2, 2, 2, 2, 2, 2, \ldots$.
    Closed form is generically $a_n = c + dn + fn^2 + gn^3$.
    When $n = 0$, we have $1 = c$.
    When $n = 1$, we have $0 = c + d + f + g$ or $g = -1 - d - f$.
    When $n = 2$, we have $0 = c + 2d + 4f + 8g$ or
    $0 = 1 + 2d + 4f + 8(-1 - d - f) = -7 - 6d - 4f$ or $f = \frac{-7-6d}{4}$.
    When $n = 3$, we have $3 = c + 3d + 9f + 27g$ or
    $3 = 1 + 3d + 9(\frac{-7-6d}{4}) + 27(-1 - d - \frac{-7-6d}{4}) = 3d + \frac{11}{2}$.
    Then $d = \frac{-5}{6}$, $f = \frac{-1}{2}$, and $g = \frac{1}{3}$, so $a_n = 1 + \frac{-5}{6}n + \frac{-1}{2}n^2 + \frac{1}{3}n^3$.

1. Let f_n be the maximum number of regions of four-dimensional space that are cut up by n three-dimensional cuts. What are f_0, f_1, f_2, f_3, f_4? And why?

   $f_0 = 1, f_1 = 2, f_2 = 4, f_3 = 8, f_4 = 16$ because there are four perpendicular directions and each of these four cuts can be perpendicular to the previous cuts.

2. If you cut a configuration with f_4 pieces with an additional cut, how many new pieces can you get?

   There can be as many as 15 new pieces, which means that the additional cut passes through all but one of the existing regions—see the answer for the next question!

3. Determine and explain a recurrence relation for f_n.

   $f_n = f_{n-1} + y_{n-1}$. Each of the n cuts is really a yam with a maximum number of planes passing through it. There are $n-1$ planes formed by intersecting the three-dimensional cuts/yams with the nth cut (yam). Each of those y_{n-1} regions of the yam represents a cutting-in-two of the region it passes through in four-dimensional space. So we add y_{n-1} to the f_{n-1} regions we already had.

4. Determine and explain a closed form for f_n.

   $f_n = \binom{n}{4} + \binom{n}{3} + \binom{n}{2} + \binom{n}{1} + \binom{n}{0}$. If we have a maximum number of regions, then the cuts are in general position. That means that any four cuts intersect in exactly one point. Consider an extra cut; if we sweep it across the four-dimensional space, the number of point-intersections it crosses will be $\binom{n}{4}$. Every time it passes an intersection, it crosses into a new region. And after it has passed through all of the intersections, it has $\binom{n}{2} + \binom{n}{1} + \binom{n}{0}$ regions passing through it because that's the number of regions *on* it of lower dimension.

5. Use induction to prove that your closed form from Problem 4 is the correct closed form for your recurrence from Problem 3.

   Here are some base cases:
   $n = 1$: $\binom{1}{4} + \binom{1}{3} + \binom{1}{2} + \binom{1}{1} + \binom{1}{0} = 0 + 0 + 0 + 1 + 1 = 2$.
   $n = 2$: $\binom{2}{4} + \binom{2}{3} + \binom{2}{2} + \binom{2}{1} + \binom{2}{0} = 0 + 0 + 1 + 2 + 1 = 4$.
   $n = 3$: $\binom{3}{4} + \binom{3}{3} + \binom{3}{2} + \binom{3}{1} + \binom{3}{0} = 0 + 1 + 3 + 3 + 1 = 8$.
   $n = 4$: $\binom{4}{4} + \binom{4}{3} + \binom{4}{2} + \binom{4}{1} + \binom{4}{0} = 1 + 4 + 6 + 4 + 1 = 16$.

   Suppose that for $n \leq k$, we have that $f_n = \binom{n}{4} + \binom{n}{3} + \binom{n}{2} + \binom{n}{1} + \binom{n}{0}$. Consider f_{k+1}. The recurrence says $f_{k+1} = f_k + y_k$. Using the inductive hypothesis, we have $f_{k+1} = (\binom{k}{4} + \binom{k}{3} + \binom{k}{2} + \binom{k}{1} + \binom{k}{0}) + y_k$, and from the closed form we calculated for y_k, we have that $f_{k+1} = (\binom{k}{4} + \binom{k}{3} + \binom{k}{2} + \binom{k}{1} + \binom{k}{0}) + (\binom{k}{3} + \binom{k}{2} + \binom{k}{1} + \binom{k}{0})$. Rearranging terms, we have $f_{k+1} = (\binom{k}{4} + \binom{k}{3}) + (\binom{k}{3} + \binom{k}{2}) + (\binom{k}{2} + \binom{k}{1}) + (\binom{k}{1} + \binom{k}{0}) + \binom{k}{0}$, and using our basic choice identity four times while noting that $\binom{k}{0} = 1 = \binom{k+1}{0}$, we get $f_{k+1} = \binom{k+1}{4} + \binom{k+1}{3} + \binom{k+1}{2} + \binom{k+1}{1} + \binom{k+1}{0}$ as desired.

Figure B.4. A graph underlying two of its spanning trees.

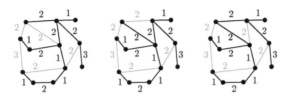

Figure B.5. An edge-weighted graph with all of its minimum-weight spanning trees.

Chapter 10, Section 10.12   page 339

1. Find two different spanning trees of the graph shown at left in Figure 10.22.

   Figure B.4 shows two different spanning trees of the relevant graph in Figure 10.22.

2. Find two different minimum-weight spanning trees of the graph shown at right in Figure 10.22. Are there more?

   Figure B.5 shows all three different minimum-weight spanning trees. You can show that these are the only three by eliminating the weight-3 edge e and then successively eliminating the highest-weight edges incident to e's vertices, etc.

3. Find, if possible, a perfect matching in each of the graphs shown in Figure 10.22.

   The left-hand graph has an odd number of vertices, so none is possible. The right-hand graph has several; the lowest-weight perfect matching has total weight 9.

4. Prove that for $n \geq 3$, every n-vertex tree has at most $n - 1$ leaves.

   We proceed by contradiction. Suppose an n-vertex tree has more than $n - 1$ leaves. It must have at least n leaves, so every vertex is a leaf. However, by the handshake lemma, the total degree (n) must equal twice the number of edges. Twice the number of edges is $2(n - 1)$ because any n-vertex tree has $n - 1$ edges. The statement $n = 2(n - 1)$ implies $n = 2$, which violates the constraints of the theorem. (As a side note, the star graph is an n-vertex tree with exactly $n - 1$ leaves.)

5. Create a binary search tree for the mini-dictionary {*block, black, brack, bract, brace, trace, race, ace, mace, maze, maize, baize*}.

   Figure B.6 shows such a binary search tree. It was created by placing *block* at the root and then adding the remaining words to the tree in the order they were listed.

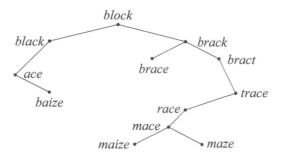

Figure B.6. A binary search tree for a mini-dictionary.

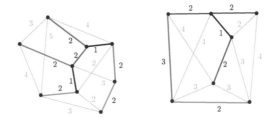

Figure B.7. Two edge-weighted graphs with minimum spanning trees highlighted.

6. Find a minimum-weight spanning tree of the graph shown at left in Figure 10.23 using Kruskal's algorithm.

Kruskal's algorithm performed on the relevant graph is shown at left in Figure B.7. Edges added later are shown in lighter grey tones. There is some choice as to which edges of a given weight are added, so your solution may differ and still be correct.

7. Create an efficient binary decision tree for identifying members of the set {coat, mittens, hat, scarf, duck, boots}.

We want to avoid questions for which all items answer *yes* (or *no*), and questions for which just one item answers *yes* (or *no*) because then we'll need more questions, so "Is it warm?" is terrible. Figure B.8 gives one possible efficient tree.

8. Prove that in any tree with at least two vertices, any two vertices are connected by a unique minimum-length path.

We proceed by contradiction. Suppose there exist two vertices v, w in a tree that are connected by at least two distinct minimum-length paths P and Q. Starting at v and heading towards w, there is some first vertex after which P, Q differ. Call this z (possibly with $z = v$). Continuing along P, Q, there is some first vertex after z that

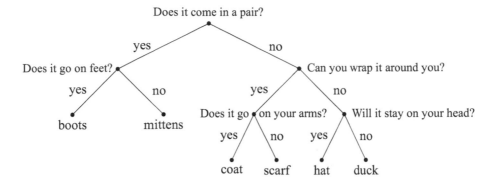

Figure B.8. This binary decision tree helps us identify warm winter wear.

P, Q have in common. Call this y. Now there are two completely distinct paths from z to y, and together these form a cycle. This contradicts tree-ness.

9. Use backtracking to find all the ways to add numbers from $\{1, 2, 3, 4, 5\}$ to get 8.

1 isn't enough. $1 + 2$ isn't enough. $1 + 2 + 3$ isn't enough. $1 + 2 + 3 + 4$ is too much. Go back.
$1 + 2 + 4$ isn't enough. $1 + 2 + 4 + 5$ is too much. Go back.
$1 + 2 + 5$ is exactly right! Keep that and go back.
$1 + 3$ isn't enough. $1 + 3 + 4$ is exactly right! Keep that and go back.
$1 + 4$ isn't enough. $1 + 4 + 5$ is too much. Go back.
$1 + 5$ isn't enough. Go back.
2 isn't enough. $2 + 3$ isn't enough. $2 + 3 + 4$ is too much. Go back.
$2 + 4$ isn't enough. $2 + 4 + 5$ is too much. Go back.
$2 + 5$ isn't enough. Go back.
3 isn't enough. $3 + 4$ isn't enough. $3 + 4 + 5$ is too much. Go back.
$3 + 5$ is exactly right! Keep that and go back.
4 isn't enough. $4 + 5$ is too much. Go back.
5 isn't enough.
Report: $8 = 1 + 2 + 5 = 1 + 3 + 4 = 3 + 5$.

10. Find a minimum-weight spanning tree of the graph shown at right in Figure 10.23 using Prim's algorithm.

Prim's algorithm performed on the relevant graph is shown at right in Figure B.7. Edges added later are shown in lighter grey tones. There is some choice as to which edges of a given weight are added, so your solution may differ and still be correct.

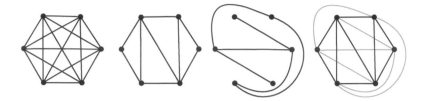

Figure B.9. Ordinary K_6; two planar graphs that make K_6 when glued at the vertices; K_6 drawn to highlight the two planar graphs that comprise it.

Chapter 11, Section 11.12   page 364

1. Compute the thickness of K_6.

   First, let's see what Theorem 11.6.7 says: $t(K_6) \geq \left\lceil \frac{15}{3\cdot6-6} \right\rceil = 2$. This isn't particularly enlightening—we already knew that K_6 is nonplanar. If we can exhibit K_6 as having thickness 2 we'll be done. See Figure B.9 for such an exhibit.

2. Check out Figure 11.14 to see an image of an annulus (or washer).

   (a) Draw a few graphs on annuli (that's the plural of *annulus*). The rule here is that you have to cover the annulus edges with graph edges (and vertices) so that you don't have partial faces.

      See Figure B.10.

   (b) Try out Euler's formula on these graphs. Does it still hold? If not, does some other formula hold?

      For the two graphs drawn in Figure B.10, we have $7 - 11 + 4 = 0$ and $14 - 27 + 13 = 0$. Euler's formula does not hold, but instead $|V(G)| - |E(G)| + |F(G)| = 0$.

   (c) Prove your conjecture.

      Notice that if we fill in the middle of an annulus with a face, and fill in the outside as well, we have a planar graph! The difference is that we have two extra faces. So, we have a one-to-one correspondence between annular graphs and

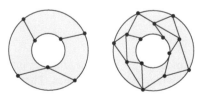

Figure B.10. Two sample annular graphs.

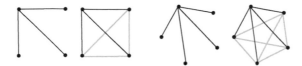

Figure B.11. Three-spoke and four-spoke stars shown together with their complements.

certain planar graphs. If we have an annular graph G and associated planar graph G', then we know $|V(G)| = |V(G')|$, $|E(G)| = |E(G')|$, and $|F(G)| = |F(G')| - 2$. From Euler's formula we have $|V(G')| - |E(G')| + |F(G')| = 2$. Substituting, we get $|V(G)| - |E(G)| + |F(G)| + 2 = 2$ or $|V(G)| - |E(G)| + |F(G)| = 0$ as desired.

3. Is the complement of any star graph planar? Are *all* complements of star graphs planar? Justify your responses.

The smallest star graphs and their complements are shown in Figure B.11. We have drawn the star graphs so that the central vertex is to the side, so that they look more like fans than stars. If we draw K_n and use this as a base for drawing K_{n+1}, what we're adding is a star graph—one vertex with n spokes! In other words, the complement of a star graph with n spokes is K_n. Therefore, the only star graphs with planar complements are those shown in the figure.

4. Can there exist a planar graph with degree sequence $(1,2,2,2,3,5,5,6)$?

The proposed graph has 8 vertices and total degree $1+2+2+2+3+5+5+6 = 26$, so it has 13 edges. Let's check Theorem 11.6.1: $13 \leq 3 \cdot 8 - 6 = 18$. So it seems possible; see Figure B.12 for such a graph with multiple edges.

5. Could the graph at left in Figure 11.15 be planar?

Nope. It has 12 vertices and 38 edges, and $38 \not\leq 3 \cdot 12 - 6 = 30$, so we have a contradiction by Theorem 11.6.1.

6. The graph at right in Figure 11.15 is definitely planar. How many faces does a planar drawing of this graph have?

It has 14 vertices and 24 edges, so $14 - 24 + f = 2$ implies it has 12 faces.

Figure B.12. A planar graph with degree sequence $(1,2,2,2,3,5,5,6)$.

Figure B.13. See how not very thick I am?   Figure B.14. I am a general n-prism graph sporting a fine Hamilton circuit.

7. Prove that a connected planar graph has exactly one face if and only if it is a tree.

First, we know that a tree has exactly one face by inspection. So we need to prove that if a planar graph has exactly one face, then it must be a tree.

Consider a planar graph with exactly one face. Then we have $v - e + 1 = 2$ or $e = v - 1$, and Theorem 10.2.1 implies that the graph must be a tree.

8. How many vertices must a 4-regular planar graph with 12 faces have?

A 4-regular planar graph with v vertices has total degree $4v$ and so has $2v$ edges. Euler's formula gives us that $v - 2v + 12 = 2$, or $v = 10$.

9. Can a planar graph with nine vertices and all faces of size 4 be k-regular for any k?

All faces are of size 4, so $2e = 4f$, which gives us $v - e + \frac{e}{2} = 2$ or $e = 2(v - 2)$. Having 9 vertices means there are 14 edges. The total degree is 28, which is not divisible by 9, so the graph cannot be regular.

10. Compute the thickness of the nonplanar Grötzsch graph, shown in Figure 11.16.

Start with Theorem 11.6.7: $t(GG) \geq \lceil \frac{20}{3\cdot 11 - 6} \rceil = 2$. If we can exhibit the Grötzsch graph as having thickness 2, we'll be done. See Figure B.13, and notice that every crossing involves one of the five grey edges. Together, the grey edges are planar, as is the remaining black-edged graph.

Chapter 12, Section 12.12   page 394

1. An n-prism graph is constructed by putting one (slightly smaller) n-cycle C_n inside another, and adding edges to join the vertices of one C_n to the other radially. (We do need $n > 3$.) See Figure 12.17 for an example. Do any n-prism graphs have Euler circuits? What about Hamilton circuits?

No Euler circuits (or even trails), by Theorem 12.3.1—all vertices have degree 3. However, as shown in Figure B.14, every n-prism graph has a Hamilton circuit.

2. List all possible orderings of ABC (how many are there?). Associate each of these orderings to a vertex of a graph. Add an edge when two orderings differ only by an adjacent transposition.

   (a) What is the degree sequence of this graph?

   We have 6 orderings: ABC, ACB, CAB, CBA, BCA, and BAC. (These are written so that each differs by the next by one adjacent transposition.) Each ordering has two possible adjacent transpositions, so the degree sequence is $(2,2,2,2,2,2)$.

   (b) Does it have an Euler circuit or trail?   Yes.

   (c) Does it have a Hamilton circuit or trail?   Yes.

   (d) Is it planar?

   Yes. The graph is a 6-cycle, so it is planar and is an Euler circuit and a Hamilton circuit.

   (e) What are the answers to the previous questions if we also consider the first and last letters to be adjacent?

   If we consider the first and last letters to be adjacent, there are three possible adjacent transpositions for each ordering; we have degree sequence $(3,3,3,3,3,3)$; and we add the edges ABC–CBA, ACB–BCA, and CAB–BAC to our 6-cycle. The resulting graph has a Hamilton circuit (the same as the cycle had) but no Euler circuit (all vertices are of odd degree); it is isomorphic to $K_{3,3}$, which is not planar.

3. Look at the graphs in Figure 10.23 on page 340. Does either have a Hamilton circuit? ... Hamilton traversal? ... Euler circuit? ... Euler traversal?

   Each graph has at least one Hamilton circuit, as shown in Figure B.15.

   The left-hand graph has four vertices of degree 3 and so has no Euler traversal by Theorem 12.3.1. The right-hand graph has two vertices of degree 3 and the rest are of degree 4, so it has an Euler trail but no Euler circuit.

 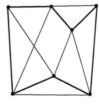

Figure B.15. We are very Hammy.

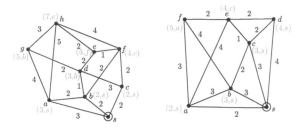

Figure B.16. **We are very Dijkstry.**

4. Again examine Figure 10.23 on page 340. For each graph, compute the shortest distance from the lower-right vertex to all other vertices. (Tip: Dijkstra is a good choice here.)

   The results of applying Dijkstra's algorithm are shown in Figure B.16.

5. Do any of the graphs in Figure 12.18 have Hamilton circuits? What about Hamilton traversals?

   All of them have Hamilton circuits, as shown in Figure B.17.

6. Do any of the graphs in Figure 12.18 have Euler circuits? What about Euler traversals?

   The left-hand graph has more than three vertices of degree 3, so by Theorem 12.3.1 there's no chance it can have an Euler traversal. The middle graph has all vertices of even degree, so has an Euler circuit. The right-hand graph has exactly two vertices of degree 3, so has an Euler trail but not an Euler circuit.

7. For which values of m, n does $K_{m,n}$ have a Hamilton circuit?

   $K_{m,n}$ has a Hamilton circuit exactly when $m = n$.

   First, note that when $m = n$ we can construct a Hamilton circuit by zig-zagging as in Figure B.18. Next, recall that in a bipartite graph, any path must alternate between

Figure B.17. **Three graphs with Hamilton circuits shown. Yup.**   Figure B.18. **A Hamilton circuit in $K_{n,n}$.**

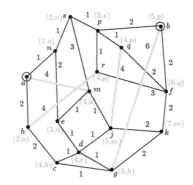

Figure B.19. A map of Altana toll plazas.

the parts. Suppose $m > n$. Our longest path (with no vertices repeated) will only reach $n + 1$ vertices of the m vertices. Even if $m = n + 1$, we only have a Hamilton path and not a Hamilton circuit.

8. The towns Gesund and Reichtum are near each other in a tourism district. In each town, all but two of the intersections are four-way stops. In Gesund, there is a five-way stop and a "T" intersection (a three-way stop), and in Reichtum there are two five-way stops. Currently, there is no direct road between Gesund and Reichtum. The tourism bureau wants to build a road so that they can create and advertise a Tour of the Towns, which will take tourists down every road of Gesund and of Reichtum without repetition. What advice can you give the tourism bureau?

   Build *two* roads, each connecting one of Reichtum's five-way stops with one of Gesund's odd-way stops. Then every intersection will have even degree, and by Theorem 12.3.1 an Euler circuit will exist.

9. In the metropolis of Altana, the Traffic Council has decreed that cars in the flying lanes must pay twice the tolls of ground-based cars (because of the additional fuel needed for flying police). What is the cheapest way to get from point a to point b? A map showing skyways in grey and ground-roads in black is shown in Figure 12.19—those dots are toll stations where you pay for the segment you've just traveled.

   Figure B.19 shows the result of running Dijkstra's algorithm on the Altana map; the cheapest toll route costs $5 and goes through n, s, and p.

10. Can you take one walk and cover every road in the map of Snakeland given in Figure 12.20 exactly once?

   By Theorem 12.3.1, no, because there are two vertices of degree 3.

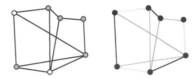

Figure B.20. A proper vertex coloring and a proper edge coloring of the graph in question.

Chapter 13, Section 13.9   page 429

1. Find the chromatic number and chromatic index of the graph shown in Figure 10.3 on page 316.

   Let the pictured graph be G. $\chi(G) = 3$; G contains K_3, so $\chi(G) \geq 3$, and we exhibit a 3-vertex coloring of G in Figure B.20. $\chi'(G) = 4$; G has a vertex of degree 4, so $\chi'(G) \geq 4$, and we exhibit a 4-edge coloring of G in Figure B.20.

2. Prove that if $\chi(G) \geq 3$, then G must contain an odd cycle.

   We proceed by contradiction. Suppose that G has no odd cycles; then by Theorem 13.5.3, G is bipartite and then by Theorem 13.5.2, G is 2-vertex-colorable. Contradiction! We know $\chi(G) \geq 3$; therefore, G must contain an odd cycle.

3. Find the chromatic number and chromatic index of each graph shown in Figure 10.22 on page 339.

   Let the left-hand graph be G and the right-hand graph be H. $\chi(G) = 3$ and $\chi'(G) = 6$; G contains K_3, so $\chi(G) \geq 3$, and G has a vertex of degree 6, so $\chi'(G) \geq 6$. We exhibit a 3-vertex coloring of G and a 6-edge coloring of G in Figure B.21. $\chi(H) = 3$ and $\chi'(H) = 4$; H contains K_3, so $\chi(H) \geq 3$, and H has a vertex of degree 4, so $\chi'(H) \geq 4$. We exhibit a 3-vertex coloring of H and a 4-edge coloring of H in Figure B.21.

4. Find the chromatic number and chromatic index of the graph shown in Figure 11.16 on page 365.

   Let the pictured graph be GG. $\chi(GG) = 4$. We know that the outer 5-cycle requires three colors, so let us suppose that there exists a 3-coloring of GG. Because GG has 5-fold rotational symmetry, we can 3-color that cycle as we please. This coloring

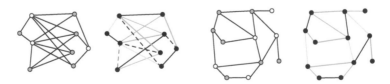

Figure B.21. Proper vertex coloring and proper edge coloring of each graph in question.

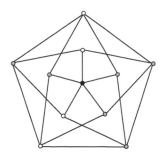 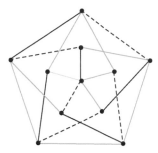

Figure B.22. A proper vertex coloring and a proper edge coloring of the Grötzsch graph.

forces the colors on the spokes of the inner star, and all three colors are present there. The middle vertex is adjacent to all of them, so it requires a fourth color. $\chi'(GG) = 5$; GG has a vertex of degree 5, so $\chi'(GG) \geq 5$. We exhibit a 4-coloring and a 5-edge coloring of GG in Figure B.22.

5. Let G be a planar graph with smallest cycle length (girth) 6. Let $v_G = |V(G)|$, $e_G = |E(G)|$, and $f_G = |F(G)|$.

(a) Develop an inequality that relates f_G to e_G.

First note that because the graph is planar, we can use Euler's formula and its consequences. The smallest cycle length is 6, so every face has size at least 6. The sum of the sizes of the faces is equal to $2e_G$. If all the faces were size 6, we would have $6f_G = 2e_G$, but some faces may be larger, and so we have $6f_G \leq 2e_G$.

(b) Use this to show that $2e_G \leq 3v_G - 6$.

Euler's formula says that $f_G = 2 - v_G + e_G$. Plugging this into $6f_G \leq 2e_G$ gives us $6(2 - v_G + e_G) = 12 - 6v_G + 6e_G \leq 2e_G$ or $6 - 3v_G + 3e_G \leq e_G$ or $2e_G \leq 3v_G - 6$.

(c) Show that G must have a vertex of degree less than 3.

We proceed by contradiction. Suppose that G has only vertices of degree 3 or more. Then the total degree of G is at least $3v_G$. By the handshaking lemma, the total degree of G is $2e_G$. Therefore $2e_G \geq 3v_G$. However, $2e_G \leq 3v_G - 6$, so we have $3v_G \leq 2e_G \leq 3v_G - 6$ or $3v_G \leq 3v_G - 6$ or $0 \leq -6$, which is a contradiction. Therefore G must have a vertex of degree less than 3.

(d) Prove that $\chi(G) \leq 3$. (Hint: use induction.)

Now we proceed by induction on the number of vertices of G. As a base case we take $v_G = 6$ so G has at least one cycle. We know this can be colored with only two colors, so $\chi(G) \leq 3$. Assume (for the inductive hypothesis) that if G has $6 \leq n \leq k$ vertices, then $\chi(G) \leq 3$. Consider G with $k + 1$ vertices and girth at least 6. We know that G must have a vertex y of degree less

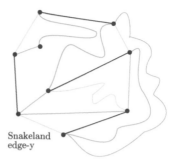

Snakeland
vertex-y

Snakeland
edge-y

Figure B.23. A proper vertex coloring and a proper edge coloring of the Snakeland map
graph.

than 3. Remove this vertex to form H. The inductive hypothesis applies to
H because either H still has girth at least 6 or H has no cycles (in which case
it is 2-vertex-colorable). So, color H with at most three colors. The vertex y
has at most two neighbors, so it can be given a third color—we restore y to
the graph, colored with this third color, and have G properly vertex colored
using at most three colors. *Finis.*

6. Without doing any actual coloring, give quick lower and upper bounds for the chromatic number and chromatic index of the graph shown in Figure 13.25.

The graph contains a K_5 and has highest degree 8, so $5 \leq \chi(G) \leq 9$. The graph has
highest degree 8, so $8 \leq \chi'(G) \leq 15$.

7. Find the chromatic number and chromatic index of the Snakeland map graph shown
in Figure 12.20 on page 395.

Let the Snakeland map graph be *ssss*. $\chi(ssss) = 3$ and $\chi'(ssss) = 4$; *ssss* contains
K_3, so $\chi(ssss) \geq 3$, and *ssss* has a vertex of degree 4, so $\chi'(ssss) \geq 4$. We exhibit a
3-vertex coloring of *ssss* and a 4-edge coloring of *ssss* in Figure B.23.

8. During the Week of Chaos at MathILy 2016, there were five timeslots for classes
and four classes offered in each timeslot. Six instructors taught three classes each,
and the director taught two classes. Create a potential class schedule.

We make a graph in which each class is a vertex, and each edge represents two
classes that cannot be taught at the same time (because an instructor cannot be in
two places at once). See Figure B.24. This gives us six disjoint triangles and a lone
edge (the director). We'll then properly color the vertices using five colors such that
each color is used exactly four times. See Figure B.25. This gives a schedule of
(1, teal): C, D, E, F;
(2, charcoal): A, B, D, G;
(3, pale teal): B, E, F, G;
(4, white): A, C, F, G;
(5, grey): A, B, C, E.

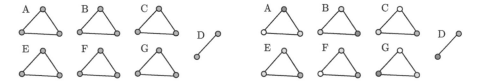

Figure B.24. No instructor can teach two or more classes at the same time.

Figure B.25. A proper coloring that uses each of five colors four times.

9. Find the chromatic number of each graph shown in Figure 12.18 on page 394.

Let the left-hand graph be G, the middle graph be Q, and the right-hand graph be H. $\chi(G) = 3$, $\chi(Q) = 4$, and $\chi(H) = 3$. G contains K_3, so $\chi(G) \geq 3$; Q contains K_4, so $\chi(Q) \geq 4$; and H contains K_3, so $\chi(H) \geq 3$. We exhibit a 3-vertex coloring of G and of H and a 4-vertex coloring of Q in Figure B.26.

10. Find the chromatic index of each graph shown in Figure 12.18 on page 394.

Let the left-hand graph be G, the middle graph be Q, and the right-hand graph be H. $\chi'(G) = 4$, $\chi'(Q) = 6$, and $\chi'(H) = 4$. G has a vertex of degree 4, so $\chi'(G) \geq 4$; Q has a vertex of degree 6, so $\chi'(Q) \geq 6$; and H has a vertex of degree 4, so $\chi'(H) \geq 4$. We exhibit a 4-edge coloring of G, a 6-edge coloring of Q, and a 4-edge coloring of H in Figure B.27.

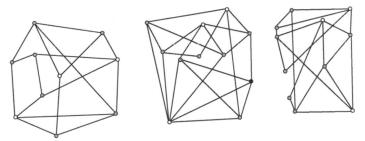

Figure B.26. A proper vertex coloring of each graph in question.

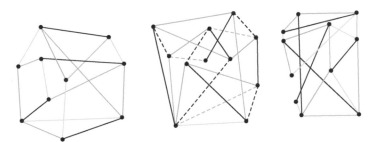

Figure B.27. A proper edge coloring of each graph in question.

1. In Lucy Worsley's *If Walls Could Talk: An Intimate History of the Home*, the author says, "The medieval death rate was one in every fifty pregnancies. Considering that it wasn't unusual for a woman to give birth a dozen times, the odds quickly mounted up for reproductive wives."
   So... what are these odds? Compute the probability of dying while pregnant for each of 1, 4, 6, and 12 pregnancies. What is the probability of dying during some one of 12 theoretical medieval pregnancies?

   One pregnancy: $\frac{1}{50} = .02$.
   Four pregnancies: If you make it to the fourth pregnancy, that means you didn't die in the first three pregnancies, so $\left(\frac{49}{50}\right)^3 \left(\frac{1}{50}\right) = .0188$.
   Six pregnancies: $\left(\frac{49}{50}\right)^5 \left(\frac{1}{50}\right) = .018$.
   A dozen pregnancies: $\left(\frac{49}{50}\right)^{11} \left(\frac{1}{50}\right) = .016$.
   But dying during the first pregnancy is exclusive of dying during the second (... or twelfth) pregnancy, so in order to compute the probability of dying during *some* pregnancy of twelve, we need to compute $\sum_{j=1}^{12} \left(\frac{49}{50}\right)^{j-1} \left(\frac{1}{50}\right) = .2153$.

2. Suppose you have a box of colored pens (fuchsia, cinnamon, tangerine, gold, lime, forest, teal, cobalt, plum) and three pencils (mechanical, yellow No. 2, printed with cupcakes).

   (a) Describe the state space of grabbing a pen and a pencil. What is the probability of each individual state?

   The state space is {fuchsia/mechanical, fuchsia/yellow No. 2, fuchsia/cupcakes, cinnamon/mechanical, cinnamon/yellow No. 2, cinnamon/cupcakes, tangerine/mechanical, tangerine/yellow No. 2, tangerine/cupcakes, gold/mechanical, gold/yellow No. 2, gold/cupcakes, lime/mechanical, lime/yellow No. 2, lime/cupcakes, forest/mechanical, forest/yellow No. 2, forest/cupcakes, teal/mechanical, teal/yellow No. 2, teal/cupcakes, cobalt/mechanical, cobalt/yellow No. 2, cobalt/cupcakes, plum/mechanical, plum/yellow No. 2, plum/cupcakes}.

   (b) What is the probability of grabbing a pen whose color begins with "f" and a mechanical pencil?

   The state space has $9 \times 3 = 27$ elements. Of those, two involve a pen whose color begins with "f" and a mechanical pencil. Thus the probability of grabbing a pen whose color begins with "f" and a mechanical pencil is $\frac{2}{27}$.

   (c) What is the probability of grabbing a pen whose color is greenish and a non-mechanical pencil?

   There are three greenish pens (lime, forest, teal) and two non-mechanical pencils (yellow No. 2, printed with cupcakes), so there are six elements in the state space that comply with the given constraints. Thus the probability of grabbing a pen whose color is greenish and a non-mechanical pencil is $\frac{6}{27} = \frac{2}{9}$.

(d) What is the probability of (grabbing a pen whose color begins with "f" and a non-mechanical pencil) or (grabbing a pen whose color is greenish and a non-mechanical pencil)?

Approach the first: In addition to the six possibilities for (grabbing a pen whose color is greenish and a non-mechanical pencil), there are two possibilities for (grabbing a pen whose color begins with "f" and a non-mechanical pencil) AND NOT (grabbing a pen whose color is greenish and a non-mechanical pencil). Thus the probability is $\frac{8}{27}$.

Approach the second: Let the event of grabbing a pen whose color is greenish and a non-mechanical pencil be G, and let the event of grabbing a pen whose color begins with "f" and a non-mechanical pencil be F. Then we seek $P(G \text{ or } F) = P(G) + P(F) - P(G \text{ and } F)$. There are four possibilities for F. There are two possibilities for grabbing a pen whose color (is greenish and begins with "f") and a non-mechanical pencil, so we have $\frac{6}{27} + \frac{4}{27} - \frac{2}{27} = \frac{8}{27}$.

3. A computer lab has 20 computers in it. On any given day, the probability that a given computer is not working is p. How many computers do you expect will be functioning when you enter the lab today? Answer the question for $p = .001$, $p = .05$.

Approach 1: We will compute the number of computers we expect to be broken and subtract from 20. Let the random variable C count the number of computers that are broken on a given day. We can write $C = \sum_{j=1}^{20} C_j$, where

$$C_j(\text{day } d) = \begin{cases} 1 & \text{computer } C_j \text{ is not working,} \\ 0 & \text{computer } C_j \text{ is working.} \end{cases}$$

$P(C_j = 1) = p$, so $\mathbb{E}[C_j] = 1P(C_j = 1) + 0P(C_j = 0) = p$. Then, by Theorem 14.7.1, $\mathbb{E}[C] = \sum_{i=1}^{20} \mathbb{E}[C_j] = \sum_{k=1}^{20} p = 20p$ broken computers. Thus we expect $20 - 20p$ computers to be functioning on any given day.

Approach 2: We will compute directly the number of computers we expect to be functioning. Let the random variable C count the number of computers that are working on a given day. We can write $C = \sum_{j=1}^{20} C_j$, where

$$C_j(\text{day } d) = \begin{cases} 1 & \text{computer } C_j \text{ is working,} \\ 0 & \text{computer } C_j \text{ is not working.} \end{cases}$$

$P(C_j = 1) = 1 - p$, so $\mathbb{E}[C_j] = 1P(C_j = 1) + 0P(C_j = 0) = 1 - p$. Then, by Theorem 14.7.1, $\mathbb{E}[C] = \sum_{i=1}^{20} \mathbb{E}[C_j] = \sum_{k=1}^{20}(1 - p) = 20 - 20p$ working computers. Thus we expect $20 - 20p$ computers to be functioning on any given day.

When $p = .001$, we expect $20 - 20(.001) \approx 20$ computers to be functioning on any given day. When $p = .05$, we expect $20 - 20(.05) \approx 19$ computers to be functioning on any given day.

4. Chips of the World come in lots of flavors. In the sale bin are
   2 bags of bacon ranch pita chips,
   1 bag of salt and vinegar potato chips,
   3 bags of hot-sauce cheese corn chips,
   5 bags of crab potato chips, and
   2 bags of peppercorn salsa pita chips.
   If you close your eyes and grab three bags of chips (one at a time, so you know you have three), what is the probability you will get…

   🐤 … all three bags of potato chips?

   There are 6 bags of potato chips out of 13 bags of chips. So the probability of the first bag grabbed being potato is $\frac{6}{13}$. Twelve bags remain, of which 5 are potato, and then 11 bags remain of which 4 are potato, so the probability is $\frac{6}{13} \cdot \frac{5}{12} \cdot \frac{4}{11}$, or $\approx .07$.

   🐤 … exactly two bags of spicy chips?

   What counts as spicy chips? Probably hot-sauce cheese corn chips and peppercorn salsa pita chips, for a total of 5 of the 13 bags of chips. There are three ways that we could get exactly two spicy-chip bags—the first or the second or the third bag isn't of spicy chips. So we have $\frac{8}{13} \cdot \frac{5}{12} \cdot \frac{4}{11} + \frac{5}{13} \cdot \frac{8}{12} \cdot \frac{4}{11} + \frac{5}{13} \cdot \frac{4}{12} \cdot \frac{8}{11}$. This simplifies to $3 \cdot \frac{8 \cdot 5 \cdot 4}{13 \cdot 12 \cdot 11}$, or $\approx .28$.

   🐤 … at least one bag of pita chips?

   To get at least one bag of pita chips, we could add the probabilities of getting exactly one, exactly two, and exactly three bags of pita chips. Or we could observe that $P(\text{grabbing at least one bag of pita chips}) = 1 - P(\text{grabbing no bags of pita chips})$. It's easy to compute the probability of grabbing no bags of pita chips—there are 4 bags of pita chips out of the 13 bags, so we get $\frac{9}{13} \cdot \frac{8}{12} \cdot \frac{7}{11}$. Our final probability is $1 - \frac{9}{13} \cdot \frac{8}{12} \cdot \frac{7}{11}$ or $\approx .71$.

   How many bags of corn chips do you expect to find in your three bags?

To compute the number of bags of corn chips we expect to find in our three bags, we need to compute expected value, which means we need to define a random variable. Let C count the number of bags of corn chips we find. By definition, $\mathbb{E}[C] = 0P(C = 0) + 1P(C = 1) + 2P(C = 2) + 3P(C = 3)$. So, we need to compute these probabilities. There are 3 bags of corn chips among our 13.

$P(C = 0) = \frac{10}{13} \cdot \frac{9}{12} \cdot \frac{8}{11}$. (Okay, that wasn't necessary. Too late.)

$P(C = 1) = 3 \cdot \frac{3 \cdot 10 \cdot 9}{13 \cdot 12 \cdot 11}$.

$P(C = 2) = 3 \cdot \frac{3 \cdot 2 \cdot 10}{13 \cdot 12 \cdot 11}$.

$P(C = 3) = \frac{3}{13} \cdot \frac{2}{12} \cdot \frac{1}{11}$.

Thus $\mathbb{E}[C] = 3 \cdot \frac{3 \cdot 10 \cdot 9}{13 \cdot 12 \cdot 11} + 2 \cdot 3 \cdot \frac{3 \cdot 2 \cdot 10}{13 \cdot 12 \cdot 11} + 3 \cdot \frac{3 \cdot 2 \cdot 1}{13 \cdot 12 \cdot 11} = \frac{9}{13}$ bags of corn chips.

5. Shoes 'R' Us has a lot of different kinds of shoes in their display case, one of each kind they sell. A shoe can be brown, black, silver, or green; it can be a low shoe, a boot, or an athletic shoe; and, it can have laces or be a slip-on.

  (a) How many different kinds of shoes does Shoes 'R' Us have in its display case?

  Shoes 'R' Us shows $4 \cdot 3 \cdot 2 = 24$ kinds of shoes in its display case.

  (b) What is the probability that a Shoes 'R' Us display shoe is brown and slip-on?

  Six shoes are brown and of those, three are slip-on, for a probability of $\frac{3}{24} = \frac{1}{8}$.

  (c) What is the probability that a Shoes 'R' Us display shoe is silver or a boot?

  There are six silver shoes in the display, and eight boots in the display case, and (check it out, here comes PIE) two of those are silver boots. Therefore, there are $6 + 8 - 2 = 12$ display shoes that are silver or boots, and the probability is $\frac{12}{24} = \frac{1}{2}$.

  (d) Given that a Shoes 'R' Us display shoe is silver, what is the probability that it is a boot?

  There are six silver shoes, of which two are boots, so the probability of a silver shoe being a boot is $\frac{1}{3}$. Or, the conditional probability formula gives $\frac{1/12}{1/4} = \frac{1}{3}$.

  (e) Given that a Shoes 'R' Us display shoe is a green athletic shoe, what is the probability that it has laces?

  There are six green shoes, of which two are also athletic. One of the green athletic shoes has laces, so the probability is $\frac{1}{2}$.

  (f) Are the properties *silver* and *boot* independent?

  To check for independence of *silver* and *boot*, we check to see whether $P(boot|silver) = P(boot)$. We already know that $P(boot|silver) = \frac{1}{3}$. And $P(boot) = \frac{1}{3}$, so these properties are independent.

6. Consider a deck of cards that is standard, except for having six suits—the two additional suits are stars and squids. (This deck exists: it is the Blue Sea Deck.) Draw a card.

(a) What is the probability that the card is a queen or a squid?

In this deck, there are 6 queens and there are 13 squids. There is a queen-of-squids card, so the total number of queens or squids is (by PIE) $6 + 13 - 1 = 18$. The total number of cards is $6 \cdot 13 = 78$. Therefore, the probability of a card being a queen or a squid is $\frac{18}{78} = \frac{3}{13}$.

(b) What is the expected value of the number on the card? (Here, Ace $= 1$, King $= 13$.)

We define the random variable X to measure the value of the number on a card. Let's use Lemma 14.3.10 to calculate the expected value. There are 78 cards, and so 78 possible states in the space S. Each of these states has a probability of $\frac{1}{78}$, so $\mathbb{E}[X] = \sum_{s \in S} X(s)P(s) = \sum_{s \in S} X(s) \cdot \frac{1}{78} = \frac{1}{78} \sum_{s \in S} X(s)$. The values of $X(s)$ range from 1 to 13, and there are 6 of each, so we now have $\frac{1}{78} \cdot 6(1 + \cdots + 13) = \frac{1}{78} \cdot 6 \cdot 91 = 7$.

7. The game *Elder Sign* has unusual dice. There are six six-sided green dice, each of which has three sides showing magnifying glasses, one side with a tentacle, one side with a skull, and one side with a scroll. There is also a six-sided yellow die with four sides showing magnifying glasses, one side with a skull, and one side with a scroll. Finally, there is a six-sided red die with three sides showing magnifying glasses, one side with a Wild sign, one side with a skull, and one side with a scroll.

(a) If you roll the six green dice, what is the expected number of magnifying glasses you'll see?

We define the random variable M to be the number of magnifying glasses we see when rolling the six green dice. This computation will be a lot easier if we define M_i to be 1 if we get a magnifying glass on the ith die, and 0 if we don't—we can note that $M = \sum_{i=1}^{6} M_i$ and use Theorem 14.7.1. $\mathbb{E}[M_i] = 0 + P(M_i = 1) = \frac{3}{6} = \frac{1}{2}$. Therefore, $\mathbb{E}[M] = \sum_{i=1}^{6} \mathbb{E}[M_i] = 6\mathbb{E}[M_i] = \frac{6}{2} = 3$.

(b) If you roll seven of the dice, what is the probability that you will roll exactly one skull?

The rolls of the dice are all independent of each other, and each die has exactly one skull, and there are seven ways to choose the die that shows a skull, so the probability is $7 \cdot \frac{1}{6} \left(\frac{5}{6}\right)^6 \approx .39$.

(c) If you roll seven of the dice, what is the probability that you will roll at least one scroll?

Observe that (probability of rolling at least one scroll) $= 1 - $ (probability of rolling no scrolls), so we compute $1 - \left(\frac{5}{6}\right)^7 \approx .72$.

8. The game of *Qwirkle* uses a bag of tiles. Each black tile has a shape on it (circle, diamond, square, crisscross, starburst, clover) that is colored (red, orange, yellow, green, blue, purple). There are three copies of each kind of tile.

(a) How many tiles are in a *Qwirkle* bag?

There are $6 \cdot 6 \cdot 3 = 108$ tiles in a *Qwirkle* bag.

(b) What is the probability that a tile drawn is red?

Of the tiles, $\frac{1}{6}$ are red, so the probability of drawing a red tile is $\frac{1}{6}$.

(c) What is the probability that a tile drawn is a sunburst?

Of the tiles, $\frac{1}{6}$ are sunbursts, so the probability of drawing a sunburst tile is $\frac{1}{6}$.

(d) What is the probability that a tile drawn is a red sunburst?

Of the red tiles, $\frac{1}{6}$ are sunbursts, so $P(\text{drawing a red sunburst tile}) = \frac{1}{36}$.

(e) What is the probability that a tile drawn is red or a sunburst?

There are 18 red tiles and 18 sunburst tiles, and 3 red sunburst tiles, so (PIE!) there are $18 + 18 - 3 = 33$ tiles that are red or sunburst. Thus the probability of drawing a red or sunburst tile is $\frac{33}{108} = \frac{11}{36}$.

(f) Is red-ness independent of sunburst-ness?

To check independence, we'll note that the conditional probability of red-given-sunburst is equal to that of sunburst-given-red (both are $\frac{1/36}{1/6} = \frac{1}{6}$) and equal to that of sunburst and of red (each $\frac{1}{6}$), so these events are independent.

9. Another *Qwirkle* qwestion: pull two tiles from the bag.

(a) What is the probability that both are blue?

We draw tiles one at a time; the first has a probability of $\frac{36}{108}$ of being blue and the second of $\frac{35}{107}$ (because only 107 tiles are left when we draw the second tile). So the probability of drawing two blue tiles is $\frac{36}{108} \cdot \frac{35}{107}$, or $\approx .11$.

(b) What is the probability that the *second* tile is blue?

If we draw two tiles and the second is blue, we either had first-tile-blue and second-tile-blue (probability $\frac{36}{108} \cdot \frac{35}{107}$), or first-tile-not-blue and second-tile-blue (probability $\frac{72}{108} \cdot \frac{36}{107}$). These events are exclusive, so we add their probabilities to obtain $\frac{36}{108} \cdot \frac{35}{107} + \frac{72}{108} \cdot \frac{36}{107} = \frac{1}{3}$.

(c) What is the probability that at least one tile is blue?

We could compute the probability of getting only the first tile blue and add that to the probability of getting only the second tile blue and add that to the probability of getting both tiles blue. Or we could compute $1 -$ (probability of getting no blue tiles). We already know the probability of getting only the second tile blue ($\frac{72}{108} \cdot \frac{36}{107}$) and the probability of getting both tiles blue ($\frac{36}{108} \cdot \frac{35}{107}$), so we might as well just add in the probability of getting only the first tile blue ($\frac{36}{108} \cdot \frac{72}{107}$) for a total of $2 \cdot \frac{72 \cdot 36}{108 \cdot 107} + \frac{36}{108} \cdot \frac{35}{107} \approx .56$.

10. What's the expected number of fixed points (items that do not move) in a permutation of n items?

We will let F count the number of fixed points in a permutation. For each of the n items, we define a random variable F_i that evaluates to 1 if the item is fixed and 0 if the item moves. There are n possibilities for where the ith item goes in a permutation, so the probability of an item staying in the same place is $\frac{1}{n}$. Thus, $\mathbb{E}[F_i] = 1P(F_i = 1) + 0P(F_i = 0) = \frac{1}{n}$. Then, by Theorem 14.7.1, $\mathbb{E}[F] = \sum_{i=1}^{n} \mathbb{E}[F_i] = \sum_{k=1}^{n} \frac{1}{n} = n\frac{1}{n} = 1$ fixed point. Hey, that seems an awful lot like Example 14.7.3 ... hmmm....

Chapter 15, Section 15.9   page 508

1. Show that $|\mathbb{Z}| = |\mathbb{Z}| + 72$.

We will exhibit a bijection between $\mathbb{Z}$ and $\mathbb{Z} \cup \mathbb{Z}_{72}$. For $k \le 0$, let $f(k) = k \in \mathbb{Z}$, for $1 \le k \le 72$ let $f(k) = k - 1 \in \mathbb{Z}_{72}$, and for $k > 72$, let $f(k) = k - 72 \in \mathbb{Z}$.
This is injective: Note that if $f(k) = f(r)$, then $f(k), f(r)$ must be both in $\mathbb{Z}$ or both in $\mathbb{Z}_{72}$. If $f(k) = f(r)$, then both are positive or both are nonpositive. If both are nonpositive, then $k = r$. If both are positive, then either $k - 72 = r - 72$, so $k = r$, or $k - 1 = r - 1$, so $k = r$.
It is also surjective: for $z \in \mathbb{Z}$, if $z \le 0$, $f(z) = z$, and if $z > 0$, $f(z + 72) = n$ and for $z \in \mathbb{Z}_{72}, f(z + 1) = z$.

2. Show that $\mathbb{Z}$ has the same cardinality as $4\mathbb{N}$.

We will exhibit a bijection between $\mathbb{Z}$ and $4\mathbb{N}$. For $k < 0$, let $f(k) = -8k$. For $k \ge 0$, let $f(k) = 8k + 4$.
Injective: Suppose $f(k) = f(r)$. Both must be divisible by 8 or not divisible by 8. If $f(k) = f(r)$ is divisible by 8, then we have $-8k = -8r$ so $k = r$. If $f(k) = f(r)$ is not divisible by 8, then we have $8k + 4 = 8r + 4$ so $k = r$.
Surjective: Every element n of $4\mathbb{N}$ is divisible by 4 and is either divisible by 8 or $\equiv 4 \pmod{8}$. If n is divisible by 8, then $f(-\frac{n}{8}) = n$, and if n is not divisible by 8, then $f(\frac{n-4}{8}) = n$.

3. Show that $\mathbb{Z}$ has the same cardinality as $\mathbb{N} \times \mathbb{N}$.

We will exhibit a bijection between $\mathbb{Z}$ and $\mathbb{N} \times \mathbb{N}$. We associate $0 \leftrightarrow (1, 1), 1 \leftrightarrow (1, 2), -1 \leftrightarrow (2, 1)$, and continue in the pattern indicated by Figure B.28.

4. Prove that $|\mathscr{P}(\mathbb{Q})| > |\mathbb{Q}|$.

Proof by contradiction: Suppose that there is a bijection f between $\mathscr{P}(\mathbb{Q})$ and $\mathbb{Q}$. Consider the set $Q \in \mathscr{P}(\mathbb{Q})$ defined as $Q = \{q \mid q \notin Q'$, where Q' is such that $f(Q') = q'\}$. Consider $f(Q)$. Is $q \in f(Q)$? Suppose $q \in f(Q)$. By definition, $q \notin f(Q)$, which is a contradiction.

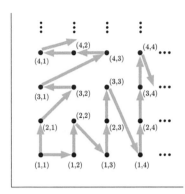

Figure B.28. A bijection between $\mathbb{Z}$ and $\mathbb{N} \times \mathbb{N}$.

5. Show that $\mathbb{W}$ has the same cardinality as $\mathbb{Z}$.

We will exhibit a bijection between $\mathbb{W}$ and $\mathbb{Z}$. When $w \in \mathbb{W}$ is odd, let $f(w) = \frac{w-1}{2} + 1$, and when $w \in \mathbb{W}$ is even, let $f(w) = \frac{-w}{2}$.

This is injective: if $f(k) = f(r)$, then either both values are positive, so that $\frac{k-1}{2} + 1 = \frac{r-1}{2} + 1$, so $k = r$, or both values are non-positive, so $\frac{-k}{2} = \frac{-k}{2}$ so $k = r$.

It is also surjective: for $n \in \mathbb{Z}$, if $n > 0$, then $f(2(n-1)+1) = \frac{2(n-1)+1-1}{2} + 1 = n$, and if $n \leq 0$, then $f(-2n) = \frac{-(-2n)}{2} = n$.

6. What is the cardinality of the set $\{\frac{p}{q} \mid p \in \mathbb{W}, q \in \mathbb{Z}\}$?

There is a bijection between this set and $\mathbb{W} \times \mathbb{Z}$, defined by $f(\frac{p}{q}) = (p, q)$. We know that $|\mathbb{W}| = |\mathbb{Z}| = |\mathbb{N}| = |\mathbb{N} \times \mathbb{N}|$. Therefore, the cardinality is $\aleph_0$.

7. What is $(\aleph_0)^3$? How about $(\aleph_0)^8$? Or $(\aleph_0)^{\aleph_0}$? Explain.

All three of these are equal to $\aleph_0$. For example, $(\aleph_0)^3 = \aleph_0 \cdot \aleph_0 \cdot \aleph_0 = |\mathbb{N} \times \mathbb{N} \times \mathbb{N}|$, and we know from Section 15.5 that there is a bijection between $\mathbb{N} \times \mathbb{N} \times \mathbb{N}$ and $\mathbb{N}$. The same holds for $(\aleph_0)^8$, and even $(\aleph_0)^{\aleph_0}$, because there is a bijection between $\bigcup_{k \in \mathbb{N}} \mathbb{N}_k$ and $\mathbb{N}$.

8. Consider the set $\mathscr{F}$ of all functions from $\mathbb{N}$ to $\mathbb{N}$. Is $\mathscr{F}$ countable or uncountable?

We will show that $\mathscr{F}$ is uncountable by showing that a proper subset of $\mathscr{F}$ is uncountable. Consider a subset $A \subset \mathbb{N}$. There exists a function $f_A : \mathbb{N} \to \mathbb{N}$ defined by

$$f_A(n) = \begin{cases} n & n \in A, \\ 1 & n \notin A. \end{cases}$$

In other words, f_A sends elements of A to themselves and sends everything else to 1. (This is neither one-to-one nor onto unless $A = \mathbb{N}$.) And, the functions f_A are in one-to-one correspondence with elements $A \in \mathscr{P}(A)$, so $|\mathscr{F}| \geq |\mathscr{P}(A)|$, and we know $\mathscr{P}(A)$ is uncountable because $|\mathscr{P}(A)| = 2^{\aleph_0} > \aleph_0$.

9. Is the total number of steps in an algorithm that does not terminate countable or uncountable?

   The steps in an algorithm are numbered, so we can count them as the algorithm proceeds and therefore the total number is countable.

10. Consider the set H of length-$\frac{1}{2}$ intervals that are contained in the interval $[0, 1]$. What is $|H|$?

   The intervals of length $\frac{1}{2}$ are in one-to-one correspondence with their left endpoints. (If you don't believe this, consider that given a left endpoint we can find the right endpoint by adding $\frac{1}{2}$.) An interval of length $\frac{1}{2}$ inside the unit interval can have a left endpoint as small as 0 and as large as $\frac{1}{2}$. So $|H| = |[0, \frac{1}{2}]|$. The elements of $[0, \frac{1}{2}]$ are in one-to-one correspondence with the elements of $[0, 1]$, so $|H| = \aleph_1$.

Chapter 16, Section 16.11   page 529

1. Given any n integers $k_1, k_2, \ldots, k_n$, show that there exist k_i, k_j such that $k_i \equiv k_j$ (mod $n - 1$).

   Compute k_1 (mod $n - 1$), k_2 (mod $n - 1$), \ldots, k_n (mod $n - 1$). These values cannot all be different; $n > n - 1$ so by the pigeonhole principle two of them must be the same. Name those two k_i, k_j and we have that $k_i \equiv k_j$ (mod $n - 1$).

2. Find a set of 13 natural numbers, each of which has a different value modulo 13 and all of which are multiples of 5.

   Well, let's see. $\{5, 10, 15, 20, 25, 30, 35, 40, 45, 50, 55, 60, 65\}$ works. Their residues are $\{5, 10, 2, 7, 12, 4, 9, 1, 6, 11, 3, 8, 0\}$.

3. Consider the equation $5x + 2y = 3$. Why can it not have any solutions with both x and y whole numbers? For which k does the equation $5x + 2y = k$ have solutions with both x and y whole numbers?

   If both x and y are whole, then the smallest possible values for $5x + 2y$ are 0 ($x = y = 0$), 2 ($x = 0, y = 1$), 4 ($x = 0, y = 2$), and 5 ($x = 1, y = 0$). None of these is equal to 3. Using $x = 0$ and all whole y, we obtain all even whole numbers for k; using $x = 1$ and all whole y, we obtain all odd numbers ≥ 5 for k. Thus we can produce all whole numbers except 1 and 3.

4. Compute $\varphi(5^k 4)$, for any positive k.

   $\varphi(5^k 4) = 5^{k-1} 4 \varphi(4) = 5^{k-1} 8$.

5. For which n is $\varphi(n) = 4$?

   Because $4 = 2^{3-1}(2 - 1) = 5 - 1 = 2^{2-1}(1)(3 - 1)$, we have $\varphi(5) = \varphi(8) = \varphi(12)$.

6. Prove that the number of natural numbers relatively prime to n and $\leq mn$ is $m\varphi(n)$.

   Recall that $GCD(k,n) = GCD(k \pmod{n}, n)$. Now notice that for every $k \pmod{n}$, there are m different numbers $\leq mn$ equivalent to $k \pmod{n}$.

7. Use the Euclidean algorithm to compute $GCD(1234, 12345)$. Now find integers k, ℓ such that $GCD(1234, 12345) = k1234 + \ell 12345$.

   $12345 \pmod{1234}$ is 5, so $GCD(1234, 12345) = GCD(5, 1234)$. Then $1234 \pmod 5 = 4$, so $GCD(1234, 12345) = GCD(5, 1234) = GCD(4, 5) = 1$. Let's see. $1 = 5 - 4$ and $4 = 1234 - 246 \cdot 5$, so $1 = 5 - 1 \cdot 1234 + 246 \cdot 5 = 247 \cdot 5 - 1 \cdot 1234$. Now, $5 = 12345 - 10 \cdot 1234$, so $1 = 247 \cdot (12345 - 10 \cdot 1234) - 1 \cdot 1234 = 247 \cdot 12345 - 2471 \cdot 1234$.

8. Prove that if $GCD(a, b) = 1$, then for any integer c, there is always a solution to $ax + by = c$, where x and y are integers. Use this fact to find a solution to the equation $2x + 3y = 4$.

   If $GCD(a, b) = 1$, then by Theorem 16.3.1 there exist integers k, ℓ such that $ka + \ell b = 1$. If we multiply through by c, we get $cka + c\ell b = c$. So, set $x = ck$ and $y = c\ell$. For the example, we know that $2(-1) + 3(1) = 1$, so $4(-1)2 + 4(1)3 = 4$. Here $x = -4$ and $y = 4$.

9. Find all x that satisfy $2x \equiv 2 \pmod 7$.

   $4 \cdot 2 = 8 \equiv 1 \pmod 7$; the equivalence becomes $x \equiv 4 \cdot 2 \equiv 1 \pmod 7$, so the desired set is all $x \equiv 1 \pmod 7$.

10. Which x satisfy both $x \equiv 1 \pmod 3$ and $x \equiv 3 \pmod 4$?

    $x = 3j + 1$ so $3j \equiv 2 \pmod 4$. Then $3 \cdot 3 \equiv 1 \pmod 4$ so $j \equiv 6 \equiv 2 \pmod 4$. Because $j = 4q + 2$, we have that $x = 3(4q + 2) + 1$, so $x \equiv 7 \pmod{12}$.

Chapter 17, Section 17.9   page 552

1. Write a runtime function for the marble-sorting algorithm given in Example 5.2.8.

   The input size is the number of marbles m in a bag. Notice that step 2 uses a different number of operations depending on the color of the marble: two operations for a red marble, three operations for a green marble, and four for a marble that is neither red nor green—we have a comparison operation for each "if." In the worst-case input, there are no red marbles and no green marbles. There are six operations per non-red–non-green marble, so the function is $6m$.

2. Consider the following algorithm that has input *list* with length n. What does the algorithm do? What is a worst-case input? What is its runtime function? What function type or complexity class is that function?

   1. Let $i = 1$.
   2. If $list_i \geq list_{i+1}$, say "nope" and exit; otherwise, continue.
   3. If $i = n - 1$, say "yup!"; otherwise, replace i with $i + 1$ and go to step 2.

The algorithm tells us whether the entries of the list are sorted in increasing order. A worst-case input is a sorted list.

Runtime function: $1 + 3(n - 1) = 3n - 2$. This is linear with $r(n) \in O(n)$.

3. Prove that $r(n) = n^2 - 16$ is in $O(n^3)$.

   We know that $n^2 - 16 \le n^2 \le n^3$, so we can choose $c = k = 1$ to satisfy the definition and see that $n^2 - 16 \in O(n^3)$.

4. Estimate the complexity of the following pointless algorithm that takes as input a natural number n.

   1. Let $i = 1$.
   2. If it's Tuesday, then continue; otherwise, go to step 5.
   3. Let $j = 2^{n+1}$ and let $m = i(2 - j)$.
   4. If $i = n$, output "Ha ha ha!"; otherwise, replace i with $i + 1$ and go to step 2.
   5. Let $j = 2i$.
   6. If $i = n$, output "Ha ha ha!"; otherwise, replace i with $i + 1$ and go to step 2.

   If it's Tuesday, then the algorithm has to do n multiplications to compute j and two operations to compute m. This happens for each value of i, and there are n iterations of i, so we have $(n + 2)n \approx n^2$ operations. If it's not Tuesday, then the algorithm does one multiplication to compute j. This happens for each value of i, for a total of n operations. Thus, the worst-case scenario is Tuesday, and the complexity is quadratic (roughly n multiplications n times).

5. Consider the following algorithm with input *list* of length n with integer elements. What does it do? What is its complexity?

   1. Let $i = 1$ and let *newlist* = {}.
   2. Append $2 \cdot list_i$ to *newlist*.
   3. If $i < n$, replace i with $i + 1$ and go to the previous step; otherwise, output *newlist*.

   This produces a new list each of whose elements is twice the value of the old list's corresponding element. The complexity is linear because the runtime is $4n + 2$.

6. Determine the complexity class of the bubble sort algorithm, whose runtime function is approximated in Section 6.12.

   Because $\binom{n}{2} = \frac{1}{2}(n^2 - n)$, bubble sort is in $O(n^2)$.

7. True or false: $\frac{1}{2}n^2 + n\log(n) \in O(n\log(n))$. Explain, and if the statement is false, make a corrected statement.

   False: for natural n, $n > \log(n)$. Thus $n^2 > n\log(n)$ and so $\frac{1}{2}n^2 + n\log(n) \in \Omega(n\log(n))$.

8. Write an algorithm that inputs a number n and outputs whether the number is even, odd, or neither (not an integer). What is its complexity?

   Answers will vary, but here is one possibility:

   1. Let $halfish = \lfloor \frac{n}{2} \rfloor$.
   2. If $2 \cdot halfish = n$, output "even!"; otherwise, continue.
   3. If $2 \cdot halfish + 1 = n$, output "odd!"; otherwise, continue.
   4. Output "this is not an integer."

   There are a whopping nine operations at most, so this algorithm is of constant complexity.

9. Consider the following algorithm. What does it do? Estimate its complexity.

   1. Input n.
   2. Let $i = 0$ and let $j = n$.
   3. If $\lfloor \frac{j}{2} \rfloor = 0$, output i; otherwise, continue.
   4. Replace j with $\lfloor \frac{j}{2} \rfloor$, replace i with $i + 1$, and go to the previous step.

   The algorithm gives the highest power of 2 that is less than or equal to n. The complexity is $O(\log_2(n))$, because the number of iterations is an integer and the logarithm bounds that from above.

10. True or false: $\sqrt{n} + n \in \Omega(n)$. Explain, and if the statement is false, make a corrected statement.

    True, because $\sqrt{n} > 0$ and so $\sqrt{n} + n > n = 1 \cdot n$.

Appendix C 🦆🦆🦆

The Greek Alphabet and Some Uses for Some Letters

Lower-case	Upper-case	Name	Lowercase uses / uppercase uses
α	A	alpha	the first letter you use / unused
β	B	beta	the second letter you use / unused
γ	Γ	gamma	curve, photon (physics) / Gamma function, Christoffel symbol (physics)
δ	Δ	delta	small amount, Dirac delta function (physics and math) / change, highest degree of a vertex in a graph
ϵ (ε)	E	epsilon	small positive number, young mathematician / unused
ζ	Z	zeta	Riemann zeta function / unused
η	H	eta	used when you have used too many common letters already / unused
θ	Θ	theta	angle / measure of algorithmic efficiency (computer science)
ι	I	iota	identity map / unused
κ	K	kappa	curvature, connectivity of a graph / unused
λ	Λ	lambda	wavelength (physics), eigenvalue / exterior product
μ	M	mu	measure, prefix for 10^{-6} / unused
ν	N	nu	neutrino (physics), frequency (physics) / unused
ξ	Ξ	xi	used as an additional variable / used occasionally in physics
o	O	omicron	unused / unused
π	Π	pi	constant $3.14159\ldots$, projection map / product
ρ	P	rho	radius (spherical coordinates), density (physics) / unused
σ	Σ	sigma	permutation / sum
τ	T	tau	torque (physics), $\frac{1-\sqrt{5}}{2}$ / unused
υ	Υ	upsilon	unused / shape of hat (not kidding!)
ϕ (φ)	Φ	phi	map, golden ratio, Euler phi function, angle (spherical coordinates) / a map
χ	X	chi	chromatic number, Euler characteristic / unused
ψ	Ψ	psi	another map, wave function (physics) / unused
ω	Ω	omega	first infinite ordinal, angular velocity / resistance, measure of algorithmic efficiency (computer science)

Appendix D

List of Symbols

A	usually a set
a	usually an element of a set
$\{a_1, a_2, \ldots, a_n\}$	finite set, and the items separated by commas are the elements of the set
$(a_1, a_2, \ldots, a_n)$	ordered n-tuple
$a_1 a_2 \ldots a_n$	a string of digits, usually forming a number
$a_1, a_2, a_3, \ldots, a_n, \ldots$	a generic integer sequence
$A_1, A_2, \ldots$	the first set, the second set, etc.
$|A|$	number of elements in (cardinality of) set A
$A \times B$	Cartesian product of sets A and B, with elements (a,b)
$\emptyset = \{\}$	empty set, or null set
$\mathbb{N}$	the natural numbers $\{1, 2, \ldots\}$
$\mathbb{Z}_2$	the binary digits $\{0, 1\}$
$\mathbb{Z}$	the integers $\{\ldots, -2, -1, 0, 1, 2, \ldots\}$
$\mathbb{Z}_n$	the integers modulo n, which as a set is $\{0, 1, \ldots, n-1\}$
$\mathbb{W}$	the whole numbers $\{0, 1, 2, \ldots\}$
$\mathbb{Q}$	the rational numbers
$\mathbb{R}$	the real numbers
$|$	"such that" or "divides" or "bar," depending on context
$\star$	a star (yes, really)
$\in$	set element, or "is an element of"; for example, $a \in A$
$\notin$	"not an element of"
$\subset$	subset, or "contained in"; for example, $A \subset B$
$\subseteq$	subset that may or may not be proper
$\subsetneq$	subset that is not equal to its superset
$\supset$	contains, superset, or subset the other way around

$\not\subseteq$	not a subset
$\mathscr{P}(A)$	the power set of A
$\overline{A}$	the complement of a set A
$B \setminus A, B - A$	the complement of A relative to B
$\cup$	union
$\cap$	intersection
$\bigcup_{i=1}^{n} A_i$	union of finitely many sets, $A_1 \cup A_2 \cup \cdots \cup A_n$
$\bigcap_{i=1}^{n} A_i$	intersection of finitely many sets, $A_1 \cap A_2 \cap \cdots \cap A_n$
$\bigcup_{i=1}^{\infty} A_i, \bigcup_{i \in \mathbb{N}} A_i$	union of a countably infinite number of sets
$\bigcap_{i=1}^{\infty} A_i, \bigcap_{i \in \mathbb{N}} A_i$	intersection of a countably infinite number of sets
$\bigcup_{\alpha \in I} A_\alpha$	the union of uncountably many sets (where I is an uncountable index set such as $[0,1]$)
$\wedge$	and
$\vee$	or
$\neg$	not
$\Rightarrow$	implies
$\rightarrow$	function direction, as in $A \rightarrow B$
$\mapsto$	an element goes somewhere, as in $a \mapsto f(a)$
$\Leftrightarrow$	if and only if
$\Longleftrightarrow$	if and only if
$(\Rightarrow)$	signals that what follows will prove this direction of implication in an if-and-only-if proof
$(\Leftarrow)$	signals that what follows will prove this direction of implication in an if-and-only-if proof
$\|$	common computer code for *or*
$\&\&$	common computer code for *and*
$==$	common computer code for $=$
$\forall$	for all
$\exists$	there exists
$f : A \rightarrow B$	usually a function, but sometimes a gipo
$f\|_C$	a map defined on a subset C of its domain
$\lfloor x \rfloor$	the floor function of x, the greatest integer less than or equal to x
$\lceil x \rceil$	the ceiling function of x, the least integer greater than or equal to x
1–1	one-to-one

$V(G)$	the vertex set of a graph G
$E(G)$	the edge set of a graph G
v_1v_2 or $\{v_1, v_2\}$	an edge of a graph
P_n	a path graph
C_n	a cycle graph
W_n	a wheel graph
K_n	the complete graph with n vertices
$K_{m,n}$	the complete bipartite graph with m vertices in one part and n vertices in the other part
φ	a function; often, a proposed isomorphism
$\star_A$	an operation on A, as in $a_1 \star_A a_2 = a_3$
$A \cong B$	A is isomorphic to B
$G \setminus e$ (or $G - e$)	the graph G but with the edge e removed and e's vertices left intact
$G \setminus v$ (or $G - v$)	the graph G but with the vertex v and all its incident edges removed
$G \setminus H$ (or $G - H$)	the graph G but with the subgraph H removed, and all edges incident to any vertex in H removed
$\overline{G}$	graph complement of G
$R(k,m)$	the Ramsey number n that indicates the smallest possible K_n that is forced to contain a monochromatic K_k or K_m
$\displaystyle\sum_{j=1}^{n} f(j), \ \Sigma_{j=1}^{n} f(j)$	the sum of the first n natural-input values of $f(j)$
$\displaystyle\sum_{s\in S} f(s), \ \Sigma_{s\in S} f(s)$	the sum of the values of $f(s)$ for all possible inputs taken from S
$deg(v)$	the degree of a vertex v in a graph G
$n\vert a$	n divides a, so that $a = kn$ for some $k \in \mathbb{Z}$
$a \equiv b \pmod{n}$	equivalence modulo n
$b \pmod{a}$	shorthand for "the smallest nonnegative integer r such that $r \equiv b \pmod{a}$"
$\sim$	a relation (comparison) between set elements, often proposed as an equivalence relation
$[a]$	the equivalence class of a, or all set elements equivalent to a under some equivalence relation $\sim$
$\binom{n}{k}$	the number of ways one can choose k things from a pile of n things
$n!$	the factorial function, which returns $n \cdot (n-1) \cdot (n-2) \cdot \cdots \cdot 3 \cdot 2 \cdot 1$
F_n	the nth Fibonacci number

w_e	the weight assigned to edge e
$w(T)$	the total weight of a tree T
$\chi(G)$	the chromatic number of a graph G
$\chi'(G)$	the chromatic index of a graph G
$\Delta(G)$	the maximum degree of any vertex in G
$P(E)$	the probability that event E occurs
$X(s)$	the value of random variable X for state s
$P(X = k)$	the probability that an event defined by the random variable X having value k occurs
$\mathbb{E}[X]$	the expected value of a random variable X
$P(E_1|E_2)$	the conditional probability that event E_1 happens, given that E_2 definitely occurs; $P(E_1|E_2) = \frac{P(E_1 \text{ and } E_2)}{P(E_2)}$
$X_1 + X_2$	the sum of two random variables, defined pointwise as $(X_1 + X_2)(s) = X_1(s) + X_2(s)$
$\aleph_0$	the cardinality of the set $\mathbb{N}$
$\aleph_1$	the smallest infinite number strictly larger than $\aleph_0$
ZFC	the axiom set we usually use for mathematics, namely the Zermelo-Fraenkel axioms plus the axiom of choice
$\varphi(n)$	the Euler phi function, which counts the natural numbers that are relatively prime to and less than n
$\text{GCD}(a,b)$	the greatest common divisor of a and b
$\text{LCM}(a,b)$	the least common multiple of a and b
$\Theta(type)$	the set of all runtime functions $r(n)$ such that there are positive real constants c_1 and c_2 and a natural number k so that whenever $n \geq k$, $c_1 \cdot type \leq r(n) \leq c_2 \cdot type$
$O(type)$	the set of all runtime functions $r(n)$ such that there is a positive real constant c and a natural number k so that whenever $n \geq k$, $r(n) \leq c \cdot type$
$\Omega(type)$	the set of all runtime functions $r(n)$ such that there is a positive real constant c and a natural number k so that whenever $n \geq k$, $r(n) \geq c \cdot type$

Bibliography

[1] Michael O. Albertson and Joan P. Hutchinson. *Discrete Mathematics with Algorithms.* Hoboken, NJ: Wiley, 1988. Available at http://www.macalester.edu/~hutchinson/book/book.html.

[2] Norman L. Biggs. *Discrete Mathematics,* Second Edition. Oxford, UK: Oxford University Press, 2003.

[3] Kenneth P. Bogart. *Combinatorics through Guided Discovery.* Manuscript, 2004. Available at http://www.math.dartmouth.edu/news-resources/electronic/kpbogart/ComboNoteswHints11-06-04.pdf.

[4] Miklós Bóna. *A Walk Through Combinatorics: An Introduction to Enumeration and Graph Theory.* River Edge, NJ: World Scientific, 2006.

[5] Gilles Brassard and Paul Bratley. *Algorithmics: Theory and Practice.* Englewood Cliffs, NJ: Prentice Hall, 1988.

[6] Gary Chartrand. *Introductory Graph Theory.* New York: Dover, 1985.

[7] Gary Chartrand and Ping Zhang. *Introduction to Graph Theory.* New York: McGraw Hill, 2005.

[8] Douglas E. Ensley and J. Winston Crawley. *Discrete Mathematics: Mathematical Reasoning and Proof with Puzzles, Patterns and Games.* Hoboken, NJ: Wiley, 2006.

[9] Edgar G. Goodaire and Michael M. Parmenter. *Discrete Mathematics with Graph Theory.* Boston: Pearson, 2006.

[10] Ronald Graham, Donald Knuth, and Oren Patashnik. *Concrete Mathematics.* Reading, MA: Addison-Wesley, 1989.

[11] Curtis Greene, Phil Hanlon, Tim Hsu, and Joan Hutchinson. *An Introduction to Cryptology and Discrete Math—Michigan Math Scholars Coursepack.* Course pack, University of Michigan, Summer 1999.

[12] Richard Hammack. *Book of Proof,* Mathematics Textbook Series. Richmond, VA: Department of Mathematics & Applied Mathematics, Virginia Commonwealth University, 2009. Available at http://www.people.vcu.edu/~rhammack/BookOfProof/.

[13] Kevin Houston. *How to Think Like a Mathematician: A Companion to Undergraduate Mathematics.* Cambridge, UK: Cambridge University Press, 2009.

[14] David J. Hunter. *Essentials of Discrete Mathematics.* Sudbury, MA: Jones and Bartlett, 2008.

[15] László Lovász and Katalin Vesztergombi. "Discrete Mathematics." Lecture notes, Yale University, Spring 1999. Available at http://www.cs.elte.hu/~lovasz/notes.html.

[16] László Lovász, József Pelikán, and Katalin Vesztergombi. *Discrete Mathematics: Elementary and Beyond.* New York: Springer, 2003.

[17] Charles Leiserson, Eric Lehman, Srinivas Devadas, and Albert R. Meyer. "6.042J/18.062J Mathematics for Computer Science, Spring 2005." *Massachusetts Institute of Technology: MIT OpenCourseWare.* http://ocw.mit.edu/, accessed 8/26/2010. License: Creative Commons BY-NC-SA.

[18] George E. Martin. *Counting: The Art of Enumerative Combinatorics.* New York: Springer-Verlag, 2001.

[19] Robert McCloskey. *Make Way for Ducklings.* New York: Puffin Books, 1941 & 1969.

[20] Ivan Niven, Herbert Zuckerman, and Hugh Montgomery. *An Introduction to the Theory of Numbers*, Fifth Edition. New York: Wiley, 1991.

[21] Daniel Pinkwater. *Ducks!* New York: Little, Brown, and Company, 1984.

[22] Jon Sczieska and Lane Smith. *Math Curse.* New York: Viking, 2007.

[23] Joseph Silverman. *A Friendly Introduction to Number Theory.* Upper Saddle River, NJ: Pearson, 2012.

[24] Richard J. Trudeau. *Introduction to Graph Theory (a.k.a. Dots and Lines).* New York: Dover, 1993.

[25] Robin J. Wilson. *Introduction to Graph Theory,* Fourth Edition. Reading, MA: Addison Wesley, 1996.

Index

Page numbers in bold face indicate definitions.

 Discrete Mathematics with Ducks
Second Edition

Designed by Charlotte Byrnes, sarah-marie belcastro, and Sarah Chow
Typeset by Charlotte Byrnes

Composed in Kabel and Times Roman using LATEX
Duck illustrations inspired by a Pekin duck and a Blue Swedish duck

T - #0456 - 071024 - C700 - 235/191/31 - PB - 9780367570705 - Gloss Lamination